Volcanic Rifted Margins

Edited by

Martin A. Menzies
Department of Geology
Royal Holloway, University of London
Egham, Surrey, TW20 OEX
United Kingdom

Simon L. Klemperer
Department of Geophysics
Stanford University
Stanford, California 94305-2115
USA

Cynthia J. Ebinger
Department of Geology
Royal Holloway, University of London
Egham, Surrey, TW20 OEX
United Kingdom

and

Joel Baker
Danish Lithosphere Centre
Øster Volgade 10
1350 Copenhagen K
Denmark

THE
GEOLOGICAL
SOCIETY
OF AMERICA

Special Paper 362

3300 Penrose Place, P.O. Box 9140 ▪ Boulder, Colorado 80301-9140 USA

2002

Published by The Geological Society of America, Inc.
3300 Penrose Place, P.O. Box 9140, Boulder, Colorado 80301
www.geosociety.org

Printed in U.S.A.

GSA Books Science Editor: Abhijit Basu
Cover graphics production by Margo Y. Good.
Cover design by James McGrow and Martin Menzies.

Library of Congress Cataloging-in-Publication Data

Volcanic rifted margins / edited by Martin A. Menzies . . . [et al.].
p. cm. — (Special paper ; 362)
Includes bibliographical references.
ISBN 0-8137-2362-0 (soft cover)
1. Volcanism. 2. Continental margins. I. Menzies, M. A. (Martin A.) II. Special papers (Geological Society of America) ; 362.
QE522 .V8827 2002
551.41—dc21

2002071278

10 9 8 7 6 5 4 3 2 1

Contents

Preface

Volcanic rifted margins form when continental lithosphere is ruptured to form new ocean basins, the main geological processes being magmatism, extension, uplift, and erosion. Volcanic rifted margins are distinguished from nonvolcanic rifted margins by the generation of excess magmatism over that required to produce normal ocean crust. Perhaps as much as 90% of the rifted continental margins around the world are volcanic rifted margins, the main exceptions occurring in eastern China, western Europe (Iberia), South Australia, and North America (Newfoundland, Alaska, Gulf of California). While a theoretical approach has helped us model the interaction between geological processes during the formation of volcanic rifted margins, the real temporal and spatial relationships are much more complex. Volcanic rifted margins can be thought of as comprising three well-defined components (Menzies et al.): (1) large igneous provinces (LIPs), comprising thick basaltic and/or silicic volcanic sequences, (2) thick high-velocity lower crust (HVLC) at the continent-ocean boundary, and (3) seaward-dipping reflector series (SDRS) with subaerial and submarine volcanic and/or sedimentary components. Menzies et al. demonstrate that LIPs formation can significantly predate breakup extension, that the geological processes associated with volcanic rifted margin formation are not synchronous, and that unequivocal evidence for kilometer-scale prevolcanic uplift is lacking in most volcanic rifted margins.

Continental rifting and magmatism

Continental rift systems provide natural laboratories where questions can be posed about the earliest stages of volcanic rifted margin formation, i.e., whether basement lineaments determine rift valley orientation, the evidence for plume involvement, whether lithosphere structure triggers melting, focuses mantle flow, or channels magmatism. Nyblade summarizes the geology and geophysics of the Tanzanian craton and evaluates the role of plumes and lithosphere architecture in the formation of an embryonic rifted margin. Whether single or multiple plumes best explain 40 m.y. of magmatism beneath Africa is investigated, but not resolved. In addition, Nyblade explores the extent to which craton edges and volcanic rifted margins are genetically linked. Watkeys describes the geology and evolution of the Lebombo volcanic rifted margin in southeastern Africa, and demonstrates how lithospheric architecture and melt production played a role in Karoo volcanism. Watkeys doubts the role of single plumes and prefers a broad zone of hot mantle due to either thermal insulation below a continent or subduction processes. He shows how dike systems can be used to unravel a complex extensional history influenced by Karoo volcanism and the opening of the South Atlantic.

An increasingly vital part of our understanding of volcanic rifted margins is accurate dating of magmatic rocks. The advent of $^{40}Ar/^{39}Ar$ methodology has been pivotal in unraveling some of the complexity of volcanic rifted margins, without which the temporal and spatial evolution of most volcanic rifted margins would be largely undefined. Miggins et al. studied uplifted fault blocks that contain prerift and synrift volcanic rocks that crop out on the flanks of Rio Grande, New Mexico, USA. The $^{40}Ar/^{39}Ar$ data are used to determine the initiation of rifting and the longevity of magmatic activity. In addition, Miggins et al. use targeted $^{40}Ar/^{39}Ar$ data to help confirm the timing of uplift and to demonstrate that a major period of uplift and extension postdated the onset of magmatism in the Rio Grande rift.

Large igneous provinces

Controversy has always surrounded the source of LIPs or flood volcanism, an integral part of volcanic rifted margins. Are plumes a necessary pre-requisite to volcanic rifted margin formation, or would normal asthenosphere provide an appropriate source region? Is the lower continental lithosphere wet enough to produce the large volumes of basaltic magmatism? Baker et al. describe the nature and possible origin of the continental mantle on the volcanic rifted margin of the southern Red Sea. No evidence for Archean or Pan-African mantle was found, perhaps due to accretion of younger mantle during Jurassic and Tertiary rifting. Continental mantle shows evidence of recent influx of mantle-derived small volume melts from the Afar plume, leading Baker et al. to conclude that continental lithosphere is unlikely to have contributed significantly to the formation of LIPs.

In a study of the passive margins of northeastern Madagascar, Melluso et al. use elemental and isotopic ratios to define potential source regions and to assist in matching conjugate margins. Comparison with volcanic rocks on the conjugate margin in India reveals that similar source areas were involved in the production of the Mahabaleshwar and Ambenali basalts of the Deccan Traps, India. Melluso et al. demonstrate involvement of mid-ocean ridge basalt mantle (asthenosphere?) with no source contribution from the Marion plume despite the conventional link to the evolution of this particular volcanic rifted margin.

Silicic rocks are a vital part of volcanic rifted margins; their presence is indicative of radical changes in the thermal state of the lithosphere and the source and/or storage level of magmas (Bryan et al.). The evolved character and eruptive style of silicic rocks mean that their potential impact on the biosphere is significantly greater than that of basaltic magmas. Bryan et al. stress the importance of the silica-rich volcanic rocks in volcanic rifted margins and point to the importance in their evolution of short-lived, high-volume, highly explosive eruptions. Details of the field character of basaltic and silicic units are dealt with by Jerram, who demonstrates the importance of a detailed understanding of the different types of volcanic units and their eruption style if we are to understand the formation of LIPs.

Atlantic margin evolution

The wealth of information on the Atlantic margins provides a basis to answer more fundamental questions, i.e., whether the continent-ocean transition is narrow or wide, whether conjugate margins are symmetrical or asymmetrical, and the nature of SDRS and the HVLC. Klausen et al. describe the coast-parallel dike swarm in Greenland, where they distinguish between fault-controlled extension and magmatic dilation by dike injection. It is apparent that magmatic extension followed tectonic extension during breakup on this volcanic rifted margin. Klausen et al. use detailed geological mapping to define a narrow continent-ocean transition zone. Mohriak et al. describe the conjugate margins of South America and Africa by integration of seismic, potential field, and borehole data. Mohriak et al. highlight the role of border faults in defining depocenters and reveal a high degree of asymmetry in some of the conjugate volcanic rifted margins. Corner et al. use potential field data on the Namibian volcanic rifted margin to discuss the rate and style of breakup and the age of volcanic-sedimentary units. Corner et al. discuss the extent to which reactivation of basement lineaments influenced rift orientation and provide some evidence that SDRS comprise both sedimentary and volcanic units. The nature of the thick (20 km) HVLC (Vp = 7.1–7.6 km/s) beneath the Namibian margin is modeled by Trumbull et al. They conclude that the velocities are consistent with highly magnesian melts, most likely olivine-rich basaltic magmas, in the lower crust. Trumbull et al. suggest that these melts may have been derived from the ancestral Tristan plume.

Volcanic rifted margins will continue to provide research targets for academics given the major interest in these margins by oil and gas companies. Future areas of interest include the nature of the lithosphere-asthenosphere in young, active volcanic rifted margins, the nature and source of volcanic SDRS, the continent to ocean transition, the temporal and spatial relationship between HVLC and SDRS, the application of geochronology to better define the prerift to synrift transition, matching conjugate margins using chronostratigraphy and thermochronology, and rifting mechanisms. Ultimately we hope to answer the question that is not yet resolved, i.e., how normal are volcanic rifted margins and what set(s) of circumstances create them during rifting.

ACKNOWLEDGMENTS

We thank the Penrose Foundation and the Geological Society of America for initially approving the proposal for the Penrose Conference at the Department of Geology, Royal Holloway University of London, with field excursions to the British Tertiary volcanic province and the Deccan Traps. Support for young scientists, keynote speakers, and overseas scientists from third world countries was provide by the International Lithosphere Programme, the International Association for Volcanology and Chemistry of the Earth's Interior, and British Petroleum plc. We thank the following reviewers for their considerable effort: U. Achauer, D. Anderson, N. Arnaud, N. Arndt, J. Cartwright, J. Chesley, J. Clemson, M. Coffin, I. Davison, R. England, R. Ernst, F. Frey, S. Holbrook, C. Jaupart, D. Jerram, N. Kusznir, G. Marsh, K. McClay, J. McHone, L. Melluso, P. Mohr, D. Peate, R. Sweeney, J. Wartho, and N. White.

Martin A. Menzies
Simon L. Klemperer
Cynthia J. Ebinger
Joel Baker

Printed in the U.S.A.

Geological Society of America
Special Paper 362
2002

Characteristics of volcanic rifted margins

Martin A. Menzies*
Department of Geology, Royal Holloway, University of London, Egham, Surrey TW20 OEX, UK
Simon L. Klemperer
Department of Geophysics, Stanford University, Stanford, California 94305-2215, USA
Cynthia J. Ebinger
Department of Geology, Royal Holloway, University of London, Egham, Surrey TW20 OEX, UK
Joel Baker
Dansk Lithosfærecenter, Oster Voldgade 10, 1350 Kobenhavn K, Denmark

ABSTRACT

Volcanic rifted margins evolve by a combination of extrusive flood volcanism, intrusive magmatism, extension, uplift, and erosion. The temporal and spatial relationships between these processes are influenced by the plate tectonic regime; the preexisting lithosphere (thickness, composition, geothermal gradient); the upper mantle (temperature and character); the magma production rate; and the prevailing climatic system. Of the Atlantic rifted margins, 75% are believed to be volcanic, the cumulative expression of thermotectonic processes over 200 m.y. Volcanic rifted margins also characterize Ethiopia-Yemen, India-Australia, and Africa-Madagascar. The transition from continental flood volcanism (or formation of a large igneous province) to ocean ridge processes (mid-ocean ridge basalt) is marked by a prerift to synrift transition with formation of a subaerial and/or submarine seaward-dipping reflector series and a significant thickness (to 15 km) of juvenile, high-velocity lower crust seaboard of the continental rifted margin. Herein we outline the similarities and differences between volcanic rifted margins worldwide and list some of their diagnostic features.

INTRODUCTION

Volcanic rifted margins (Fig. 1) are produced where continental breakup is associated with the eruption of flood volcanism during prerift and/or synrift stages of continental separation (Fig. 2) (Mutter et al., 1982; White et al., 1987; Holbrook and Kelemen, 1993; Eldholm and Grue, 1994; Courtillot et al., 1999). These margins are easily distinguished from nonvolcanic margins, like the Iberian margin, that do not contain such a large amount of extrusive and/or intrusive igneous rock and that may exhibit unusual features, such as unroofed mantle peridotites (e.g., Pickup et al., 1996; Louden and Chian, 1999). Mapping of flood basalt provinces and subsurface seismic volcanic-stratigraphic analyses show that volcanic rifted margins border the northern, central, and southern Atlantic Ocean, the southern Red Sea, the east coast of Africa, circum-Madagascar, the east and west coasts of India, the western and eastern coasts of Australia, and possibly parts of Antarctica (Coffin and Eldholm, 1992, 1994; Mahoney and Coffin, 1997; Planke et al., 2000) (Fig. 1). The initiation of a flood basalt province (or of a large igneous province [LIP]) (Fig. 2) is commonly a prerift phenomenon and takes the form of subaerial basaltic and/or silicic volcanism (e.g., Cox, 1988; Renne et al., 1992; Menzies et al., 1997a; Larsen and Saunders, 1998). The prerift to synrift transition is marked by a structural change, in some cases a magmatic hiatus, erosion of newly formed rift mountains, and the formation of high-velocity lower crust (HVLC), and a seaward-dipping reflector series (SDRS) (Mutter et al., 1982; White et al., 1987; Eldholm and Grue, 1994; Planke et al., 2000) (Fig. 2). SDRS comprise subaerial and submarine volcanic rocks and

*E-mail: menzies@gl.rhul.ac.uk.

Menzies, M.A., Klemperer, S.L., Ebinger, C.J., and Baker, J., 2002, Characteristics of volcanic rifted margins, *in* Menzies, M.A., Klemperer, S.L., Ebinger, C.J., and Baker, J., eds., Volcanic Rifted Margins: Boulder, Colorado, Geological Society of America Special Paper 362, p. 1–14.

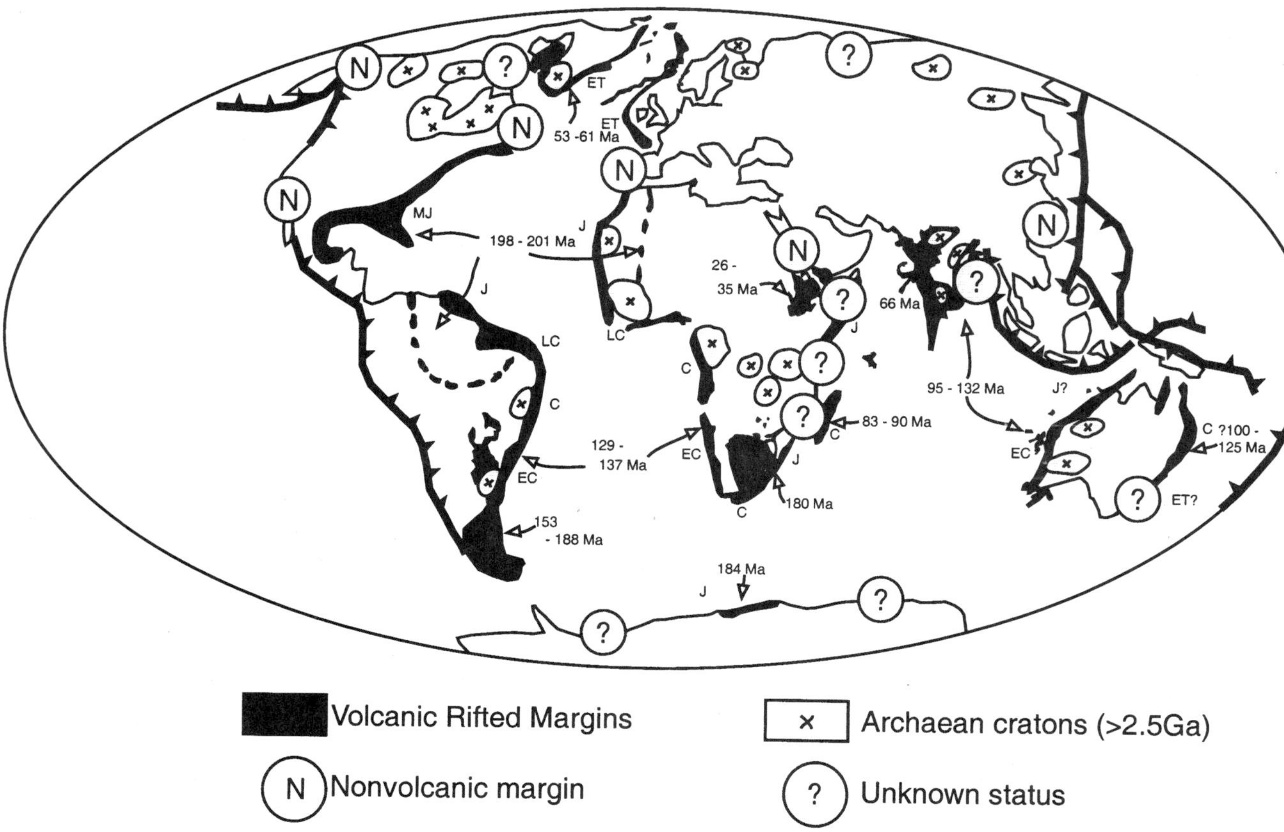

Figure 1. Distribution of volcanic rifted margins (<200 Ma), nonvolcanic rifted margins (N), and rifted margins of unknown status (?). Approximate age of oldest oceanic crust is shown adjacent to volcanic rifted margins along with age of large igneous provinces (LIPs) on the continental margin (J, Jurassic; C, Cretaceous; T, Tertiary; E, early, M, middle, L, late). Archean cratons (>2.5 Ga) are shown to help illustrate the relationship (or lack thereof) between LIPs and cratonic edges. After Mahoney and Coffin (1997), Planke et al. (2000), Press and Siever (2000).

probably variable amounts of sedimentary detritus shed from the volcanic rifted margin during uplift and tectonic denudation of the kilometer-scale rift mountains. The formation of SDRS is associated with the establishment of thicker than normal oceanic crust, seaward of the rifted margin at the continent to ocean transition (Fig. 2). Eventually stretching and heating lead to effective rupture of magmatically modified continental lithosphere, and seafloor spreading commences. This early oceanic crust may be thicker than normal owing to hotter asthenosphere associated with the plume and/or steep gradients at the lithosphere-asthenosphere boundary (e.g., Boutilier and Keen, 1999) (Fig. 2). The interval between the first expression of volcanic rifted margin formation on the prerift continental margin and the formation of true ocean floor can be tens of millions of years (Fig. 3).

In this paper we focus on evidence from a few of the better-known volcanic rifted margins, Ethiopia-Yemen, the Atlantic margins, and the Australia-India conjugate margins. Miocene to recent volcanic rifted margins (<30 Ma) exist in northeastern Africa (Hoffmann et al., 1997; George et al., 1998; Ebinger and Casey, 2001) and southwestern Arabia (e.g., Baker et al., 1996a; Menzies et al., 1997a, 1997b), where the conjugate margins are separated by ocean spreading centers in the southern Red Sea. The southern Red Sea and eastern Gulf of Aden are the youngest, hottest, and most active volcanic rifted margins. They are characterized by active volcanism, high heat flow, and shallow earthquakes (e.g., Davison et al., 1994; Manighetti et al., 1998; Ebinger and Casey, 2001). On the volcanic rifted margins of the southern Red Sea one can observe the temporal transition from prerift flood volcanism to synrift domino fault-block terranes (Yemen) and subaerial seaward-dipping reflector series (Ethiopia) (Fig. 1). Cretaceous-Tertiary volcanic rifted margins occur in the North Atlantic (i.e., western Greenland, Norway, and the United Kingdom) (e.g., Larsen and Jakobsdottir, 1988; Larsen and Saunders, 1998; Saunders et al., 1997; Klausen and Larsen, this volume), around peninsular India, and western Australia (e.g., Kent et al., 1997) (Fig. 1). In Brazil and Namibia, volcanic rifted margins are related to the opening of the South Atlantic (Peate, 1997), beginning with Parana and Etendeka flood volcanism (Hawkesworth et al., 1992; Renne et al., 1992, 1996a, 1996b; Peate, 1997; Mohriak et al., this volume; Corner et al., this volume). One of the spatially most extensive volcanic rifted

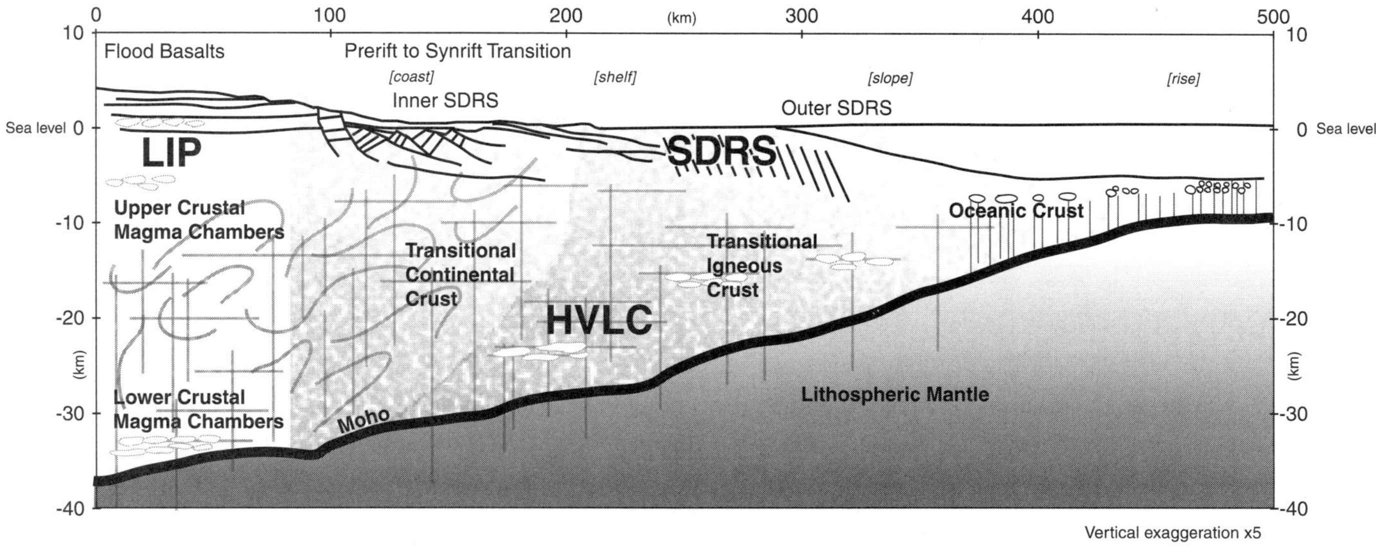

Figure 2. Schematic volcanic rifted margin based on data from Ethiopia-Yemen and the Atlantic margins. Volcanic rifted margin is characterized by a subaerial flood basalt province (i.e., large igneous province [LIP]) with upper and lower crustal magmatic systems; prerift to synrift transition with extended continental crust, and formation of a subaerial inner seaward-dipping reflector series (SDRS); development of high-velocity lower crust (HVLC) in the transition from continental to oceanic domains; formation of submarine outer seaward-dipping reflectors (SDRs), and an ocean basin (i.e., mid-ocean ridge basalt). In this example volcanism has extrusive episodes that are prerift (flood basalts), synrift (inner subaerial SDRS) and synrift and/or postrift (outer submarine SDRS).

margins, the Central Atlantic magmatic province, formed ca. 200 Ma (Holbrook and Kelemen, 1993; McHone, 1996; Hames et al., 2000) and records the initial breakup of Pangea. This volcanic rifted margin has been reduced to erosional remnants of intrusive and/or extrusive complexes (McHone, 1996) and offshore SDRS (Benson, 2002) spread over 10^6 km^2 (Fig. 1). Older intraplate flood basalt provinces erupted in the Permian-Triassic occur in Siberia, Russia, and Emeishan, China. However, their relationship to rifted margins is unknown, and they are not discussed herein. In addition, we do not consider relics of Precambrian flood basalt provinces that are apparent as dike swarms and unroofed plutonic complexes, because links to continental breakup are even more elusive (Mahoney and Coffin, 1997, and references therein).

We use these better known volcanic rifted margins as we discuss some of the controversies surrounding the origin of volcanic rifted margins (whether associated with active or passive rifting) and describe some of their diagnostic features, i.e., common association of silicic volcanism with the dominant flood basalts; crustal architecture of HVLC and SDRS at the continent-ocean transition; temporal relation between extension and magmatism; and rift-margin uplift and mountain building.

Passive continental rifted margins: Plate-driven and plume-driven processes

Traditionally, passive (plate driven) and active (plume driven) rifting models were invoked as an explanation for the formation of nonvolcanic and volcanic passive margins, respectively. Passive or plate-driven rifting models required that continental breakup was initiated by extensional forces, followed by surface uplift and magmatism related to the passive upwelling of normal asthenospheric material. In this case melts would be generated by shallow decompression melting processes. In contrast, active or plume-driven rifting models required deeper melt generation and subsequent interaction with the continental lithosphere. In these models the expectation is that volcanic rifted margins formed by kilometer-scale surface uplift prior to LIP formation and extension. However, recent research on volcanic rifted margins indicates that such simple chronologies do not apply to many rifted margins, suggesting that their formation is also not simple. The timing of uplift and extension relative to LIP formation is complex, requiring more detailed observations in individual provinces (Table 1).

Lithospheric thinning is an uncontested requirement for volcanic rifted margin formation. Extensional forces large enough to initiate rifting are generated by the presence of hot, low-density asthenosphere and subsequent heating of mantle lithosphere (Crough, 1978). More controversial is the mechanism responsible for the production of large volumes of basaltic volcanism at the Earth's surface, which in the majority of cases is spatially and temporally related to continental breakup (Watkeys, this volume). Adiabatic decompression melting due to active upwelling of normal asthenosphere is triggered by lithospheric thinning, as observed at ocean ridges. This occurs without a thermal anomaly and/or plume and may have led to considerable melt production on rifted margins (e.g., Holbrook and Kelemen, 1993; Boutilier and Keen, 1999; Korenaga et al.,

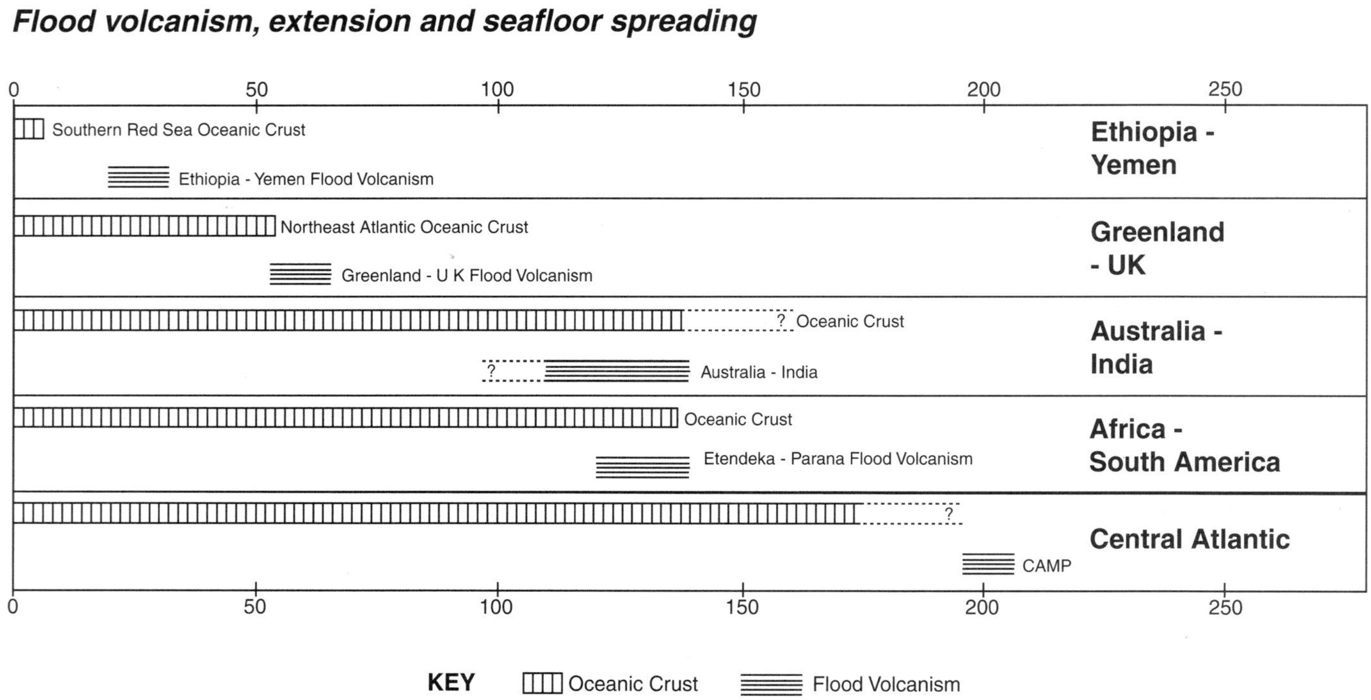

Figure 3. Formation of flood basalt provinces (large igneous provinces [LIPs]) on volcanic rifted margins and timing of formation of oceanic crust. Subaerial or submarine seaward-dipping reflector series (SDRS) characterize prerift to synrift (continent to ocean) transition and predate oldest oceanic crust on that volcanic rifted margin. Note that in most instances flood volcanism or LIP formation precedes breakup and formation of ocean crust. High-velocity lower crust and SDRS tend to form after the main flood basalt episode and before the youngest ocean crust. See text and table 1 for references.

2000). Alternatively, thermal anomalies or plume processes are believed to be a vital prerequisite for the generation of large volumes of melt (e.g., White and McKenzie, 1989; Sleep, 1996; Ernst and Buchan, 1997). In such cases the mantle potential temperature is elevated above that of the normal asthenosphere (~1300°C). However, there is considerable controversy over whether plumes initiate rifting, or rifting focuses plume activity (e.g., King and Anderson, 1998; White and McKenzie, 1989; Ebinger and Sleep, 1998; Nyblade, this volume). Controversy also continues with regard to the geometry of plumes; their temperature; their depth of origin; and their chemical identity (e.g., Turcotte and Emerman, 1983; Richards et al., 1989).

Geophysical and geochemical evidence, claimed as proof for the origin of plumes in the deep or shallow mantle, is equivocal. Mantle tomography models show distinct low-velocity zones at the core-mantle boundary, but their continuity with upper mantle low-velocity zones may be ambiguous. Ray dispersion complicates the simultaneous resolution of the width and depth of particular features in the upper mantle (Shen et al., 1998). However, as more and more receiver-function studies are undertaken in plume provinces, an exciting new plume detection method has evolved that allows direct measurement of the 410 km and 670 km discontinuities. As a result, the locations of plume stems can be mapped (Wolfe et al., 1997; Shen et al., 1998). For petrologists and geochemists the controversy surrounds the identification of volcanic rocks with unequivocal primary mantle signatures. Low magnesian volcanic rocks that have undergone low-pressure fractionation are obviously inappropriate probes of high-pressure mantle processes. Even highly magnesian unfractionated volcanic rocks can inherit the chemical signature of the lithosphere (crust and mantle) because of their higher temperature. However, advances have been made toward the identification of geochemical criteria that may help resolve mantle and/or crustal characteristics (e.g., Baker et al., 2000; Breddam et al., 2000; Melluso et al., this volume; Baker et al., this volume).

What may be stated, with some certainty, is that large-scale thermal anomalies, plumes, and/or hotspots exist in the Earth, and their distribution, temporal evolution, and spatial extent are highly variable. Variations may arise because of preplume lithospheric structure and differences in the velocity of plates over plumes. Slow moving plates may show larger volumes of melting than fast moving plates. In addition, their longevity and genesis may manifest as uplift, subsidence, and extension of the lithosphere and associated magmatism, occurring to different degrees and in various sequences. In volcanic rifted margins melts are produced by variations in pressure and temperature, and these can be achieved, respectively, by lithospheric thinning and thermal anomalies.

TABLE 1. CHARACTERISTICS OF VOLCANIC RIFTED MARGINS

	LIP: Present-day thickness of sub-aerial volcanic rocks	LIP: Period of eruption of 70%–80% of the basaltic rocks	LIP: Age of silicic volcanic rocks: pre-basaltic, syn-basaltic or post-basaltic eruptions	LIP and tectonics: pre-rift, syn-rift or post rift?	LIP: Estimated scale of pre-magmatic uplift	Presence and/or absence of a seaward dipping reflector series	Presence or absence of high velocity (~ 7.4 km/s) lower crust	HVLC: Presence of >10 km of new mafic igneous crust
1a Ethiopa	>2 km	Basalt-rhyolite (29–31 Ma) (base not dated)	Synchronous with basaltic eruptions 26–31 Ma	Pre-rift magmatism	Not known — buried by rift activity	Sub-aerial 'inner' SDRS	Yes	Not known
1b Yemen	>2 km: originally ca. 4 km with ca. 2 km lost to erosion	Basaltic eruptions 29–31 Ma	Post-basaltic eruptions 26–29 Ma	Pre-rift magmatism	10–100 m (marine to continental transition in sediments)	denuded remnant of 'inner' SDRS and buried 'outer' SDRS	Yes	Not known
2a Greenland	5–7 km	53–56 Ma	Intrusions, no volcanics reported	Pre-rift, syn-rift and post rift magmatism	100s meters	SDRS — no sediments reported	Yes	Yes
2b United Kingdom Tertiary Volcanic Province	ca. 1 km: heavily denuded margin	58–61 Ma and 53–56 Ma	Syn-basaltic eruptions from 58–61 Ma and absent 53–56 Ma	Pre-rift magmatism then syn-rift and post-rift (i.e., SDRS)	Unknown but volcanics erupted onto sub-aerially weathered marine sediments	SDRS — mostly volcanic rocks	Yes — Rockall (5 km thick)	Yes — under continent-ocean transition
3a India	1 km: Limited data; ca. 1 km (SDRS)	Basalt/rhyolite ca. 95–118 Ma	Syn-basalts	Not known	Not known	SDRS — Sylhet province?	Not known	Not known
3b Australia	1–2 km: >1 km (SDRS) and 1 km (Wallaby Plateau)	Bunbury 123 –132 Ma	Syn-basalts	Syn-rift and post-rift magmatism	Not known	SDRS — Wallaby Plateau	7.2–7.3 km/s Exmouth Plateau	Not known
4a Brazil Parana	1.8 km: originally ca. 4.8 km with 3 km lost to erosion	Basalt eruptions 129–133 Ma	Syn-basalts (and post-) basalts	? Pre-rift and syn-rift magmatism	? Possible pre-rift elevation ca. 500 m. (timing unclear)	Yes	Not known	Not known
4b Namibia-Etendeka	0.9 km: originally ca. 3.9 km with 3 km lost to erosion	Basalt eruptions 131–133 Ma	Syn-basalts (and post-) basalts	? Pre-rift and syn-rift magmatism	? Possible pre-rift elevation ca. 500 m. (timing unclear)	SDRS mixture of volcanics and sedimentary rocks (Corner et al., this volume)	Yes	Yes
5 Central Atlantic Magmatic Province	1 km	198–201 Ma	No silicic volcanic rocks	Post-rift magmatism (S.E. USA) and syn-rift magmatism (N.E. USA and Africa)	?? 0.9–2.6 km syn-rifting or pre-volcanic	SDRS volcanics with some interbedded sediments	Yes, significant igneous intrusions	No, normal oceanic crust (7–8 km)

Note: LIP, large igneous province; SDRS, seaward dipping reflector series; HVLC, high velocity lower crust.

Examples of source references for specific volcanic rifted margins: Ethiopia and Yemen: Berckhemer et al. (1975); Davison et al. (1994); Baker et al. (1996a,b); Menzies et al. (1997a,b); Egloff et al. (1997); Al'Subbary et al. (1998); George et al. (1998); Hoffmann et al. (1997); Baker et al. (2000); Ebinger and Casey (2001); Ukstins et al. (2002); Baker et al. (this volume). Greenland and UK: Roberts et al. (1979); Mutter et al. (1982); White et al. (1987); White and MacKenzie (1989); Brodie and White (1994); Saunders et al. (1997); Larsen and Saunders (1998); Jolley (1997); Korenaga et al. (2000); Planke et al. (2000); Klausen et al. (this volume). India and Australia: Von Stackleberg et al. (1980), Von Rad and Thurow (1992); Storey et al. (1992); Colwell et al. (1994); Exon and Colwell (1994); Milner et al. (1995); Frey et al. (1996); Kent et al. (1997). Parana and Etendeka: Hawkesworth et al. (1992); Renne et al. (1992); Gallagher et al. (1994); Turner et al. (1994); Renne et al. (1996a, 1996b); Gladczenko et al. (1997); Peate (1997); Clemson et al. (1999); Davison (1999); Jerram et al. (1999); Stewart et al. (1996); Hinz et al. (1999) and refs therein; Bauer et al. (2000); Corner et al. (this volume); Mohriak et al. (this volume); Trumbull et al. (this volume); Watkeys et al. (this volume); Central Atlantic Magmatic Province: McBride (1991); Holbrook and Kelemen (1993); McHone (1996); Lizarralde and Holbrook (1997); Withjack et al. (1998); Hames et al. (2000); Benson (2001); McHone and Puffer (2001); Schlische et al. (2001).

LIP continental basaltic and silicic flood volcanism: Shallow and deep sources

The birth of volcanic rifted margins (Table 1) is associated with the subaerial eruption of basaltic rocks and the minor eruption of submarine pillow lavas (e.g., Jolley 1997; Planke et al., 2000) (Fig. 2). Whereas basaltic volcanism normally dominated the evolution of the LIP, silicic volcanism may have contributed significantly to the total volume of the volcanic pile (e.g., Peate, 1997; Bryan et al., this volume; Jerram, this volume). LIPs, which characterize all volcanic rifted margins, are rarely thicker than 2 km (Table 1) because they represent the erosional remnants of earlier sequences estimated to have been as much as 2–3 times as thick at the time of eruption (Table 1) (Cox, 1980; Mahoney and Coffin, 1997). These estimates of the original erupted thickness on the continental margin take into account the amount of subaerial volcanism that has been eroded by synrift or postrift processes. However, the erupted thickness differs from the actual melt thickness produced during volcanic rifted margin formation, which must include igneous intrusives added to the continental crust as dike-sill complexes and plutonic centers. Today these may be evident as unroofed magma chambers extending the length of rifted margins (e.g., Namibia, Scotland), exposed dike swarms (e.g., Saudi Arabia), or overthickened HVLC. HVLC is never exposed at the surface, but is frequently reported from seismic data across volcanic rifted margins.

Many geochronological methods have been applied to volcanic rifted margins (e.g., Rb-Sr, K-Ar, Ar-Ar), but major advances in argon-argon dating using K-rich phenocryst phases (e.g., sanidine, amphiboles) and lasers have led to an improved understanding of the genesis of silicic and basaltic volcanic rocks in volcanic rifted margins (Renne et al., 1992, 1996a, 1996b; Turner et al., 1994; Hames et al., 2000; Ukstins et al., 2002; Miggins et al., this volume). There is considerable debate, however, about the age of individual provinces (see Peate, 1997, for review). In the majority of volcanic rifted margins, dating indicates that the main pulse (i.e., 70%–80%) of subaerial continental margin volcanism, both basaltic and silicic, occurred over a relatively short period of time ranging from 1 to 4 m.y. (Table 1).

In some volcanic rifted margins, basaltic volcanic rocks are dominant (e.g., Central Atlantic magmatic province, Greenland), while in others silicic volcanic rocks can constitute a significant part of the volcanic stratigraphy (e.g., northeastern Africa, South America, Africa) (Table 1). Silicic volcanism can occur early during the main basaltic episode or after the main basaltic eruptions (e.g., Ethiopia and Parana). Extrusive silicic rocks do not exist in all volcanic rifted margins, but may occur as silicic intrusives (e.g., Greenland; Table 1). The coexistence of basaltic and silicic volcanic rocks or the eventual switch from basaltic to silicic volcanism reveals the complexity of magmatic processes within volcanic rifted margins. Overall the complex relationships vary from basalt-dominated volcanic rifted margins, bimodal basalt-rhyolite volcanic rifted margins, to intermixed basalt-rhyolite volcanic rifted margins (Table 1). Crustal magma chambers play a pivotal role in the formation of silicic magmas, as does melting of the lower crust, perhaps fueled by basaltic underplating (Cox, 1980, 1988). In Yemen, the silicic volcanism that postdated basaltic volcanism and lasted >3 m.y. (Baker et al., 1996a) is believed to have originated by processes of assimilation and fractional crystallization of mantle-derived melts. Individual silicic volcanic units can be geochemically linked to nearby intrusive centers, often unroofed as granite-syenite-gabbro complexes (e.g., UK Atlantic margin, Yemen). These are presumed to have acted as source regions for the silicic volcanic rocks. In contrast, the origin of silicic volcanic rocks from Etendeka-Paraná erupted during the lifetime of the flood basalt province (Peate, 1997) may relate more to the formation of large-scale crustal melts. While several igneous complexes in Namibia have been identified as sources for the volcanic rocks on the basis of similar ages, the extensive synrift lava cover in many other examples may hide the identity of associated plutonic complexes. In other volcanic rifted margins (e.g., Greenland, Paraná, Yemen) the presence of a monotonous basalt stratigraphy on the rifted margin, and a paucity of plutonic rocks, may indicate that the plutonic rocks are offshore (e.g., Deccan), or are preserved on the conjugate margin (e.g., Yemen). It is possible that the basalt stratigraphy that dominates the volcanic rifted margins in Brazil, Ethiopia, and Greenland was inextricably linked to igneous centers now preserved in their conjugate margins, Namibia, Yemen, and Scotland, respectively.

Along the youthful northeastern African margins, silicic volcanic rocks were explosively erupted, typically venting 10^2–10^3 km^3 of magma (Ukstins et al., 2002). In the Deccan Traps and the North Atlantic Tertiary volcanic province, the presence of ash layers in the volcanic stratigraphy may indicate silicic volcanism (Deccan) or alkaline volcanism (Greenland) between periods of basaltic volcanism (e.g., Heister et al., 2001). In other volcanic rifted margins (e.g., Etendeka) (Peate, 1997), individual silicic eruptive units have thicknesses of ca. 100 m, aerial extents >8000 km^2, and volumes of 3000 km^3. These silicic units are comparable in volume to individual mafic lava units from LIPs like the Columbia River. Plinian eruption columns associated with the emplacement of voluminous ignimbrites in these volcanic rifted margins could have injected large amounts of aerosols into the atmosphere, and so affected global climate more than basaltic eruptions of similar volume.

Eruption rates in volcanic rifted margins have not been adequately defined by volume-time studies of individual eruptive units, but as a first approximation, thickness-time relationships reveal a marked decline in eruption rate from the mafic to the felsic eruptive stages of volcanic rifted margins (e.g., Hawkesworth et al., 1992; Baker et al., 1996a). This is consistent with the requirement for longer time periods to allow basaltic magmas to pond in shallow magma chambers and to evolve toward silicic derivatives by a combination of fractionation processes and assimilation of surrounding basement and/or roof rocks.

Continent-ocean transition: HVLC and SDRS

Voluminous subaerial flood volcanism on a continental margin lasting for millions of years requires a well-established magma transfer system within the crust and shallow mantle (Fig. 2). Cox (1980) first alluded to the potentially important contribution of sill-dike complexes to crustal growth during flood volcanism. Shallow (i.e., caldera structures) and deeper crustal magma chambers are a requirement of many models where the mineralogy and chemistry of mafic magmas indicate fractionation at lower crustal pressures and temperatures. The presence of plagioclase, clinopyroxene, and olivine phenocrysts in basaltic rocks alludes to fractional crystallization processes in lower crustal magma chambers, and, in many instances the geochemistry of these rocks reveals crustal contamination probably occurring concomitantly with evolution of the magmas in shallow or deep crustal chambers (e.g., Cox, 1980; Hooper, 1988). Even more extreme fractionation processes are apparent in the rhyolites found within volcanic rifted margins. Such rocks contain quartz, mica, and amphibole phenocrysts indicative of high-level processes. While some authors argue for an inextricable link between underplating and basin inversion (Brodie and White, 1994), there are few reports of kilometer-scale, underplated, high-velocity layers spatially limited below many basins that could be analogues of the well-documented HVLC at volcanic rifted margins (Lizarralde and Holbrook, 1997; Korenaga et al., 2000).

Characteristic features of volcanic rifted margins are zones of HVLC (Fig. 2) between stretched continental crust and normal thickness oceanic crust (e.g., Kelemen and Holbrook, 1995; Boutilier and Keen, 1999; Korenaga et al., 2000; Benson, 2001; Trumbull et al., this volume). Most likely the HVLC was emplaced during the breakup stage or, if it was a synrift feature, was associated with mantle upwelling (e.g., Kelemen and Holbrook, 1995; Boutilier and Keen, 1999). In southeast Greenland crustal thicknesses, at equivalent positions on the continental margin, vary from 30–40 km thick close to the thermal anomaly (i.e., track of Iceland hotspot) to 18 km 500–1000 km from the anomaly (Korenaga et al., 2000) (Fig. 2). In some volcanic rifted margins, the continent-ocean transition can be abrupt (e.g., Namibia) with entirely new HVLC formed seaward of almost unchanged, perhaps slightly thinned, continental crust. In this case the generation of additional igneous material may have more to do with extension and decompression melting than plumes and/or hotspots.

Current models for volcanic rifted margins are largely based on the results of geophysical surveying and scientific drilling in the northeastern Atlantic although few deep wells are available to calibrate interpretations (e.g., Korenaga et al., 2000). Scientific drilling in the northeastern Atlantic and industry drilling off Namibia (Kudu Field) show that lavas were erupted subaerially (e.g., Mutter et al., 1982; Clemson et al., 1999). SDRS, first recognized along the North Atlantic margin, mark the synrift stage in continental breakup and as such are characteristic of volcanic rifted margins (Roberts et al., 1979; Mutter et al., 1982; White et al., 1987; Larsen and Jakobsdottir, 1988; Korenaga et al., 2000; Benson, 2001). Volcanic rifted margins have thick sequences of seaward-dipping volcanic-sedimentary strata above, or seaward of, the region of HVLC, and extending landward to the ocean-continent transition zone (e.g., Mutter et al., 1982; Clemson et al., 1999). Reflector packages within these SDRS diverge downward and dip oceanward 20° or more (Fig. 2). Planke et al. (2000) divided the SDRS into "inner" and "outer" packages (Fig. 2) on the basis of studies of the North Atlantic margins (Fig. 1). The inner SDRS were subaerially emplaced flows, the geometry of which was affected by basin architecture. They proposed that this phase of volcanism occurred during subaerial seafloor spreading or syntectonic infilling of rift basins. The outer SDRS are believed to represent sheet flows in marine basins, and have similarities to subaerial flows. Submarine eruptions (i.e., pillowed flows and hyaloclastites) characterize this developmental stage. SDRS are synrift phenomena and are distinct from flood basalts; they straddle the continent-ocean boundary and can include subaerial and submarine volcanic and sedimentary rock types.

On the Namibian margin, modeling of magnetic data from seismic profiles suggests that the SDRS is a mixture of volcanic and sedimentary rocks. Presumably some portion of the SDRS must comprise sedimentary rocks, given that the volcanic stratigraphy on the uplifted margin can be reduced in thickness during synrift erosional processes (Gallagher et al., 1994). However, whether these sediments are argillaceous or arenaceous depends on the nature of the material removed from the margin (e.g., metamorphic, sedimentary, or igneous rocks). On the Norwegian volcanic rifted margin, seismic sections have been interpreted as representing a transition from subaerial to submarine volcanic deposits that comprise lavas and volcaniclastic sedimentary rocks (Planke et al., 2000). If we take the Yemen margin as an indication of what might constitute seaward-dipping reflector series, it is clear that a significant proportion (at least 50%) of these features must be sedimentary in nature. A sediment-budget analysis of the Red Sea margin (Davison et al., 1994) in Yemen indicated that several kilometers of basaltic and/or silicic volcanic rock were removed from the volcanic rifted margin during classic synrift extension. This erosional period would have contributed to the SDRS constructed on the stretched continental crust and embryonic oceanic crust.

Several volcanic rifted margins show an abrupt termination of the SDRS against a high-velocity structural high, which may be a late synrift intrusion (e.g., Planke et al., 2000), a fault, or an abandoned spreading ridge marking the ocean-continent boundary (e.g., Korenaga et al., 2000). Ebinger and Casey (2001) provided a mechanism for synrift emplacement of some SDRS via the development of high-strain neovolcanic zones and the abandonment of crustal detachments. Formation of SDRS on volcanic rifted margins is synchronous with the prerift to synrift transition on the continental margin. SDRS typically postdate flood volcanism on the rifted margin, and their formation may be synchronous with a hiatus in magmatism, a change in magmatic source area, and a peak in denudation. Because this is a situation that would not be associated with the generation of melt, it is likely that strain localization and focused extension accelerated melt generation. Although SDRS may predate ocean crust formation at a mid-ocean ridge, they are transitional between rifted continental margin processes and ocean ridge processes (Fig. 2). The continent-ocean transition is difficult to determine, and therefore considerable controversy surrounds the nature of the crust beneath many SDRS. The petrology and geochemistry of both SDRS and the HVLC hold a vital clue to a major change in the source of magmas, from one that fed a LIP to one that produced oceanic crust. SDRS and HVLC are two principal diagnostics of volcanic rifted margins (Fig. 2).

Breakup extension: Pre-LIP, syn-LIP, or post-LIP?

The relationships between the timing of LIP formation and rifting leading to ocean-floor formation are complex. This may in part be explained by the fact that some volcanic rifted margins are proximal, others distal, to plume heads and/or stems, so it is unlikely that volcanic rifted margins will show the same relationships. It is also complicated by the possibility that magma sources for volcanic rifted margins may reside either in the deep mantle (i.e., plumes) or the shallow mantle (i.e., asthenospheric small-scale convection). The temporal relationship between magmatism and extension may differ greatly if, as we believe, in deep-sourced plumes enhanced temperatures triggered melt production, whereas asthenospheric melts are decompression melts triggered by lithospheric thinning. We envisage plume-derived magmatism occurring at any stage in the development of a rifted continent (prerift, synrift, or postrift), whereas magmatism derived from the shallow mantle would largely be synrift or postrift.

Another problematic aspect of understanding the relationship between extension and magmatism is defining the timing of rifting and/or extension. Extension may be fault controlled or via dike injection (Klausen and Larsen, this volume), and may be identified as the appearance of the first fault, the first volcanic rock, or the first depocenter. Is the onset of extension the timing of the initiation of continental extension, or is breakup marked by the formation of seafloor sensu stricto? Tens of millions of years can pass between the initiation of LIP formation (prerift) and the generation of seafloor, so it is important to understand absolute and relative timing of the geological processes leading to the formation of new seafloor. Any generalization about the apparent synchroneity of magmatism, extension, and uplift ignores the reality that, with the technology available, we can resolve the relative and absolute timing of these processes and so better understand rift processes.

In Figure 3 the relationship between flood volcanism (i.e., LIP formation) and the formation of oceanic crust is summarized for many of the volcanic rifted margins that formed in the past 200 m.y. (see also Courtillot et al., 1999). The age of the oldest oceanic crust adjacent to the volcanic rifted margin in question can be used as a minimum age of seafloor spreading because it is conceivable that this is not the oldest ocean floor, but merely the oldest seafloor for which samples exist (Fig. 3). The age of oceanic crust can be compared with the age of flood volcanism on the volcanic rifted margin to better understand the relationship between extension and magmatism.

In Ethiopia-Yemen, magmatism is dated by Ar-Ar methods as 31–26 Ma (Baker et al., 1996a; Hoffman et al., 1997; Ukstins et al., 2002). Extension (leading to the formation of domino fault-block terranes) is defined by Ar-Ar and fission-track dating of hanging-wall and footwall lithologies (Menzies et al., 2001). Extension in Yemen (i.e., southern Red Sea margin) began in the late Oligocene (ca. 26 Ma), coincident with a marked hiatus in extrusive activity and significant tectonic erosion and/or crustal cooling dated by fission-track methods and validated by Ar-Ar dating of unconformities as 19–25 Ma (Baker et al., 1996a; Menzies et al., 1997a). On the conjugate margin in Ethiopia, extension occurred along the length of the western escarpment ≥25 Ma, indicating that rifting occurred after the onset of flood basaltic volcanism ca. 31 Ma (Ukstins et al., 2002). Volcanic rocks were erupted from isolated centers located along the western escarpment in Ethiopia (Kenea et al., 2001; Ukstins et al., 2002). We conclude that much of the Ethiopian-Yemeni flood volcanism was prerift in character. While the timing will not be the same for all volcanic rifted margins, the southern Red Sea is an illustration of how breakup and the continent-ocean transition can be protracted.

In the case of the North Atlantic (Greenland-UK) (Fig. 1), LIP formation lasted from 61 to 53 Ma (e.g., Eldholm and Grue, 1994; Saunders et al., 1997) and the oldest oceanic crust indicates that extension must have taken place before 52 Ma (Fig. 3). From this it appears that volcanism in the North Atlantic straddled breakup with a prerift (LIP) and a synrift stage (SDRS) (Larsen and Saunders, 1998). Such a protracted period of volcanism may explain the attenuated, heavily intruded nature of the broad continent-ocean transition.

The details of the timing are less well known for Australia-India (Fig. 3). Volcanism on the Indian and Australian margins occurred between 100 and 130 Ma (e.g., Kent et al., 1997), and breakup between Australia, India, and Antarctica was 125–133 Ma (Fig. 3). It appears that volcanism on the rifted margin was synchronous with continental breakup, but that volcanism continued (sporadically?) during formation of oceanic crust.

In the Paraná-Etendeka volcanic rifted margins (Fig. 1), oceanic crust located off Africa is slightly older than that known off South America. The age of the oceanic crust indicates that extension occurred ca. 135 Ma, overlapping with the Paraná-Etendeka LIP (Peate, 1997). Because the main pulse of basaltic magmatism occurred ca. 130–133 Ma, it can be inferred that the LIP was largely prerift to synrift. A prerift stage is supported by the fact that the main volcanic units can be traced, and the volcanic stratigraphies matched, from the Etendeka across the Atlantic Ocean to the Paraná of Brazil (e.g., Milner et al., 1995; Mohriak et al., this volume). Synrift magmatism is supported by offshore valley systems that appear to be filled with extrusive lavas with later deformation and faulting-controlled emplacement of the volcanic units (cf. Clemson et al., 1999). Alternatively, both these observations could be explained by a synrift model for the magmatic activity. Initial pulses of magmatism would fill topographic lows, as described by Clemson et al. (1999), and further synrift activity would mantle the filled topography such that units were traceable from South America to Africa, as reported by Milner et al. (1995).

The relationships for the Central Atlantic magmatic province (Fig. 3) appear more complex, probably because of the size of the province and the extent to which it has been eroded. Continental magmatism has been dated as 198–201 Ma (Hames et al., 2000). However, along the eastern margin of North America the relationship between magmatism and tectonics is variable (J. McHone, 2001, personal commun.). In southeastern North America, volcanic rocks of the Central Atlantic magmatic province appear to postdate both the cessation of rifting by ca. 10 Ma, and uplift and/or erosion. This should be contrasted with Central Atlantic magmatic province magmatism in northeastern North America and northwestern Africa, where magmatism is synrift and rifting continued for ~25 m.y. after magmatism followed by Middle to Late Jurassic uplift (J. McHone 2001, personal commun.). SDRS from offshore northeastern United States are thought to have been emplaced ca. 175 Ma (Withjack et al., 1998; Benson, 2002; Schlische et al., 2002), and flood volcanism appears to be synrift or postrift. This contrasts with the North Atlantic margins (Greenland, UK) where a significant prerift flood volcanic stage is evident. However, there may be a bias in the rock record. In the Central Atlantic magmatic province, onshore intrusive rocks are used to define the timing of magmatism on the rifted margin. However, in deeply eroded volcanic rifted margins, like the Central Atlantic magmatic province, these hypabyssal and/or plutonic rocks may bias the dating toward the synrift stage. We use the Yemen volcanic rifted margin as an illustration of how hypabyssal and/or plutonic rocks may be largely synrift in age, despite a prerift history of 4–5 m.y. of flood basalt volcanism unrepresented in these exposed hypabyssal and/or plutonic rocks. In Yemen the original subaerial volcanic stratigraphy has an age of 31–26 Ma, and is known to be prerift (Baker et al., 1996a; Menzies et al., 1997a, 1997b). Hypabyssal and plutonic rocks underlying or intruding the volcanic rifted margin have ages that are primarily younger than 25 Ma (Chazot et al., 1998, and references therein) and so intrusive activity, as exposed, is largely synrift. It appears that peak extension (and erosion 19–26 Ma) was associated with a possible extrusive hiatus, but with significant intrusive activity exemplified by the dike swarms and granite-gabbro-syenite laccoliths. This synrift intrusive stage is confirmed by <25 Ma dikes that are parallel to the Red Sea margin and that occur in Saudi Arabia and Egypt (Chazot et al., 1998, and references therein). If these hypabyssal rocks were all that remained of the Yemen volcanic rifted margin, their ages (21–25 Ma) would be biased toward dating the peak of extension (<25 Ma), possibly the period of formation of the SDRS (<25 Ma), but definitely not the onset of flood volcanism (31 Ma) or the true age of the magmatic period (31–19 Ma).

The relationship between magmatism and faulting is complex, and synchroneity between the two processes, a model driven expectation, appears to be unsupported in many volcanic rifted margins. Magmatism may have predated rifting by several million years as in the Ethiopia-Yemen volcanic rifted margins, postdated rifting as in some of the Central Atlantic volcanic rifted margins, or straddled the prerift to synrift transition as in the North Atlantic volcanic rifted margins of Greenland-UK. To resolve this issue, as in all volcanic rifted margins, accurate dating of extrusive and intrusive volcanic rocks on the continental margin and the ocean floor and the timing of extension are needed. The volcanic part of volcanic rifted margins is complex, with the possibility, as seen in Yemen, of prerift (i.e., flood basalts), synrift (i.e., hypabyssal and/or plutonic rocks), and posterosional (i.e., alkaline volcanic rocks) stages (Chazot et al., 1998). Selection of any of those rock types in trying to understand the relationship between magmatism and tectonics may predetermine the outcome. The magmatic source for volcanic rifted margins may be deep (i.e., plumes) or shallow (i.e., asthenosphere). Shallow melt production from decompression melting should be synrift or postrift (e.g., Central Atlantic magmatic province). Deep melt production can exploit already established rift systems (e.g., India-Australia) or help generate new rift systems (e.g., Yemen-Ethiopia) and so may be prerift, synrift, or postrift.

Rift margin mountains: Prevolcanic or synvolcanic uplift and erosion

Many rifted continental margins are bordered by eroded mountain ranges, evident as topographic highs proximal to the rifted margin. This can be seen in the Paraná-Etendeka, northeastern African, Greenland-Scotland, and the Deccan Traps (Fig. 2). The highest points in Arabia and the UK are atop relic mountain ranges that border the volcanic rifted margin of the southern Red Sea and the North Atlantic, respectively. The continental margins of eastern Brazil and western India are bordered by steep scarps facing the rift valley, whereas the rift shoulder is characterized by less dramatic slope development (Gallagher et al., 1994). In most of these volcanic rifted margins no kilometer-scale mountain range existed prior to rifting and magmatism, so an important part of the evolution of volcanic rifted margins is mountain building and erosion. The juvenile nature of the mountains and/or landscape is evident from the drainage patterns, which were influenced by mountain building and margin uplift. Cox (1989) drew attention to the drainage patterns on volcanic rifted margins and the fact that, in many cases, major rivers flowed away from the present-day coastline and followed a lengthy inland course because of the uplifted rift margins. Some of the best examples are the Rio de la Plata–Parana (Brazil), which has its headwaters <250 km from the Atlantic margin; the river follows a course of several thousand kilometers to the west and south away from the present coastline, reaching the Atlantic Ocean at Buenos Aires. Similarly the Blue Nile has headwaters in the rift mountains of the Ethiopian highlands within 500 km of the Red Sea. However, these waters flow westward and northward for several thousand kilometers to reach the Mediterranean Sea at Alexandria.

To help bracket the period of uplift and mountain building, limitations have to be placed on the prevolcanic rifted margin paleoenvironment and the onset of erosion. Paleoenvironmental clues are in the prerift sedimentary rocks that underlie the earliest volcanic rocks of volcanic rifted margins. However, caution has to be exercised in interpreting the prerift sedimentary rocks, because in many cases they are very difficult to date and therefore cannot be definitively shown to relate in space and time to the volcanic rifted margin sensu strictu. Whereas paleoenvironmental analysis helps us determine the approximate time when mountain building began, the timing of crustal cooling or denudation may be used to define the minimum time when topography existed on the margin. On most volcanic rifted margins (Table 1) the magnitude of uplift prior to volcanism is on a scale of hundreds of meters, typically measured by locating marine horizons within the volcanic stratigraphy. Around the Red Sea, as in many other volcanic rifted margins, marine sediments are several kilometers above present sea level (Davison et al., 1994), revealing significant rift shoulder uplift.

In Yemen a paleoshoreline existed (>31 Ma) close to the present location of a 4-km-high mountain range. Paleocurrent information and the maturity of the prevolcanic sediments in Yemen require a hinterland on what is now the opposite side of the rift in the Danakil horst, Eritrea (Al'Subbary et al., 1998). The prevolcanic sedimentary rocks imply that the continental masses were close to sea level in the southern Red Sea and, by inference, Eritrea (Al'Subbary et al., 1998). The predominance of subaerial volcanic rock units (rather than submarine flows or hyaloclastites) also indicates a subaerial continental environment at 31 Ma; some rift-related uplift must have occurred before that time. Possibly the initiation of uplift is recorded in changes to the orientation of the paleoshoreline, along with a shallow marine to continental transition that occurred prior to volcanism. These changes indicate that uplift of that surface (ca. 31 Ma) was tens to hundreds of meters. However, the exact age of these sediments relative to the period of formation of the volcanic rifted margin is unknown. In the unlikely event that these sedimentary rocks are considerably older than the volcanic rifted margin, the paleoenvironmental changes would not relate to the evolution of the volcanic rifted margin.

Fission-track ages date crustal cooling and hence rapid tectonic denudation as having occurred between 19 and 26 Ma (Menzies et al., 1997a) on the Red Sea margin. We presume that Oligocene-Miocene denudation required greater topography than the paleoshoreline inferred to exist at 31 Ma, hence such topography was generated, and the period of uplift and exhumation is bracketed, in the late Oligocene (26–31 Ma). Independent verification of this period of denudation is to be found in unconformities in the volcanic stratigraphy that formed between 19 and 26 Ma (Baker et al., 1996a). In contrast to the largely synvolcanic uplift in Yemen, uplift on a scale of hundreds of meters is believed to have preceded volcanism in the North Atlantic province (e.g., Larsen and Saunders, 1998). In western Greenland major unconformities beneath the volcanic rocks are associated with fluvial peneplanation and valley incision, indicating a period of prevolcanic uplift and erosion. However, in eastern Greenland during the same time the landscape was close to sea level and, in northwest Scotland-Faeroes, the prevolcanic landscape was a low-relief, vegetated land surface (Jolley, 1997). Furthermore, in northwest Scotland subaerially weathered marine sediments (chalk) underlie the lowermost flood volcanic rocks (G. Fitton, 2001, personal commun.). The prerift North Atlantic volcanic rifted margin (Brodie and White, 1994) could be classified as a low-relief land surface with some incised river systems. The proximity of that land surface to sea level is revealed by studies in Norway (Planke et al., 2000), where the margin is believed to have developed in the continental to oceanic transition.

In Namibia and Brazil (Etendeka-Parana) the basal volcanic rocks overlie and are interbedded with continental eolian sandstones, which occasionally overlie fluvial deposits. A large eolian erg system is reported intercalated with the lowermost flood basalts (Jerram et al., 1999), but how far above sea level it was formed is not known. On this volcanic rifted margin uplift and doming prior to rifting could be argued for due to the lack of Upper Karoo sediments (Clemson et al., 1999). This may be consistent with conclusions based on fission-track data that argue for a prerift elevation of ~500 m in southeastern Brazil (Gallagher et al., 1994).

The degree of preservation of volcanic rifted margins is inextricably linked to climate, elevation, and the amount and/or rate of erosion. The youngest volcanic rifted margins survive as 3–4-km-high mountain ranges in the desert climate of northeastern Africa, and Cretaceous-Tertiary volcanic rifted margins are characterized by ~5–7-km-thick volcanic sections in the mountain ranges of subpolar Greenland, deeply eroded rift mountains in the west maritime climate of the UK, and major scarp retreat in tropical India and Brazil. The western Ghat escarpment (Deccan) is believed to have an erosional, rather than a tectonic origin, and scarp retreat is believed to be the major determinant of landscape with the original continental margin ~75 km west of its present location. Erosion over 200 m.y. has reduced the subaerial portion of the Central Atlantic magmatic province in the eastern United States and western Africa to a dike swarm. However, submarine equivalents to these margins have survived offshore as SDRS. Just as in the relationship between magmatism and rifting, we have a complex history of uplift and denudation operating on a variety of prevolcanic landscapes.

Volcanic rifted margins: Summary

Volcanic rifted margins evolved in response to local thermotectonic conditions, and consequently marked differences can be found in the temporal and spatial relationships between tectonics, magmatism, uplift, and erosion. Of all passive continental margins around the world, 90% are volcanic rifted margins to varying degree, the exceptions being continental margins in eastern China, Iberia, the northern Red Sea, South Australia, the Newfoundland Basin–Labrador Sea, and possibly the Gulf of California. Although the Arctic and Antarctic margins have largely unknown status, parts of the Antarctic margin are believed to be volcanic (Fig. 1).

A considerable variation exists in volcanic rifted margins. The prevolcanic environment can vary from shallow marine-continental (e.g., North Atlantic), fluvial-continental (e.g., Yemen-Ethiopia) to eolian continental (e.g., Etendeka). Flood volcanism can be thick (e.g., 7 km, Greenland) or relatively thin (e.g., 1.5 km, Deccan). Volcanism can be represented by predominantly basaltic volcanic rocks at the base and by mainly silicic volcanic rocks at the top (e.g., Yemen), or silicic volcanic rocks may be found throughout the volcanic stratigraphy (e.g., Ethiopia) or be essentially absent (e.g., Deccan). Processes related to volcanic rifted margin can vary in time and space. Magmatism can predate breakup extension by several million years (e.g., Yemen-Ethiopia), magmatism and breakup can be synchronous (e.g., Greenland–North Atlantic Tertiary volcanic province), or magmatism can postdate breakup by several million years (e.g., Australia-India). Magmatism in volcanic rifted margins may have originated by decompression melting associated with lithospheric thinning and/or upwelling of thermal anomalies modified by melting of lithospheric rocks. In most volcanic rifted margins prevolcanic uplift can vary from tens of meters (Yemen) to hundreds of meters (North and South Atlantic), but it appears that kilometer-scale prevolcanic uplift is not as widespread as some models have predicted.

The magmatic and structural evolution of individual volcanic rifted margins is complex and may not fit simple models. This may be due to the geology, age, and thickness of the prerift lithosphere and proximity to plume heads, which are potentially variable in temperature, longevity, and dimensions. There appears to be a continuous gradation from volcanic rifted margins to nonvolcanic rifted margins; a possible continuum is evident in the southern Red Sea. Much remains to be learned about the extent to which plumes drive, or are focused by, lithospheric extension and the exact geophysical and geochemical nature of plumes. Perhaps we can better understand the process of formation of volcanic rifted margins by comparing and contrasting their geological characteristics with nonvolcanic rifted margins where continental rifting occurs without thermally enhanced mantle (e.g., Newfoundland, Iberia), or the formation of intraplate large igneous provinces in ocean basins (e.g., Ontong-Java oceanic plateau) and continents (e.g., Siberian flood basalts) where widespread rifting is absent. Clearly a single active rifting model cannot explain the formation of all volcanic rifted margins around the world.

Volcanic rifted margins: Characteristics

The following characteristics are sufficiently common in volcanic rifted margins to be diagnostic.

1. Flood volcanism may have reached 4–7 km thickness prior to erosion, which has reduced several margins to thicknesses of 1–2 km.
2. Basaltic and silicic volcanic rocks are erupted subaerially. Of the exposed subaerial basaltic rocks, 70%–80% occurred in <3 m.y. The eruption of silicic volcanic rocks can occur during or after eruption of the basaltic rocks and can last for as many as 5 m.y.
3. Magmatism and rifting are not necessarily synchronous. Magmatism can occur before, during, or after rifting. In some volcanic rifted margins a magmatic hiatus coincided with the peak of extension.
4. Rift mountains are uplifted and rapidly eroded by synrift (or postrift) processes.
5. Seaward-dipping reflectors comprise a mixture of volcanic flows, volcaniclastic deposits (e.g., hyaloclastites), and nonvolcanic sediments. Formation of the SDRS postdates flood volcanism.
6. HVLC (~7.4 km/s) forms in the continent-ocean transition by igneous processes and reaches considerable thicknesses (10–15 km). The exact relationship between the formation of the high-velocity lower crust and continental (flood) or oceanic (mid-ocean ridge basalt) volcanism is unknown.

ACKNOWLEDGMENTS

We acknowledge the financial support of the Penrose Foundation, British Petroleum, the International Lithosphere Program, and the International Association for Volcanology and Chemistry of the Earth's Interior. We thank Steve Holbrook for his constructive comments on an earlier version of this manuscript. The manuscript has also benefited from comments by Millard Coffin and from input by Hans Christian Larsen, Godfrey Fitton, Scott Bryan, Dave Peate, and Greg McHone. We also thank those who freely shared ideas at the Penrose "Volcanic rifted margin" meeting at the Department of Geology, Royal Holloway in March–April 2000 and during the Mahabaleshwar and Mull field trips.

REFERENCES CITED

Al'Subbary, A.K., Nichols, G.N., Bosence, D.W.J., and Al-Kadasi, M. 1998, Pre-rift doming, peneplanation or subsidence in the southern Red Sea?: Evidence from the Medj-zir Formation (Tawilah Group) of western Yemen, *in* Purser, B., and Bosence, D., eds., Sedimentation and tectonics in rift basins: Red Sea–Gulf of Aden case: London, Chapman and Hall, p. 119–134.

Baker, J., Snee, L., and Menzies, M.A., 1996a, A brief Oligocene period of flood volcanism in Yemen: Implications for the duration and rate of continental flood volcanism at the Afro-Arabian triple junction: Earth and Planetary Science Letters, v. 138, p. 39–56.

Baker, J., Thirlwall, M.F., and Menzies, M.A., 1996b, Sr-Nd-Pb isotopic and trace element evidence for crustal contamination of a mantle plume: Oligocene flood volcanism in western Yemen: Geochimica et Cosmochimica Acta, v. 60, p. 2559–2581.

Baker, J., MacPherson, C.G., Menzies, M.A., Thirlwall, M.F., Al-Kadasi, M., and Mattey, D.P., 2000, Resolving crustal and mantle contributions to continental flood volcanism, Yemen: Constraints from mineral oxygen isotope data: Journal of Petrology, v. 41, no. 12, p. 1805–1820.

Bauer, K., Neben, S., Schrekenberger, B., Emmerman, R., Hinz, K., Jokat, W., Schulze, A., Trumbull, R.B., and Weber, K., 2000, Deep structure of the Namibia continental margin as derived from integrated geophysical studies: Journal of Geophysical Research, v. 105, p. 25829–25853.

Benson, R.N., 2002, Chronology of CAMP basalts and seaward-dipping reflectors of the North American Atlantic continental margin, *in* Hames, W.E., ed., The Central Atlantic Magmatic Province: American Geophysical Union Monograph (in press).

Boutilier, R., and Keen, C., 1999, Small-scale convection and divergent plate boundaries: Journal of Geophysical Research, v. 104, p. 7389–7403.

Breddam, K., Kurz, M.D., and Storey, M., 2000, Mapping out the conduit of the Iceland Mantle plume with helium isotopes: Earth and Planetary Science Letters, v. 176, p. 45–55.

Brodie, J., and White, N., 1994, Sedimentary basin inversion caused by igneous underplating: Northwest European continental shelf: Geology, v. 22, p. 147–150.

Chazot, G., Menzies, M.A., and Baker, J., 1998, Pre-, syn- and post-rift volcanism on the south-western margins of the Arabian Plate, *in* Purser, B., and Bosence, D., eds., Sedimentation and tectonics in rift basins: Red Sea–Gulf of Aden case: London, Chapman and Hall, p. 50–55.

Clemson J., Cartwright, J., and Swart, J., 1999, The Namib rift: A rift system of possible Karoo age, offshore Namibia, *in* Cameron, N.R., Bate, R.H., and Clure, V.S., eds., The oil and gas habitats of the South Atlantic: Geological Society [London] Special Publication 153, p. 381–402.

Coffin, M.F., and Eldholm, O., 1992, Volcanism and continental break-up: A global compilation of large igneous provinces, *in* Storey, B.C., Alabaster, T., and Pankhurst, R.J., eds., Magmatism and the causes of continental break-up: Geological Society [London] Special Publication 68, p. 21–34.

Coffin, M.F., and Eldholm, O., 1994, Large igneous provinces: Crustal structure, dimensions and external consequences: Reviews of Geophysics, v. 32, p. 1–36.

Colwell, J.B., Symonds, P.A., and Crawford, A.J., 1994, The nature of the Wallaby (Cuvier) Plateau and other igneous provinces of the West Australian Margin, *in* Exon, N., ed., Geology of the outer North West Shelf, Australia: Australian Geological Survey Organisation, AGSO Journal of Australian Geology and Geophysics, v. 15, no. 1, p. 137–156.

Courtillot, V., Jaupart, C., Manighetti, I., Tapponier, P., and Besse, J., 1999, On causal links between flood basalts and continental break-up: Earth and Planetary Science Letters, v. 166, p. 177–195.

Cox, K.G., 1980, A model for flood basalt volcanism: Journal of Petrology, v. 21, p. 629–650.

Cox, K.G., 1988, The Karoo Province, *in* McDougall, J.D., ed., Continental flood basalts: Amsterdam, Kluwer Publishers, p. 239–271.

Cox, K.G., 1989, The role of mantle plumes in the development of continental drainage patterns: Nature, v. 342, p. 873–877.

Crough, S.T., 1978, Thermal origin of mid-plate hotspot swells: Geophysical Journal of the Royal Astronomical Society, v. 55, no. 2, p. 451–469.

Davison, I., 1999, Tectonics and hydrocarbon distribution along the Brazilian S. Atlantic margins, *in* Cameron, N., Bate, R., and Clure, V., eds., The oil and gas habitats of the S. Atlantic: Geological Society [London] Special Publication 153, p. 133–152.

Davison, I., Al-Kadasi, M., Al-Khirbash, S., Al-Subbary, A., Baker, J., Blakey, S., Bosence, D., Dart, C., Owen, L., Menzies, M., McClay, K., Nichols, G., and Yelland, A., 1994, Geological evolution of the southern Red Sea rift margin—Republic of Yemen: Geological Society of America Bulletin, v. 106, p. 1474–1493.

Ebinger, C., and Casey, M., 2001, Continental break-up in magmatic provinces: An Ethiopian example: Geology, v. 29, p. 527–530.

Ebinger, C., and Sleep, N., 1998, Cenozoic magmatism in Africa: One plume goes a long way: Nature, v. 395, p. 788–791.

Eldholm, O., and Grue, K., 1994, North-Atlantic volcanic margins: Dimensions and production rates: Journal of Geophysical Research, v. 99, p. 2955–2968.

Ernst, R.E., and Buchan, K.L., 1997, Giant radiating dike swarms: Their use in identifying pre-Mesozoic large igneous provinces and mantle plumes: *in* Mahoney, J., and Coffin, M.F., eds., Large igneous provinces: Continental oceanic and planetary flood volcanism: American Geophysical Union Geophysical Monograph 100, p. 297–334.

Exon, N., and Colwell, J.B., 1994, Geological history of the outer North West Shelf of Australia: A synthesis, *in* Exon, N., ed., Geology of the outer North West Shelf, Australia: Australian Geological Survey Organisation, AGSO Journal of Australian Geology and Geophysics, v. 15, no. 1, p. 177–190.

Frey, F.A., McNaughton, N.J., Nelson, D.R., de Laeter, J.R., and Duncan, R.T.A., 1996, Petrogenesis of the Bunbury Basalt, Western Australia: Between the Kerguelen plume and Gondwana lithosphere?: Earth and Planetary Science Letters, v. 144, no. 1–2, p. 163–183.

Gallagher, K., Hawkesworth, C.J., and Mantovani, M., 1994, The denudation history of the inshore continental margin of SE Brazil inferred from apatite fission track data: Journal of Geophysical Research, v. 99, p. 18117–18145.

George, R., Rogers, N., and Kelley, S., 1998, Earliest magmatism in Ethiopia: Evidence for two mantle plumes in one flood basalt province: Geology, v. 26, p. 923–926.

Gladczenko, T., Hinz, K., Eldholm, O., Meyer, H., Neben, S., and Skogseid, J., 1997, South Atlantic volcanic margins: Journal of the Geological Society of London, v. 154, p. 465–470.

Hames, W.E., Renne, P.R., and Ruppel, C., 2000, New evidence for geologically-instantaneous emplacement of earliest Jurassic Central Atlantic magmatic province basalts on the North American margin: Geology, v. 28, p. 859–862.

Heister, L.E., O'Day, P.A., Brooks, C.K., Neuhoff, P.S, and Bird, D.K., 2001, Pyroclastic deposits within the East Greenland Tertiary basalts: Journal of the Geological Society of London, v. 158, p. 269–284.

Hawkesworth, C.J., Gallagher, K., Kelley, S., Mantovani, M.S.M., Peate, D.W., Regelous, M., and Rogers N.W., 1992, Parana magmatism and the opening of the South Atlantic, *in* Storey, B., Alabaster, A., and Pankhurst, R., eds., Magmatism and the causes of continental break-up: Geological Society [London] Special Publication 68, p. 221–240.

Hinz, K., Neben, S., Schreckenberger, B., Roeser, H.A., Block, M., Goncalves de Souza, K., and Meyer, H., 1999, The Argentine continental margin north of 48 deg N: Sedimentary successions, volcanic activity during break-up: Marine and Petroleum Geology, v. 16, no. 1, p. 1–25.

Hofmann, C., Courtillot, V., Feraud, F., Rochette, P., Yirgu, G., Ketefo, E., and Pik, R., 1997, Timing of the Ethiopian flood basalt event: Implications for plume birth and global change: Nature, v. 389, p. 838–840.

Holbrook, W.S., and Kelemen, P.B., 1993, Large igneous province on the US Atlantic margin and implications for magmatism during continental breakup: Nature, v. 364, p. 433–436.

Hooper, P., 1988, Crystal fractionation and recharge (RFC) in the American bar flows of the Imnaha Basalt, Columbia River basalt group: Journal of Petrology, v. 29, p. 1097–1118.

Jerram, D.A., Mountney, N., Holzforster, F., and Stollhofen, H., 1999, Internal stratigraphic relationships in the Etendeka Group in the Huab Basin, NW Namibia: Understanding the onset of flood volcanism: Journal of Geodynamics, v. 28, p. 393–418.

Jolley, D.W., 1997, Palaeosurface palynofloras of the Skye lava field and the age of the British Tertiary volcanic province, *in* Widdowson, M., ed., Palaeosurfaces: Recognition, reconstruction and palaeo-environmental interpretation: Geological Society [London] Special Publication 120, p. 67–94.

Kelemen, P., and Holbrook, S., 1995, Origin of thick, high-velocity igneous crust along the U.S. East Coast margin: Journal of Geophysical Research, B, Solid Earth and Planets, v. 100, no. 6, p. 10077–10094.

Kenea, N., Ebinger, C., and Rex, D.C., 2001, Late Oligocene volcanism and extension in the southern Red Sea hills, Sudan: Journal of the Geological Society of London, v. 158, p. 285–294.

Kent, R.W., Saunders, A.D., Kempton, P.D., and Ghose, N.C., 1997, Rajmahal basalts, eastern India: Mantle sources and melt distribution at a volcanic rifted margin, *in* Mahoney, J., and Coffin, M.F., eds., Large igneous provinces.: Continental oceanic and planetary flood volcanism: American Geophysical Union Geophysical Monograph 100, p. 145–182.

King, S., and Anderson, D., 1998, Edge-driven convection: Earth and Planetary Science Letters, v. 160, p. 289–296.

Knox, R., Nyblade, A., and Langston, C., 1999, Upper mantle *S* velocities beneath Afar and western Saudi Arabia from Rayleigh wave dispersion: Geophysical Research Letters, v. 25, p. 4233–4236.

Korenaga, J., Holbrook, S., Kent, G., Kelemen, P., Detrick, R., Larsen, H.-C., Hopper, J., and Dahl-Jensen, T., 2000, Crustal structure of the southeast Greenland margin from joint refraction and reflection seismic tomography: Journal of Geophysical Research, v. 105, p. 21591–21614.

Larsen, H.-C., and Jakobsdottir, S., 1988, Distribution, crustal properties and significance of seaward-dipping sub-basement reflectors off E. Greenland, *in* Morton, A.C., and Parson, L.M., eds., Early Tertiary volcanism and the opening of the NE Atlantic: Geological Society [London] Special Publication 39, p. 95–114.

Larsen, H.-C., and Saunders, A., 1998, Scientific results, Ocean Drilling Program, Leg 152: Tectonism and volcanism at the southeast Greenland rifted margin: A record of plume impact and later continental rupture: College Station, Texas, Ocean Drilling Program, p. 503–533.

Lizarralde, D., and Holbrook, W.S., 1997, U.S. mid-Atlantic Margin structure and early thermal evolution: Journal of Geophysical Research, v. 102, p. 22855–22875.

Louden, K., and Chian, D., 1999, The deep structure of non-volcanic rifted continental margins: Philosophical Transactions of the Royal Society of London, v. 357, p. 767–804.

Mahoney, J.J., 1988, Deccan traps, *in* MacDougall, J.D., ed., Continental flood basalts: Amsterdam, Kluwer Publishers, p. 151–194.

Mahoney, J., and Coffin, M.F., editors, 1997, Large igneous provinces: Continental oceanic and planetary flood volcanism: American Geophysical Union Geophysical Monograph 100, 438 p.

Manighetti, I., Tapponnier, P., Courtillot, V., Gruszow, S., and Gillot, P.-Y., 1998, Propagation of rifting along the Arabia-Somalia plate boundary: Journal of Geophysical Research, v. 102, p. 2681–2710.

McBride, J.H., 1991, Constraints on the structure and tectonic development of the Early Mesozoic South Georgia rift, southeastern United States: Seismic reflection data processing and interpretation: Tectonics, v. 10, p. 1065–1083.

McHone, J.G., 1996, Broad-terrane Jurassic flood basalts across northeastern North America: Geology, v. 24, p. 319–322.

Menzies, M.A., Gallagher, K., Hurford, A., and Yelland, A., 1997a, Red Sea volcanic and the Gulf of Aden non-volcanic margins, Yemen: Denudational histories and margin evolution: Geochimica et Cosmochimica Acta, v. 61, p. 2511–2528.

Menzies, M.A., Baker, J., Chazot, G., and Al'Kadasi, M., 1997b, Evolution of the Red Sea volcanic margin, western Yemen, *in* Mahoney, J., and Coffin, M.F., eds., Large igneous provinces: Continental oceanic and planetary flood volcanism: American Geophysical Union Geophysical Monograph 100, p. 29–43.

Menzies, M.A., Baker, J., and Chazot, G., 2001, Cenozoic plume evolution and flood basalts in Yemen: A key to understanding older examples, *in* Ernst, R.E., and Buchan, K.L., eds., Mantle plumes: Their identification through time: Geological Society of America Special Publication 352, p. 23–36.

Milner, S.C., Duncan, A.R., Whittingham, A.M., and Ewart, A., 1995, Trans-Atlantic correlation of eruptive sequences and individual silicic volcanic units within the Parana-Etendeka igneous province: Journal of Volcanology and Geothermal Research, v. 69, no. 3–4, p. 137–157.

Mutter, J., Talwani, M., and Stoffa, P., 1982, Origin of seaward-dipping reflectors in oceanic crust off the Norwegian margin by subaerial seafloor spreading: Geology, v. 10, p. 353–357.

Peate, D., 1997, The Paraná-Etendeka Province, *in* Mahoney, J., and Coffin, M.F, eds., Large igneous provinces: Continental oceanic and planetary flood volcanism": American Geophysical Union Geophysical Monograph 100, p. 217–245.

Pickup, S., Whitmarsh, R.B., Fowler, C.M.R., Reston, T.J., 1996, Insight into the nature of the ocean-continent transition off West Iberia from a deep multichannel seismic reflection profile: Geology, v. 24, no. 12, p. 1079–1082.

Planke, S., Symonds, P.A., Alvestad, E., and Skogseid, J., 2000, Seismic volcano-stratigraphy of large-volume basaltic extrusive complexes on rifted margins: Journal of Geophysical Research, B, Solid Earth and Planets, v. 105, no. B8, p. 19335–19351.

Press, F., and Siever, R., 2000, Understanding Earth (third edition): New York, W.H. Freeman and Co., 573 p.

Renne, P., Ernesto, M., Pacca, I.G., Coe, R.S., Glen, J.M., Prevot, M., and Perrin, M., 1992, The age of Paraná flood volcanism, rifting of Gondwanaland and the Jurassic-Cretaceous boundary: Science, v. 258, p. 975–979.

Renne, P., Deckart, K., Ernesto, M., Feraud, G., and Piccirillo, E., 1996a, Age of the Ponta Grossa dyke swarm (Brazil), and implications to the Parana flood volcanism: Earth and Planetary Science Letters, v. 144, p. 199–211.

Renne, P., Glen, J.M., Milner, S.C., and Duncan, A.R., 1996b, Age of Etendeka flood volcanism and associated intrusions in south-western Africa: Geology, v. 24, no. 7, p. 659–662.

Richards, M.A., Duncan, R.A., and Courtillot, V.E., 1989, Flood basalts and hotspot tracks—Plume heads and tails: Science, v. 246, p. 105–107.

Roberts, D.G., Montadert, L., and Searle, R.C., 1979, The western Rockall Plateau: Stratigraphy and structural evolution, Initial reports of the Deep Sea Drilling Project, Volume 48: Washington, D.C., U.S. Government Printing Office, p. 1061–1088.

Saunders, A.D., Fitton, J.G., Kerr, A.C., Norry, M.J., and Kent, R.W., 1997, The North Atlantic igneous province, *in* Mahoney, J., and Coffin, M.F., eds., Large igneous provinces: Continental oceanic and planetary flood volcanism: American Geophysical Union Geophysical Monograph 100, p. 45–94.

Schlische, R.W., Withjack, M.O., and Olsen, P.E., 2002, Relative Timing of CAMP, rifting, continental breakup, and basin inversion: Tectonic significance, *in* Hames, W.E., ed., The Central Atlantic magmatic province: American Geophysical Union Geophysical Monograph (in press).

Shen, Y., Solomon, S.C., Bjarnasson, I.T., and Wolfe, C.J., 1998, Seismic evidence for a lower-mantle origin of the Iceland Plume: Nature, v. 395, no. 6697, p. 62–65.

Sleep, N.H., 1996, Lateral flow of hot plume material ponded at sub-lithospheric depths: Journal of Geophysical Research, v. 101, p. 28065–28083.

Stewart, K., Turner, S., Kelley, S., Hawkesworth, C.J., Kirstein, L., and Mantovani, M., 1996, 3-D, $^{40}Ar/^{39}Ar$ geochronology in the Parana continental flood basalt province: Earth and Planetary Science Letters, v. 143, p. 95–109.

Storey, M., Kent, R.W., Saunders, A.D., Hergt, J., Salters, V.J.M., Whitechurch, H., Sevigny, J.H., Thirlwall, M.F., Leat, P., Ghose, N.C., and Gifford, M., 1992, Scientific Results, Ocean Drilling Program, Leg 120: Lower Cretaceous volcanic rocks on continental margins and their relationship to the Kerguelen Plateau: College Station, Texas, Ocean Drilling Program, p. 33–53.

Turcotte, D.L., and Emerman, S.H., 1983, Mechanisms of active and passive rifting: Tectonophysics, v. 94, p. 39–50.

Turner, S., Regelous, M., Kelley, S., Hawkesworth, C., and Mantovani, M., 1994, Magmatism and continental break-up in the South Atlantic: High precision ^{40}Ar-^{39}Ar geochronology: Earth and Planetary Science Letters, v. 121, p. 333–348.

Ukstins, I., Renne, P., Wolfenden, E., Baker, J., Ayalew, D., and Menzies, M.A., 2002, Matching conjugate volcanic rifted margins: $^{40}Ar/^{39}Ar$ chronostratigraphy of pre- and syn-rift bimodal flood volcanism in Ethiopia and Yemen: Earth and Planetary Science Letters, v. 198, p. 289–306.

Von Rad, U., and Thurow, J., 1992, Bentonitic clay as indicators of early Neocomian post-breakup volcanism off Northwest Australia: Proceedings of the Ocean Drilling Program, Exmouth Plateau, covering Leg 122 of the cruises of the drilling vessel JOIDES Resolution, Singapore, Republic of Singapore, sites 759–764: College Station, Texas, Ocean Drilling Program, p. 213–232.

White, R., and McKenzie, D., 1989, Magmatism at rift zones: The generation of continental margins and flood basalts: Journal of Geophysical Research, v. 94, p. 7685–7729.

White, R.S., Spence, G.D., Fowler, S.R., McKenzie, D.P., Westbrook, G.K., and Bowen, A.N., 1987, Magmatism at rifted continental margins: Nature, v. 330, p. 439–444.

Withjack, M.O., Schlische, R.W., and Olsen, P.E., 1998, Diachronous rifting, drifting, and inversion on the passive margin of central eastern North America: An analog for other passive margins: American Association of Petroleum Geologists Bulletin, v. 82, p. 817–835.

Wolfe, C.J., Bjarnason, I.T., VanDecar, J.C., Solomon, S.C., 1997, Seismic structure of the Iceland Mantle plume: Nature, v. 385, no. 6613, p. 245–247.

Manuscript Accepted by the Society October 15, 2001

Printed in the U.S.A.

Geological Society of America
Special Paper 362
2002

Crust and upper mantle structure in East Africa: Implications for the origin of Cenozoic rifting and volcanism and the formation of magmatic rifted margins

Andrew A. Nyblade
Department of Geosciences, Penn State University, University Park, Pennsylvania 16802, USA

ABSTRACT

Crust and upper mantle structure in East Africa, together with the tectonic history of the region, is used to evaluate geodynamic processes commonly associated with the formation of magmatic rifted margins. Cenozoic rifting and volcanism in East Africa represent the earliest stage in the development of a rifted continental margin, and East Africa is one of the few places where geodynamic processes may be active that could lead to the development of a magmatic rifted margin. The Precambrian tectonic framework of East Africa is characterized by an Archean craton (Tanzania Craton) surrounded by Proterozoic mobile belts. An extensive nonmagmatic rift system associated with the separation of Madagascar from Africa developed in the mobile belts during the Permian-Cretaceous. In the Cenozoic, two rift branches (Western Rift and Eastern Rift) formed in the mobile belts. Volcanism is present in both branches, but most of it is concentrated in the Eastern Rift. Crustal structure away from the Cenozoic rifts is typical for Precambrian crust. Similarly, uppermost mantle structure across East Africa does not appear to have been altered, except beneath the rift valleys. Deeper in the upper mantle a thick (~200 km) lithospheric keel is found under the Tanzania Craton, and a broad (200–400 km wide) thermal anomaly extending to a depth of at least 400 km is beneath the Eastern Rift. Several models cannot fully account for the tectonic history of East Africa and/or the structure of the crust and upper mantle, including edge flow in the convecting mantle around the keels of Archean cratons, broad (>500 km wide) thermal upwellings originating in the lower mantle, a relatively stationary plume head, and a plume head that flows outward along topography on the underside of the lithosphere. Consequently, these models are not strong candidates for the origin of volcanism and rifting in East Africa. Models with two or more plume heads, however, can explain the relevant observations, and therefore a multiple plume head explanation for the rifting and volcanism in East Africa is favored. These findings further call into question the viability of nonplume models for the formation of magmatic rifted margins, particularly models invoking edge flow in the convecting mantle around cratonic keels.

Nyblade, A.A., 2002, Crust and upper mantle structure in East Africa: Implications for the origin of Cenozoic rifting and volcanism and the formation of magmatic rifted margins, *in* Menzies, M.A., Klemperer, S.L., Ebinger, C.J., and Baker, J., eds., Volcanic Rifted Margins: Boulder, Colorado, Geological Society of America Special Paper 362, p. 15–26.

INTRODUCTION

The origin of magmatic rifted continental margins is highly contended. Mantle plume models have long been advocated (e.g., White and McKenzie, 1989; Richards et al., 1989; Hill et al., 1992), but nonplume models invoking passive rifting above warmer than normal regions of the sublithospheric mantle have also been argued for vigorously (e.g., Anderson, 1994; King and Anderson, 1995, 1998).

Evaluating models for the formation of magmatic rifted margins is difficult because there are few, if any, magmatic rifted margins developing today where crust and upper mantle structure can be investigated in an active tectonic setting. Cenozoic rifting and volcanism in East Africa represent the earliest stage in the development of a rifted continental margin, and East Africa is one of the few places where geodynamic processes that could lead to the development of a magmatic rifted margin may be active. Therefore, in this chapter crust and upper mantle structure in East Africa, together with the tectonic history of the region, is used to evaluate geodynamic processes commonly associated with the formation of magmatic rifted margins.

Specifically, two questions concerning the formation of magmatic rifted margins are addressed. (1) Is flow in the convecting mantle around the edges of Archean cratons a viable mechanism for generating magmatic rifted margins? Many magmatic rifted margins are similar to the East African rift system in that they developed adjacent to an Archean craton (e.g., Deccan–Chagos–Lacadive Ridge, North Atlantic volcanic province, Karoo; Coffin and Eldholm, 1992). Even if the East African rift system never develops into a passive continental margin, its special tectonic setting nonetheless affords an excellent opportunity to evaluate models for magmatic rifted margins invoking convective flow in the mantle around the edges of cratons. (2) Can plume models adequately explain crust and upper mantle structures found beneath a possibly embryonic magmatic rift? Plume models for the formation of magmatic rifted margins are favored by many. Several different plume models have been proposed for East Africa, and by evaluating these models it might be possible to gain further insights into the nature of mantle plumes and their role in the formation of magmatic rifted margins.

CRUST AND UPPER MANTLE STRUCTURE IN EAST AFRICA

Geology

The Precambrian basement of East Africa consists of the Archean Tanzania Craton, which is in the center of the region, and a number of Early to Late Proterozoic mobile belts surrounding it (Fig. 1). The Cenozoic rift valleys have developed almost exclusively within the mobile belts, largely skirting the cratonic nucleus.

The Tanzania Craton consists mainly of granites, gneisses, and amphibolites; some greenstone belts are in the region north of 4.5°S. The youngest dates for the craton (ca. 2500 Ma) come from granitic rocks near its eastern margin (Cahen et al., 1984). To the east of the Tanzania Craton is the Mozambique Belt, which has mainly north- to south-striking structures formed by multiple collisional events dated between 1200 Ma and 450 Ma (Cahen et al., 1984; Shackleton, 1986; Key et al., 1989). Nd and Sr model ages and Pb isotopes indicate that much of the Mozambique Belt crust initially formed in the Late Archean and Early Proterozoic (Moller et al., 1998). The Tanzania Craton is bordered to the southeast and southwest by the Early Proterozoic Usagaran and Ubendian Belts, respectively (Lenoir et al., 1994; Theunissen et al., 1996). The northern part of the Ubendian Belt is truncated west of the Tanzania Craton by the northeast-trending Late Proterozoic Kibaran Belt (Cahen et al., 1984). To the north of the craton is the Late Proterozoic Ruwenzori Belt (Cahen et al., 1984).

Extensive late Paleozoic to Mesozoic rifts can be found throughout parts of East Africa and are generally oriented northeast-southwest or northwest-southeast. The present outlines of these rifts, now reduced from their original sizes by uplift and erosion, exceed the size of the Cenozoic rifts (Fig. 2). Two episodes of rifting are well documented (Kreuser, 1995). Karoo rifting began during the Early Permian with the initial separation of Madagascar from Africa and terminated in the Early Jurassic, leaving a rift system that extended from the coastal region of Kenya southward through Tanzania, Mozambique, Malawi, Zambia, and Zimbabwe. Shortly afterward, a new phase of rifting commenced in the Early Jurassic along the entire coast of East Africa and terminated at 165 Ma with the final separation of Madagascar from the African continent.

There is little volcanism associated with the Paleozoic and Mesozoic rifts in East Africa (and also Madagascar). The only known Paleozoic or Mesozoic volcanic units are some Late Cretaceous sills in the Anza graben (Bosworth and Morley, 1994) and a few Early Jurassic sills in the Luwegu trough (Fig. 2) (Hankel, 1987). Locally, the sills in the Luwegu trough attain thicknesses of ~120 m (Spence, 1957) and are similar in age and composition to the extensive Karoo flood basalts of southern Africa (Kreuser, 1995; Hankel, 1987). In addition to these sills, a number of volcanic plugs and dikes, mostly carbonatitic, have been found in some of the Karoo basins, but they are predominantly early Tertiary in age (Hankel, 1987; Bosworth and Morley, 1994).

The Cenozoic rift system in East Africa consists of two branches, the Western Rift and the Eastern Rift (Fig. 1). Extension within the Eastern Rift in Kenya has led to the formation of a narrow (50–80 km wide) rift graben commonly referred to as the Kenya or Gregory rift. At the terminus of the Eastern Rift in northeastern Tanzania, the graben structures found in Kenya give way to a much wider zone (~300 km) of block faulting (Dawson, 1992; Ebinger et al., 1997; Foster et al., 1997). The Western Rift is composed of numerous en echelon fault-bounded basins (Ebinger, 1989). Many of these rift basins to the

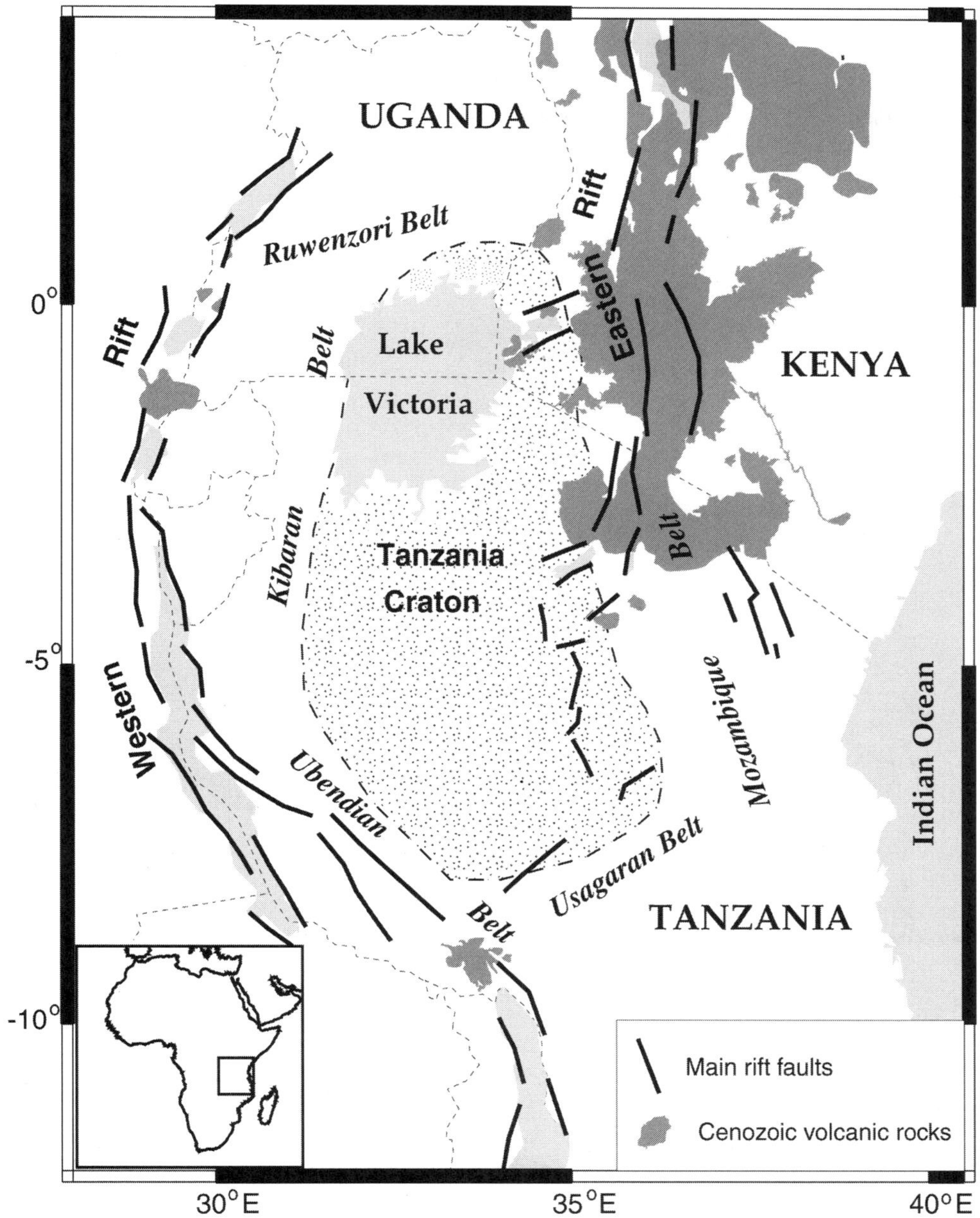

Figure 1. Schematic map of East Africa showing political boundaries, large rift lakes, Precambrian terranes, Cenozoic rift system, and Cenozoic volcanic provinces.

south and southwest of the Tanzania Craton developed within or adjacent to Karoo rifts, and in some cases Karoo-aged faults may have been reactivated.

The timing of Cenozoic volcanism in East Africa is fairly well defined. George et al. (1998) provided a summary of magmatism in the Eastern Rift showing a clear southward progression in onset age. Magmatism commenced at 40–45 Ma in southern Ethiopia, 30–35 Ma in northern Kenya, 15 Ma in central Kenya, and 8 Ma in northern Tanzania. In contrast to the Eastern Rift, there are only a few volcanic centers in the Western Rift (Fig. 1). Magmatism began ca. 12 Ma in the northern eruptive centers around Lake Kivu and at 8 Ma in the Rungwe Province in southern Tanzania (Ebinger et al., 1989; Pasteels et al., 1989; Kampunzu et al., 1998). Rift faulting in the Eastern and Western Rifts probably commenced about the same time as the magmatism (Ebinger, 1989).

Crustal structure

Until recently, investigations of crustal structure in East Africa focused primarily on the Eastern and Western Rifts. Early studies used seismic refraction data and observations from teleseismic and regional earthquakes to examine crustal structure (Bonjer et al., 1970; Griffiths et al., 1971; Long et al., 1972; Mueller and Bonjer, 1973; Bram and Schmeling, 1975; Nolet and Mueller, 1982; Hebert and Langston, 1985), yielding estimates of Moho depths of 40–48 km beneath unrifted crust, and of 20–32 km under the rift valleys. Work on crustal structure in

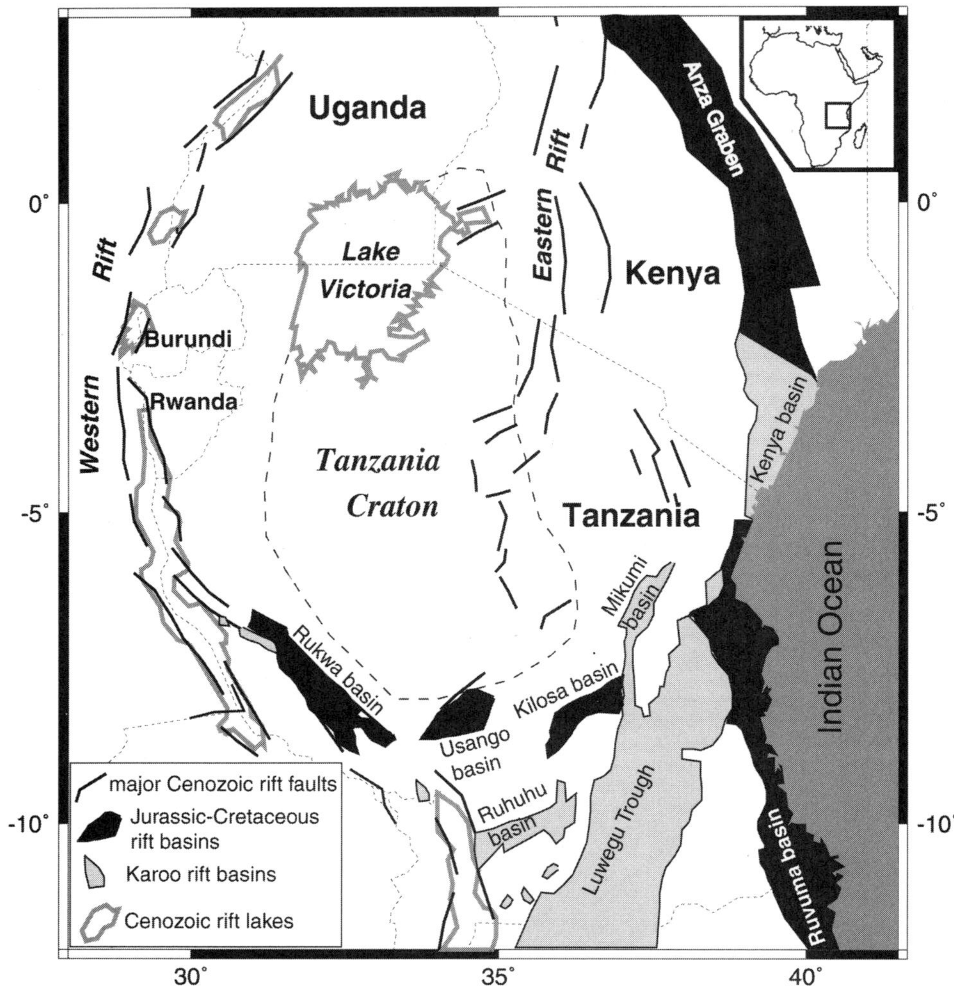

Figure 2. Schematic map of East Africa showing locations of Permian-Cretaceous rift basins, main Cenozoic rift faults, and Archean Tanzania Craton (Permian-Cretaceous rifts taken from Kreuser, 1995; Bosworth and Morley, 1994).

and around the Kenya Rift was undertaken by the Kenya Rift International Seismic Project (KRISP) using seismic refraction profiles (Prodehl et al., 1994; Fuchs et al., 1997, and references therein). The KRISP group found that along the axis of the rift Moho depth shallows from 35 km in southern Kenya to ~20 km beneath northern Kenya. Away from the rift, crustal thicknesses of 34–40 km beneath the Tanzania Craton and 35–42 km beneath the Mozambique Belt were obtained.

A more recent investigation of crustal structure in East Africa by Last et al. (1997) using teleseismic receiver functions and Rayleigh wave dispersion measurements has provided additional information about crustal structure in the mobile belts and craton. Earthquakes used in this study were recorded by the Tanzania Broadband Seismic Experiment (Nyblade et al., 1996). For the Tanzania Craton, Last et al. (1997) obtained Moho depths of 37–42 km, a mean crustal shear velocity ($\overline{V}s$) of 3.79 km/s, and Poisson's ratios of 0.24–0.26. For the Mozambique Belt, Last et al. (1997) obtained a $\overline{V}s$ of 3.74 km/s, Moho depths between 36 and 39 km, and Poisson's ratios between 0.24 and 0.27. Results obtained by Last et al. (1997) from the Ubendian Belt indicate a $\overline{V}s$ of ~3.74 km/s and Moho depths between 40 and 45 km. Based on a comparison of these results and the KRISP results to global averages for Precambrian crust, it can be concluded that crustal structure away from the narrow rift valleys has not been modified to any significant extent by the Cenozoic tectonism in East Africa.

Upper mantle structure

Uppermost mantle structure beneath the Kenya Rift was investigated by the KRISP group using seismic refraction profiles (Prodehl et al., 1994; Fuchs et al., 1997, and references therein). The pattern of Pn velocities beneath and adjacent to the Kenya Rift is fairly simple (Fig. 3). Low Pn velocities of 7.5–7.8 km/s are found under the axis of the rift, and Pn velocities of 8.1–8.3 km/s are found under the unrifted areas of the Mozambique Belt and Tanzania Craton. The transition from low to high Pn velocities is abrupt and coincides with the main rift border faults, indicating that thermal modification of uppermost mantle structure is confined in Kenya to the rift proper. Models derived from inverting teleseismic travel time residuals, however, suggest that deeper in the upper mantle under Kenya the zone of modified

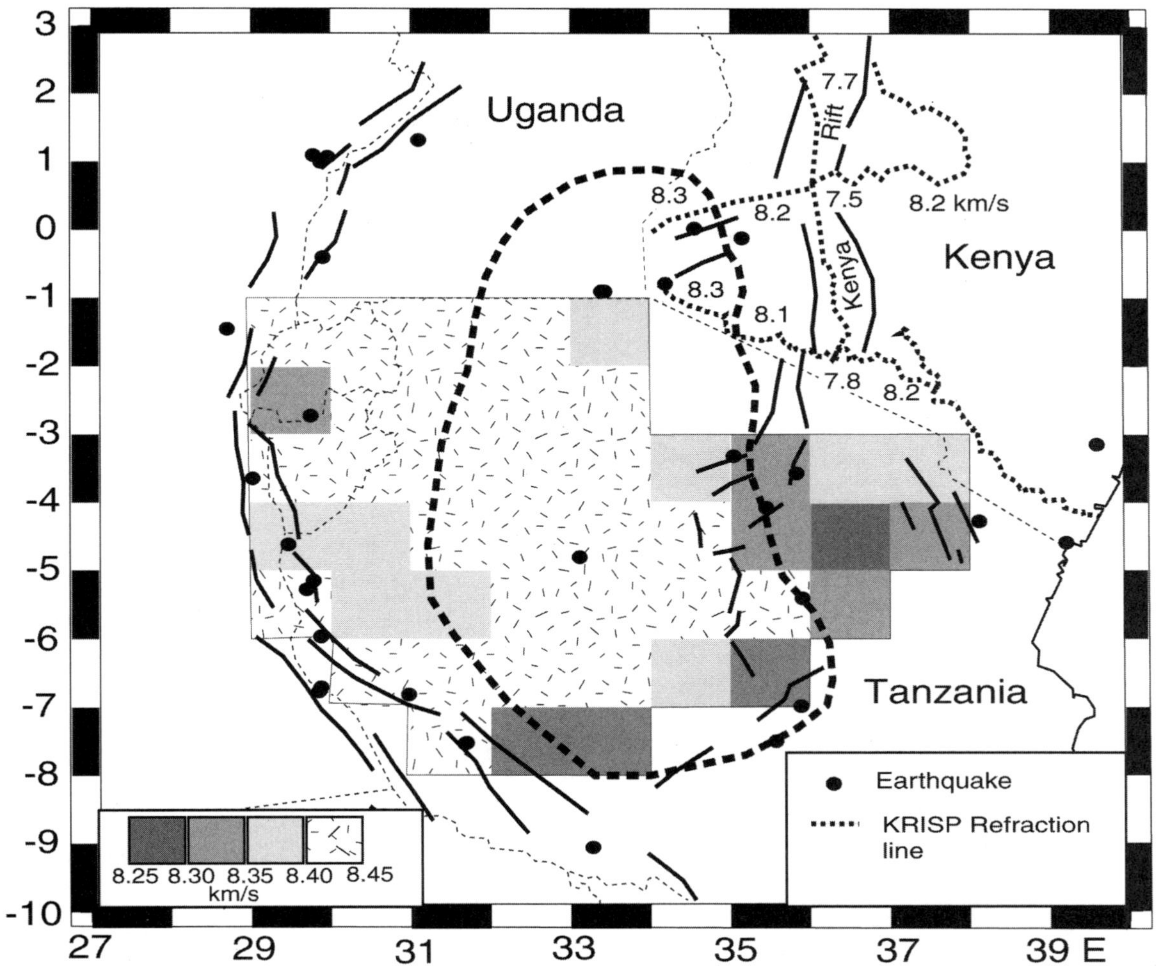

Figure 3. Pn velocities across East Africa from Brazier et al. (2000) and KRISP group. Geological features as in Figure 1. Numbers next to KRISP refraction lines give Pn velocities (in km/s).

lithosphere may extend over a broader region (Achauer et al., 1994; Slack et al., 1994; Green et al., 1991). Nolet and Mueller (1982) examined mantle structure beneath the Western Rift by simultaneously inverting teleseismic body wave traveltimes and surface wave phase and group velocities. They found a thin (~20 km), high-velocity lid beneath the Western Rift underlain by a low-velocity channel.

Upper mantle structure beneath the craton and rifted mobile belts in Tanzania has been investigated in a number of studies using data from the Tanzania Broadband Seismic Experiment (Nyblade et al., 1996). The P wave traveltimes from regional earthquakes were inverted for long wavelength (>100 km) Pn velocity variations beneath Tanzania by Brazier et al. (2000) using a generalized inverse algorithm. Brazier et al. found Pn velocities of 8.40–8.45 km/s beneath the center of the Tanzania Craton, 8.30–8.35 km/s beneath the Mozambique Belt where the Eastern Rift terminates, and 8.35–8.40 km/s beneath the Western Rift (Fig. 3). These velocities are high for Precambrian lithosphere and thus indicate that there are no broad (>100 km wide) thermal anomalies in the uppermost mantle beneath the craton or rifted mobile belts in Tanzania.

Structure deeper in the upper mantle beneath the Mozambique Belt and craton in Tanzania was imaged by inverting relative traveltimes from teleseismic P and S waves for upper mantle seismic velocity variations (Ritsema et al., 1998). The patterns of P- and S-wave velocity variation obtained are similar, and so only the S-wave velocity model is discussed here (Fig. 4A). The model shows higher than average velocities beneath the Tanzania Craton and predominantly lower than average velocities beneath the rifted mobile belts surrounding the craton. The low-velocity region under the Eastern Rift extends vertically to depths greater than 400 km and laterally over a region ~300 km wide. The lithospheric keel beneath the craton, as defined by the relatively fast velocities, extends to a depth of ~200 km (the continuation of fast velocities to 300–400 km depth in Fig. 4A is due to limited vertical resolution). Between depths of 200 and 300 km the low-velocity structure associated with the rifts begins to extend westward under the fast structure

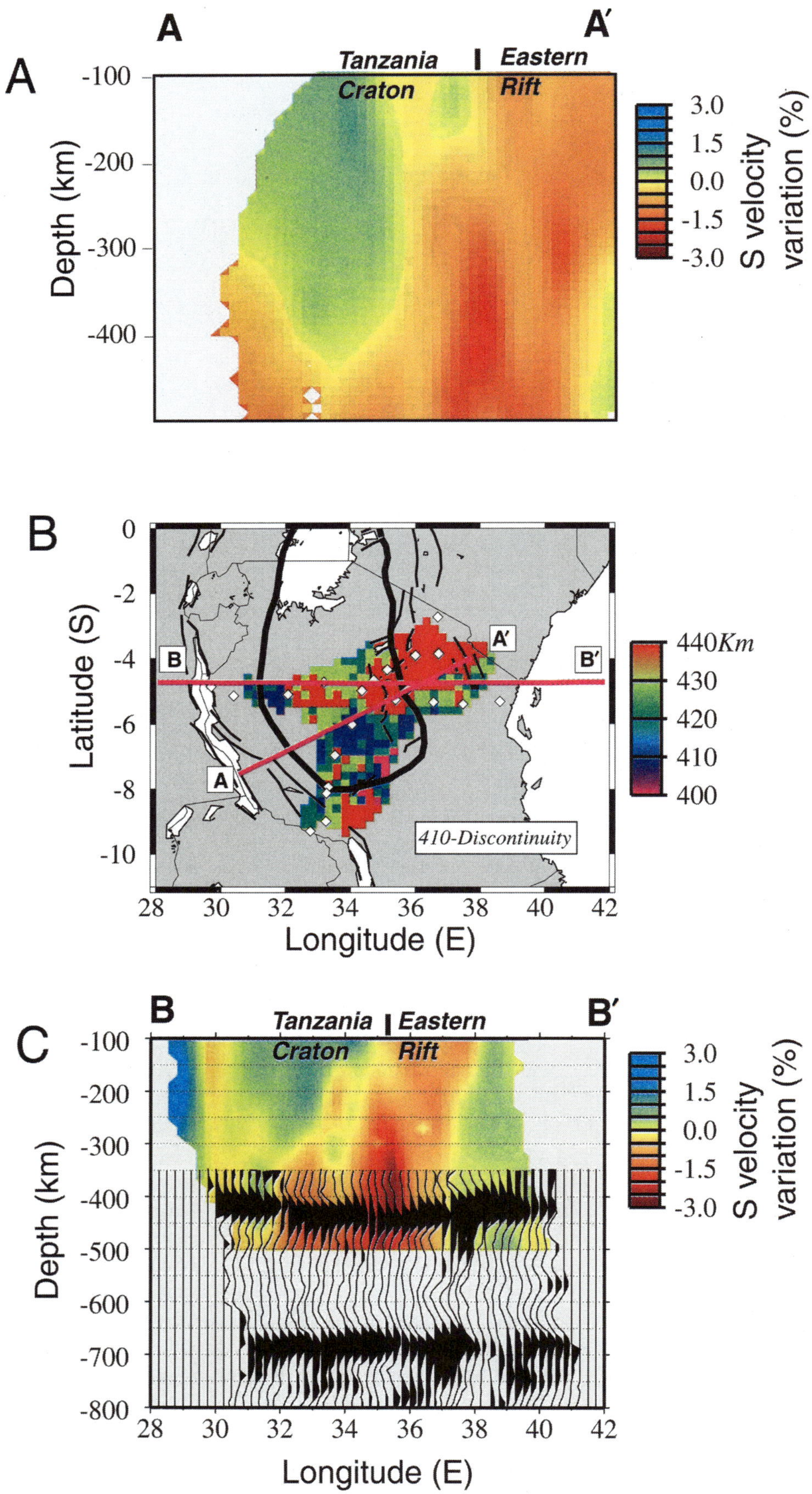

Figure 4. A: Vertical slice through S wave velocity model for East Africa from Ritsema et al. (1998). Geographic location of vertical slice is shown in B. Uncertainties in horizontal and vertical dimensions of velocity structures are ~50 and 100 km, respectively. Velocity structure above 100 km and below 500 km is poorly resolved and therefore is not shown. B: Depth to 410 km discontinuity beneath Tanzania from stacking receiver functions from Owens et al. (2000). Uncertainties in depth estimates shown are ±5 km. Open diamonds show seismic station locations, and A–A′ and B–B′ give locations of figures in A and C. Geological features are as in Figure 1. C: Stacked receiver functions showing broad depression of 410 km discontinuity and flat 660 km discontinuity superimposed on S wave velocity model of Ritsema et al. (1998). Traces begin at 350 km depth to highlight arrivals from 410 and 660 km discontinuities. Geographic location of profile is shown in B.

of the cratonic lithosphere. Ritsema et al. (1998) were not able to determined if the low-velocity structure under the Mozambique Belt extends to shallower depths because velocity structure above 100 km depth could not be resolved using the teleseismic data. However, the results discussed by Brazier et al. (2000) show clearly that the low velocities at depth within the upper mantle do not reach the base of the crust, at least not broadly across the entire mobile belt.

Further information on the structure of the upper mantle beneath Tanzania comes from a study of the topography on the 410 and 660 km discontinuities (Owens et al., 2000). The 410 and 660 km discontinuities are generally interpreted as transitions in the α-phase to the β-phase of (Mg, Fe)SiO_4 and from γ-(Mg, Fe)SiO_4 to perovskite + magnesiowustite, respectively. The Clapeyron slopes of the equilibrium phase boundaries indicate that the depth of the 410 km discontinuity should be deflected downward in regions of warmer temperature and the 660 km discontinuity should be deflected upward (Bina and Helffrich, 1994). Therefore, topography on the discontinuities provides information about upper mantle thermal structure in the 400–700 km depth range and can assist in the interpretation of velocity variations obtained from tomographic inversions.

Topography on the 410 and 660 km discontinuities beneath Tanzania was estimated by geographically stacking receiver functions (Owens et al., 2000). Results show that the transition zone (e.g., the region between the discontinuities) is thinned by 30–40 km over an area 200–400 km wide beneath the Eastern Rift (Fig. 4, B and C). The transition-zone thinning, which corresponds to a temperature increase of ~200–300 K, is primarily due to a wide depression of the 410 km discontinuity and coincides directly with the low-velocity anomaly seen beneath the Eastern Rift (Fig. 4, B and C). In comparison to the 410 km discontinuity, little relief is observed on the 660 km discontinuity (Fig. 4C).

The coincidence of the depressed 410 km discontinuity and the low-velocity region beneath the Eastern Rift (Fig. 4C) indicates that at least some of the S-wave velocity variation between the craton and Eastern Rift seen in the model of Ritsema et al. (1998) is due to temperatures beneath the Eastern Rift elevated by 200–300 K. According to laboratory measurements of the temperature derivatives of wave speeds in olivine (Isaak, 1992), a 200–300 K temperature increase in the upper mantle would reduce S wave velocities by ~2%.

In addition to seismic investigations, heat flow and gravity studies place important constraints on crust and upper mantle structure in East Africa. The ensemble of heat-flow data from East Africa (Nyblade et al., 1990; Nyblade, 1997; Wheildon et al., 1994), when compared to the global distribution of heat flow in tectonically undisturbed Precambrian terrains of similar age, indicates that heat flow from areas of East Africa outside of the rift valleys is not elevated. Heat flow from within the rift valleys, although probably elevated, is not well determined by observations. The lack of a broad heat-flow anomaly across East Africa is consistent with the seismic structure of the crust and upper mantle described here, which shows that the crust and uppermost mantle away from the Kenya Rift has not been modified to any great extent by the Cenozoic tectonism.

Constraints on crust and upper mantle structure from gravity studies come primarily from regional estimates of effective elastic plate thickness (Te). Te estimates in areas of East Africa away from the rift valleys are typical for Precambrian lithosphere, and thus support the findings from seismic and heat-flow data for relatively unperturbed crust and uppermost mantle structure away from the rifts and for modified lithospheric structure beneath the axes of the rifts (Ebinger et al., 1989, 1991; Petit and Ebinger, 2000; Upcott et al., 1996).

ORIGIN OF CENOZOIC RIFTING AND VOLCANISM IN EAST AFRICA

Several plume and nonplume models for the origin of the Cenozoic rifting and volcanism in East Africa have been proposed over the past few decades, models similar to those proposed for the early formation of magmatic rifted margins. In this section, models relevant to the two questions posed in the introduction about the formation of magmatic rifted margins are evaluated using the tectonic setting and crustal and upper mantle structure reviewed in the preceding sections. To summarize, the main characteristics of crust and upper mantle structure that viable models must account for are (1) unmodified crust away from the rift valleys; (2) a lithospheric keel beneath the Tanzania Craton that is at least 200 km thick; and (3) a broad (200–400 km wide) thermal anomaly in the upper mantle beneath the Eastern Rift extending from the shallow upper mantle to depths of ≥400 km.

EDGE model

The EDGE model was proposed by Anderson (1994) and King and Anderson (1995, 1998). In this model magmatic rifting is controlled by the presence of suture zones in the lithosphere coinciding with asymmetries in the thickness of the lithosphere between older (i.e., cratonic) and younger (i.e., mobile belt) lithosphere (Fig. 5, A and B). The difference in heat flux through the older, thicker lithosphere and the thinner, younger lithosphere drives small-scale convection, drawing mantle from under the thicker lithosphere to the suture zone. The mantle material rises as it approaches the boundary to the thinner lithosphere without requiring large stretching factors, enabling the material to cross into the melting zone relatively easily and produce significant quantities of basaltic melt. Extensional stresses across the suture are required to focus the flow at the suture, but this model does not necessarily require stretching to thin the lithosphere extensively and allow mantle from below the lithosphere to melt extensively by adiabatic decompression. Instead, the small-scale flow brings deeper, fertile mantle from beneath the thicker lithosphere to shallower depths under the thinner lithosphere, where it melts without significant amounts of lithospheric stretching.

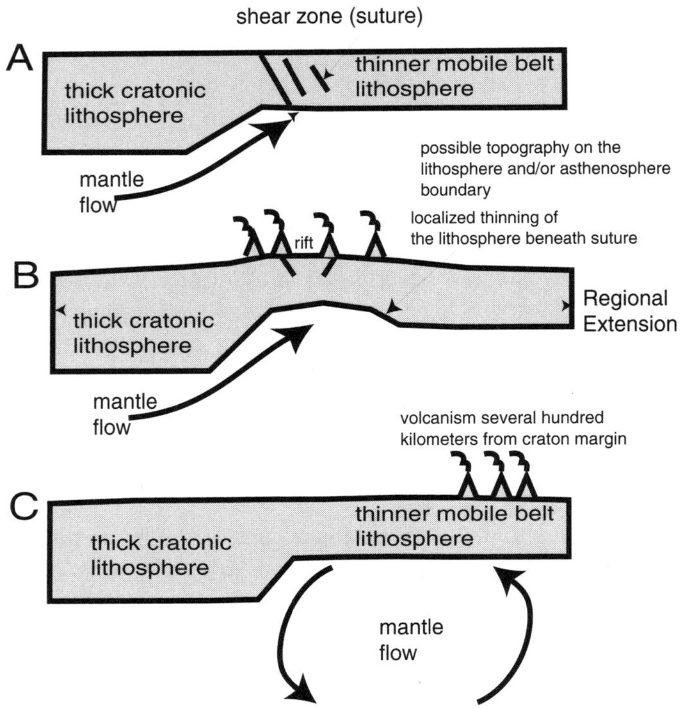

Figure 5. A, B: EDGE model of King and Anderson (1995, 1998). C: Modified EDGE model of King and Ritsema (2000). See text for further explanation.

The application of this model to East Africa might appear to be straightforward. The Tanzania Craton provides the requisite thick lithosphere, and the small amount of lithospheric stretching needed to initiate the development of the magmatic rift comes from the far-field extensional stresses linked to the opening of the Afar triple junction. However, there are at least two observations that this model does not explain.

The first observation is the lack of magmatism in the Western Rift. The amount of crustal extension in the Western Rift is similar to the Eastern Rift (Ebinger, 1989; Morley, 1988), and so there is little reason not to expect the same volume of melting beneath the Western Rift as the Eastern Rift. It could be argued that there should be more melting beneath the Western Rift than the Eastern Rift because the Western Rift is between two cratons (the Tanzania and Congo Cratons). According to the EDGE model, flow in the mantle from beneath both cratons should be channeled toward the Western Rift.

The other observation that is not easily reconciled with the EDGE model is that rifting in East Africa during the late Paleozoic and Mesozoic was nonmagmatic. If convective flow in the mantle around the cratonic keel led to the magmatism in the Cenozoic rifts, then it is difficult to explain why it would not have led to a similar amount of magmatism during the development of the late Paleozoic and Mesozoic rifts. There should have been more magmatism in the older rifts because of the greater amount of lithospheric extension, extension that culminated with the separation of Madagascar from East Africa. The inability of the EDGE model to account for these observations calls into question its viability as a comprehensive model for the Cenozoic tectonism in East Africa.

Modified EDGE model

In another model for African volcanism invoking edge-driven convection, King and Ritsema (2000) make the convection cell flow in the opposite direction (Fig. 5C) by removing the long-wavelength thermal variation in the convecting upper mantle imposed in the earlier models of King and Anderson (1995, 1998). In the King and Ritsema model, East African volcanism is attributed to warm upwelling mantle several hundred kilometers from the craton boundary. It is clear that this model does not provide a viable explanation for East African volcanism because the volcanism occurs adjacent to the craton margin, not several hundred kilometers away from it.

Broad thermal upwelling from the lower mantle

On the basis of tomographic images of the mantle beneath Africa that show a broad S-wave velocity anomaly in the upper mantle beneath Tanzania possibly connecting with low-velocity structure in the lower mantle beneath southern Africa, it has been suggested that a broad (i.e., >500 km wide) thermal upwelling extends from the core-mantle boundary all the way to the uppermost mantle beneath the rift system (Lithgow-Bertelloni and Silver, 1998; Ritsema et al., 1999).

Although the depth extent of the thermal anomaly and the wide depression of the 410 km discontinuity beneath northern Tanzania (Fig. 4) are consistent with a broad thermal upwelling, the flat 660 km discontinuity under Tanzania is not easily explained by a broad thermal upwelling. The γ-(Mg, Fe)SiO_4 to perovskite + magnesiowustite phase transformation should occur at depths shallower than 660 km if temperatures at that depth are elevated (e.g., Shen et al., 1998). Moreover, the average thickness of the transition zone beneath Tanzania (253 km; Owens et al., 2000) is consistent with estimates of the global average transition-zone thickness (e.g., Flanagan and Shearer, 1998; Chevrot et al., 1999), indicating that there is no broad thinning of the transition zone, as would be expected if a broad thermal anomaly was throughgoing across the transition zone. Although the lower mantle low-velocity structure beneath southern Africa may somehow be linked geodynamically to the upper mantle low-velocity structure beneath East Africa, there appears to be little evidence to support a broad throughgoing mantle thermal anomaly beneath Tanzania.

Stationary plume head model

The existence of a nearly stationary plume head beneath East Africa has been proposed by many investigators (e.g., Simiyu and Keller, 1997, Green et al., 1991; Smith, 1994; Slack et al., 1994; Zeyen et al., 1997; Burke, 1996; George et al., 1998). Most of the plume head models assume a plume structure that is similar to the starting plume model of Griffiths and Campbell

(1991). The crustal and upper mantle structure in East Africa can be attributed to a nearly stationary plume head under two conditions (Fig. 6, A and B) (Nyblade et al., 2000): (1) the plume head must have come up under the eastern side of the Tanzania Craton in central Kenya and then flowed small distances (north-south) along the craton margin and underneath the craton, and (2) the plume head must be several hundred kilometers in diameter so that the bottom of it is across the 410 km discontinuity.

In this interpretation (Fig. 6, A and B), the thermal structure beneath the Eastern Rift is caused by buoyant (warm) plume head material that has migrated around and laterally along the eastern side of the cratonic keel, modifying the mantle lithosphere beneath the Eastern Rift. This plume head is distinct from the Afar plume. Plume head temperatures are estimated to be 100–300 K above ambient mantle temperatures (McKenzie and Bickle, 1988; Campbell and Griffiths, 1990; Farnetani and Richards, 1994), sufficient to reduce S-wave velocities by a few percent. The wide depression of the 410 km discontinuity beneath northern Tanzania is caused by the bottom of the plume crossing the 410 km discontinuity. Fluid dynamic studies of plume heads suggest that they could be several hundred kilometers in diameter (Griffiths and Campbell, 1991). If a plume head of this size impinged on thick (200–250 km) cratonic lithosphere, it is possible that the bottom of the plume head might extend to depths of ≥~400 km, giving rise to a depression of the 410 km discontinuity that is several hundred kilometers across. The 660 km discontinuity beneath Tanzania (Fig. 4C) is not disrupted by the plume tail in this interpretation because the tail is to the north beneath central Kenya, where the rifting and volcanism are centered (Fig. 6B).

There may be a number of problems with the model in Figure 6, and so a more or less stationary plume head model might not be a valid model for East Africa. One potential problem is that the starting plume head model was developed to explain the rapid eruption of flood basalts over an interval of 1–3 m.y. Volcanism in northern Tanzania has been ongoing for the past 8 m.y. and over a 20–25 m.y. interval in Kenya. The longer duration of volcanism in northern Tanzania and Kenya compared to most flood basalt provinces could perhaps be accounted for with a stationary plume head model if the plume head has a smaller temperature anomaly (<100 K) than plume heads that form flood basalt provinces. However, the depression of the 410 km discontinuity beneath northern Tanzania indicates temperatures elevated by 200–300 K, which is at the high end of the anomalous temperature range typically associated with plumes. Thus, the small melt volume in East Africa erupted over several millions of years and the large depression of the 410 km discontinuity are not easily reconciled within the framework of a stationary plume head model.

Another potential problem with a stationary plume head model is that some numerical studies of plumes (e.g., Sleep, 1996, 1997) suggest that the plume head material may flow rapidly away from the plume head center channeled by topography on the underside of the lithosphere. If correct, then these studies would indicate that the bottom of the plume head, as shown in Figure 6, would probably not extend across the 410 km discontinuity.

Runny plume head models

On the basis of the numerical flow calculations of Sleep (1996, 1997), Ebinger and Sleep (1998) proposed a "runny" plume head model for East Africa. They argued that a single plume head impinged on the lithosphere in southern Ethiopia, and that the plume material flowed ~1000 km outward from this location channeled by topography on the lithosphere-asthenosphere boundary (Fig. 6C). The channeled plume material is ~50–100 km thick.

Considering that warm material spreading outward from the plume center in the Ebinger and Sleep model is only 50–100 km thick, it is not easy to explain with their model the >400 km deep thermal anomaly beneath northern Tanzania, unless for some unknown reason the plume material has ponded there beneath the lithosphere. George et al. (1998) suggested that a plume head originally impinged on the lithosphere beneath southern Ethiopia, where the oldest volcanism is found, and that the plume tail is now located beneath Kenya because of the northward motion of the African plate. This model has the same problem as the Ebinger and Sleep model in that it requires some way to get the warm material from the plume to pond beneath northern Tanzania in order to explain the depth extent of the upper mantle thermal anomaly.

SUMMARY AND CONCLUSIONS

In summary, all of the models discussed in the preceding section may be flawed in one way or another. It is not obvious how the nonplume models could be modified to explain the relevant observations. However, a plume origin for the rifting and volcanism in East Africa can be made to work by invoking more than one plume head. For example, the tectonic development of the East African rift system along with the structure of the crust and upper mantle could be accounted for by combining the effects of a runny plume head arriving beneath southern Ethiopia ca. 40 Ma, as proposed by Ebinger and Sleep (1998), with a new plume head rising through the upper mantle beneath Kenya and northern Tanzania, as proposed by Nyblade et al. (2000). A plume origin for the Cenozoic rifting and volcanism in East Africa, whether it is one or more plumes, is strongly supported by many geochemical and/or petrologic studies of the volcanics and mantle xenoliths from various parts of East Africa. (A review of the findings reported in these studies or even a complete listing of references is beyond the scope of this chapter.)

In conclusion, the two questions raised in the introduction are addressed. (1) Is flow in the convecting mantle around the edges of Archean cratons a viable mechanism for generating magmatic rifted margins? (2) Can plume models adequately explain crust and upper mantle structures found beneath a possibly embryonic magmatic rift? The first question is easier to address than the second. As mentioned in the introduction, the development of

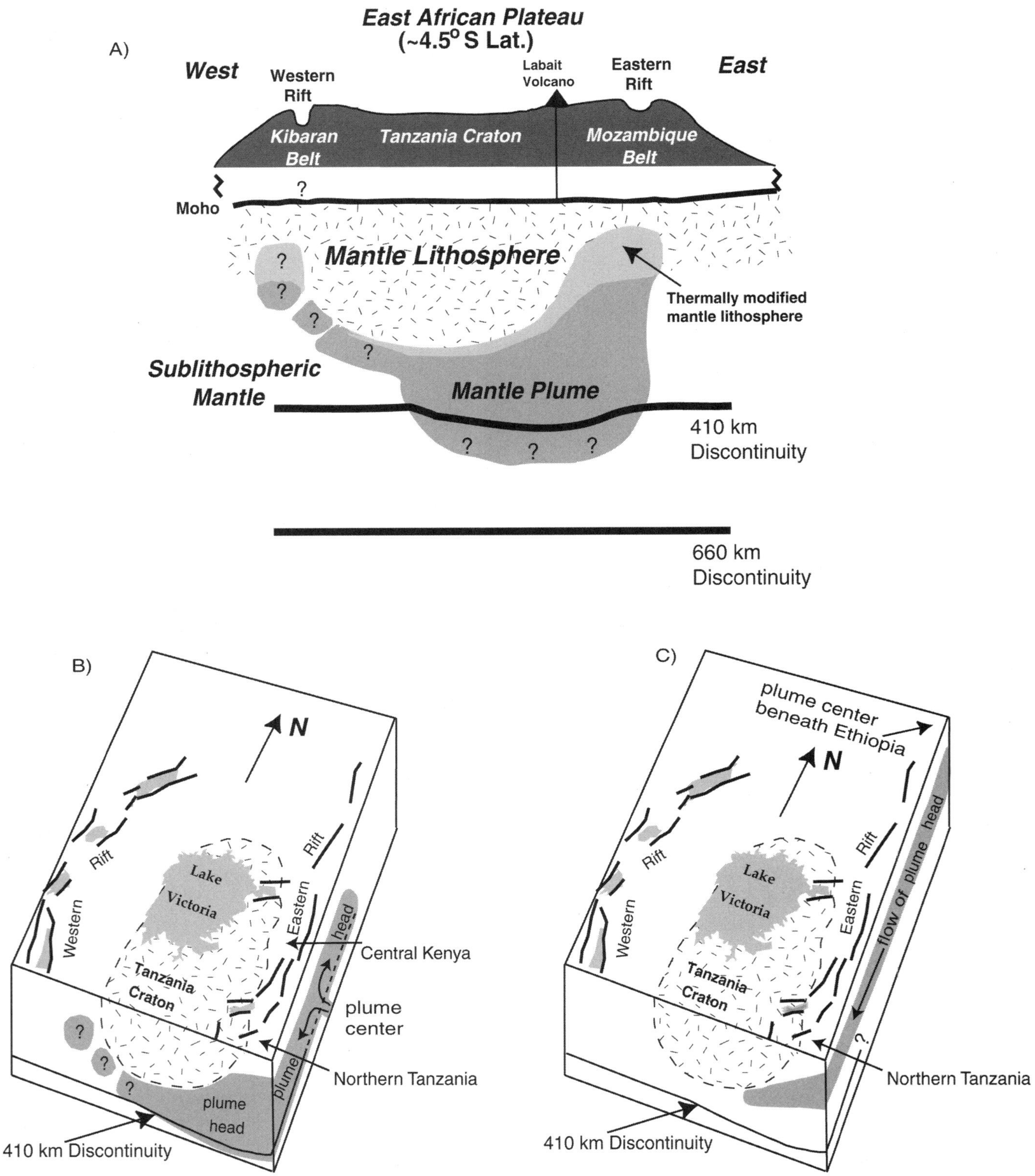

Figure 6. A: Schematic cross section at ~4.5 s showing plume head beneath eastern margin of Tanzania Craton. Question marks beneath Eastern and Western Rifts and at bottom of plume head indicate that structures illustrated there are poorly determined. B: Schematic three-dimensional diagram of model in A showing plume head centered beneath central Kenya and flow of plume head material around and beneath cratonic keel. C: Sketch diagram showing "runny" plume head model of Ebinger and Sleep (1998) with plume head centered to north beneath southern Ethiopia.

the East African rift system along a craton boundary is similar to the tectonic setting in which many magmatic rifted margins developed. The proximity of many magmatic rifted margins to craton boundaries motivated Anderson (1994) and King and Anderson (1995, 1998) to propose the EDGE model.

Because the EDGE model, as originally proposed, as well as the modified version of King and Ritsema (2000), cannot account for many of the structural features found in the crust and upper mantle beneath East Africa, as well as aspects of the tectonic history of the region, these models are not viewed as strong candidates for explaining the origin of magmatic rifted margins. This conclusion does not necessarily preclude the existence of the convective flow pattern in the mantle in the EDGE models, but it does imply that if the flow pattern exists, then, in all likelihood, it does not influence strongly the formation of magmatic rifts.

Addressing the second question raised in the introduction is more difficult. Several plume models have been proposed for East Africa, but none of the models invoking a single plume can explain fully the structural features found in the crust and upper mantle beneath East Africa along with the tectonic development of the region. Because single-plume models cannot explain all of the relevant observations, it is not easy to draw clear inferences about the nature of plumes from this study or to comment on their possible role in the formation of magmatic rifted margins.

Nonetheless, the findings of this study may provide some insights about the source region of plumes: (1) the source region must be able to generate multiple plumes within the time span of volcanism in East Africa (~40–45 million years), and (2) the source region is likely to be in the lower mantle. The observation that the transition zone beneath Tanzania is not thinner than normal suggests a lower mantle origin for the plumes. If the source region was at the base of the upper mantle, then the 660 km discontinuity should be elevated broadly beneath East Africa (cf. Shen et al., 1998), which it is not. Because East Africa is above the edge of a large low-velocity region in the lower mantle, the most plausible plume source region is within or along the boundary of the low-velocity region in the lower mantle centered under southern Africa.

ACKNOWLEDGMENTS

This study was funded by the National Science Foundation (grant EAR-9304555). Reviews by U. Achauer, C. Ebinger, M. Menzies, and an anonymous reader greatly improved this paper.

REFERENCES CITED

Achauer, U., and the KRISP Teleseismic Working Group, 1994, New ideas of the Kenya rift based on the inversion of the combined dataset of the 1985 and 1989/90 seismic tomography experiments: Tectonophysics, v. 236, p. 305–330.

Anderson, D.L., 1994, The sublithospheric mantle as the source of continental flood basalts: The case against the continental lithosphere and plume head reservoirs: Earth and Planetary Science Letters, v. 123, p. 269–280.

Brazier, R.A., Nyblade, A.A., Langston, C.A., and Owens, T.J., 2000, Pn wave velocities beneath the Tanzania craton and adjacent rifted mobile belts, East Africa: Geophysical Research Letters, v. 27, p. 2365–2368.

Bina, C., and Helffrich, G., 1994, Phase transition Clapeyron slopes and transition zone seismic discontinuity topography: Journal of Geophysical Research, v. 99, p. 15853–15860.

Bonjer, K.P., Fuchs, K., and Wohlenburg, J., 1970, Crustal structure of the East African rift system from spectral response ratios of long period body waves: Zeitschrift für Geophysik, v. 36, p. 287–297.

Bosworth, W., and Morley, C.K., 1994, Structure and stratigraphic evolution of the Anza rift, Kenya: Tectonophysics, v. 236, p. 93–115.

Bram, K., and Schmeling, B.D., 1975, Structure of the crust and upper mantle beneath the western rift of East Africa, derived from investigations of near earthquakes, *in* Pilger, A., and Rosler, A., eds., Afar between continental and oceanic rifting: Stuttgart, Germany, Schweizerbart, p. 138–142.

Burke, K., 1996, The African Plate: South African Journal of Geology, v. 99, p. 339–410.

Cahen L., Snelling, N.J., Delhal, J., and Vail, J.R., 1984, The geochronology and evolution of Africa: New York, Oxford University Press, 512 p.

Campbell, I.H., and Griffiths, R.W., 1990, Implications of mantle plume structure for the evolution of flood basalts: Earth and Planetary Science Letters, v. 99, p. 79–93.

Chevrot, S., Vinnik, L., and Montagner, J.-P., 1999, Global-scale analysis of the mantle Pds phases: Journal of Geophysical Research, v. 104, p. 20203–20219.

Coffin, M.F., and Eldholm, O., 1992, Volcanism and continental break-up: A global compilation of large igneous provinces, *in* Storey, B.C., Alabaster, T., and Pankhurst, R.J., eds., Magmatism and the causes of continental break-up: Geological Society [London] Special Publication 68, p. 17–30.

Dawson, J.B., 1992, Neogene tectonics and volcanicity in the north Tanzania sector of the Gregory rift valley: Contrasts with the Kenya sector: Tectonophysics, v. 204, p. 81–92.

Ebinger, C.J., 1989, Tectonic development of the western branch of the East African rift system: Geological Society of America Bulletin, v. 101, p. 885–903.

Ebinger, C.J., and Sleep, N.H., 1998, Cenozoic magmatism throughout east Africa resulting from impact of a single plume: Nature, v. 395, p. 788–791.

Ebinger, C.J., Karner, G.D., and Weissel, J.K., 1991, Mechanical strength of the extended continental lithosphere: Constraints from the Western rift system, East Africa: Tectonics, v. 10, p. 1239–1256.

Ebinger, C.J., Bechtel, T.D., Forsyth, D.W., and Bowin, C.O., 1989, Effective elastic plate thickness beneath the East African and Afar plateaus and dynamic compensation of the uplifts: Journal of Geophysical Research, v. 94, p. 2883–2901.

Ebinger, C.J., Poudjom, Y., Mbede, E., Foster, F., and Dawson, J.B., 1997, Rifting Archean lithosphere: The Eyasi-Manyara-Natron rifts, east Africa: Journal of the Geological Society, London, v. 154, p. 947–960.

Farnetani, C.G., and Richards, M.A., 1994, Numerical investigations of the mantle plume initiation model for flood basalt events: Journal of Geophysical Research, v. 99, p. 13813–13833.

Flanagan, M.P., and Shearer, P.M., 1998, Global mapping of tomography on transition zone velocity discontinuities: Journal of Geophysical Research, v. 103, p. 2673–2901.

Foster, A., Ebinger, C., Mbede, E., and Rex, D., 1997, Tectonic development of the northern Tanzanian sector of the East African rift system: Journal of the Geological Society, London, v. 154, p. 689–700.

Fuchs, K., Altherr, B., Muller, B., and Prodehl, C., 1997, Structure and dynamic processes in the lithosphere of the Afro-Arabian rift system: Tectonophysics, v. 278, p. 1–352.

George, R., Rogers, N., and Kelley, S., 1998, Earliest magmatism in Ethiopia: Evidence for two mantle plumes in one flood basalt province: Geology, v. 26, p. 923–926.

Green, W.V., Achauer, U., and Meyer, R.P., 1991, A three-dimensional seismic image of the crust and upper mantle beneath the Kenya rift: Nature, v. 354, p. 199–203.

Griffiths, R.W., and Campbell, I.H., 1991, Interaction of mantle plume heads with the earth's surface and onset of small-scale convection: Journal of Geophysical Research, v. 96, p. 18295–18310.

Griffiths, D., King, R., Khan, M., and Blundell, D., 1971, Seismic refraction line in the Gregory Rift: Nature, v. 229, p. 69–71.

Hankel, O., 1987, Lithostratigraphic subdivisions of the Karoo rocks of the Luwegu basin (Tanzania) and their biostratigraphic classification based on microfloras, marcofloras, fossil woods and vertebrates: Geologische Rundundschau, v. 76, p. 539–565.

Hebert, L., and Langston, C.A., 1985, Crustal thickness estimates at AAE (Addia Ababa, Ethiopia) and NAI (Nairobi, Kenya) using *P*-wave conversions: Tectonophysics, v. 111, p. 299–327.

Hill, R.I., Campbell, I.H., Davies, G.R., and Griffiths, R.W., 1992, Mantle plumes and continental tectonics: Science, v. 256, p. 186–193.

Isaak, D., 1992, High-temperature elasticity of iron-bearing olivines: Journal of Geophysical Research, v. 96, p. 1871–1885.

Kampunzu, A.B., Bonhomme, M.G., and Kanika, M., 1998, Geochronology of volcanic rocks and evolution of the Cenozoic Western Branch of the East African Rift System: Journal of African Earth Sciences, v. 26, p. 441–461.

Key, R.M., Charsley, T.J., Hackman, B., Wilkinson, A.F., and Rundle, C.C., 1989, Superimposed upper Proterozoic collision controlled orogenesis in the Mozambique orogenic belt of Kenya: Precambrian Research, v. 44, p. 197–225.

King, S.D., and Anderson, D.L., 1995, An alternative mechanism of flood basalt formation: Earth and Planetary Science Letters, v. 136, p. 269–279.

King, S.D., and Anderson, D.L., 1998, Edge-driven convection: Earth and Planetary Science Letters, v. 160, p. 289–296.

King, S.D., and Ritsema, J., 2000, African hot-spot volcanism: Small-scale convection in the upper mantle beneath cratons: Science, v. 290, p. 1137–1140.

Kreuser, T., 1995, Rift to drift evolution in Permian-Jurassic basins of East Africa, *in* Lambiase, J.J., ed., Hydorcarbon habitat in rift basins: Geological Society [London] Special Publication 80, p. 297–315.

Last, R.J., Nyblade, A.A., Langston, C.A., Owens, T.J., 1997, Crustal structure of the East African plateau from receiver functions and Rayleigh wave phase velocities: Journal of Geophysical Research, v. 102, p. 24469–24483.

Lenoir, J.L., Liegeois, J.P., Theunissen, K., and Klerkx, J., 1994, The Paleoproterozoic Ubendian shear belt in Tanzania: Geochronology and structure, Journal of African Earth Sciences, v. 19, no. 3, p. 160–184.

Lithgow-Bertelloni, C., and Silver, P.G., 1998, Dynamic topography, plate driving forces and the African superswell: Nature, v. 395, p. 269–272.

Long, R.E., Backhouse, R.W., Maguire, P.K.H., and Sundarlingham, K., 1972, The structure of East Africa using the surface wave dispersion and Durham seismic array data: Tectonophysics, v. 15, p. 165–178.

McKenzie, D., and Bickle, M.J., 1988, The volume and composition of melt generated by extension of the lithosphere: Journal of Petrology, v. 29, p. 625–679.

Moller, A., Mezger, K., and Schenk, V., 1998, Crustal age domains and the evolution of the continental crust in the Mozambique Belt of Tanzania: Combined Sm-Nd, Rb-Sr, and Pb-Pb isotopic evidence: Journal of Petrology, v. 39, p. 749–783.

Morley, C.K., 1988, Variable extension in Lake Tanganyika: Tectonics, v. 7, p. 785–801.

Mueller, S., and Bonjer, K.P., 1973, Average structure of the crust and upper mantle in East Africa: Tectonophysics, v. 20, p. 238–253.

Nolet, G., and Mueller, S., 1982, A model for the deep structure of the East African rift system from the simultaneous inversion of teleseismic data: Tectonophysics, v. 84, p. 151–178.

Nyblade, A.A., 1997, Heat flow across the east African plateau: Geophysical Research Letters, v. 24, p. 2083–2086.

Nyblade, A.A., Birt, C., Langston, C.A., Owens, T.J., and Last, R.J., 1996, Seismic experiment reveals rifting of craton in Tanzania: EOS (Transactions, American Geophysical Union), v. 77, p. 520–521.

Nyblade, A.A., Pollack, H.N., Jones, D.L., Podmore, F., and Mushayandedbvu, M., 1990, Terrestrial heat flow in east and southern Africa: Journal of Geophysical Research, v. 95, p. 17371–17384.

Nyblade, A.A., Owens, T.J., Gurrola, H., Ritsema, J., and Langston, C., 2000, Seismic evidence for a deep upper mantle thermal anomaly beneath East Africa: Geology, v. 28, p. 599–602.

Owens, T.J., Nyblade, A.A., Gurrola, H., and Langston, C.A., 2000, Mantle transition zone structure beneath Tanzania, East Africa: Geophysical Research Letters, v. 27, p. 827–830.

Pasteels, P., Villeneuve, M., De Paepe, P., and Klerkx, J., 1989, Timing of the volcanism of the southern Kivu province: Implications for the evolution of the western branch of the East African rift system: Earth and Planetary Science Letters, v. 94, p. 353–363.

Petit, C., and Ebinger, C., 2000, Flexure and mechanical behavior of cratonic lithosphere: Gravity models of the East African and Baikal rifts: Journal of Geophysical Research, v. 105, p. 19151–19162.

Prodehl, C., Keller, G.R., and Khan, M.A., 1994, Crustal and upper mantle structure of the Kenya Rift: Tectonophysics, v. 236, p. 1–483.

Richards, M., Duncan, R., and Courtillot, V., 1989, Flood basalts and hotspot tracks: Plume heads and tails: Science, v. 246, p. 103–108.

Ritsema, J., van Heijst, H., and Woodhouse, J.H., 1999, Complex shear wave velocity structure imaged beneath Africa and Iceland: Science, v. 286, p. 1925–1928.

Ritsema, J., Nyblade, A.A., Owens, T.J., and Langston, C.A., 1998, Upper mantle seismic velocity structure beneath Tanzania, East Africa: Implications for the stability of cratonic lithosphere: Journal of Geophysical Research, v. 103, p. 21201–21213.

Shackleton, R.M., 1986, Precambrian collision tectonics in Africa, *in* Coward, M.P., and Ries, A.C., eds., Collision tectonics: Geological Society [London] Special Publication 19, p. 329–349.

Shen, Y., Solomon, S.C., Bjarnason, I.T., and Wolfe, C.J., 1998, Seismic evidence for a lower-mantle origin of the Iceland plume: Nature, v. 395, p. 62–65.

Simiyu, S.M, and Keller, G.R., 1997, An integrated analysis of lithospheric structure across the east African plateau based on gravity anomalies and recent seismic studies: Tectonophysics, v. 278, p. 291–313.

Slack, P.D., Davis, P.M. and the KRISP Teleseismic Working Group, 1994, Attenuation and velocity of P-waves in the mantle beneath the east African rift, Kenya: Tectonophysics, v. 236, p. 331–358.

Sleep, N., 1996, Lateral flow of hot plume material ponded at sublithospheric depths: Journal of Geophysical Research, v. 101, p. 28065–28083.

Sleep, N., 1997, Lateral flow and ponding of starting plume material: Journal of Geophysical Research, v. 102, p. 10001–10012.

Smith, M., 1994, Stratigraphic and structural constraints on mechanisms of active rifting in the Gregory Rift, Kenya: Tectonophysics, v. 236, p. 3–22.

Spence, J., 1957, The geology of part of the eastern province of Tanganyika: Geological Survey of Tanganyika, Bulletin 84.

Theunissen, K., Klerkx, J., Melnikov, A., and Mruma, A., 1996, Mechanisms of inheritance of rift faulting in the western branch of the East African Rift, Tanzania: Tectonics, v. 15, p. 776–790.

Upcott, N.M., Mukasa, R.K., Ebinger, C.J., and Karner, C.D., 1996, Along-axis segmentation and isostacy in the western rift, East Africa: Journal of Geophysical Research, v. 101, p. 3247–3268.

Wheildon, J., Morgan, P., Williamson, K.H., Evans, T.R., and Swanberg, C.A., 1994, Heat flow in the Kenya rift zone: Tectonophysics, v. 236, p. 131–149.

White, R., and McKenzie, D., 1989, Magmatism at rift zones: The generation of volcanic continental margins and flood basalts: Journal of Geophysical Research, v. 94, p. 7685–7729.

Zeyen, H., Volker, F., Wehrle, V., Fuchs, K., Sobolev, S.V., and Altherr, R., 1997, Styles of continental rifting: Crust-mantle detachment and mantle plumes: Tectonophysics, v. 278, p. 329–352.

Manuscript Accepted by the Society October 15, 2001

Printed in the U.S.A.

Geological Society of America
Special Paper 362
2002

Development of the Lebombo rifted volcanic margin of southeast Africa

M.K. Watkeys
School of Geological and Computer Sciences, University of Natal, Durban 4041, South Africa

ABSTRACT

The north-south–trending Lebombo region of southeast Africa has many of the elements of a classic rifted volcanic margin, including a monocline, seaward-dipping basalts, and the mid-ocean ridge basalt–like Rooi Rand sheeted dike swarm. At the northern end it also has the appearance of a classic triple junction, the Sabi monocline being the other successful rift arm and the Okavango dike swarm being the failed arm. Consequently the region has been interpreted as developing due to the impact of a plume into a stable region with the simultaneous development of the three arms. However analysis of the geology reveals that this is not the case. There is evidence for tectonic activity at least 70 m.y. before the arrival of the plume, dextral transtension taking place in the Limpopo Belt basement during deposition of Permian-Triassic Karoo sediments. The Jurassic Karoo dolerite dike swarms display a number of trends, some of which are influenced by preexisting structures. These dikes intruded during different but overlapping times ca. 183 Ma in a linked sequence that does not correspond to a synchronous development of the classic triple junction shape. The initial 183 Ma nephelinite and picrite volcanism and dikes occurred at the northern end of the Lebombo, along or adjacent to the east-northeast–trending Limpopo Belt. It was prevented from spilling southward due to the presence of a long-lived paleohigh on the northeast Kaapvaal craton. With the onset of low MgO basaltic volcanism, dikes injected along a west-northwest trend to form the Okavango dike swarm, which extends from the northern end of the Lebombo. Dilation of this swarm caused older fractures to open and resulted in the intrusion of the northeast-trending Olifants River swarm. Dilation also resulted in movement along the Agulhas-Falklands Fracture Zone that caused the coastal faulting in KwaZulu-Natal at the southern end of the Lebombo. This induced east-west extension across the Lebombo, which then completed the sequence of fracture events by linking back to the site of initial volcanism at the northern end. Subsequent thermal doming of the Kaapvaal craton enabled both north-south and east-west dikes to intrude at the same time. Further extension on the Lebombo involved the intrusion of the Rooi Rand sheeted dike swarm, followed by extrusion of rhyolites produced by partial melting of underplated basaltic material. During this event, the main monoclinal flexing and faulting took place, which affected earlier normal faults. There was no further significant east-west extension along the Lebombo even during the actual breakup of Gondwana, which occurred 40 m.y. later with the opening of the South Atlantic and the development of Cretaceous volcanism east of the Lebombo. In this breakup event the Falkland Plateau was pulled out along the southeast coast of Africa so that the extension direction was virtually north-south.

Watkeys, M.K., 2002, Development of the Lebombo rifted volcanic margin of southeast Africa, *in* Menzies, M.A., Klemperer, S.L., Ebinger, C.J., and Baker, J., eds., Volcanic Rifted Margins: Boulder, Colorado, Geological Society of America Special Paper 362, p. 27–46.

Consequently, in terms of the Karoo event, the Lebombo is a failed rifted margin, because full continental breakup did not occur. However, in terms of its overall development, it may be regarded as a doubly rifted volcanic margin, having been involved in the 183 Ma Karoo event and the 135 Ma Cretaceous breakup of Gondwana, the extension directions in these two events being at right angles to each other.

INTRODUCTION

The rifted volcanic margin of southeast Africa is marked by the Lebombo monocline, a north-south–trending structure more than 700 km long, along the eastern edge of the Archean Kaapvaal craton (Fig. 1). Together with the Sabi monocline and the Okavango dike swarm, it has the classic shape of a triple junction (Burke and Dewey, 1972), the dike swarm representing the failed arm (Reeves, 1978). The Lebombo has some typical characteristics of a generalized model of rifted volcanic margins, including a monoclinal flexure induced by coast-parallel faulting (Figs. 2 and 3). This has resulted in a seaward dip of the voluminous Karoo volcanic rocks, which are covered by the Cenozoic sequences of the Mozambique coastal plains. These volcanics may be interpreted as the equivalents of the seaward-dipping reflectors of many submerged rifted volcanic margins elsewhere in the world; in this region they have been uplifted due to the post-Gondwana breakup history of southern Africa.

It has long been recognized in South Africa that Karoo volcanism did not result in actual continental breakup, and that the Lebombo should be likened to the early stages of rifting along the east coast of Greenland (du Toit, 1929, 1937). Seafloor spreading occurred ~40 m.y. later as part of a Cretaceous volcanic event. Therefore, the Lebombo might be regarded as a doubly rifted volcanic margin. Cox (1970) pointed out that there was basement control on the Mesozoic tectonic patterns seen in the region, but subsequent work has tended to concentrate on geochemical aspects (viz. Erlank, 1984). With the emergence of the mantle plume theory (Campbell and Griffiths, 1990; Hill, 1991), this has become the most popular explanation for the volcanism and rifting (viz. Cox, 1992; White, 1997).

Any model for continental breakup is, by necessity, a generalization and should vary according to the local geology. In southeast Africa, the rifting events were not superimposed on an isotropic medium, but on a heterogeneous lithosphere consisting of an Archean nucleus enclosed by Mesozoic-Proterozoic mobile belts. This regional framework has played an important control on the Mesozoic pattern, and this chapter is aimed at returning to Cox's (1970) theme concerning how preexisting structures might have influenced the rifting process. This involves examining the sediments beneath the volcanic rocks and the dikes feeding those extrusions. A model is provided for the Gondwana breakup sequence around southeast Africa, summarizing events from the time of Karoo volcanism until the mid-Cretaceous, when rifting finally ceased and drifting took over as all contact was lost between the continental crust of Africa and South America.

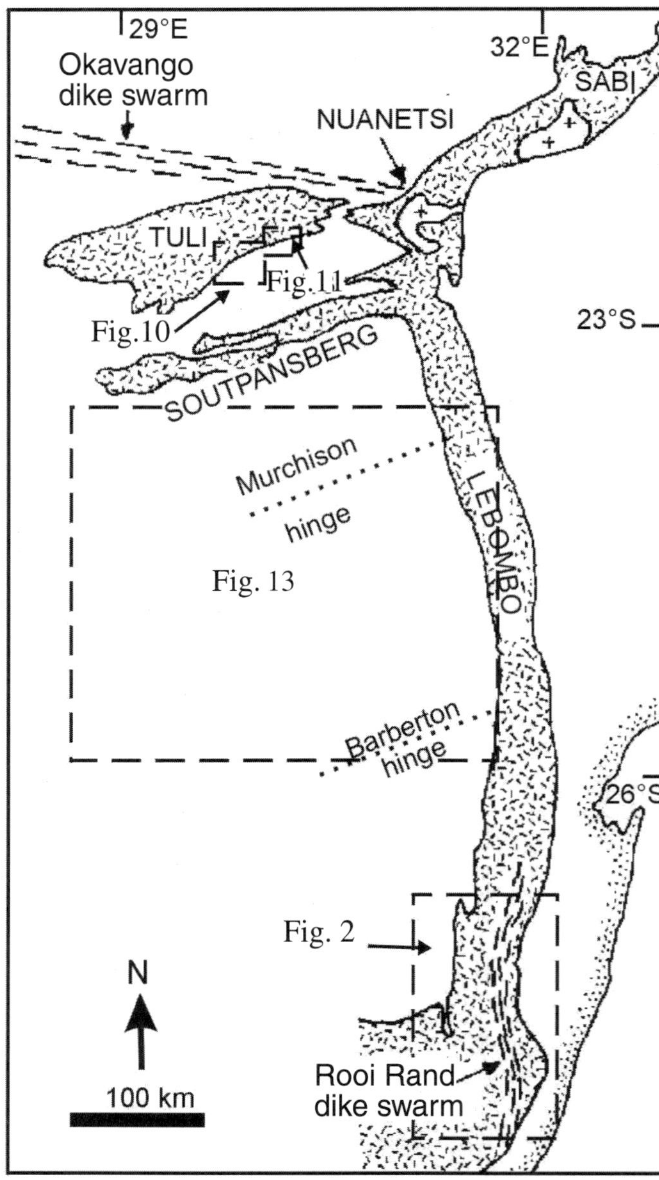

LEBOMBO AND ENVIRONS

The Lebombo monocline occurs along the eastern edge of the Archean Kaapvaal craton (Fig. 1). At the southern end of the Lebombo, the Kaapvaal craton is overthrust by the Mesozoic-Proterozoic Natal metamorphic province (Matthews, 1981). In

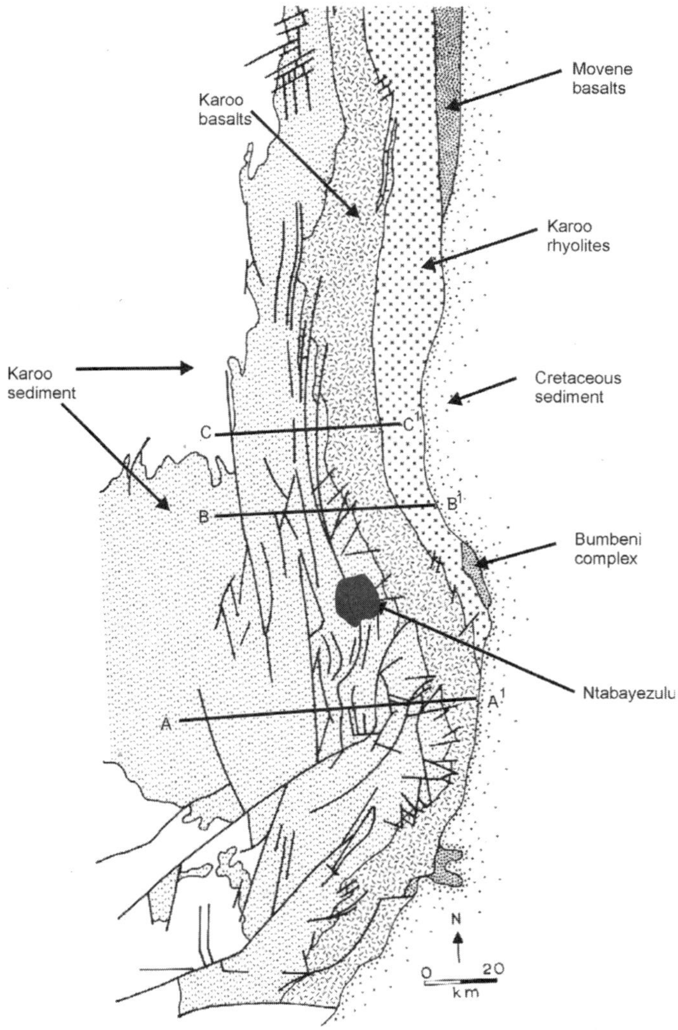

Figure 2. Simplified geological map of southern Lebombo (see Fig. 1 for locality). Lines of cross section are shown in Figure 3.

that province, a region of Mesozoic coastal faulting that lacks downfaulted volcanics is known as the Natal monocline (King, 1972; Von Veh and Andersen, 1990; Watkeys and Sokoutis, 1998). The northern end of the Lebombo coincides with the termination of the Kaapvaal craton against the Limpopo Belt. Farther north, the Karoo Supergroup is preserved in four downfaulted troughs, the Soutpansberg trough, the Tuli trough, the Nuanestsi trough, and the Sabi monocline (Fig. 1).

These regions represent more than 1200 km strike length of continuous exposure of the Karoo Supergroup, which consists of Carboniferous to Jurassic sediments, capped by Jurassic volcanics. Karoo sedimentation was essentially continuous from the underlying Ordovician to Devonian Cape Supergroup, which is found at the southern and western part of South Africa (Veevers et al., 1994). The two supergroups therefore represent a Gondwana succession deposited on continental crust over 300 m.y.; the succession was abruptly terminated by the voluminous outpourings of Karoo basalts. This prolonged period of relative tectonic stability in southern Africa was interrupted in the Late Permian and Early Triassic by the development of the Cape fold belt along the southern and western parts of the subcontinent (Dingle et al., 1983). This event is related to the final stages of amalgamation of Gondwana, when the Patagonia microplate collided with South America and southern Africa.

Sedimentary stratigraphy of the Lebombo and environs

The sediments of the Karoo Supergroup of the Lebombo and environs may be correlated with the succession in the main Karoo basin to the southwest (Tankard et al., 1982; Smith, 1990; Veevers et al., 1994). Because of this correlation, and in order to avoid confusion with the proliferation of stratigraphic nomenclature across international boundaries as well as within South Africa (South African Committee for Stratigraphy, 1980), only the broad terms will be used here in giving the overall description of the Karoo Supergroup along the Lebombo.

The sediments may be subdivided into two successions. The basal Dwyka Group is absent or only very thinly developed along the Lebombo, so the lower succession here consists of Permian to Triassic Ecca and Beaufort Groups, the upper succession being the Molteno, Elliott, and Clarens Formations. The two successions are separated by an unconformity that can be dated indirectly as being younger than 205 Ma. This is the date obtained for the Dokolwayo kimberlite in Swaziland (Allsopp and Roddick, 1984), diamonds from which have been found in Molteno sediments (Turner and Minter, 1985). When the thicknesses of the Karoo sediments along the Lebombo are plotted on a cross section (Fig. 4), it is apparent that there was a paleohigh toward the northern Lebombo, particularly for the lower succession. The sediments deposited south of this are the northeastern facies of the main Karoo basin, whereas sediments to the north are part of the fault-bounded Zambezian tectonic-sedimentary terrain of Rust (1975). There is also a change in paleocurrent between the lower and upper sedimentary successions. Whereas the lower succession drained south and southwest into the main Karoo basins (Tankard et al., 1982; Dingle et al., 1983), the upper succession has eastward-flowing paleocurrents (Turner and Minter, 1985).

The paleohigh along the Lebombo has been well documented (viz. Tankard et al., 1982). In the Ecca Group, it has been ascribed to the Permian-Carboniferous glaciation retreating to a highland as southern Africa moved away from the south pole. In the Beaufort Group, it is due to being at the northern edge of a downflexed foreland basin associated with the development of the Cape fold belt (Catuneanu et al., 1998). However, there are also inflections or hinges in the paleoslope, where there are two other important east-northeast–trending structures in the basement: the Barberton and Murchison Archean greenstone belts (Fig. 4). It shows that the Kaapvaal craton was not behaving as a coherent body during deposition of the Karoo sediments, but was undergoing reactivation along older tectonic lineaments, so that the Archean was already being segmented into separate blocks (Fig. 4).

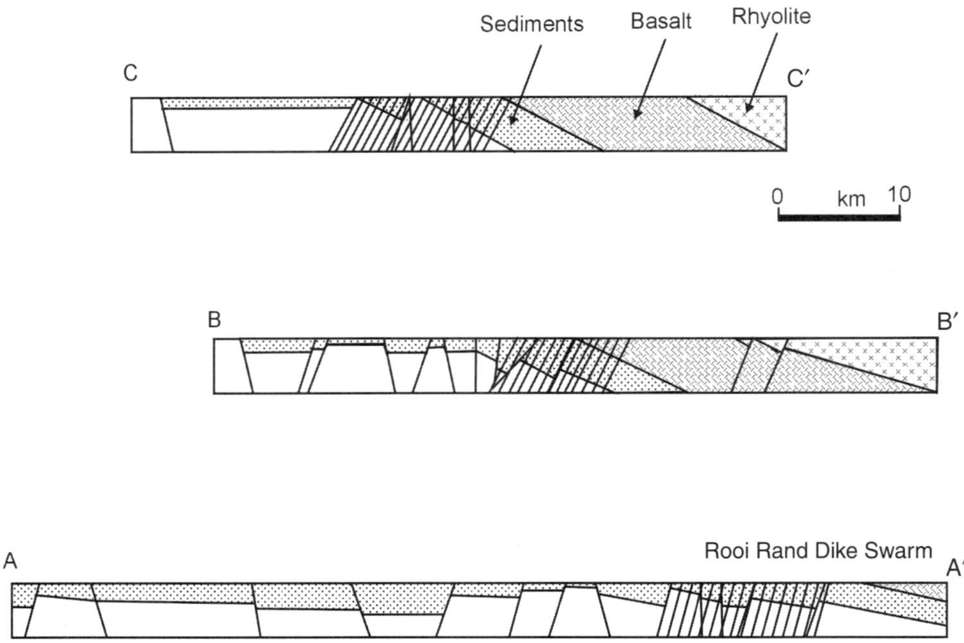

Figure 3. Simplified cross sections of southern Lebombo (see Fig. 2 for localities), showing zone of normal faulting in west and faulted monocline in east, and Rooi Rand sheeted dike swarm intruding at flexure.

Such segmentation is not unexpected farther north in the downfaulted troughs of the Soutpansberg, Bubye, Tuli, Nuanetsi, and Sabi regions (Figs. 1 and 4). The Beaufort Group is absent north of the Bubye trough due to fault-controlled deposition. A plot of the thicknesses of sediments (Fig. 5) reveals a paleohigh centered on the Nuanetsi igneous province, which intrudes the sediments and volcanics of the region. In places the sediments are completely absent and the basalts directly overlie basement (Broderick, 1979).

One possible explanation for this paleohigh is thermal uplift of the area during intrusion of the Nuanetsi igneous province. However, this seems unlikely because the Karoo sediments show an onlapping relationship with this paleohigh, indicating that it was present as early as the Permian-Carboniferous. If this was due to thermal uplift, it implies a long-lived event in excess of 100 m.y. The paleocurrents in the braided river deposits of the Ecca Group flow away from this paleohigh; coal seam caps show a fining away from the paleohigh (Fig. 6). It is difficult to explain the presence of this paleohigh as a regional feature or as having been on a margin of a foreland basin. It appears more likely that it was a horst.

Volcanic stratigraphy of the Lebombo and environs

The volcanic rocks of the Karoo Supergroup have been reviewed by a number of workers (viz. Bristow and Saggerson, 1983; Eales et al., 1994; Cox, 1988), and the stratigraphy is summarized in Figure 4. The exact thicknesses along the Lebombo are difficult to establish due to the lack of distinct stratigraphic markers and the probability of repetition due to faulting. Consequently, the values shown in Figure 7 represent the maximum thicknesses. They reveal, however, that the paleohigh that affected the deposition of the sediments also influenced the distribution and thicknesses of the volcanics.

The earliest manifestations of volcanism are in the north, i.e., the eruption of the Mashikiri nephelinites onto an active eolian landscape represented by the Clarens Formation. They are as thick as 170 m and are confined to the northern Lebombo, the Soutpansberg trough, and narrow exposures along the northern parts of the Sabi trough (Bristow, 1984a). They have been dated as ca. 183 Ma (Duncan et al., 1997) and consist of undersaturated (SiO_2 40.4%–45.0%) and strongly nepheline-normative with MgO in the range 2.6%–12%. It is likely that the nephelinites represented the first low-degree melts from a carbonated mantle (Bristow, 1984a).

The Letaba picrites overlie the nephelinites. The picrites are indistinguishable in age from the nephelinites (Duncan et al., 1997) and are found in the same regions as the nephelinites, but with a much wider distribution, including the southern part of the Tuli trough (Bristow, 1984b). These rocks, which may be as thick as 4 km, have a range in MgO (9%–23%) and require parental liquids in excess of 13.5% MgO (Bristow, 1984b). The picrites also contain very variable K and incompatible element contents (Cox et al., 1984). The low-K picrite end member (Ellam and Cox, 1989) may have been derived from a mantle not compositionally distinct from the asthenosphere (Ellam and Cox, 1991; Sweeney et al., 1991; Ellam et al., 1992).

The voluminous low MgO continental flood basalts that followed and covered most of southern Africa appear to have erupted ca. 183 Ma (Duncan et al., 1997; Marsh et al., 1997), synchronous with emplacement of an extensive network of subvolcanic dolerite dikes and sills (Encarnación et al., 1996).

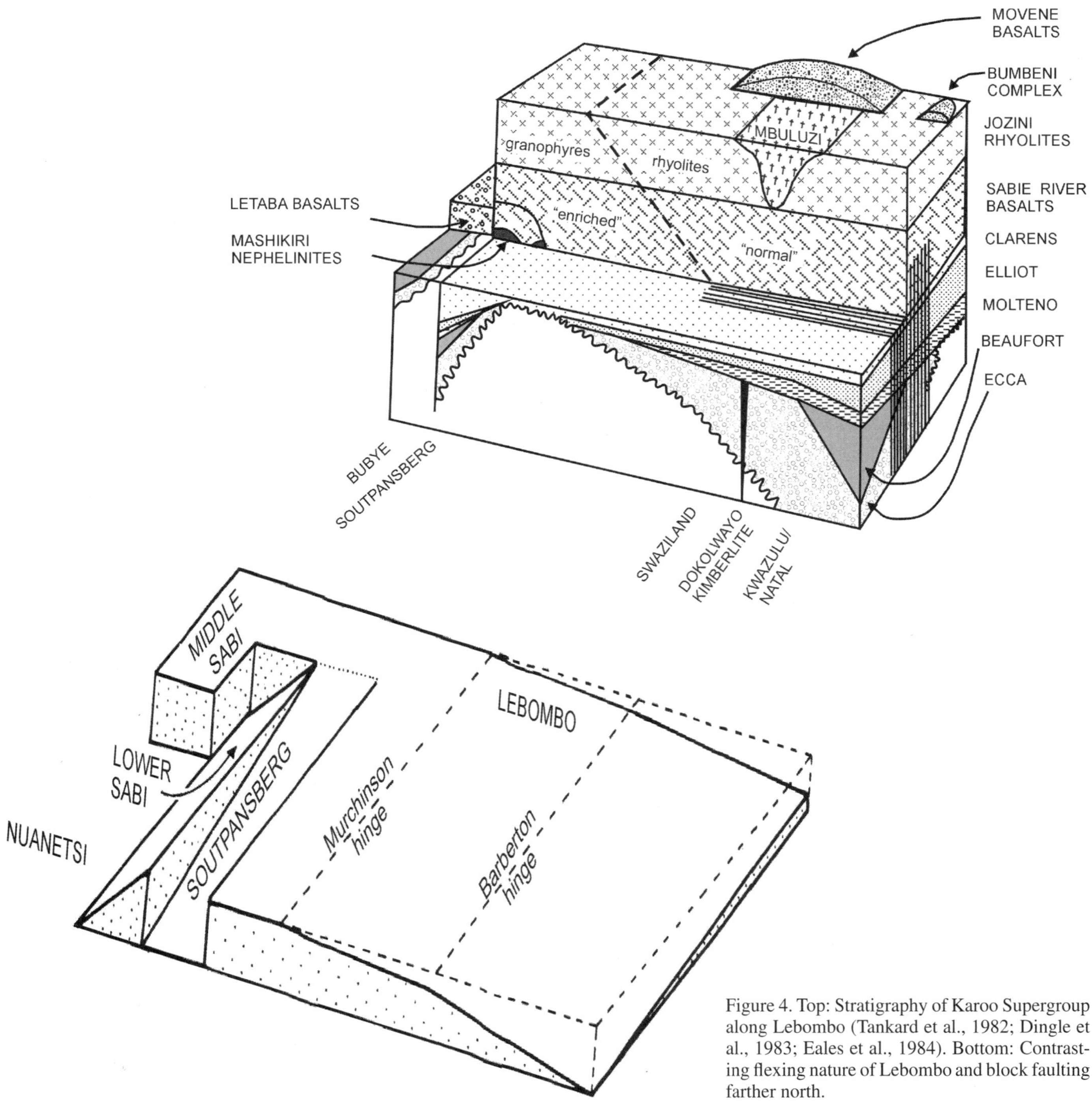

Figure 4. Top: Stratigraphy of Karoo Supergroup along Lebombo (Tankard et al., 1982; Dingle et al., 1983; Eales et al., 1984). Bottom: Contrasting flexing nature of Lebombo and block faulting farther north.

Along the Lebombo these basalts are termed the Sabie River Basalt Formation (Cox and Bristow, 1984). They have ages that are virtually indistinguishable from the underlying volcanics and, on the basis of a paleomagnetic reversal (Hargraves et al., 1997), probably erupted within the space of 0.5 m.y. (Duncan et al., 1997). The north-south divide across the paleohigh along the Lebombo is represented in these rocks not only in thicknesses, but in composition: the basalts along the northern Lebombo and farther north are relatively enriched in incompatible elements (high Ti-Zr) compared to those in the south (low Ti-Zr) (Cox et al., 1967). This feature is found across the Karoo Province (Cox, 1983; Duncan et al., 1984). Rather than ascribe this directly to a plume, Sweeney and Watkeys (1990) explained it as being due to lithospheric control, the enriched basalts being derived from sub-Archean lithospheric mantle and the normal basalts from a post-Archean mantle. This slightly naive suggestion was super-

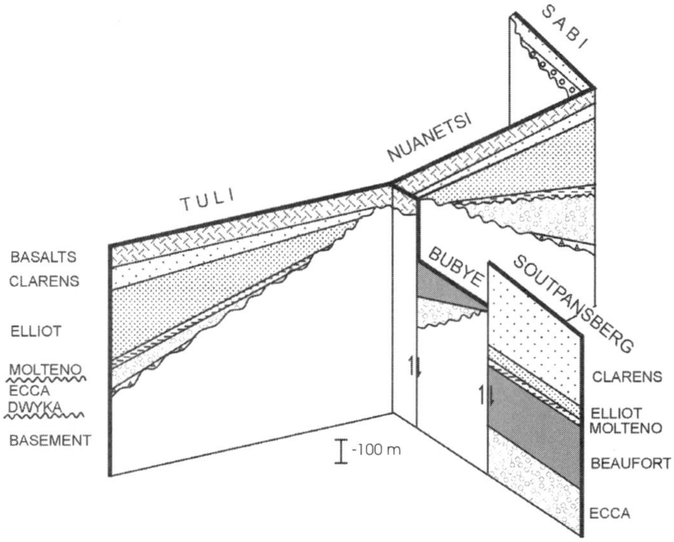

Figure 5. Fence diagram showing thicknesses of Karoo Supergroup sediments in Soutpansberg, Bubye, Tuli, Nuanetsi, and Sabi troughs (compiled from Swift et al., 1953; Swift, 1962; Thompson, 1975; Broderick, 1979; Watkeys, 1979; Light and Broderick, 1998). See Figures 1 and 4 for localities of troughs.

seded by a model proposing that ~30%–40% of the incompatible trace element component in the basalts might have been derived from the lithospheric mantle (Sweeney et al., 1994).

The low MgO basalts may be related to the underlying picrite basalts by crystal fractionation processes (Cox et al., 1984), the low-K picrites being a significant (63 wt%) component of the parent to the low-MgO, high Ti-Zr basalts, and the remaining contribution being from the high-K picrites (Sweeney et al., 1991). The low-Ti-Zr basalts may have been derived from a thermal asthenospheric plume but equilibrated with refractory mantle lithosphere before eruption (Sweeney et al., 1991). It is only in the late-stage evolved high-Fe basalts, found locally in the upper parts of the basalt sequence in the central Lebombo, that there might be the influence of a compositionally distinct mantle plume (Sweeney et al., 1994).

Various layered intrusions are scattered along the Lebombo (Saggerson and Logan, 1970). The largest of these is Ntabayezulu in the southern Lebombo (Fig. 2), which occurs along the line of flexure and was intruded by the Rooi Rand sheeted dike swarm (Fig. 3). This swarm extends from northern Kwa-Zulu Natal into southern and central Swaziland (Saggerson et al., 1983). It intrudes the basalts but not the overlying rhyolites (Fig. 4), is aligned approximately north-south, being subparallel to the axis of the Lebombo monocline, and varies from 10 to 22 km in width. It consists of sheeted dolerite dikes generally dipping steeply toward the west with variable amounts of intervening country rock (or wall-rock screens). Crustal extension may be as much as 40% across this swarm. The geochemical evidence indicates that the dike swarm is not the feeder to the Sabie River Basalt Formation, being mid-ocean ridge basalt–like in character (Armstrong et al., 1984; Duncan et al., 1990). Detailed mapping and geochemical studies of the various phases of dike intrusion reveal that there is not a random geochemical pattern or a simple geochemical evolution (Meth, 1996). All of the phases have unradiogenic Pb isotopic compositions, so they appear to lack the HIMU (high $^{238}U/^{204}Pb$ ratio) mantle component characteristic of plume involvement (Watkeys et al., 2001).

The top of the Sabie River Formation contains some interbedded rhyolites and dacites that predate the main eruption of rhyolites, ca. 178 Ma, along the Lebombo and in the Nuanetsi

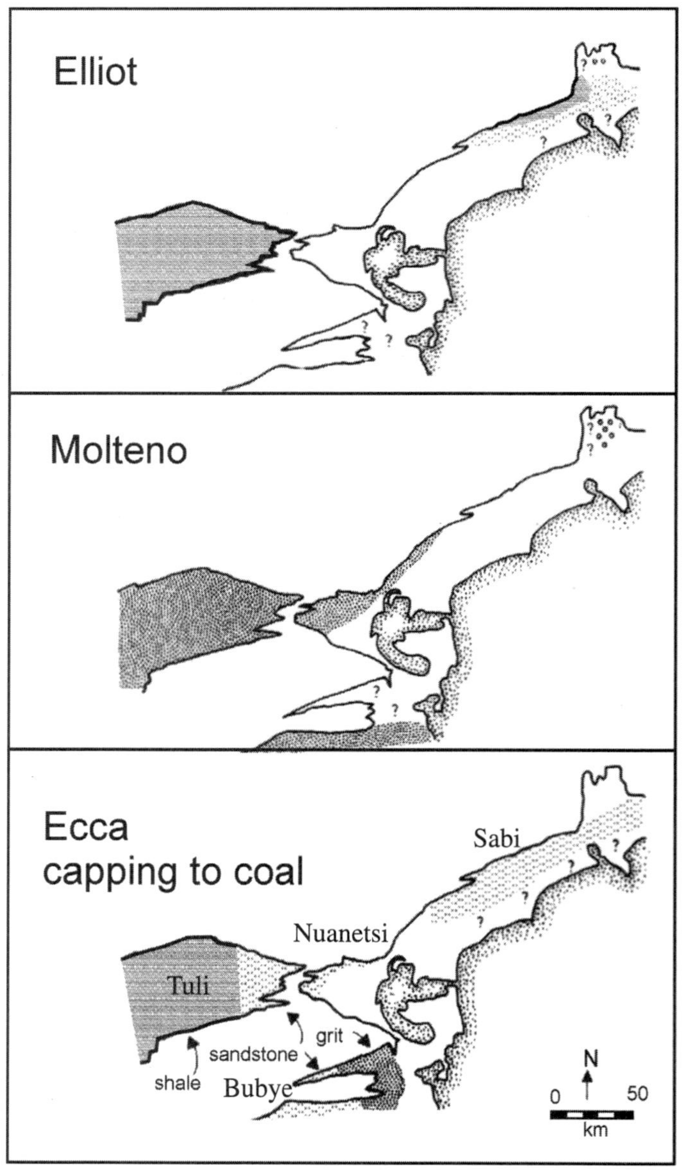

Figure 6. Distribution of Ecca Group and Molteno and Elliot Formations in Bubye, Tuli, Nuanetsi, and Sabi troughs (compiled from Swift et al., 1953; Swift, 1962; Thompson, 1975; Broderick, 1979; Watkeys, 1979; Light and Broderick, 1998). See Figures 1 and 4 for localities of troughs.

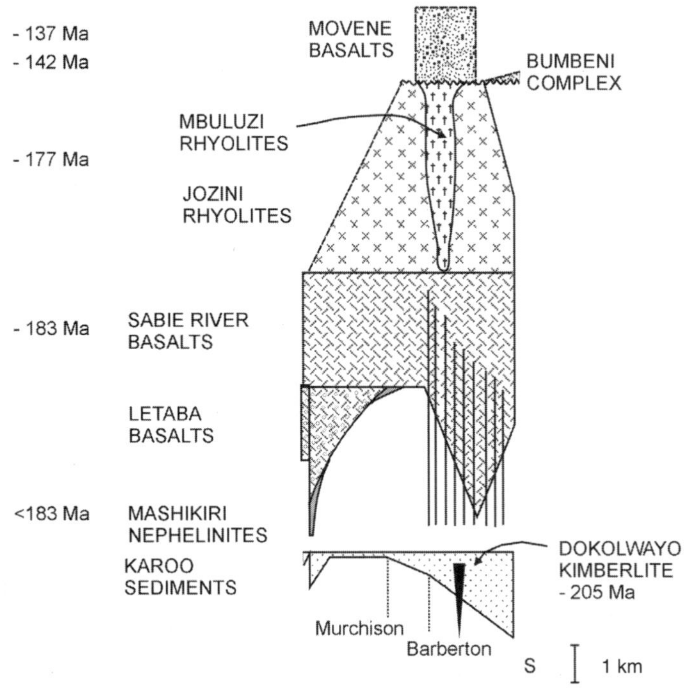

Figure 7. Thicknesses of Karoo Supergroup along Lebombo (Tankard et al., 1982; Dingle et al., 1983; Eales et al., 1984; Bristow, 1984a, 1984b; Cox and Bristow, 1984; Cleverley et al., 1984).

trough (Cleverly et al., 1984; Duncan et al., 1997). The Lebombo rhyolites are as thick as 5 km and are subdivided into the older Jozini Formation and younger Mbuluzi Formation (Fig. 4). The former consists of a monotonous sequence of high-temperature ash-flow tuffs showing little chemical variation along the Lebombo, but also showing evidence of the north-south divide as it occurs in the north as granophyres (Fig. 4). The Mbuluzi rhyolites are confined to Swaziland, south of the Barberton hinge, and are distinguished by the presence of quartz phenocrysts and a lower Zr content. The Nuanetsi rhyolites, which are as thick as 1.8 km and contain some interbedded basalt flows, are geochemically distinct from the Lebombo rhyolites (Cleverly et al., 1984; Betton et al., 1984).

The Lebombo and Nuanetsi rhyolites are considered to derive from partial melting of hot underplated basaltic rocks (Cleverly et al., 1984; Harris and Erlank, 1992) with some subtle crustal contamination (Betton et al., 1984). The Mkutshane Beds, which are well below the main rhyolite sequence and are interbedded with the basalts, are an exception; these are either crustal melts or are highly contaminated by crust (Cleverly et al., 1984).

The youngest of the Karoo events is the Nuanetsi igneous province (Cox et al., 1965), which intrudes the rhyolites of the Nuanetsi trough (Fig. 1). The largest intrusion is a transgressive acid sill, called the Main granophyre, which occurs near the base of the rhyolites and may extend to a similar body in the lower Sabi region. Ring complexes associated with this province are generally aligned east-northeast, and consist of early gabbros and later granite, granophyres, and microgranites; nepheline syenites occur in the 177 Ma Marangudzi Complex (Foland and Henderson, 1976).

Post-Karoo igneous activity and sedimentation

The top of the Karoo rhyolites represents the end of the Karoo volcanic event and marks a major hiatus. Volcanism associated with Gondwana breakup in this region commenced ca. 146 Ma with the Ntabankosi (formerly Kuleni) rhyolites, which intruded the Jozini rhyolites ca. 146 Ma (Allsopp et al., 1984). These are overlain by conglomerates of the Msunduze Formation and then by ~50 m of trachybasalts and trachyandesites of the Mpilo Formation (Wolmarans and du Preez, 1986). These may be equivalent to the Movene Formation basalts found overlying the Jozini rhyolites farther north (Du Preez and Wolmarans, 1986).

The Mpilo basalts are overlain and intruded by the rocks of the Bumbeni Complex, which is 14 km long and 5 km wide, and consists of at least 360 m of pyroclastics, rhyolites, and trachytes (Bristow and Duncan, 1983), which have been dated as 133 ± 4 Ma (Allsopp et al., 1984). Most of the volcanics related to this event are hidden beneath Cretaceous, Tertiary, and Quaternary sediments of the southeast Africa coastal plain (Dingle et al., 1983; Watkeys et al., 1993).

TECTONIC CONTROLS ON THE RIFTING EVENTS

Limpopo region

In light of the fact that both the oldest and the youngest Karoo igneous events took place in the Limpopo Belt, it is remarkable that actual rifting did not take place along that region, but rather at a high angle to it, particularly because the Limpopo Belt has been a zone of weakness and reactivation since its formation in the Archean (Watkeys, 1983). The belt consists of exhumed high-grade polymetamorphic and highly deformed rocks representing a continental collision between the Kaapvaal and Zimbabwean cratons in the Late Archean. The Southern and Northern marginal zones represent high-grade cratonic material, whereas the Central zone is a volcanic-sedimentary sequence that has been thrust northward onto the Zimbabwean craton. The east-northeast alignment of the belt reflects a strong fabric in this orientation that subsequently became an important fracture direction that developed in the Proterozoic.

The Karoo Supergroup is preserved in the Central zone in a number of downfaulted troughs (Figs. 1 and 8). The lower sedimentary succession was deposited on a basement with well-developed east-northeast fractures that controlled a series of half-grabens, indicating that the region was undergoing north-northwest–south-southeast extension. The ultimate control on the positioning of these grabens appears to be inversion of the southward-dipping thrust along which the Central zone had

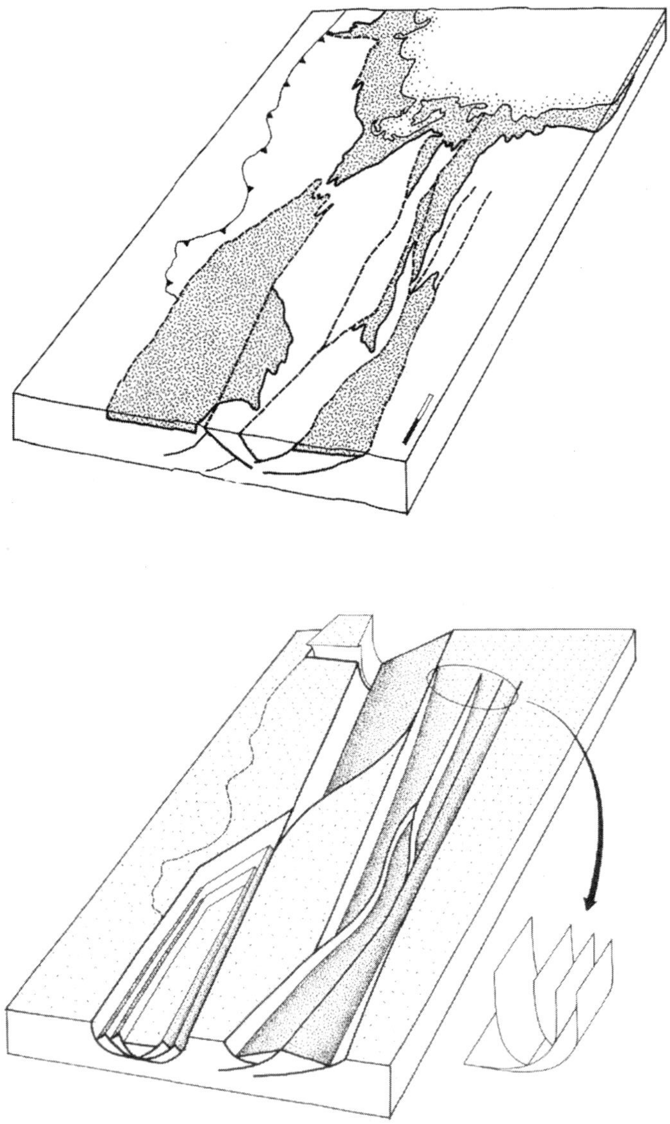

Figure 8. Top: Soutpansberg, Bubye, Tuli, Nuanetsi, and Sabi troughs viewed from southwest, showing main Karoo faults and position of thrust (teeth in upper plate) along northern edge of Central zone, Limpopo Belt. Scale bar is 20 km long. See Figure 1 for regional locality. Bottom: Soutpansberg, Bubye, Tuli, Nuanetsi, and Sabi troughs showing half-grabens (exaggerated) and upthrown block in Sabi region. Scale bar is 20 km long. See Figure 1 for regional locality.

been thrust northward. Extension along such a structure will result in graben development on either side of a horst (Fig. 9). This is considered the most likely reason for the presence of a paleohigh in the vicinity of the future site of the Nuanetsi igneous province.

Following the retreat of the Permian-Carboniferous Dwyka glaciation, the Ecca Group, and in the Bubye trough, the Beaufort Group, was deposited into these half-grabens during a period of extension. This was not major stretching of the crust but rather a slight extension at right angles to the dominant east-northeast trend in the basement; this resulted in the formation of normal faults along this trend, so that there was downfaulting in the Limpopo Belt relative to the Kaapvaal craton. This event ceased in late Beaufort time due to compression related to the formation of the Cape fold belt being transmitted across the Kaapvaal craton. With the cessation of this event, extension commenced again with deposition of the upper sedimentary succession, which onlaps the basement (Fig. 9). This scenario explains the lack of Beaufort sediments in places and the presence of the unconformity between the lower and upper sedimentary successions.

In addition to extension, there are also features indicative of dextral strike-slip deformation (Reading, 1980). Along the southern edge of the Tuli trough, the Karoo sediments both unconformably overlie the basement and are faulted against it (Thompson, 1975; Watkeys, 1979) (Fig. 10). The offset of the unconformity combined with the dip of the strata and the dip and strike of the faults yield a reconstruction that is representative of a dextral strike-slip system. Later reactivation of the faults by normal movement has resulted in slickenlines that are not related to the strike-slip event.

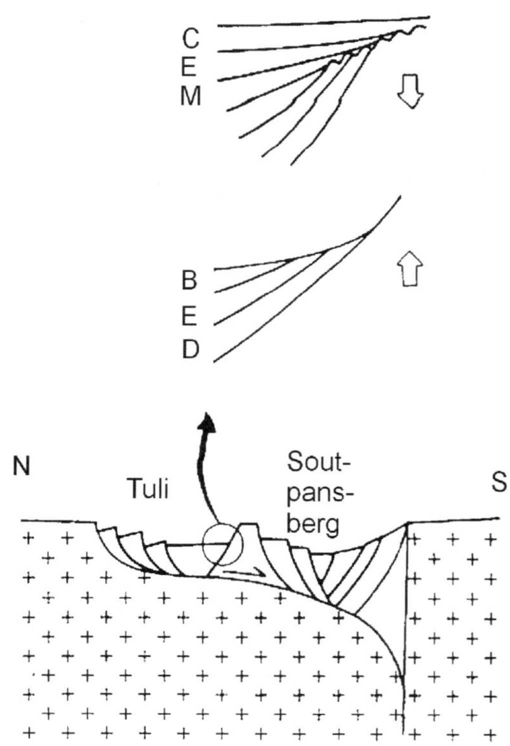

Figure 9. Development of graben and horst caused by inversion of Limpopo Belt Central zone thrust (see Fig. 8), and schematic representation of development of stratigraphy due to relative movement of adjacent craton. Lower sedimentary succession: D, Dwyka; E, Ecca; B, Beaufort. Upper sedimentary succession: M, Molteno; E, Elliot; C, Clarens.

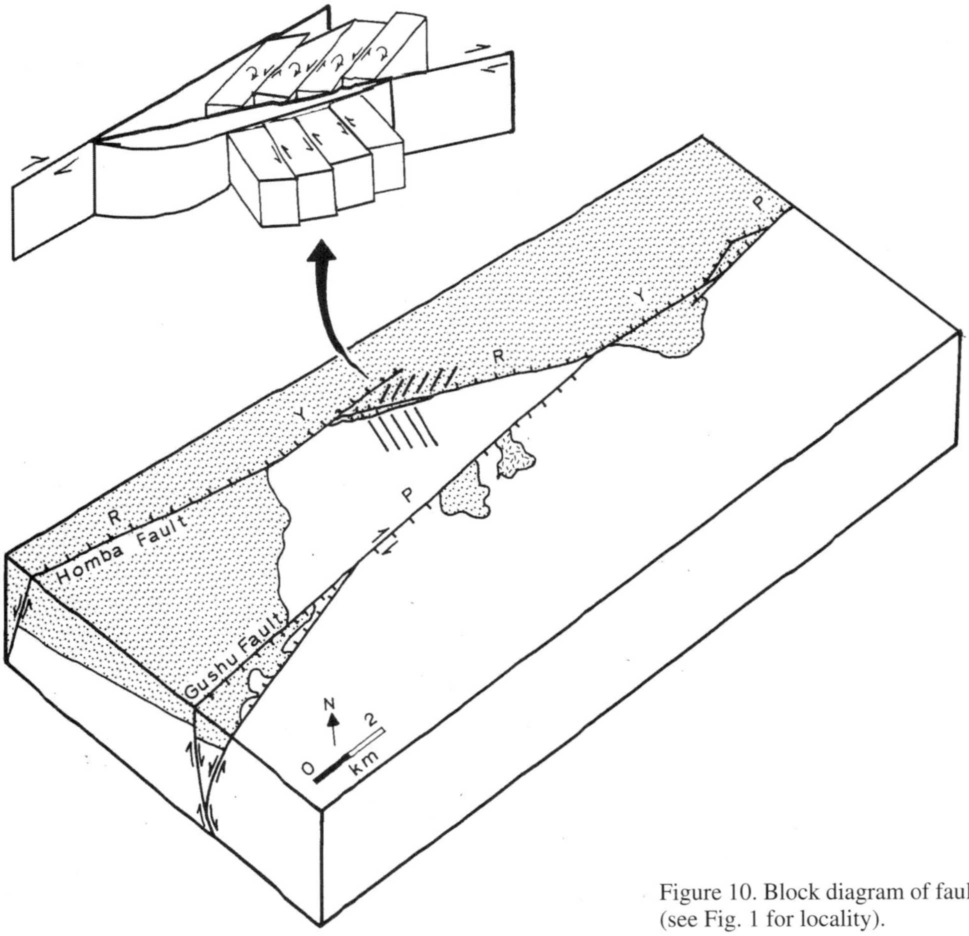

Figure 10. Block diagram of fault pattern along southern end of Tuli trough (see Fig. 1 for locality).

The Homba fault dips north and is downthrown to the north (Fig. 10). At its eastern end, a segment of Molteno Formation sediments is trapped as a wedge between the northeast-trending Y shear and the east-northeast–trending R shear. Within this wedge, the Molteno sediments have undergone domino-style faulting and slight rotation around a vertical axis in order to accommodate the extension taking place in this small releasing bend. This rotation is probably taking place on the Ecca shales near the unconformity. On the south side of the R shear, the basement has also been segmented into small blocks by R′ shears. A sliver of Ecca shales parallel to the R shear is trapped between these two sets of blocks, and is probably assisting with the dissipation of the terminations of the two sets of faults in the blocks to the north and south. This R shear is a connecting splay with another Y shear that forms the Gushu fault, which dips south and is downthrown to the south.

The dikes in the adjacent basement include ages that range from Early Proterozoic to Karoo. Selecting those that appear to be Karoo in age yields a number of different orientations that fit the expected pattern of T, R, R′, and P shears, suggesting that they may be infilling these fractures (Fig. 11). There are also north-south dikes that cut this pattern, as does the large Masasanye dike, which has with a more or less east-west trend and is probably related to the alkalic intrusions associated with the Nuanetsi igneous province. This dike displays very clear segmentation, indicating vertical or subvertical magma flow direction during intrusion rather than being the result of horizontal injection away from that center.

On a larger scale, the general rhomb-like shape and nature of the faults of the Tuli and Nuanetsi troughs also indicate that a dextral strike-slip system has been in operation (Fig. 12). It accounts for the horst of basement south of the Tuli trough and southwest of the Nuanetsi trough. The large synclinal structure within the Tuli trough (Cox et al., 1965; Thompson, 1975) is probably the result of sagging due to north-south stretching during the transtensional event. The axis of this syncline is coincident with a zone of faulting in the adjacent basement that continues into the Nuanetsi trough as a normal fault (Broderick, 1979). In this area, the dextral strike-slip faults are associated with a right-stepping fault pattern, resulting in a releasing bend and formation of a basin in each case. Farther east, in the vicinity of the Sabi Valley, there appears to be a left step, so that a restraining bend has developed. This will result in an upthrown block, and may explain the different nature of the Karoo sedi-

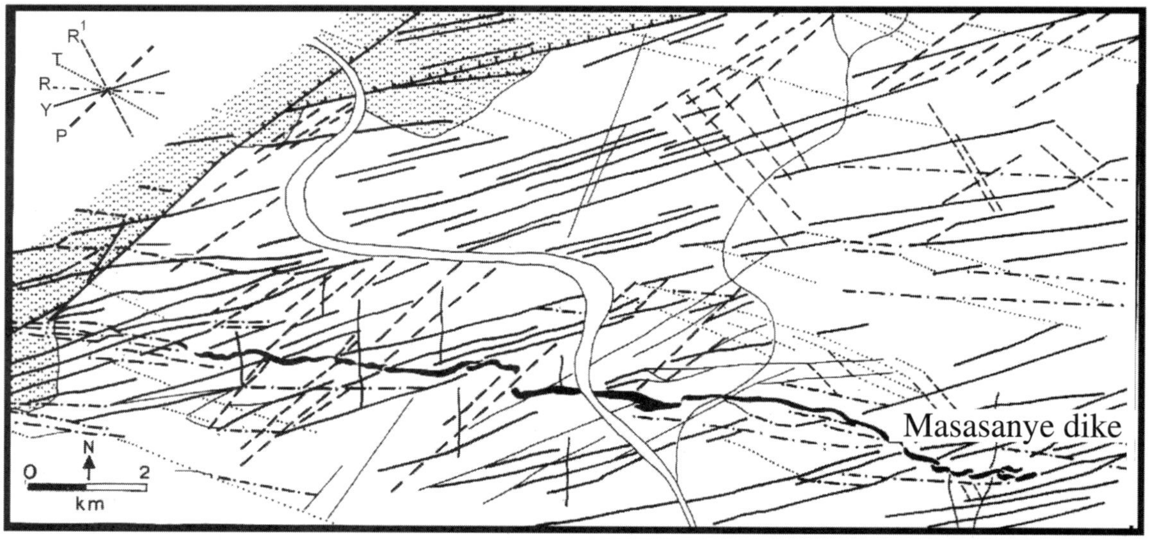

Figure 11. Karoo dikes in part of Limpopo Belt basement south of Tuli trough (see Fig. 1 for locality) and interpretation of fracture directions in terms of dextral strike-slip system.

ments in this region; i.e., they are more immature, proximal sediments derived from an adjacent uplifted terrain (Swift et al., 1953; Swift, 1962) (Fig. 8).

Dike swarms

Dike swarms are good indicators of the relative age of fracturing and the paleostress directions at that time. However, a problem in southern Africa is that there are swarms dating to the Archean, and many of the older directions were reactivated in younger events (Hunter and Reid, 1987; Wilson et al., 1987; Uken and Watkeys, 1997). Without isotopic dating, the exact age of a dike intruding the basement is difficult to ascertain; however, the Karoo dikes tend to be distinctive, not having undergone alteration and, in places, the low-grade metamorphism that has affected older dikes. When swarms are found cutting the Karoo Supergroup, it is not always certain whether this is a reactivated trend present in the unexposed basement or a new trend. Geophysical mapping of dike swarms yields excellent results and has been undertaken in southern Africa (Mabu, 1995;

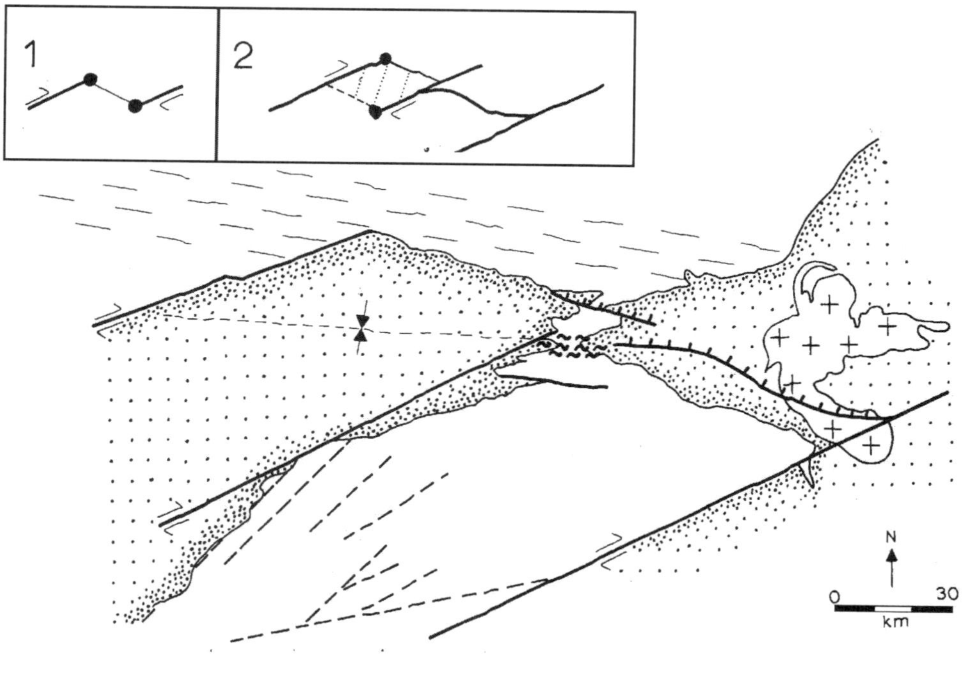

Figure 12. Simplified sketch of normal and strike-slip faults at eastern end of Tuli trough and western end of Nuanetsi trough, which contains Nuanetsi igneous province. Inset shows development of these faults in terms of dextral strike-slip releasing bend model (see Fig. 1 for regional locality).

Gomez, 2000), but the interpretation of relationships between dikes is often equivocal (Reeves, 2000) and requires more detailed geological control.

The simplified dike pattern for the northeast Kaapvaal craton west of the central and northern Lebombo is shown in Figure 13. This was obtained using geophysics, satellite imagery, and other aerial photographs as well as geological mapping, supplementing the map obtained by Uken and Watkeys (1997). The pre-Karoo dike swarms include the east-west swarm, ~100 km wide in the northernmost part of the map, some of the east-west dikes in the center of the map, and some of the northeast-southwest and northwest-southeast dikes. All of these directions have been utilized by dolerite dikes that have been observed intruding Karoo sediments and volcanics, with the additional development of a north-south direction.

West-northwest–trending swarm. The west-northwest–trending swarm extending from the southeast corner of the map is ~100 km wide. It utilizes a zone that was originally initiated in the Archean and that has been reutilized many times (Uken and Watkeys, 1997). The 205 Ma Dokolwayo kimberlite in Swaziland is along the east-southeast extension of this zone. The zone is not heavily intruded by dikes; they tend to concentrate in three narrower zones along either side of the overall zone as well as in the middle of the zone.

This trend is parallel to the Okavango dike swarm, also termed the North Botswana dike swarm by Reeves (2000) (Fig. 1). The main part of the swarm extends from the Nuanetsi igneous province across southern Zimbabwe and northern Botswana toward the Okavango swamps, a lesser density of dikes being on either side. The intersection of this swarm with the Lebombo and Sabi monoclines is where a postulated plume head would be positioned (viz. Ernst et al., 1996). Uken and Watkeys (1997) considered this swarm to be Karoo in age. The dike density decreases with stratigraphic height when the swarm crosses the basalts of the Tuli trough, indicating that the dikes are probably feeders for that succession. In addition, these dikes are the only potential feeders for the Karoo basalts under the Kalahari sands and at Victoria Falls. The north-northeast–south-southwest extension that they indicated is similar to the extension direction required for the fold to develop in the Tuli trough (Fig. 12). However, Reeves (2000) interpreted the swarm as being Cretaceous, pointing out that the swarm yields a pole of rotation with a small circle that corresponds to the Agulhas-Falklands Fracture Zone that operated as a transform fault during the opening of the South Atlantic. Ar-Ar dating has revealed the swarm to be 175–179 Ma (Le Gall et al., 2001), Karoo in age. However, this date does not invalidate Reeves's (2000) observation concerning the interaction of the dilation of this dike swarm and tectonic activity along the Agulhas-Falklands Fracture Zone. Only the age of movement changes, and the Karoo date corresponds nicely to the model of Watkeys and Sokoutis (1998), with respect to both the timing and the sense of movement along that fracture zone, in order to explain some of the coastal faulting south of the Lebombo.

Figure 13. Simplified sketch of dikes of northeast Kaapvaal craton, west of central and northern Lebombo (see Fig. 1 for locality).

Northeast-trending dikes. This swarm is as wide as 150 km and was termed the Olifants River swarm by Uken and Watkeys (1997), whereas Reeves (2000) calls it the Orange River swarm. It extends southwest from the Olifants River area at the northern end of the Lebombo, where dolerite dikes of this orientation intrude the Karoo sediments, nephelinites, picrites, and basalts. In the basement, it is clear that the dolerite dikes follow a Late Archean dike direction before they change to a north-northeast trend. This "knee bend" partly coincides with where the dolerite dikes intrude the Proterozoic cover overlying the Archean basement, but is also apparent in Archean basement exposed to the east of this cover, so it does not seem to be a change in stress pattern between basement and cover. Farther south, the dikes rotate back to a more northeast trend where they intrude the Karoo sediments and then feed the sills. If this swarm were to continue along this trend farther southwest, it would intersect the large remnant of Karoo basalts preserved around Lesotho. Overall, they appear to indicate northwest-southeast extension.

East-west–trending dikes. These occur in a zone ~150 km wide along the Lebombo that may widen toward the west, and indicative of north-south extension of the craton. In the basement areas it is apparent that this swarm consists of dikes of varying ages, probably extending to the Proterozoic (Uken and Watkeys, 1997). Unlike the northeast-tending Olifants River swarm, where the dolerite dikes intruded the Karoo rocks of the Lebombo, they seem to cut through most of the basalt pile.

North-south–trending dikes. This appears to be a new trend induced by the Karoo event. The main concentration of these is along the Lebombo, where they cut through the entire basalt sequence and across the strike of the rocks. There are other north-south–trending dikes farther west on the Kaapvaal craton, but they tend to be widely spaced. This direction is also present in the Limpopo Belt, and reflects a craton-wide east-west extension late during the basalt event.

RIFTING OF THE MARGIN OF SOUTHEAST AFRICA

It is apparent from the preceding description of the geology that there are several preexisting structural controls that influenced the rifting events in the Karoo. Before discussing these it is necessary to restore southeast Africa into a Gondwana context so that the overall evolution of the region can be linked to the breakup events.

Refitting Africa, Antarctica, and South America

There is no completely satisfactory refit between Africa and Antarctica. Martin and Hartnady (1986) provided a longitudinally defined but latitudinally flexible fit, the continental margin of Drönning Maud Land, Antarctica, being juxtaposed with that of southern Mozambique and northern KwaZulu-Natal. De Wit et al. (1988) had a looser fit on their map of Gondwana, which is adopted in this chapter, while Lawver and Scotese (1987), Groenewald et al. (1991), and Roeser et al. (1996) all proposed tighter fits. The problem with the tight fits is the elimination of the Mozambique Ridge (Mougenot et al., 1991), which consists of both continental and oceanic material (Scrutton, 1976; Roeser et al., 1996). A possible solution to maintaining both a tight fit and a continental Mozambique Ridge is to have the ridge initially move southward as part of the Antarctic plate (Martin and Hartnady, 1986). However, in the absence of evidence for such an event, the looser fit of de Wit et al. (1988) is adopted here (Fig. 14).

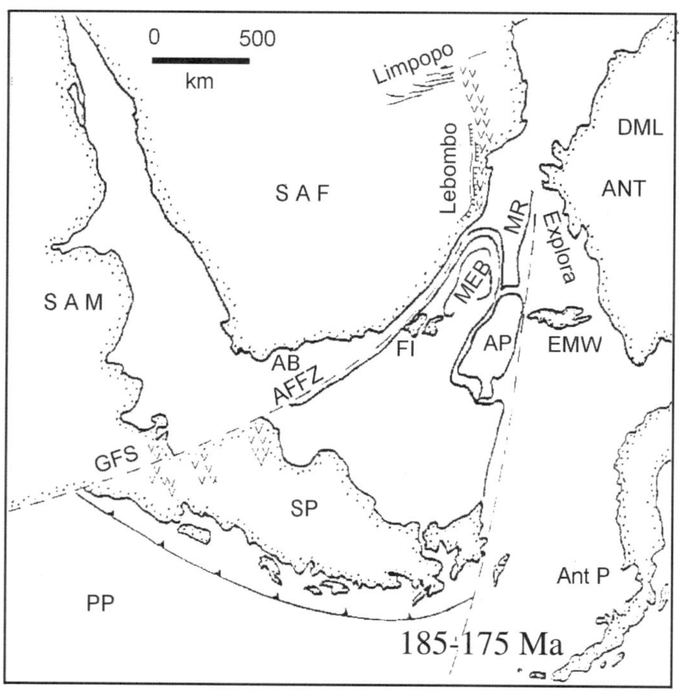

Figure 14. Southeast Africa and environs during stage 1 (180–175 Ma). SAM, South America; GFS, Gastre fault system; PP, Proto-Pacific Ocean; SP, southern Patagonia; SAF, southern Africa; AB, Agulhas Bank; AFFZ, Agulhas-Falklands Fracture Zone; AP, Agulhas Plateau; FI, Falkland Islands; MEB, Maurice Ewing Bank; MR, Mozambique Ridge; ANT, Antarctica; Ant P, Antarctic Peninsula; DML, Drönning Maud Land; EWM, Ellsworth Island.

With regard to South America, it is now generally accepted that the Falkland Islands and associated Maurice Ewing Bank should be positioned against the southeast coast of Africa, as first proposed by Adie (1952). From the earliest work (Halle, 1912), a correlation with the geology of the Eastern Cape Province in South Africa was obvious and straightforward, providing that the Falklands are rotated by almost 180° for the geology to match correctly. This rotation has been confirmed by paleomagnetic data (Mitchell et al., 1986) and is considered to have occurred in the Middle to Late Jurassic (Ben-Avraham et al., 1993). However, after rotation the Falkland Plateau is too long to fit back into the southern Natal Valley, which is offshore between southern KwaZulu-Natal and the Eastern Cape. This problem is resolved only when more than 400 km extension of the Falkland Plateau is removed (Marshall, 1994).

Breaking apart Africa, South America, and Antarctica

The paleogeography at the time of Karoo volcanism is shown in Figure 14. This volcanism was the first stage of the following five stages of breakup.

1. The first stage (180–175 Ma) is the rifting associated with Karoo volcanism.
2. The second stage (175–155 Ma) involves linking a fracture system across Gondwana from the proto-Pacific Ocean to the Lebombo. Strike-slip motion along this system resulted in microplate rotation.
3. The third stage (155–135 Ma) involved further linking of a fracture system across Gondwana from Tethys past the Lebombo to the proto-Pacific Ocean. Strike-slip motion along this system split Gondwana into two plates: East Gondwana (Antarctica, Madagascar, India, and Australia) and West Gondwana (South America and Africa).
4. In the fourth stage (135–115 Ma), the two trans-Gondwana fracture systems linked enabling East and West Gondwana to drift apart. South America and Africa separation is related to the arrival of the Tristan da Cuñha plume beneath a preexisting rift.
5. The fifth stage (115–90 Ma) involved the split between Antarctica and Australia as well as the final severing of any continental link between South America and Africa.

Stage 1 (180–175 Ma): Fracturing associated with Karoo volcanism

It is apparent that, prior to Karoo volcanism, there was some fracturing in the region. Dextral transtensional movement during Karoo influenced sedimentation in the Limpopo Belt. The upper sedimentary succession of the Lebombo shows eastward-flowing paleocurrents that might indicate the beginnings of downwarping of that region. At this time, the Gastre fault system of South America was active ca. 220–208 Ma (Rapela and Pankhurst, 1992) and, via the link along the Agulhas-Falklands Fracture Zone, may have been influencing the region.

The earliest Karoo volcanics were erupted through the Central zone of the Limpopo Belt, which was a zone of reactivation since the Early Proterozoic. The dominantly east-northeast–trending Karoo dikes of this region reflect the strong basement control (Fig. 15). The picritic dikes are only found intruding this trend and the associated less common northeast trend, whereas the low-MgO basaltic dikes occur not only in both these trends, but also in the slightly younger west-northwest Okavango trend. The latter trend is present not just in the Okavango swarm, but also farther afield on both the Zimbabwe and Kaapvaal cratons (Fig. 15). In the Kaapvaal craton, it represents direction first initiated in the Early Proterozoic.

The northeast-trending Olifants River dike swarm, with its distinctive "knee bend," is another swarm that appears to be regionally associated with the nephelinites and picrites as it extends out from this area of initial Karoo volcanism (Fig. 15). It is impossible to establish a relative age relationship with the east-northeast–trending Limpopo dikes because they do not intersect. When they intersect west-northwest–tending dikes on the Kaapvaal craton, age relationships seem to indicate that the west-northwest set is slightly older, but that there was some overlap in time.

Thus, it appears that after injection of three major dike swarms, the 120° triple junction pattern had not yet emerged. This only occurred in the next stage when the north-south dikes developed, defining the eventual line of the Lebombo (Fig. 15). It is at this time that east-west extension took place, resulting in the development of the normal faulting pattern observed on the western section of the Lebombo. The Explora wedge (Hinz and Krause, 1982; Hübscher et al., 1996) probably developed at this time as the conjugate margin to this rifting (Fig. 14).

The north-south dikes do not appear to overlap in time with the older swarms, but seem to overlap with the east-west dikes (Fig. 15). Although they are not observed in contact, the east-west dikes must predate the north-south Rooi Rand dike, which represents the final stage of basaltic volcanism in the Karoo. Synchronous injection of north-south and east-west dikes occurred in the Rooi Rand; thin east-west dikes cut across the chill margins of the north-south dikes, but then merged into what must have been the still-liquid center of the dike.

According to the available dating (Duncan et al., 1997), there can be little doubt that all of these events are close in time, possibly within 1 m.y., and certainly within 5 m.y. Now that the Okavango Dike swarm is known to be Karoo in age, they can also be neatly linked by integrating the conclusion of Reeves (2000), with regard to the pole of rotation of the Okavango swarm, with the model of Watkeys and Sokoutis (1998) for the coastal faulting south of the Lebombo.

The initial dike intrusions were along the east-northeast trend within the Limpopo Belt. This earliest event utilized an old weakness to erupt the nephelinites and picrites, as well as the

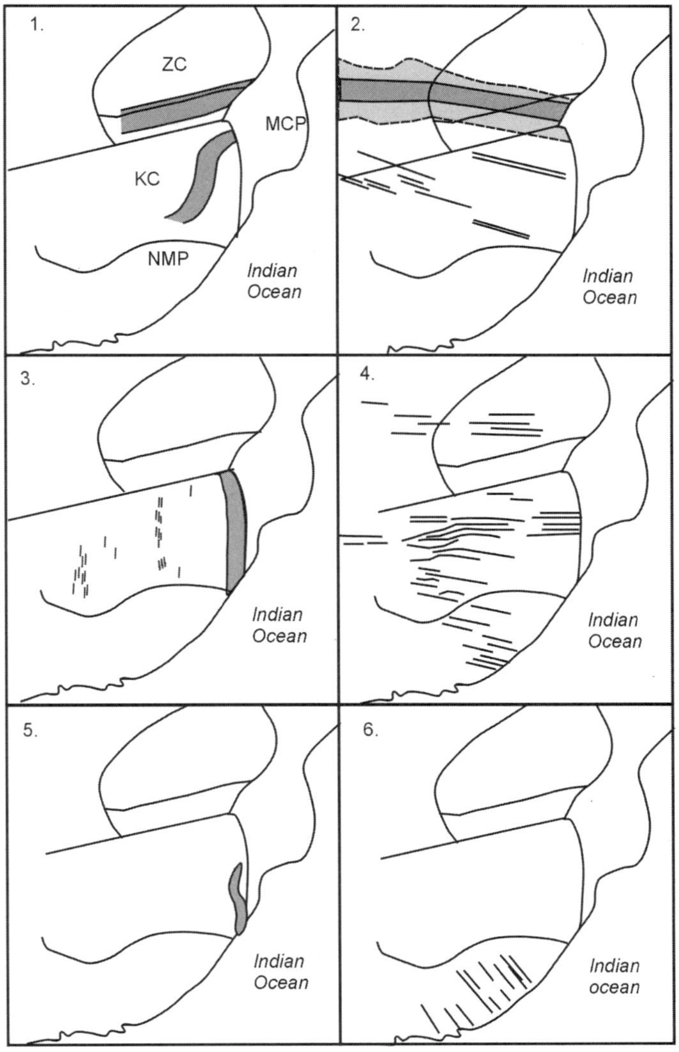

Figure 15. Simplified maps showing main dike swarms related to Karoo and Cretaceous volcanism along margin of southeast Africa together with main regional crustal blocks (ZC, Zimbabwe craton; KC, Kaapvaal craton; MCP, Mozambique coastal plain; NMP, Natal metamorphic province). 1: East-northeast–trending Limpopo swarm and northeast-trending Olifants River swarm. 2: West-northwest Okavango swarm, with central core of dense dikes flanked by zone of lesser dike density, and other west-northwest dikes. 3: North-south–trending dikes. 4: East-west–trending dikes. Rooi Rand dikes swarm. 6: Northwest-trending dikes.

early low-MgO basalts. Further major outpourings of the continental flood basalts involved a change to the west-northwest trend. Although this is a reactivated trend on the Kaapvaal craton, it appears to be a completely new fracture direction with respect to the Okavango swarm. The pole of rotation for this swarm (Reeves, 2000) yields a small circle that coincides with the Agulhas-Falklands Fracture Zone, which defines the coastline of southeastern Africa south of the Lebombo (Fig. 14). It also closely coincides with the northeast trend of the Olifants River dike swarm and completely coincides with the north-northeast trend caused by the "knee bend."

The Olifants River dike swarm does not intersect the Okavango swarms, but the "knee bend" occurs where it intersects a parallel west-northwest–trending swarm on the Kaapvaal craton (Fig. 13). Away from the intersection, the slightly oblique nature of preexisting northeast-trending fractures with respect to the ideal small circle would have resulted in oblique opening of such fractures as the west-northwest–trending swarm dilated. This would make the northeast trend more susceptible to magma injection. This explains not only the orientation of the dikes, but also the field relationships whereby the west-northwest dikes tend to be first, but overlap with the later Olifants River dikes. On the map (Fig. 15) the Olifants River dike swarm appears to be a zone linking the Okavango swarm with the other west-northwest dikes on the Kaapvaal craton (Fig. 15). The Olifants River swarm emanates from the region of early Karoo volcanism, where magma production must have been at its maximum at that time, but the dikes intruded vertically or subvertically. Thus they may have provided a zone at depth along which later magma could penetrate beneath the Kaapvaal craton.

The Agulhas-Falklands Fracture Zone is also a small circle to the Okavango pole (Reeves, 2000), so opening of the Okavango swarm would have caused movement along it. Watkeys and Sokoutis (1998) deduced that sinistral movement along this fault was responsible for the development of the coastal faulting of KwaZulu-Natal beyond the southern end of the Lebombo. They pointed out that the Lebombo is not on the Agulhas-Falklands Fracture Zone small circle, so any movement along the fracture zone would cause the Lebombo to undergo extension; they considered this to be responsible for the east-west extension. They postulated that it might have been caused by movement of the Falklands Plateau, induced by compression from the subduction zone along the western end of Gondwana. However, rotation related to dilation of the Okavango swarm is a better explanation of this movement, appealing to a known geological event of the correct age.

This east-west extension formed the sinistral east-northeast strike-slip faults of Cox (1992). The extension also overlaps with the north-south extension observed across the craton (Fig. 15). This may be due to the entire region beginning to dome over a thermal high, resulting in intrusion of north-south and east-west dikes (Fig. 15). Continued extension along the Lebombo allowed further mantle upwelling, which involved the intrusion of the Rooi Rand dike swarm at the southern end of the Lebombo (Fig. 15). This swarm does not appear to have injected horizontally from a plume head. It terminates both northward and southward along the southern Lebombo, so it seems that local centers of intrusion are more likely. However, even extensive lateral flow from such centers seems unlikely, because individual dikes have steeply plunging terminations; some are even blunt ended (Kattenhorn and Watkeys, 1995). It also lacks the HIMU signature typical of a plume. Continued upwelling of the hot mantle eventually caused partial melting in underplated basaltic material, eruption of the Karoo rhyolites, and downflexing caused by crustal thinning. Only after this event did the faulted monoclinal structure become fully developed, as seen at the eastern end of the cross sections in Figure 3; however, initial downwarping may have initiated during late Karoo sedimentation, as suggested by the paleocurrent data (Turner and Minter, 1985).

Stage 2 (175–155 Ma): Rotation of microplates

After cessation of Karoo volcanism, the South African stratigraphic record is particularly sparse until the Late Jurassic because of a regional uplift and erosion event that also occurred in the Patagonia and the Antarctic Peninsula (Dalziel, 1983). However, there is evidence of tectonism on the western Maurice Ewing Bank, i.e., extension associated with rotation of the Falkland Islands and intrusion of dikes that took place in the Middle to Late Jurassic (Taylor and Shaw, 1989). No oceanic crust formed on the Falkland Plateau at this time (Richards et al., 1986), but spreading took place in the Weddell Sea (Storey et al., 1996). Because rotation requires simple shear, the most likely cause of this deformation is strike-slip movement along the Gastre fault–Agulhas Falklands Fracture Zone as Patagonia moved to the southwest (Rapela and Pankhurst, 1992) (Fig. 16). This may be related to a 30° clockwise rotation of the Antarctic Peninsula with respect to East Antarctica, so that it was aligned with southern Patagonia by the end of the Jurassic (Grunow et al., 1987).

Stage 3 (155–135 Ma): Strike-slip movement between East and West Gondwana

The onset of seafloor spreading in the Somali Basin (Scotese et al., 1988) caused dextral-strike slip movement along one of its transform faults, the Davie Fracture Zone, which is now off the coast of Tanzania and northern Mozambique (Martin and Hartnady, 1986). This transform fault formed part of a fracture system that propagated across Gondwana via the Mozambique Basin and the Weddell Sea, which may have opened in the east (Leitchenker et al., 1996) and been subducted in other areas (Storey et al., 1996).

Strike-slip movement along this system from ca. 151 Ma to ca. 133 Ma split Gondwana into two plates. East Gondwana (Madagascar, India, Antarctica, and Australia) began to move southward with respect to West Gondwana (Africa and South America) and started to partially fragment. This movement may have reactivated the east-northeast faults of the Limpopo region and north-south faults of the Lebombo (Fig. 16). The right-stepping pattern of the regional fracture system, combined with the dextral movement, opened up a strike-slip basin now buried under the Mozambique coastal plains. The crust here is probably continental, but highly extended; a large component of basalt is represented by the sparse outcrops of the Movene and Mpilo basalts east of the Lebombo. As the two Gondwana plates slid past each other, movement was initiated along this proto-Pacific linked system and gave rise to a rifted landscape on the Agulhas Bank, which formed a number of depocenters in the Outeniqua Basin bounded by arcuate faults trending east-southeast to south-southeast (Bate and Malan, 1992). Similar offshore Late Jurassic sedimentation took place along the Drönning Maud Land margin of Antarctica (Jacobs et al., 1996). In South Africa, this period of deposition onshore is represented by the lower part of the Uitenhage Group preserved in a number of discontinuous half-grabens (Algoa, Oudsthoorn, Heidelberg and/or Riversdale) (Dingle et al., 1983).

This movement along the Gastre fault–Agulhas-Falklands Fracture Zone may also have been responsible for a north-south zone of incipient rifting, which propagated northward in the South Atlantic during the Late Jurassic and Early Cretaceous. This is recorded by the development of the Orange and Walvis basins separated by the Luderitz arch (Dingle et al., 1983). The rift seems to have followed the grain of the Pan-African basement, the normal faults involving early inversion of old thrusts (Light et al., 1992, 1993). It does not represent onset of South America–Africa separation, but its development and position ultimately gave rise to the South Atlantic in the next stage of development.

Stage 4 (135–115 Ma): Extraction of the Falkland Plateau from the Natal Valley

The eruption of the Bumbeni Complex at the southern end of the Lebombo is virtually synchronous with other events around the southern African coast, such as the intrusion of the Cape Peninsula dolerites of the southwest Cape (Reid et al., 1991) and the Mehlberg dike in the northern Richtersveld along the South Atlantic coast (Reid and Rex, 1994). The reason for this event is the opening of the South Atlantic and the onset of extraction of the Falkland Plateau from the Natal Valley ca. 133.5 Ma (Goodlad et al., 1982). This resulted in the formation of a mid-ocean

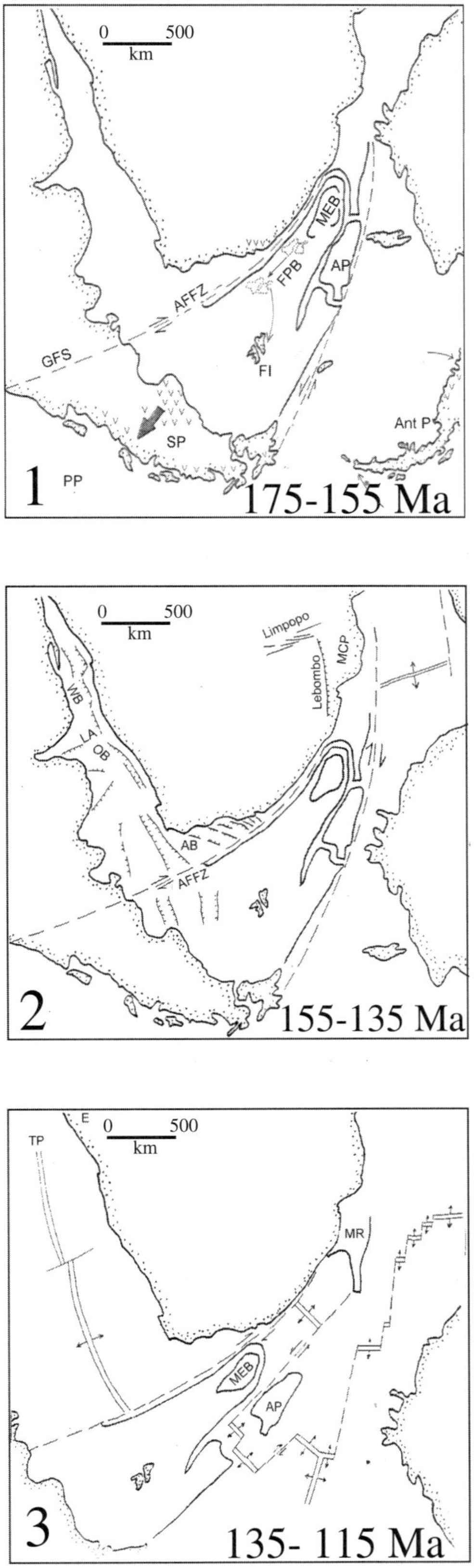

Figure 16. Top: Southeast Africa and environs during stage 2 (175–155 Ma). AFFZ, Agulhas-Falklands Fracture Zone; Ant P, Antarctic Peninsula; AP, Agulhas Plateau; FI, Falkland Islands; FPB, Falkland Plateau Basin; GFS, Gastre fault system; MEB, Maurice Ewing Bank; PP, Proto-Pacific Ocean; SP, Southern Patagonia. Middle: Southeast Africa and environs during stage 3 (155–135 Ma). AB, Agulhas Bank; LA, Luderitz arch; MCP, Mozambique coastal plain; OB, Orange River Basin; WB, Walvis Basin. Bottom: Southeast Africa and environs during stage 4 (135–115 Ma). E, Etendeka basalts; MR, Mozambique Ridge; TP, Tristan plume.

ridge normal to the southeast coast of Africa at about the present position of Durban, so that the Bumbeni Complex may effectively be a volcanic center on the northern side of this rift. The Falkland Plateau moved past the southeast coast of Africa along the Agulhas-Falklands Fracture Zone, at this stage an intracontinental transform fault ~1200 km long offsetting spreading ridges in the South Atlantic and Natal Valley (Fig. 16). This caused further coast-parallel faulting of KwaZulu-Natal south of the Lebombo (Von Veh and Andersen, 1990) and diking events around the coast of South Africa. The northwest-trending dikes of southern KwaZulu-Natal and the Eastern Cape (Fig. 15) may be of this age, being related to the southwestern movement of the mid-ocean ridge along the Agulhas-Falklands Fracture Zone, but could also be related to the Karoo age activity along this zone.

The driving force behind this event was the arrival of the Tristan da Cuñha plume under the northwest Paraná basin of South America at 137 Ma (Turner et al., 1994), evidence of which is present in the upper mantle of the region (Vandecar et al., 1995). This resulted in the production of voluminous basaltic melts preserved in the Paraná continental flood basalt province (Hawkesworth et al., 1992). Gondwana moved westward over the plume until it was positioned beneath a preexisting rift and seafloor spreading began, leaving the 132 Ma Etendeka volcanics of Namibia on the African plate (Renne et al., 1996).

Stage 5 (115–90 Ma): Strike slip of the Falkland Plateau past the Agulhas Bank

The final split in the South Atlantic involved a triple junction jump to the southern continental part of Agulhas Plateau, which was intruded by oceanic crust ca. 93 Ma (Tucholke et al., 1981). It was this Cenomanian-Turonian breaching of contact between continental crust of Africa and South America that enabled the entire Atlantic Ocean to become linked (Fairhead, 1988).

Most of the geological record of South Africa during this stage is submerged on the continental shelf in the Mesozoic basins (Dingle et al., 1983), the notable exception being the coastal plain of Maputaland, east of the Lebombo (Watkeys et al., 1993). In this region, the Makatini Formation was deposited from the late Barremian until the time of the Aptian unconformity. This hiatus coincides with the mid-Albian change in the rotation pole between South America and Africa, which is the time that Madagascar reached its present position. After this break, deposition of the Mzinene Formation took place until another hiatus occurred that corresponds to the closely spaced Cenomanian and Turonian unconformities identified offshore. Although these Cretaceous sediments display some evidence of minor faulting, it is apparent that by this time rifting along the Lebombo had completely ceased.

CONCLUSIONS

The rifting of the margin of southeastern Africa was caused by two major events: Karoo volcanism ca. 183 Ma that did not cause continental breakup, and the opening of the South Atlantic ca. 135 Ma. Karoo volcanism was influenced by two factors, the production of melts in the mantle and the lithospheric architecture. The latter is the framework into which the mantle melts intrude, and there is clear evidence of the control it played in influencing the position of Karoo initial volcanism and diking. The Limpopo Belt was important in providing a weak zone for the initial magmas to pass through to the surface, not only in being a zone of reactivation since the Proterozoic, but also because it was undergoing some extension during Karoo time.

The cause of the melting, however, is less clear. It may have been caused by heating the mantle, by "wetting" the mantle, or by stretching the lithosphere (Gallagher and Hawkesworth, 1992). There is no evidence of wet melting in the Karoo magmas. Although the initial nephelinite event hints at the possibility of melting induced by the introduction of carbonate fluids, this does not seem to be a satisfactory explanation for the picrites and tholeiitic flood basalts. Although there was some extension at the site of initial magmatism, there was not nearly enough to cause mantle melting. Therefore melting must have been due to heating, which raises the issue of a plume (White and McKenzie, 1989).

While plumes are a plausible explanation for continental flood basalts (Campbell and Griffiths, 1990; Hill, 1991; White and McKenzie, 1995), they are not the sole reason for continental breakup (White, 1992); this is certainly true for Gondwana (Storey, 1995). In the case of Karoo volcanism, there is no compelling geochemical reason for invoking the type of single plume with a head and tail as envisaged by Campbell and Griffiths (1990) and White (1997), but it is clear that the mantle must have been heated.

The model that is preferred here involves the upwelling of anomalously hot sublithospheric mantle across a broad zone. This heating may have been the result of thermal insulation of the mantle beneath a supercontinent (Anderson, 1982) or induced by subduction processes taking place along the western margin of Gondwana (Cox, 1978, 1992; Storey et al., 1992). One possibility in the latter case could be the brief coupling and then disengagement of overlying elongate convection cells in a layered mantle convection scenario. Such an event would increase the upwelling of the upper cell and then cut off the heat source from below this cell. This eliminates the necessity for having an initial hot cell in the upper mantle, as envisaged by Anderson et al. (1992). This would explain the elongate shape of the Karoo-Ferrar superplume (Storey, 1995; Storey and Kyle, 1997), and eliminates the requirement of a Karoo hotspot trail, which has always been a contentious issue. In contrast, there is much more compelling evidence for the Tristan hotspot trail being the result of a single plume with a head and tail that caused the opening of the South Atlantic in the Early Cretaceous (Turner et al., 1994; Vandecar et al., 1995).

The interpretation of the final rifting pattern along the Lebombo and Sabi monoclines as representing a triple junction produced by a plume is an oversimplification. There are elements of the 120° triple junction expected (Fahrig, 1987), but an examination of the field evidence reveals that it did not develop synchronously. Initial magmatism was along the east-northeast Limpopo trend, and this may be due not just to it being a zone of weakness and faulting, but also to a fundamental difference in the lithosphere beneath this region when compared to the adjacent cratons. Whatever the case, the presence of a long-lived paleohigh to the south prevented these volcanics from flooding southward. The initial volcanic centers for the volcanism in the Eastern Cape and Lesotho (Marsh et al., 1997) also occur at the southern edge of the Kaapvaal craton, in contact with the Proterozoic Namaqua-Natal mobile belt. This seems to indicate a difficulty in fracturing the craton during the early stages of volcanism and a deflection of activity toward the margins and adjacent weak zones, or zones with a younger and warmer lithosphere.

The first time a new trend developed is with intrusion of the west-northwest Okavango dike swarm, which cut across all preexisting structures in the Limpopo Belt and Zimbabwe craton. At this stage the "plume power" is controlling the fracture pattern. The dilation of this dike swarm induced movement along small circles that assisted intrusion of the northeast Olifants River swarm into preexisting fractures. It also caused movement along the Agulhas-Falklands Fracture Zone, resulting in the development of the north-south Lebomb. This took the fracturing back to the original site of volcanism and resulted in the 120° triple junction pattern. Regional doming of the Kaapvaal craton over the thermal high caused by the plume produced a combination of east-west and north-south dikes. The available dating indicates that all of these diking events took place ca. 183 Ma.

The geological control on two arms of the triple junction is apparent: the west-northwest–trending Okavango dike swarm cuts across all other structures and the Sabi monocline developed along the east-northeast–trending Limpopo Belt. It is the intersection of these two features and the contact of the Limpopo Belt with the Kaapvaal craton that appear to have controlled the actual position of the triple junction. The Lebombo developed at 120° to these two arms. Why the major east-west rifting in the Karoo took place along the Lebombo instead of along the Limpopo tend is not obvious. It may be that the slower deformation in the Limpopo region allowed the strain to be distributed over a wider zone, whereas rapid brittle failure of the Kaapvaal resulted in a narrow distinct zone of fracturing that cut through the crust. Alternatively, the Lebombo may be parallel to a Proterozoic mobile belt now hidden beneath the Cenozoic cover of the Mozambique coastal plain.

The dilation direction of the west-northwest–trending Okavango swarm is the best indication of the dominant horizontal extension direction imposed by the mantle on the lithosphere at this time. The swarm cuts across all other preexisting structures and is parallel to the actively subducting western margin of Gondwana, so any mantle cells induced by subduction will have been elongated parallel to this margin. The Okavango swarm dilation direction is subparallel to the Limpopo Belt, so strike-slip movement was more likely in that region, but the dilation direction is oblique to the Lebombo, which would have more likely given rise to extension.

Full continental separation does not appear to have taken place, despite the intrusion of the Rooi Rand dike swarm. This is probably because the plume was a relatively short lived event that did not penetrate any region of previous recent crustal thinning, and did not continue long enough to initiate seafloor spreading. The development of oceanic crust adjacent to the Lebombo region is related to younger events that resulted in extension at right angles to the Lebombo; these events took place farther north in the Somali basin and then on the other side of Africa, in the South Atlantic.

ACKNOWLEDGMENTS

I thank Goonie Marsh and Russell Sweeney for reviews of an earlier draft of this paper.

REFERENCES CITED

Adie, R.J., 1952, The position of the Falkland Islands in a reconstruction of Gondwanaland: Geological Magazine, v. 89, p. 401–410.

Allsopp, H.L., and Ruddick, J.C., 1984, Rb-Sr and ^{40}Ar-^{39}Ar age determinations on phlogopite micas from the pre–Lebombo Group Dololwayo kimberlite pipe: Geological Society of South Africa Special Publication 13, p. 267–272.

Allsopp, H.L., Manton, W.I., Bristow, J.W., and Erlank, A.J., 1984, Rb-Sr geochronology of Karoo felsic volcanics: Geological Society of South Africa Special Publication 13, p. 273–280.

Anderson, D.L., 1982, The chemical composition and evolution of the mantle, *in* Akimoto, S., and Manghnani, M.H., eds., High-pressure research in geophysics: Advances in Earth and Planetary Sciences, v. 12, p. 301–318.

Anderson, D.L., Zhang, Y.-S., and Tanimoto, T., 1992, Plume heads, continental lithosphere and tomography, *in* Storey, B.C., Alabaster, T., and Pankhurst, R.J., eds., Magmatism and the causes of continental break-up: Geological Society[London] Special Publication 68, p. 99–124.

Armstrong, R.A., Bristow, J.W., and Cox, K.G., 1984, The Rooi Rand dyke swarm, southern Lebombo: Geological Society of South Africa Special Publication 13, p. 77–86.

Bate, K.J., and Malan, J.A., 1992, Tectonostratigraphic evolution of the Algoa, Gamtoos and Pletmos basins, offshore South Africa, *in* de Wit, M.J., and Ransome, I.G.D., eds., Inversion tectonics of the Cape Fold Belt, Karoo, and Cretaceous Basin of Southern Africa: Rotterdam, Balkema, p. 61–73.

Ben-Avraham, Z., Hartnady, C.J.H., and Malan, J.A., 1993, Early extension between the Agulhas Bank and the Falkland Plateau due to the rotation of the Lafonia microplate: Earth and Planetary Science Letters, v. 117, p. 43–58.

Betton, P.J., Armstrong, R.A., and Manton, W.I., 1984, Variations in the lead isotopic composition of Karoo magmas: Geological Society of South Africa Special Publication 13, p. 331–340.

Bristow, J.W., 1984a, Nephelinites of the north Lebombo and south-east Zimbabwe: Geological Society of South Africa Special Publication 13, p. 87–104.

Bristow, J.W., 1984b, Picritic rocks of the north Lebombo and south-east Zimbabwe: Geological Society of South Africa Special Publication 13, p. 105–124.

Bristow, J.W., and Duncan, A.R., 1983, Rhyolite dome formation and Plinean activity in the Bumbeni Complex, southern Lebombo: Transactions of the Royal Society of South Africa, v. 86, p. 273–279.

Bristow, J.W., and Saggerson, E.P., 1983, A general account of Karoo vulcanicity in southern Africa: Geologische Rundschau, v. 72, p. 1015–1060.

Broderick, T.J., 1979, Explanation of the geological map of the country south of Nuanetsi, Nuanetsi and Beitbridge districts: Short Report—Rhodesia Geological Survey, v. 46, 108 p.

Burke, K., and Dewey, J.F., 1972, Plume-generated triple junctions: Key indicators in applying plate tectonics to old rocks: Journal of Geology, v. 81, p. 406–433.

Campbell, I.H., and Griffiths, R.W., 1990, Implications of mantle plume structure for the origin of flood basalts: Earth and Planetary Science Letters, v. 99, p. 79–93.

Catuneanu, O., Hancox, P.J., and Rubridge, B.S., 1998, Reciprocal flexural behaviour and contrasting stratigraphies: A new basin development model for the Karoo retroarc foreland system, South Africa: Basin Research, v. 10, p. 417–439.

Cleverly, R.W., Betton, P.J., and Bristow, J.W., 1984, Geochemistry and petrogenesis of the Lebombo rhyolites: Geological Society of South Africa Special Publication 13, p. 171–194.

Cox, K.G., 1970, Tectonics and volcanism of the Karoo period and their bearing on the postulated fragmentation of Gondwanaland, *in* Clifford, T.N., and Gass, I.G., eds., African magmatism and tectonics: Edinburgh, Oliver and Boyd, p. 211–235.

Cox, K.G., 1978, Flood basalts, subduction and the break-up of Gondwanaland: Nature, v. 274, p. 47–49.

Cox, K.G., 1983, The Karoo Province of southern Africa: Origin of trace element enrichment patterns in the Karoo, *in* Hawkesworth, C.J., and Norry, M., eds., Continental basalts and mantle xenoliths: Nantwich, Cheshire, UK, Shiva Press, p. 139–157.

Cox, K.G., 1988, The Karoo Province, *in* MacDougall, J.D., ed., Continental flood basalts: Dordrecht, Kluwer, p. 239–271.

Cox, K.G., 1992, The Karoo igneous activity, and the early stages of the break-up of Gondwanaland, *in* Storey, B.C., Albaster, T., and Pankhurst, R.J., eds., Magmatism and the causes of continental break-up: Geological Society [London] Special Publication 68, p. 137–148.

Cox, K.G., and Bristow, J.W., 1984, The Sabie River basalt formation of the Lebombo monocline and south-east Zimbabwe: Geological Society of South Africa Special Publication 13, p. 125–148.

Cox, K.G., MacDonald, R., and Hornung, G., 1967, Geochemical and petrogenetic provinces in the Karoo basalts of southern Africa: American Mineralogist, v. 52, p. 1451–1474.

Cox, K.G., Bristow, J.W., Taylor, S.R., and Erlank, A.J., 1984, Petrogenesis of the basic rocks of the Lebombo: Geological Society of South Africa Special Publication 13, p. 149–170.

Cox, K.G., Johnson, R.L., Monkman, L.J., Stillman, C.J., Vail, J.R., and Wood, D.N., 1965, The geology of the Nuanetsi igneous province: Royal Society of London Philosophical Transactions, ser. A, v. 257, p. 71–218.

Dalziel, I.W.D., 1983, The evolution of the Scotia Arc: A review, *in* Oliver, R.L., James, P.R., and Jago, J.B., eds., Antarctic Earth Science: Canberra, Australian Academy of Science, p. 283–288.

De Wit, M., Jeffery, M., Bergh, H., and Nicolaysen, L., 1988, Geological map of sectors of Gondwana reconstructed to their disposition ca.150 Ma, scale 1:10 000 000: American Association of Petroleum Geologists and University of Witwatersrand, Johannesburg, 2 sheets.

Dingle, R.V., Siesser, W.G., and Newton, A.R., 1983, Mesozoic and Tertiary geology of southern Africa: Rotterdam, Balkema, 375 p.

Du Preez, J.W., and Wolmarans, L.G., 1986, Die geologie van die gebeid Kosibaai: Toeligting van Blad 2632: Geological Survey of South Africa, 19 p.

Du Toit, A.L., 1929, The volcanic belt of the Lebombo: A region of tension: Transactions of the Royal Society of South Africa, v. 18, p. 189–218.

Du Toit, A.L., 1937, Our wandering continents: Edinburgh, Oliver and Boyd, 366 p.

Duncan, A.R., Erlank, A.J., and Marsh, J.S., 1984, Regional geochemistry of the Karoo igneous province: Geological Society of South Africa Special Publication 13, p. 355–388.

Duncan, A.R., Armstrong, R.A., Erlank, A.J., Marsh, J.S., and Watkins, R.T., 1990, MORB-related dolerites associated with the final phases of Karoo flood basalt volcanism in southern Africa, *in* Parker, A.J., Rickwood, P.C., and Tucker, D.H., eds., Mafic dykes and emplacement mechanisms: Rotterdam, Balkema, p. 119–129.

Duncan, R.A., Hooper, P.R., Rehacek, J., Marsh, J.S., and Duncan, A.R., 1997, The timing and duration of the Karoo igneous event, southern Gondwana: Journal of Geophysical Research, v. 102, p. 18127–18138.

Eales, H.V., Marsh, J.S., and Cox, K.G., 1984, The Karoo igneous province—An introduction: Geological Society of South Africa Special Publication 13, p. 1–26.

Ellam, R.M., and Cox, K.G., 1989, A Proterozoic lithospheric source for Karoo magmatism: Earth and Planetary Science Letters, v. 93, p. 207–218.

Ellam, R.M., and Cox, K.G., 1991, An interpretation of K picrite basalts in terms of interaction between asthenospheric magmas and mantle lithosphere: Earth and Planetary Science Letters, v. 105, p. 330–342.

Ellam, R.M., Carlson, R.W., and Shirley, S.B., 1992, Evidence from Re-Os isotopes for plume-lithosphere mixing in Karoo flood basalt genesis: Nature, v. 359, p. 718–721.

Encarnación, J., Flemming, T.H., Elliot, D.H., and Eales, H.V., 1996, Synchronous emplacement of Ferrar and Karoo dolerites and the early breakup of Gondwana: Geology, v. 24, p. 535–538.

Erlank, A.J., editor, 1984, Petrogenesis of the volcanic rocks of the Karoo Province: Geological Society of South Africa Special Publication 13, 395 p.

Ernst, R.E., Buchan, K.L., West, T.D., and Palmer, H.C., 1996, Diabase (dolerite) dyke swarms of the world (first edition): Geological Survey of Canada Open-File Report 3241, 104 p.

Fahrig, W.F., 1987, The tectonic settings of continental mafic dyke swarms, *in* Halls, H.C., and Fahrig, W.F., eds., Mafic dyke swarms: Geological Association of Canada Special Paper 34, p. 331–348.

Fairhead, J.D., 1988, Mesozoic plate tectonic reconstructions of the central South Atlantic Ocean: The role of the West and Central African rift systems: Tectonophysics, v. 155, p. 181–191.

Foland, K.A., and Henderson, C.M.B., 1976, Application of age and Sr isotope data to the petrogenesis of the Marungudzi ring complex, Rhodesia: Earth and Planetary Science Letters, v. 29, p. 291–301.

Gallagher, K., and Hawkesworth, C.J., 1992, Dehydration melting and the generation of continental flood basalts: Nature, v. 358, p. 57–59.

Gomez, S.C., 2000, A database of dykes in eastern and southern Africa supported by aeromagnetic survey interpretation [M.S. thesis]: Delft, The Netherlands, International Institute for Aerospace and Earth Sciences (ITC), Department of Earth Resources Surveys, 75 p.

Goodlad, S., Martin, A.K., and Hartnady, C.J.H., 1982, Mesozoic magnetic anomalies in the southern Natal Valley: Nature, v. 295, p. 686–688.

Groenewald, P.B., Grantham, G.H., and Watkeys, M.K., 1991, Geological evidence for a Proterozoic to Mesozoic link between southeastern African and Dronning Maud Land, Antarctica: Journal of the Geological Society of London, v. 148, p. 1115–1123.

Grunow, A.M., Kent, D.V., and Dalziel, I.W.D., 1987, Mesozoic evolution of the Weddell Sea Basin: New paleomagnetic constraints: Earth and Planetary Science Letters, v. 86, p. 12–26.

Halle, T.G., 1912, On the geological structure and history of the Falkland Islands: Bulletin of the Geological Institute, University of Upsala, v. 11, p. 115–229.

Hargraves, R.B., Rehacek, J., and Hooper P.R., 1997, Palaeomagnetism of Karoo igneous rocks in southern Africa: South African Journal of Geology, v. 100, p. 195–212.

Harris, C., and Erlank, A.J., 1992, The production of large-volume, low-δO^{18} rhyolites during rifting of Africa and Antarctica: The Lebombo monocline, southern Africa: Geochimica et Cosmochimica Acta, v. 56, p. 3561–3570.

Hawkesworth, C.J., Gallacher, K., Kelley, S., Mantovani, M., Peate, D.W., Regelous, M., and Rogers, N.W., 1992, Paraná magmatism and the opening of the South Atlantic, *in* Storey, B.C., Alabaster, T., and Pankhurst, R.J., eds., Magmatism and the causes of continental break-up: Geological Society [London] Special Publication 68, p. 221–240.

Hill, R.I., 1991, Starting plumes and continental break-up: Earth and Planetary Science Letters, v. 104, p. 398–416.

Hinz, R., and Krause, W., 1982, The continental margin of Queen Maud Land/Antarctica: Seismic sequences, structural elements and geological development: Geologische Jahrbuchs Reihe E, v. 23, p. 17–41.

Hübscher, C., Jokat, W., and Miller, H., 1996, Crustal structure of the Antarctic continental margin in the eastern Weddell Sea, *in* Storey, B.C., King, E.C., and Livermore, R.A., eds., Weddell Sea tectonics and Gondwana break-up: Geological Society [London] Special Publication 108, p. 165–174.

Hunter, D.R., and Reid, D.L., 1987, Mafic dykes swarms in southern Africa, *in* Halls, H.C., and Fahrig, W.F., eds., Mafic dyke swarms: Geological Association of Canada Special Paper 34, p. 445–456.

Jacobs, J., Kaul, N., and Weber, K., 1996, The history of denudation and resedimentation at the continental margin of western Dronning Maud Land, Antarctica, during break-up of Gondwana, *in* Storey, B.C., King, E.C., and Livermore, R.A., eds., Weddell Sea tectonics and Gondwana break-up: Geological Society [London] Special Publication 108, p. 191–199.

Kattenhorn, S.A., and Watkeys, M.K., 1995, Blunt-ended dyke segments: Journal of Structural Geology, v. 11, p. 1525–1542.

King, L.C., 1972, The Natal Monocline: Explaining the origin and scenery of Natal: Durban, South Africa, University of Natal, 112 p.

Lawver, L.A., and Scotese, C.R., 1987, A revised reconstruction of Gondwanaland, *in* McKenzie, G.D., ed., Gondwana six: Structure, tectonics and geophysics: American Geophysical Union Geophysical Monograph 40, p. 17–23.

Le Gall, B., Gilbert, F., Tchoso, G., Bertrand, H., Tiecelin, J.-J., Kampunzu, H., Dyment, J., and Maia, M., 2001, First precise age data on the giant lower Jurassic Okavango dike swarm (N. Botswana): European Geophysical Society XXVI General Assembly, Nice, Abstracts, GRA3, p. 693.

Leitchenkov, G.L., Miller, H., and Zatzepin, E.N., 1996, Structure and Mesozoic evolution of the eastern Weddell Sea, Antarctica: History of early Gondwana break-up, *in* Storey, B.C., King, E.C., and Livermore, R.A., eds., Weddell Sea tectonics and Gondwana break-up: Geological Society [London] Special Publication 108, p. 175–190.

Light, M.P.R., and Broderick, T.J., 1998, The Geology of the country east of Beitbridge: Bulletin—Zimbabwe Geological Survey, v. 87, 121 p.

Light, M.P.R., Maslanyi, R.J., and Banks, N.L., 1992, New geophysical evidence for extensional tectonics on the divergent margin offshore Namibia, *in* Storey, B.C., Albaster, T., and Pankhurst, R.J., eds., Magmatism and the causes of continental break-up: Geological Society [London] Special Publication 68, p. 257–270.

Light, M.P.R., Maslanyi, R.J., Greenwood, R.J., and Banks, N.L., 1993, Seismic sequence stratigraphy and tectonics offshore Namibia, *in* Williams, G.D., and Dobb, A., eds., Tectonics and seismic sequence stratigraphy: Geological Society [London] Special Publication 71, p. 163–191.

Mabu, S.M., 1995, Aeromagnetic mapping and interpretation of mafic dyke swarms in southern Africa [M.S. thesis]: Delft, The Netherlands, International Institute for Aerospace and Earth Sciences (ITC), Department of Earth Resources Surveys, 63 p.

Marsh, J.S., Hooper, P.R., Rehacek, J., Duncan A.R., and Duncan R.A., 1997, Stratigraphy and age of Karoo basalts of Lesotho and implications for correlations within the Karoo Igneous Province, *in* Mahoney, J.J., and Coffin, M.F., eds., Large igneous provinces: Continental, oceanic, and planetary flood volcanism: American Geophysical Union Geophysical Monograph 100, p. 247–272.

Marshall, J.E.A., 1994, The Falkland Islands: A key element in Gondwana paleogeography: Tectonics, v. 13, p. 499–514.

Martin, A.K., and Hartnady, C.J.H., 1986, Plate tectonic development of the south west Indian Ocean: A revised reconstruction of East Antarctica and Africa: Journal of Geophysical Research, v. 91, p. 4767–4786.

Matthews, P.E., 1981, Eastern or Natal sector of the Namaqua-Natal mobile belt in southern Africa, *in* Hunter, D.R., ed., Precambrian of the Southern Hemisphere: Amsterdam, Elsevier, p. 705–715.

Meth, D.L., 1996, The geology and geochemistry of the Rooi Rand dyke swarm [M.S. thesis]: Durban, South Africa, University of Natal, 190 p.

Mitchell, C., Taylor, G.K., and Cox, K.G., 1986, Are the Falkland Islands a rotated microplate?: Nature, v. 319, p. 131–139.

Mougenot, D., Gennesseaux, M., Hernadndez, J., Lepvrier, C., Malod, J.-A., Raillard, S., Vanney, J.-R., and Villneuve, M., 1991, La ride du Mozambique (Océan Indien): Un fragment continental individualisé lors du coulissement de l'Amérique et de l'Antarctique le long de l'Afrique de l'Est?: Comptes Rendus de l'Academie des Sciences, Serie 2, Sciences de la Terre et des Planetes, v. 312, p. 655–662.

Rapela, C.W., and Pankhurst, R.J., 1992, The granites of northern Patagonia and the Gastre Fault System in relation to the break-up of Gondwana, *in* Storey, B.C., Albaster, T., and Pankhurst, R.J., eds., Magmatism and the causes of continental break-up: Geological Society [London] Special Publication 68, p. 209–220.

Reading, H.G., 1980, Characteristics and recognition of strike-slip fault systems, *in* Ballance, P.F., and Reading, H.G., eds., Sedimentation in oblique slip mobile zones: International Association of Sedimentologists Special Publication 4, p. 7–26.

Reeves, C.V., 1978, A failed Gondwana spreading axis in southern Africa: Nature, v. 273, p. 222–223.

Reeves, C.V., 2000, The geophysical mapping of Mesozoic dyke swarms in southern Africa and their origin in the disruption of Gondwana: Journal of African Earth Sciences, v. 30, p. 499–513.

Reid, D.L., Erlank, A.J., and Rex, D.C., 1991, Age and correlation of the False Bay dolerite dyke swarm, south-western Cape, Cape Province: South African Journal of Geology, v. 94, p. 155–158.

Reid, D.L., and Rex, D.C., 1994, Cretaceous dykes associated with the opening of the South Atlantic: The Mehlberg dyke, northern Richtersveld: South African Journal of Geology, v. 97, p. 135–145.

Renne, P.R., Glen, J.M., Milner, S.C., and Duncan, A.R., 1996, Age of the Etendeka flood volcanism and associated intrusions in southwestern Africa: Geology, v. 24, p. 659–662.

Richards, P.R., Gatliff, R.W., Quinn, M.F., Williamson, J.P., and Fannin, N.G.T., 1996, The geological evolution of the Falkland Islands continental shelf, *in* Storey, B.C., King, E.C., and Livermore, R.A., eds., Weddell Sea tectonics and Gondwana break-up: Geological Society [London] Special Publication 108, p. 105–128.

Roeser, H.A., Fritsch, J., and Hinz, K., 1996, The development of the crust off Dronning Maud Land, East Antarctica: *in* Storey, B.C., King, E.C., and Livermore, R.A., eds., Weddell Sea tectonics and Gondwana break-up: Geological Society [London] Special Publication 108, p. 243–264.

Rust, I.C., 1975, Tectonic and sedimentary framework of Gondwana basins in southern Africa, *in* Campbell, K.S.W., ed., Gondwana geology: Canberra, Australian National University Press, p. 537–564.

SACS (South African Committee for Stratigraphy), 1980, Stratigraphy of South Africa. 1. Kent, L.E., compiler, Lithostratigraphy of the Republic of South Africa, South West Africa/Namibia, and the Republics of Bophuthatswana, Transkei and Venda: Handbook—Geological Survey of South Africa, Department of Mineral and Energy Affairs, v. 8, p. 535–564.

Saggerson, E.P., and Logan, C.T., 1970, Distribution controls of layered and differentiated mafic intrusions in the Lebombo volcanic subprovince: Geological Society of South Africa Special Publication, v. 1, p. 721–733.

Saggerson, E.P., Bristow, J.W., and Armstrong, R.A., 1983, The Rooi Rand dyke swarm: South African Journal of Science, v. 79, p. 365–369.

Scotese, C.R., Gahagan, L.M., and Larson, R.L., 1988, Plate tectonic reconstructions of the Cretaceous and Cenozoic ocean basins: Tectonophysics, v. 155, p. 27–48.

Scrutton, R.A., 1976, Crustal structure at the continental margin south of South Africa: Geophysical Journal of the Royal Astronomical Society, v. 44, p. 601–623.

Smith, R.M.H., 1990, A review of the stratigraphy and sedimentary environments of the Karoo Basin of South Africa: Journal of African Earth Science, v. 10, p. 117–137.

Storey, B.C., 1995, The role of mantle plumes in continental break-up: Case histories from Gondwanaland: Nature, v. 377, p. 301–308.

Storey, B.C., and Kyle, P.R., 1997, An active mantle mechanism for Gondwana breakup: South African Journal of Geology, v. 100, p. 707–716.

Storey, B.C., Alabaster, T., Hole, M.J., Pankhurst, R.J., and Wever, H.E., 1992, Role of subduction-plate boundary forces during initial stages of Gondwana break-up: Evidence from the proto-Pacific margin of Antarctica, *in* Storey, B.C., Alabaster, T., and Pankhurst, R.J., eds., Magmatism and the causes of continental break-up: Geological Society [London] Special Publication 68, p. 149–163.

Storey, B.C., Vaughan, A.P.M., and Millar, I.L., 1996, Geodynamic evolution of the Antarctic Peninsula during Mesozoic times and its bearing on Weddell Sea history, *in* Storey, B.C., King, E.C., and Livermore, R.A., eds., Weddell Sea tectonics and Gondwana break-up: Geological Society [London] Special Publication 108, p. 87–103.

Sweeney, R.J., and Watkeys, M.K., 1990, A possible link between Mesozoic lithospheric architecture and Gondwana flood basalts: Journal of African Earth Sciences, v. 10, p. 707–716.

Sweeney, R.J., Duncan, A.R., and Erlank, A.J., 1994, Geochemistry and petrogenesis of central Lebombo basalts of the Karoo Igneous Province: Journal of Petrology, v. 35, p. 95–125.

Sweeney, R.J., Falloon, T.J., Green, D.H., and Tatsumi, Y., 1991, The mantle origins of the Karoo picrites: Earth and Planetary Science Letters, v. 107, p. 256–271.

Swift, W.H, 1962, The geology of the Middle Sabi Valley: Bulletin—Southern Rhodesia Geological Survey, v. 52, 30 p.

Swift, W.H., White, W.C., Wiles, J.W., and Worst, B.G., 1953, The geology of the Lower Sabi Coalfield: Bulletin—Southern Rhodesia Geological Survey, v. 40, 94 p.

Tankard, A.J., Jackson, M.P.A., Eriksson, K.A., Hobday, D.K., Hunter, D.R., and Minter, W.E.L., 1982, Crustal evolution of southern Africa: 3.8 billion years of Earth history: New York, Springer-Verlag, 523 p.

Taylor, G.K., and Shaw, J., 1989, The Falkland Islands: New palaeomagnetic data and their origin as a displaced terrane from southern Africa, *in* Hillhouse, J.W., ed., Deep structure and past kinematics of accreted terranes: American Geophysical Union Geophysical Monograph 50, p. 59–72.

Thompson, A.O., 1975, The Karoo Rocks in the Mazunga area, Beitbridge District: Short Report—Rhodesia Geological Survey, v. 40, 79 p.

Tucholke, B.E., Houtz, R.E., and Barrett, D.M., 1981, Continental crust beneath the Agulhas Plateau, southwest Indian Ocean: Journal of Geophysical Research, v. 86, p. 3791–3806.

Turner, B.R., and Minter, W.E.L., 1985, Diamond-bearing upper Karoo fluvial sediments in NE Swaziland: Journal of the Geological Society of London, v. 142, p. 765–776.

Turner, S., Kelley, S., Hawkesworth, C.J., and Mantovani, M., 1994, Magmatism and continental break-up in the South Atlantic: High precision ^{40}Ar-^{39}Ar geochronology: Earth and Planetary Science Letters, v. 121, p. 333–348.

Uken, R., and Watkeys, M.K., 1997, An interpretation of mafic dyke swarms and their relationships with major mafic magmatic events on the Kaapvaal Craton and Limpopo Belt: South African Journal of Geology, v. 100, p. 314–348.

Vandecar, J.C., James, D.E., and Assumpçao, M., 1995, Seismic evidence for a fossil mantle plume beneath South America and implications for plate driving forces: Nature, v. 378, p. 25–31.

Veevers, J.J., Cole, D.I., and Cowan, E.J., 1994, Southern Africa: Karoo Basin and Cape Fold Belt: Geological Society of America Memoir, v. 184, p. 223–279.

Von Veh, M.W., and Andersen, N.J.B., 1990, Normal-slip faulting in the coastal areas of northern Natal and Zululand, South Africa: South Africa Journal of Geology, v. 93, p. 574–582.

Watkeys, M.K., 1979, Explanation of the geological map of the country west of Beitbridge: Short Report—Zimbabwe-Rhodesia Geological Survey, 45, 96 p.

Watkeys, M.K., 1983, Brief explanatory notes on the provisional geological map of the Limpopo Belt and environs: Geological Society of South Africa Special Publication 8, p. 5–8.

Watkeys, M.K., and Sokoutis, D., 1998, Transtension in south-east Africa during Gondwana break-up, *in* Holdsworth, R.E., Strachan, R., and Dewey, J.F., eds., Continental transpressional and transtensional tectonics: Geological Society [London] Special Publication 135, p. 203–214.

Watkeys, M.K., Light, M.P.R., and Broderick, T.J., 1983, A retrospective view of the Central Zone of the Limpopo Belt, Zimbabwe: Geological Society of South Africa Special Publication 8, p. 65–80.

Watkeys, M.K., Mason, T.R., and Goodman, P.S., 1993, The role of geology in the development of Maputaland, South Africa: Journal of African Earth Science, v. 16, p. 205–221.

Watkeys, M.K., Harmer, R.E., and Meth, D.L., 2001, The Rooi Rand dyke swarm, Lebombo, South Africa: Implications for the Karoo plume, Abstract, Fourth International Dyke Conference, Ithala, South Africa, p. 31.

White, R.S., 1992, Magmatism during and after continental break-up, *in* Storey, B.C., Alabaster, T., and Pankhurst, R.J., eds., Magmatism and the causes of continental break-up: Geological Society [London] Special Publication 68, p. 1–16.

White, R.S., 1997, Mantle plume origin for the Karoo and Ventersdorp flood basalts: South African Journal of Geology, v. 100, p. 271–282.

White, R., and McKenzie, D.P., 1989, Magmatism at rift zones: The generation of volcanic margins and flood basalts: Journal of Geophysical Research, v. 94, p. 7685–7729.

White, R., and McKenzie, D.P., 1995, Mantle plumes and flood basalts: Journal of Geophysical Research, v. 100, p. 17543–17585.

Wilson, J.F., Jones, D.L., and Kramers, J.D., 1987, Mafic dyke swarms in Zimbabwe, *in* Halls, H.C., and Fahrig, W.F., eds., Geological Association of Canada Special Paper 34, p. 433–444.

Wolmarans, L.G., and du Preez, J.W., 1986, The geology of the St. Lucia area: Explanation of sheet 27½32: Geological Survey of South Africa, 42 p.

Manuscript Accepted by the Society October 15, 2001

Printed in the U.S.A.

Geological Society of America
Special Paper 362
2002

Extension and uplift of the northern Rio Grande Rift: Evidence from $^{40}Ar/^{39}Ar$ geochronology from the Sangre de Cristo Mountains, south-central Colorado and northern New Mexico

Daniel P. Miggins*
Ren A. Thompson
Charles L. Pillmore
Lawrence W. Snee
U.S. Geological Survey, Box 25046, Denver Federal Center, Denver, Colorado 80225, USA
Charles R. Stern
University of Colorado, Department of Geological Sciences, Box 399, Boulder, Colorado 80309, USA

ABSTRACT

New $^{40}Ar/^{39}Ar$ dating and geologic mapping of middle to late Cenozoic volcanic rocks and basin-fill sedimentary rocks preserved in the San Luis Valley and adjacent Sangre de Cristo Mountains give new insights into the timing of rift volcanism and opening of the northern Rio Grande Rift. The age, orientation, and elevation of these preserved sections delimit the timing of early rift volcanism, establishment of protorift or paleorift basins, and the uplift rate of the Sangre De Cristo crustal block along the eastern rift margin.

We use offset geomorphic surfaces along with newly dated volcanic rocks to determine the exhumation or uplift rate of the Culebra Range relative to the adjacent San Luis Valley. On a first-order scale, we determined an exhumation rate of 58 m/m.y. based on the preserved section of 25 Ma volcanic rocks in the Culebra Range and age-equivalent volcanic rocks preserved in the San Luis Hills. A second uplift rate based on a 400 m offset section of Servilleta basalt dated as 4.7 Ma yields an exhumation rate of 87 m/m.y. Previous age data, exhumation and uplift rates, and field observations from the San Luis Valley region indicate that a major period of uplift and subsequent block faulting occurred during the middle Miocene (15 Ma), ~10 m.y. after the initiation of rifting.

INTRODUCTION

The Rio Grande Rift in the western United States is a large north-south–trending Cenozoic continental rift that extends northward from New Mexico through Colorado (Fig. 1). The rift is bounded on the west by the southern limbs of the Rocky Mountains and Colorado Plateau and on the east by the Front Range, the eastern limb of the southern Rocky Mountains, and the Great Plains, effectively separating the tectonically active Cordillera from the stable craton to the east. Establishing the temporal and

*E-mail: dmiggins@usgs.gov.

Miggins, D.P., Thompson, R.A., Pillmore, C.L., Snee, L.W., and Stern, C.R., 2002, Extension and uplift of the northern Rio Grande Rift: Evidence from $^{40}Ar/^{39}Ar$ geochronology from the Sangre de Cristo Mountains, south-central Colorado and northern New Mexico, *in* Menzies, M.A., Klemperer, S.L., Ebinger, C.J., and Baker, J., eds., Volcanic Rifted Margins: Boulder, Colorado, Geological Society of America Special Paper 362, p. 47–64.

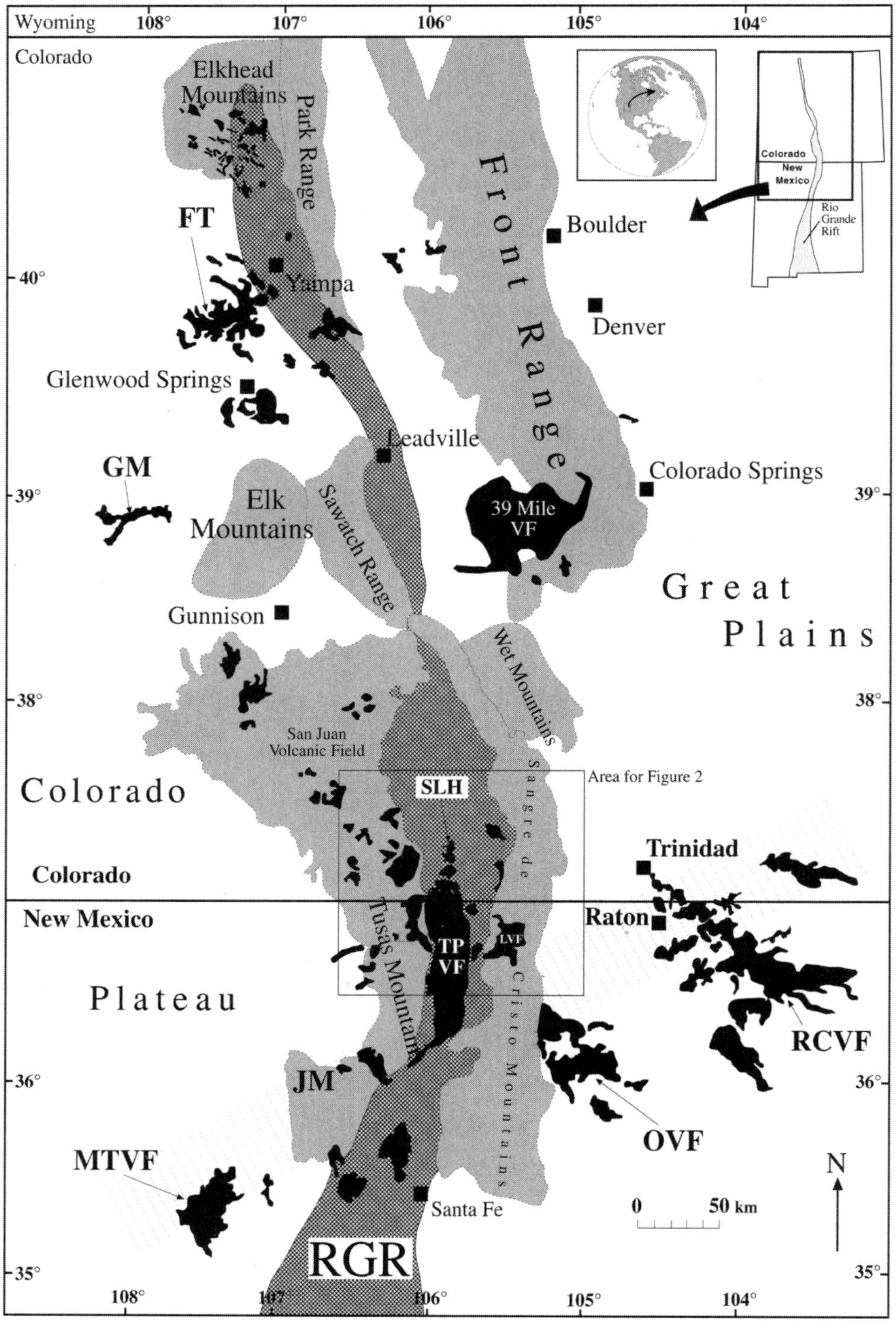

Figure 1. Regional map showing locations of major prerift and synrift volcanic fields in southern Colorado and northern New Mexico and their relationship to the Rio Grande Rift (RGR) and major uplifts of southern Rocky Mountains. Our study area is located at the Colorado–New Mexico border. RCVF = Raton Clayton volcanic field; OVF = Ocate volcanic field; JM = Jemez Mountains; MTVF = Mount Taylor volcanic field; TPVF = Taos Plateau volcanic field; LVF = Latir volcanic field; SLH = San Luis Hills; GM = Grande Mesa; FT = Flat Tops; 39 Mile VF = 39 Mile volcanic field. Cross-hachured field indicates general trend of Jemez lineament. Modified from Thompson et al. (1991) after Tweto (1979).

spatial association of rift volcanism and the tectonic and structural evolution of its rift valleys is critically important to the study of the mechanisms of continental rift systems.

Previous studies of the Rio Grande Rift have established the broad geochronologic framework for rift volcanism (Stormer, 1972a, 1972b; Lipman and Mehnert, 1979; O'Neill and Mehnert, 1988a; Stroud, 1997; Appelt, 1998; Penn, 1994; Lipman et al., 1986; Leat et al., 1988; Perry et al., 1987; Thompson et al., 1991; Gibson et al., 1992), a framework that can be used in conjunction with more detailed studies in aerially restricted segments of the rift to define both the timing of rift initiation and the uplift or exhumation history of rift flanks.

We present new field observations and $^{40}Ar/^{39}Ar$ data for a sequence of middle to late Cenozoic volcanic rocks preserved in uplifted fault blocks on the eastern flank of the Rio Grande Rift at the latitude of the Colorado–New Mexico border. The preserved fault blocks in the southern Sangre de Cristo Mountains and San Luis Valley contain both prerift- and synrift-related volcanic rocks. The age, orientation, and elevation of these preserved sections delimit the timing of early rift volcanism, establishment of protorift or paleorift basins, and the uplift rate of the Sangre de Cristo crustal block along the rift margin.

These data are interpreted in light of previously reported age data for regional Cenozoic volcanic rocks along an east-west transect across the rift at ~36°N.

GEOLOGIC SETTING

The central valley of the Rio Grande Rift in northern New Mexico (Taos Plateau) and southern Colorado (San Luis Valley) is composed of a series of predominantly down-to-the-west half-grabens, the most prominent being the Culebra graben (Brister and Gries, 1994) bounded on the east by high-angle faults against the Culebra Range of the Sangre de Cristo Mountains (Fig. 2). The western margin forms an east-dipping flexure or hinge composed of middle to late Tertiary volcanic rocks, and interbedded basin-fill volcaniclastic sediments shed eastward from the Oligocene San Juan volcanic field.

Contained wholly within the Culebra graben is an intrarift horst, exposed as the San Luis Hills, a prominent geomorphic plateau in the center of the San Luis Valley (Fig. 2). West of the San Luis Hills, exposed basin-fill deposits are principally pebble to cobble gravels of the middle Tertiary Los Pinos Formation that have a western provenance in the San Juan volcanic field. To the east, temporally equivalent sediments of the Santa Fe Group shed from the Sangre de Cristo Mountains fill basins formed against basin-bounding faults with thicknesses to 3500 m (Keller et al., 1984).

The San Luis Hills are the surface expression of a major intrarift horst within the central depression of the San Luis Valley in the northern Rio Grande Rift (Thompson and Machette, 1989; Thompson et al., 1991). For most of its length, the horst is confined to the subsurface (Kleinkopf et al., 1970; Brister and Gries, 1994). Two additional exposures occur ~55 km south of the Colorado–New Mexico border at Brushy Mountain and Timber Mountain (Thompson et al., 1986) (Fig. 2). At all three localities, middle Tertiary volcanic rocks are exposed with a notable paucity of Santa Fe Group basin-fill sediments, suggesting that the horst maintained a position of positive relief throughout extensional deformation and subsidence of basins on either side. The San Luis

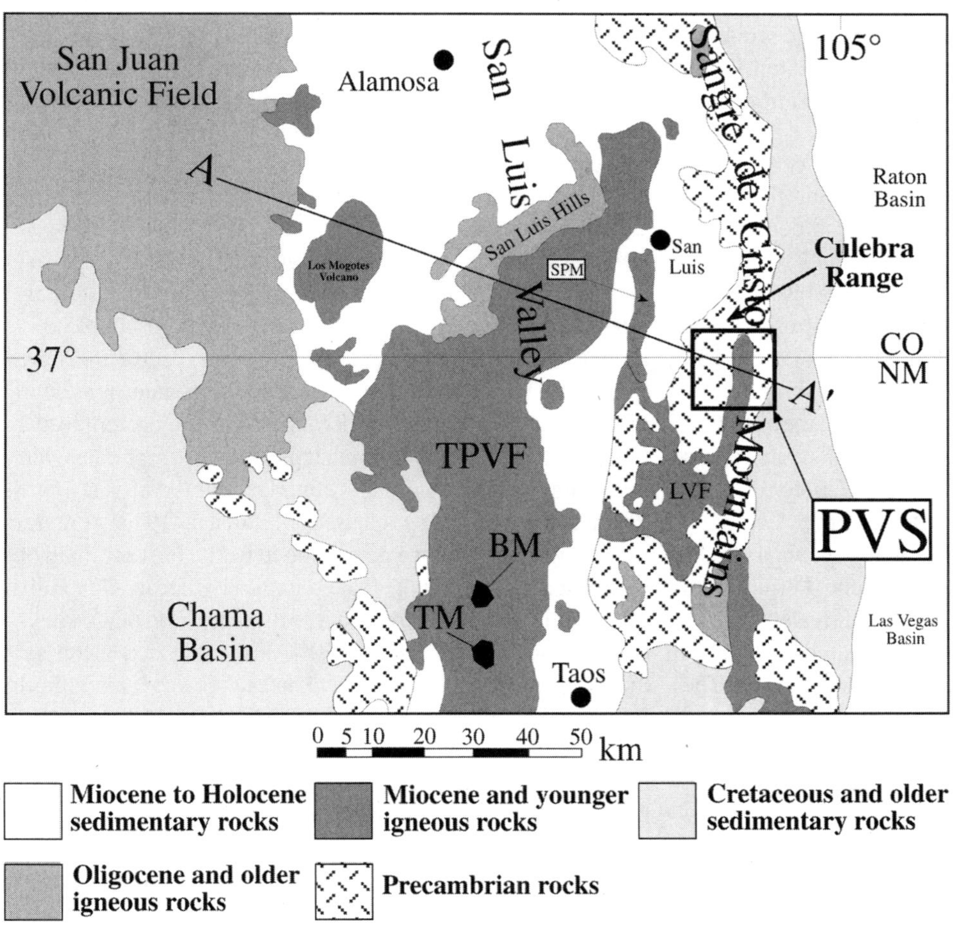

Figure 2. Generalized geologic map showing relationship of study area relative to the San Luis Valley and Sangre De Cristo Mountains, south-central Colorado and northern New Mexico. TPVF = Taos Plateau volcanic field; SPM = San Pedro Mesa; PVS = preserved volcanic sequence; LVF = Latir volcanic field; TM = Timber Mountain; BM = Brushy Mountain; CO, Colorado; NM = New Mexico. Modified from Keller et al. (1984).

Hills are bounded by undissected basin-fill sediments of the San Luis Valley that terminate to the east at a major west-dipping fault system along the west side of the Sangre de Cristo Mountains (Cordell, 1978; Personius and Machette, 1984).

In the San Luis Hills, middle Tertiary andesitic to dacitic volcanic eruptions were followed closely in time by subvolcanic intrusion of cogenetic plutons (Thompson et al., 1986). Subsequent uplift and erosion unroofed these plutons and remnants of the volcanic carapace and provided the irregular topography onto which 26 Ma basalts were erupted. These basalts filled local paleovalleys to thicknesses in excess of 100 m, resulting in the plateau surface observed today. The basalts are correlated with the oldest of the Hinsdale Formation basalts in the San Juan Mountains (Lipman and Mehnert, 1975), where they are generally thought to represent early stages of basaltic volcanism associated with rifting. The intermediate composition volcanic and plutonic rocks beneath the basaltic cover are temporally and compositionally equivalent to the Oligocene volcanic rocks of the San Juan volcanic field, which represent remnants of the southeastern margin of this complex. This volcanic sequence probably underlies basin-fill sediments of both the Los Pinos Formation and Santa Fe Group on the west and east sides of the horst, respectively.

High-angle normal faulting on the east side of the San Luis Valley displaced prerift sedimentary rocks, synrift volcanic rocks, and associated basin fill in the Culebra Range of northern New Mexico. The volcanic rocks of the Culebra Range sequence are preserved in the hanging wall of the northward extension of the Blue Lake thrust fault (Lipman and Reed, 1989) in the upper reaches of the Costilla River drainage.

During Laramide thrusting, Proterozoic rocks of the crystalline basement were pushed eastward, overturning the Pennsylvanian and Permian red arkosic sandstones and conglomerates of the Sangre de Cristo Formation. A thick overlying sedimentary sequence comprising formations of Mesozoic age including the Chinle to Pierre Shale crop out in overturned beds to the east of the preserved volcanic sequence (PVS) in the Underwood area (Fig. 2). Thrust relations are rarely observed because of poor exposures and a rotational fault along the east margin of the Underwood area that forms a half-graben that juxtaposes Proterozoic rocks against Sangre de Cristo rocks.

Following uplift and thrusting during the Laramide orogeny, widespread erosion of Proterozoic and younger rocks characterized geologic processes of the Eocene Period. Red clays and conglomerates considered to be equivalent to the Vallejo Formation of Upson (1941) were deposited in local irregularities and depositional basins on the eroded surface. The Vallejo Formation appears to be overlain by unconsolidated gravels and sediments considered equivalent to the lower part of the Santa Fe Group. The gravels comprise mostly Proterozoic cobbles and boulders of quartzite and gneiss. Near the base of this section are aerially restricted outcrops of the distal outflow of the Amalia Tuff, a 26–25 Ma silicic ash-flow tuff associated with the collapse of the Questa caldera in the Latir volcanic field, 40 km to the south (Fig. 2). The tuff was erupted onto an erosional surface of probable Eocene age. The presence of local accumulation of Santa Fe Group sediments beneath the Amalia Tuff suggests local accumulation of early-rift deposits in rift-related subbasins marking the early stages of extension. Overlying the Amalia Tuff, preserved in disparate rotated fault blocks of the Culebra Range, are interbedded basaltic to dacitic volcanic rocks and Santa Fe Group sediments with an aggregate thickness of ~300 m. These deposits occur at elevations as high as 1100 m above the current valley floor. The individual faults that accommodate this exhumation of the Culebra Range are complex (A. Wallace, 2002, written commun.), but were largely responsible for blocking out the geometry of the western margin of the present rift valley.

As extension progressed, range-front faulting migrated westward, as evidenced by west-dipping normal faults in the San Luis Valley that offset late Tertiary volcanic rocks of the 4.8–4.4 Ma Servilleta Formation and middle Pleistocene alluvium of the modern rift valley (Personius and Machette, 1984; Thompson and Machette, 1989). Basaltic lava flows preserved in the footwall of these Quaternary faults are offset as much as 400 m and exhibit local rotation as much as 25°.

$^{40}Ar/^{39}Ar$ GEOCHRONOLOGY

We collected 22 samples for $^{40}Ar/^{39}Ar$ dating. Both groundmass concentrations and mineral separates are used in this study. Given the degree of uncertainty associated with analysis of groundmass separates from basaltic rocks, special care was taken to remove xenoliths and phenocrysts, common sources of inaccuracy in groundmass dating. Of the 22 samples analyzed, 16 are from the preserved volcanic sequence of the Culebra Range and 6 are from basaltic lava flows in the San Luis Valley. Location coordinates are given in Table 1.

Most of the interpreted $^{40}Ar/^{39}Ar$ dates in Table 1 are plateau dates. Incremental heating data for each sample are presented in Table 2. Incremental heating age spectra are shown in Figure 3, and the interpreted dates are presented in a diagrammatic section in Figure 4. The term "plateau" refers to two or more contiguous temperature steps with apparent dates that are indistinguishable at the 95% confidence interval and represent ≥50% of the total $^{39}Ar_K$ released (Fleck et al., 1977). A preferred date represents what we determine to be the best estimate of the apparent age for a sample that contains no plateau. Generally we restrict this term to a portion of the age spectrum that shows near concordancy but comprises <50% of the released argon, or the included fractions of argon have dates that overlap within three standard deviations of the weighted mean.

Isochron analysis (York, 1969) for those samples given a preferred date was used to assess if nonatmospheric argon components were trapped in any samples and in some cases to define an apparent age of the sample. A total gas date, analogous to conventional K-Ar date, is calculated for each sample by weight averaging all dates of all gas fractions for the sample.

TABLE 1. SUMMARY OF INTERPRETED ^{40}AR/^{39}AR AGE-SPECTRUM RESULTS FOR VOLCANIC ROCKS FROM THE SANGRE DE CRISTO MOUNTAINS AND SAN LUIS VALLEY, SOUTH-CENTRAL COLORADO

Sample	Location	Figure	Min/GC	Latitude (°N)	Longitude (°W)	Type of date	Apparent age (Ma)	Initial $^{40}Ar/^{36}Ar$
Preserved Volcanic Sequence								
DM-98-25A	Laharic flow					Isochron	29.63 ± 0.17	304 ± 2.46
	Underwood area	4A	Biotite	36°56′17″	105°13′26″	Preferred	29.53 ± 0.08	
76P-2-DM	Laharic flow Underwood area	4B	Hornblende	36°59′24″	105°13′37″	Plateau	31.07 ± 0.05	
P80-14A-DM	Dead Horse Hill-lower andesite flow breccia	4C	GC	36°58′44″	105°11′18″	Plateau	29.32 ± 0.05	
P80-14E-DM	West of Dead Horse Hill Auto breccia andesite	4D	GC	36°58′43″	105°11′43″	Plateau	29.86 ± 0.06	
P80-14B-DM	Dead Horse Hill Tuffaceous sandstone	4E	Sanidine	36°58′17″	105°11′40″	Plateau	25.71 ± 0.25	
P80-14C-DM	Bernal trail Andesite flow					Isochron	26.05 ± 0.91	264 ± 9.18
		4F	GC	36°56′20″	105°12′30″	Preferred	25.35 ± 0.09	
P80-26-DM	Amalia Tuff unwelded Base of Hoo Doo's	4G	Sanidine	36°56′03″	105°13′18″	Plateau	24.96 ± 0.11	
B-S-1-DM	Flatiron basalt flow					Isochron	15.08 ± 0.15	309 ± 2.79
	#1 Creek	4H	GC	36°59′17″	105°15′53″	Preferred	15.08 ± 0.06	
B-S-2-DM	Flatiron basalt flow #1 Creek	4I	GC	37°00′04″	105°16′03″	Plateau	15.16 ± 0.03	
B-S-4-DM	Flatiron basalt flow #1 Creek	4J	GC	37°00′38″	105°16′05″	Plateau	14.97 ± 0.09	
P83-23B-DM	Basalt flow Head of #1 Creek	4K	GC	37°01′32″	105°16′17″	Plateau	15.11 ± 0.03	
DM-98-26-2	Skeleton Mesa basalt flow (top)	4L	GC	36°59′06″	105°11′47″	Preferred	14.53 ± 0.02	
						Isochron	14.42 ± 0.11	301 ± 0.03
P80-14D-DM	Skeleton Mesa basalt flow (top)	4M	GC	36°59′07″	105°11′50″	Plateau	14.22 ± 0.03	
DM-98-22-2	Devil's Park cinder cone	4N	GC	37°01′45″	105°13′38″	Plateau	11.87 ± 0.03	
DM-98-16*	Basalt flow related to Devil's Park cinder cone	4O	GC	37°02′12″	105°14′38″	Isochron	11.80 ± 0.06	321.09 ± 10.35
P80-15-DM	Horblende andesite #1 Creek	4P	Hornblende	36°58′33″	105°15′13″	Plateau	11.42 ± 0.04	
San Luis Valley Basalts								
DM-99-52A	Basalt flow east of Sanchez Resevoir	4Q	GC	37°03′53″	105°20′25″	Plateau	10.74 ± 0.10	
DM-99-52B	Basalt flow east of Sanchez Resevoir	4Q	GC	37°03′53″	105°20′25″	Plateau	11.98 ± 0.10	
DM-99-51	Basalt flow south of San Luis	4R	GC	37°04′17″	105°20′54″	Plateau	4.37 ± 0.17	
DM-99-54	Basalt flow northeast of San Luis	4S	GC	37°19′13″	105°22′35″	Plateau	4.55 ± 0.15	
DM-99-53	San Luis Basalt flow (Pedro Mesa)	4T	GC	37°12′40″	105°25′41″	Plateau	4.59 ± 0.02	
DM-99-53A	San Luis Basalt flow (Pedro Mesa)	4U	GC	37°15′58″	105°26′16″	Isochron	4.75 ± 0.10	289.24 ± 9.45

Note: GC, groundmass concentrate.

TABLE 2. ^{40}AR/^{39}AR INCREMENTAL HEATING DATA FOR SAMPLES FROM SOUTHERN COLORADO AND NORTHERN NEW MEXICO

Data from MAP 215

DM-98-25A Welded tuff boulder, lower andesite unit
Biotite; 254.5 mg; J-value = .003842 ± 0.1%

Temperature (°C)	$^{40}Ar_R$ [2*] (Ma at ± 2σ)	$^{39}Ar_K$	$^{40}Ar_R/^{39}Ar_K$ [3†]	$^{39}Ar/^{37}Ar$ [4§]	$\%^{40}Ar_R$	$\%^{39}Ar$	Apparent age[5**]
700	0.0142	0.0024	5.77	23	16.1	0.1	39.6 ± 1.0
800	.0541	.0114	4.720	25	9.9	.6	32.4 ± 0.6
900	.3398	.0775	4.385	127	72.7	3.9	30.14 ± 0.07
1000	.98531	.22592	4.361	301	87.8	11.3	29.98 ± 0.06
1050	.89598	.20915	4.284	358	92.2	10.5	29.45 ± 0.08
1100	1.3130	.30588	4.293	267	93.2	15.3	29.51 ± 0.07
1150	1.1004	.25678	4.285	171	90.5	12.9	29.46 ± 0.09
1200	.85466	.19804	4.316	61	88.7	9.9	29.67 ± 0.06
1300	2.6853	.62489	4.297	47	90.4	31.3	29.54 ± 0.05
1525	.36465	.08292	4.397	69	85.3	4.2	30.22 ± 0.18

Total gas date: 29.66 ± 0.07 Ma
Preferred date (steps 5–9): 29.53 ± 0.08 Ma; Isochron date (steps 1–10): 29.63 ± 0.17 Ma; $(^{40}Ar/^{36}Ar)_i$ = 304 ± 2.46

76P-2-DM Welded tuff boulder, lower andesite unit
Hornblende; 352.45 mg; J-value = .006725 ± 0.1%

Temperature (°C)	$^{40}Ar_R$ [2*] (Ma at ± 2σ)	$^{39}Ar_K$	$^{40}Ar_R/^{39}Ar_K$ [3†]	$^{39}Ar/^{37}Ar$ [4§]	$\%^{40}Ar_R$	$\%^{39}Ar$	Apparent age[5**]
700	0.0233	0.00486	4.80	0.74	1.5	0.1	57.3 ± 0.2
800	.0401	.00997	4.02	.25	3.0	.1	48.2 ± 0.4
900	.0411	.01305	3.15	.17	25.0	.1	37.8 ± 0.2
950	.01314	.00536	2.45	.11	13.5	.1	29.5 ± 0.4
1000	.08446	.03034	2.78	.23	40.6	.3	33.5 ± 1.1
1050	2.4492	.94468	2.593	.31	84.2	10.6	31.18 ± 0.05
1075	4.3840	1.6944	2.587	.30	90.4	19.1	31.12 ± 0.05
1100	4.5748	1.7744	2.578	.30	90.3	20.0	31.01 ± 0.05
1125	4.9436	1.9153	2.581	.30	90.7	21.6	31.05 ± 0.05
1150	3.3686	1.3027	2.586	.30	86.2	14.7	31.10 ± 0.05
1200	1.7411	.67002	2.599	.28	75.5	7.6	31.25 ± 0.1
1250	.8912	.34042	2.618	.28	61.4	3.8	31.5 ± 0.1
1300	.2436	.09155	2.661	.30	36.2	1.0	32.00 ± 0.05
1400	.2068	.07711	2.68	.24	37.7	.9	32.2 ± 0.2

Total gas date: 31.18 ± 0.06 Ma
Plateau date (steps 6–10): 31.07 ± 0.05 Ma

P80-14A-DM Dead Horse Hill Lower Andesite flow breccia
GC; 200.0 mg; J-value = .00664 ± 0.1%

Temperature (°C)	$^{40}Ar_R$ [2*] (Ma at ± 2σ)	$^{39}Ar_K$	$^{40}Ar_R/^{39}Ar_K$ [3†]	$^{39}Ar/^{37}Ar$ [4§]	$\%^{40}Ar_R$	$\%^{39}Ar$	Apparent age[5**]
600	0.1381	0.10283	1.34	2.02	4.6	0.8	16.01 ± 0.30
700	.71069	.3517	2.021	3.82	27.5	2.8	24.04 ± 0.04
800	1.9828	.81405	2.436	4.42	89.5	6.5	28.9 ± 0.05
850	3.2038	1.2850	2.493	4.79	82.9	10.3	29.6 ± 0.05
900	3.2032	1.2932	2.477	5.06	96.0	10.4	29.4 ± 0.05
950	5.9363	2.4044	2.469	5.15	97.5	19.3	29.3 ± 0.05
1000	5.7538	2.3309	2.468	4.80	97.1	18.7	29.3 ± 0.05
1050	2.8296	1.1466	2.468	4.16	94.9	9.2	29.3 ± 0.05
1150	3.7898	1.5367	2.466	3.71	87.6	12.3	29.30 ± 0.03
1250	2.1948	.91094	2.409	1.64	83.0	7.3	28.6 ± 0.04
1350	.7254	.30057	2.414	0.20	73.9	2.4	28.7 ± 0.18

Total gas date: 29.02 ± 0.05 Ma
Plateau date (steps 6–9): 29.32 ± 0.05 Ma

P80-14E-DM West of Dead Horse Hill, auto breccia andesite
GC; 243.93 mg; J-value = .00662 ± 0.1%

Temperature (°C)	$^{40}Ar_R$ [2*] (Ma at ± 2σ)	$^{39}Ar_K$	$^{40}Ar_R/^{39}Ar_K$ [3†]	$^{39}Ar/^{37}Ar$ [4§]	$\%^{40}Ar_R$	$\%^{39}Ar$	Apparent age[5**]
500	0.26709	0.16405	1.63	3.80	4.3	0.9	19.3 ± 0.3
700	1.3920	.65166	2.136	5.60	20.0	3.5	25.33 ± 0.09
800	3.9344	1.5950	2.467	6.14	88.8	5.8	29.22 ± 0.05
900	11.6527	4.6298	2.517	6.55	92.9	24.6	29.81 ± 0.05
1000	12.9068	5.9095	2.184	6.06	97.6	31.4	29.90 ± 0.07
1100	8.1166	3.2703	2.482	4.84	94.1	17.4	29.40 ± 0.05
1200	4.1968	1.7106	2.453	4.28	83.4	9.1	29.06 ± 0.05
1300	2.1447	.8792	2.439	1.42	75.8	4.7	28.90 ± 0.1

Total gas date: 28.10 ± 0.06 Ma
Plateau date (steps 3–8): 29.86 ± 0.06 Ma

TABLE 2. (continued)

Temperature (°C)	$^{40}Ar_R$[2*] (Ma at ± 2σ)	$^{39}Ar_K$	$^{40}Ar_R/^{39}Ar_K$[3†]	$^{39}Ar/^{37}Ar$[4§]	%$^{40}Ar_R$	%^{39}Ar	Apparent age[5**]
Data from MAP 215							
P80-14B-DM Dead Horse Hill, Tuffaceous sandstone							
Sanidine; 11.28 mg; J-value = .00663±0.1%							
700	0.01548	0.00224	6.90	1.18	5.6	0.2	81 ± 10
1000	.25681	.1118	2.296	26.4	31.3	11.1	27.2 ± 0.8
1100	.2291	.1056	2.169	25.0	77.9	10.5	25.78 ± 0.5
1200	.4806	.2237	2.148	34.8	89.8	22.2	25.5 ± 0.2
1300	.80745	.3723	2.169	29.7	95.0	37.0	25.76 ± 0.04
1350	.4167	.1916	2.175	22.7	95.0	19.0	25.8 ± 0.2
Total gas date: 26.01 ± 0.25 Ma							
Plateau date (steps 3–6): 25.71 ± 0.25 Ma							
P80-26-DM Base of Hoo Doo's, ash-flow tuff							
Sanidine; 31.7 mg; J-value = .006684 ± 0.1%							
750	0.0191	0.0089	2.16	488	4.9	0.2	26 ± 3
850	.04447	.0196	2.266	17.4	8.2	.5	27.1 ± 0.6
950	.12136	.05720	2.122	16.6	71.2	1.5	25.4 ± 0.3
1050	.31357	.14835	2.114	33.1	84.1	3.9	25.3 ± 0.2
1150	.75983	.36110	2.104	33.3	84.4	9.6	25.2 ± 0.1
1200	.53757	.26010	2.067	29.7	94.0	6.9	24.8 ± 0.2
1250	1.1269	.54203	2.079	69.1	96.2	14.4	24.9 ± 0.1
1300	1.7012	.81663	2.084	58.9	98.0	21.7	24.96 ± 0.07
1350	1.9833	.95112	2.085	65.6	98.4	25.3	24.97 ± 0.08
1400	1.2510	.60096	2.082	39.6	97.5	16.0	24.93 ± 0.09
Total gas date: 24.99 ± 0.11 Ma							
Plateau date (steps 5–10): 24.96 ± 0.11 Ma							
P80-14C-DM Bernal andesite flow							
GC; 252.89 mg; J-value = .006687 ± 0.1%							
700	2.4594	1.4027	1.75	6.30	32.6	30.9	21.03 ± 0.04
800	1.5643	.78793	1.985	3.62	53.3	17.3	23.79 ± 0.04
900	3.6770	1.7372	2.117	.62	61.3	38.2	25.35 ± 0.04
1000	.62711	.2962	2.117	.14	73.1	6.5	25.4 ± 0.1
1100	.39104	.19197	2.037	.11	69.5	4.2	24.4 ± 0.1
1200	.14994	.08092	1.853	.13	33.4	1.8	22.2 ± 0.6
1300	.04494	.03137	1.433	.01	26.3	.7	17.2 ± 1.1
1400	.02547	.01478	1.723	.00	19.5	.3	21 ± 3
Total gas date: 23.58 ± 0.08 Ma							
Preferred date (steps 3–4): 25.35 ± 0.09 Ma							
B-S-1-DM Flat Iron basalt							
GC; 251.16 mg; J-value = .006771 ± 0.1%							
500	0.0786	0.0394	1.99	0.94	6.1	0.5	24.2 ± 0.6
700	.01547	.00787	1.966	1.13	11.6	.1	24 ± 5
800	.85866	.63236	1.358	1.56	43.1	8.4	16.51 ± 0.03
900	2.7003	2.1300	1.268	2.58	69.2	28.4	15.42 ± 0.03
1000	2.6043	2.0948	1.243	2.21	85.7	27.9	15.12 ± 0.03
1100	1.2271	.97542	1.258	1.06	75.1	13.0	15.30 ± 0.05
1200	.8014	.65482	1.224	.90	69.6	8.7	14.89 ± 0.05
1300	.87006	.70602	1.232	.77	67.5	9.4	14.99 ± 0.08
1400	.31507	.25625	1.23	.05	30.8	3.4	14.96 ± 0.07
Total gas date: 15.37 ± 0.04 Ma							
Preferred date (steps 5–9): 15.08 ± 0.06 Ma							
B-S-2-DM Flat Iron basalt							
GC; 245.57 mg; J-value = .0067 ± 0.1%							
700	0.16215	0.09925	1.63	1.06	7.4	1.4	19.6 ± 0.3
800	.32890	.23966	1.372	1.49	32.5	3.3	16.5 ± 0.2
900	.94044	.71391	1.317	1.82	71.9	9.7	15.85 ± 0.05
1000	4.1034	3.2569	1.260	2.57	74.9	44.4	15.16 ± 0.03
1100	1.5277	1.2132	1.259	1.21	76.3	16.5	15.16 ± 0.02
1200	.84327	.68323	1.234	.86	69.5	9.3	14.86 ± 0.06
1300	.91399	.75738	1.207	.81	65.8	10.3	14.53 ± 0.05
1400	.44939	.37154	1.210	.09	37.3	5.1	14.6 ± 0.1
Total gas date: 15.21 ± 0.05 Ma							
Plateau date (steps 4–5): 15.16 ± 0.03 Ma							

(*continued*)

TABLE 2. (continued)

Temperature (°C)	$^{40}Ar_R$[2*] (Ma at ± 2σ)	$^{39}Ar_K$	$^{40}Ar_R/^{39}Ar_K$[3†]	$^{39}Ar/^{37}Ar$[4§]	%$^{40}Ar_R$	%^{39}Ar	Apparent age[5**]
			Data from MAP 215				
B-S-4-DM Flat Iron basalt							
GC; 245.7 mg; J-value = .006724 ± 0.1%							
700	0.0773	0.0373	2.07	0.51	6.8	0.9	25.0 ± 0.8
800	.21914	.16433	1.334	.71	36.0	3.9	16.1 ± 0.2
900	.78843	.61277	1.287	.98	40.8	14.4	15.54 ± 0.02
1000	1.58312	1.2769	1.240	1.13	82.1	30.1	14.98 ± 0.03
1100	1.2930	1.0408	1.242	.76	82.9	24.5	15.01 ± 0.08
1200	.46049	.37181	1.239	.51	67.5	8.8	14.96 ± 0.05
1300	.50473	.40861	1.235	.48	53.9	9.6	14.9 ± 0.2
1400	.41352	.33470	1.235	.06	31.8	7.9	14.92 ± 0.09
Total gas date: 15.19 ± 0.07 Ma							
Plateau date (steps 4–8): 14.97 ± 0.09 Ma							
P80-23B-DM, Basalt flow at head of #2 creek							
GC; 218.48 mg; J-value = .006628 ± 0.1%							
500	0.05115	0.02372	2.16	1.47	6	0.2	25.6 ± 0.3
600	.08643	.04498	1.921	2.12	22.8	0.4	22.8 ± 0.9
700	.37667	.28308	1.331	2.4	62	2.4	15.84 ± 0.09
800	.85914	.68521	1.254	2.29	82.1	5.8	14.93 ± 0.04
900	4.7506	3.7440	1.269	2.54	85.1	31.8	15.11 ± 0.02
1000	5.8005	4.5663	1.270	2.60	96.4	38.8	15.13 ± 0.02
1100	1.2695	1.0020	1.269	2.53	83	8.5	15.09 ± 0.03
1200	.85212	.67315	1.266	2.13	74.2	5.7	15.07 ± 0.05
1350	.89703	.73391	1.222	.14	70.7	6.2	14.56 ± 0.05
Total gas date: 15.13 ± 0.03 Ma							
Plateau date (steps 5–8): 15.11 ± 0.03 Ma							
DM-98-26-2, Dead Horse Hill, top basalt flow							
GC; 219.4 mg; J-value = .003558 ± 0.1%							
500	0.00024	0.00198	0.12	1.51	0.1	0.0	.8 ± 1.7
600	.38829	.16638	2.334	1.45	23.3	3.3	14.92 ± 0.03
700	1.4413	.63075	2.285	1.33	72.6	12.6	14.61 ± 0.02
800	3.6316	1.5904	2.283	1.60	92.2	31.7	14.60 ± 0.02
900	3.7955	1.6868	2.250	1.91	96.3	33.6	14.39 ± 0.02
1000	1.3324	.60096	2.217	1.8	91.1	12.0	14.18 ± 0.02
1100	.37734	.17915	2.106	1.38	70.4	3.6	13.47 ± 0.02
1200	.2932	.14816	1.979	.18	51.0	3.0	12.66 ± 0.05
1350	.02608	.01338	1.949	.02	11.2	.3	12.47 ± 0.10
Total gas date: 14.38 ± 0.03 Ma							
Prefered date (steps 4–5): 14.53 ± 0.02 Ma; Isochron date (steps 2–6): 14.42 ± 0.10 Ma: ($^{40}Ar/^{36}Ar$)i = 301.03 ± 0.03							
P80-14D-DM, Dead Horse Hill, Skeleton Mesa basalt flow							
GC; 257.58 mg; J-value = .006685 ± 0.1%							
500	0.02842	0.02021	1.41	1.32	3.6	0.2	16.8 ± 1.1
700	.21585	.20224	1.067	1.37	37.2	1.8	12.8 ± 0.6
800	.34190	.53740	1.194	1.40	77.0	4.7	14.3 ± 0.1
900	1.9417	1.6198	1.199	1.49	68.8	14.2	14.40 ± 0.02
1000	5.2739	4.4503	1.185	1.65	95.5	39.0	14.24 ± 0.02
1100	3.3307	2.8196	1.181	1.72	93.6	24.7	14.19 ± 0.04
1200	1.5530	1.3256	1.172	1.79	82.5	11.6	14.07 ± 0.03
1350	.50317	.44963	1.119	.17	42.1	3.9	13.45 ± 0.08
Total gas date: 14.18 ± 0.05 Ma							
Plateau (steps 4–7): 14.22 ± 0.03 Ma							

TABLE 2. (continued)

Temperature (°C)	$^{40}Ar_R$ [2*] (Ma at ± 2σ)	$^{39}Ar_K$	$^{40}Ar_R/^{39}Ar_K$ [3†]	$^{39}Ar/^{37}Ar$ [4§]	$\%^{40}Ar_R$	$\%^{39}Ar$	Apparent age [5**]
			Data from MAP 215				
DM-98-22-2, Devil's Park cinder cone							
GC; 231.08 mg; J-value = .003415 ± 0.1%							
500	0.1282	0.04009	3.20	2.13	22.4	0.7	19.6 ± 0.6
700	.79011	.40601	1.946	2.41	81.8	6.9	11.95 ± 0.02
800	2.4462	1.2669	1.931	2.52	77.6	21.6	11.86 ± 0.02
900	3.7532	1.9379	1.937	2.76	97.4	33.1	11.89 ± 0.02
1000	2.8905	1.4978	1.930	2.83	96.0	25.6	11.85 ± 0.02
1100	.78726	.40582	1.940	3.17	90.0	6.9	11.91 ± 0.06
1200	.42029	.22809	1.843	.45	77.7	3.9	11.32 ± 0.05
1300	.12839	.07294	1.760	.05	53.2	1.2	10.81 ± 0.09
Total gas date: 11.90 ± 0.03 Ma							
Plateau date (steps 3–6): 11.87 ± 0.03 Ma							
P80-15-DM, # 1 Creek hornblende andesite							
Hornblende; 92.62 mg; J-value = .006737 ± 0.1%							
850	0.036	0.0126	2.89	0.14	2.8	0.7	35 ± 4
1000	.0117	.00897	1.301	.64	13.1	.5	16 ± 4
1100	.0038	.0047	.804	.60	4.5	.2	9.8 ± 2.3
1150	.0099	.0068	1.453	.30	12.9	.4	18 ± 4
1250	.83555	.88629	.943	.24	76.9	46.1	11.42 ± 0.02
1350	.92112	.97854	.941	.24	81.2	50.9	11.40 ± 0.03
1400	.04533	.02596	1.746	.22	89.1	1.3	21.1 ± 0.4
Total gas date: 11.74 ± 0.06 Ma							
Preferred date (steps 5–6):11.42 ± 0.04 Ma							
DM-98-16, Devil's Park cinder cone–related basalt flow							
GC; 257.6 mg; J-value = .002108 ± 0.25%							
750	2.8	70.4	.3433	3.263	1.43	470	12.4 ± 0.1
850	17.1	92.8	2.100	3.102	1.66	6469	11.76 ± 0.02
900	18.9	95.5	2.328	3.129	2.09	53356	11.86 ± 0.03
950	15.3	96.3	1.887	3.137	2.19	5862	11.89 ± 0.03
1000	10.5	97.1	1.294	3.161	1.90	2506	11.98 ± 0.02
1050	7.2	97.2	.8853	3.159	2.22	807	11.97 ± 0.03
1100	5.9	97.2	.7273	3.144	2.42	518	11.92 ± 0.04
1150	5.3	93.6	.6467	3.098	1.97	191	11.74 ± 0.07
1250	11.3	91.1	1.393	3.065	.30	165	11.62 ± 0.03
1650	5.7	80.9	.7071	3.107	.05	106	11.78 ± 0.09
Total gas date: 11.85 Ma							
Preferred date (Steps 3–7): 11.87 ± 0.03 Ma; Isochron age (steps 1–10): 11.80 ± 0.06 Ma; $^{40}Ar/^{36}Ar)i$ = 321.09 ± 10.35							
DM-99-51, Basalt flow in Santa fe Gravels, San Luis Valley							
GC; 235.5 mg; J-value = .002200 ± 0.25%							
600	2.4	4.8	.06846	1.115	1.48	28	4.4 ± 0.4
700	7.7	10	.2208	1.156	1.06	42	4.6 ± 0.2
750	6.8	13.1	.1944	1.079	.67	53	4.3 ± 0.2
800	8.6	21.7	.2460	1.233	.41	81	4.9 ± 0.3
850	9.8	34.3	.2830	1.292	.27	144	5.1 ± 0.2
900	13.4	42.9	.3861	1.129	.21	283	4.5 ± 0.3
1000	27.0	50.7	.7770	1.108	.18	442	4.4 ± 0.1
1200	24.3	32.1	.6980	1.097	.15	100	4.3 ± 0.1
Total gas date: 4.51 Ma							
Plateau date (steps 6–8): 4.37 ± 0.17 Ma							

(continued)

TABLE 2. (continued)

Temperature (°C)	$^{40}Ar_R$[2*] (Ma at ± 2σ)	$^{39}Ar_K$	$^{40}Ar_R/^{39}Ar_K$[3†]	$^{39}Ar/^{37}Ar$[4§]	$\%^{40}Ar_R$	$\%^{39}Ar$	Apparent age[5**]
			Data from MAP 215				
DM-99-52A, Basalt flow, San Luis Valley							
GC; 239.0 mg; J-value = .002211 ± 0.25%							
700	1.6	6.9	.1329	2.037	.39	45	8.1 ± 0.6
750	2.2	16.3	.1900	2.519	.54	111	10.0 ± 0.4
800	4.6	45.9	.3944	2.747	.75	485	10.9 ± 0.3
850	8.6	62.4	.7314	2.743	.04	1130	10.9 ± 0.1
900	13.1	67.8	1.112	2.701	.37	1686	10.74 ± 0.05
950	16.3	56.4	1.387	2.747	.37	976	10.92 ± 0.06
1000	9.9	72.2	.8422	2.705	.88	1888	10.8 ± 0.1
1200	33.5	81.2	2.847	2.702	.62	495	10.75 ± 0.03
1350	6.4	82.9	.5474	2.632	.07	854	10.5 ± 0.3
1450	3.0	81.9	.2541	2.673	.06	688	10.6 ± 0.5
1650	.8	54.9	.06934	2.562	.06	252	10.2 ± 1.0
Total gas date: 10.71 Ma							
Plateau date (steps 6–11): 10.74 ± 0.10 Ma							
DM-99-52B, Basalt flow, San Luis Valley							
GC; 255.2 mg; J-value = .002225 ± 0.25%							
600	1.8	13.4	0.1134	3.311	0.67	26	13.2 ± 0.8
700	1.7	58.1	.1079	3.411	.74	81	13.6 ± 0.9
800	10.3	89.2	.6423	3.108	.67	496	12.4 ± 0.1
850	16.9	95.6	1.048	3.006	.71	5343	12.0 ± 0.1
900	21.4	96.8	1.332	2.988	.58	18827	12 ± 0.1
950	18.0	96.4	1.121	3.006	.49	3824	12.02 ± 0.08
1000	10.5	93.4	.6528	3.026	.41	1127	12.1 ± 0.1
1100	12.8	85.0	.7967	2.94	.47	475	11.8 ± 0.1
1200	6.4	64.1	.4007	2.708	.14	79	10.8 ± 0.3
Total gas date: 12.00 Ma							
Plateau date (steps 4–9): 11.98 ± 0.10 Ma							
DM-99-53, San Pedro Mesa, basalt flow, San Luis Valley							
GC; 1.000 g; J-value = .0002250 ± 0.25%							
800	5.5	17.6	0.04438	10.74	.12	31	4.83 ± 0.08
850	11.5	48.6	.09299	10.54	.18	128	4.74 ± 0.03
900	16.3	62.3	.1316	10.414	.23	232	4.68 ± 0.03
950	19.9	66.2	.1603	10.255	.30	317	4.61 ± 0.02
1000	19.6	67.7	.1578	10.161	.34	340	4.57 ± 0.02
1100	17.0	55.7	.1373	10.185	.29	183	4.58 ± 0.04
1250	8.9	29.2	.07183	9.659	.07	33	4.3 ± 0.1
1350	1.3	11.5	.01074	10.282	.01	21	4.6 ± 0.4
Total gas date: 4.61 Ma							
Plateau date (steps 4–8): 4.59 ± 0.02 Ma							
DM-99-53A, San Pedro Mesa, basalt flow, San Luis Valley							
GC; 1.000 g; J-value = .000250 ± 0.25%							
800	0.9	22.6	0.007081	10.309	0.13	33	4.6 ± 0.5
850	5.4	65.5	.04288	10.732	.15	170	4.83 ± 0.07
900	8.3	79.3	.06551	10.854	.20	350	4.88 ± 0.05
950	17.1	76.9	.1347	10.599	.28	412	4.77 ± 0.03
1000	21.2	76.9	.1671	10.337	.40	431	4.65 ± 0.02
1100	28.9	72.8	.2280	10.129	.59	311	4.56 ± 0.02
1350	16.0	38.7	.1264	9.552	.05	53	4.30 ± 0.05
1450	1.6	37.7	.01264	8.79	.02	79	4 ± 0.3
1650	.5	38.1	.003847	11.034	.02	60	5 ± 0.7
Total gas date: 4.60 Ma							
Isochron date (steps 1–7): 4.75 ± 0.10 Ma; $(^{40}Ar/^{36}Ar)i$ = 289.24 ± 9.45							

TABLE 2. (continued)

Temperature (°C)	$^{40}Ar_R$ [2*] (Ma at ± 2σ)	$^{39}Ar_K$	$^{40}Ar_R/^{39}Ar_K$ [3†]	$^{39}Ar/^{37}Ar$ [4§]	%$^{40}Ar_R$	%^{39}Ar	Apparent age[5**]
			Data from MAP 215				
DM-99-54, Basalt flow in Santa fe Gravels, San Luis Valley							
GC; 1.000 g; J-value = .000250 ± 0.25%							
700	5.2	6.9	.03561	9.133	0.19	13	4.11 ± 0.20
800	3.2	53.4	.02187	10.85	.19	121	4.9 ± 0.3
850	9.1	71.8	.06219	10.844	.25	253	4.9 ± 0.1
900	15.2	74.4	.1038	10.617	.34	372	4.77 ± 0.06
950	16.4	73.0	.1123	10.316	.42	351	4.64 ± 0.06
1000	17.8	70.5	.1216	10.179	.44	278	4.58 ± 0.06
1100	17.6	51.5	.1201	10.076	.36	118	4.53 ± 0.05
1200	10.0	30.5	.06815	9.834	.23	37	4.4 ± 0.1
1350	5.6	5.8	.03823	9.738	.01	10	4.4 ± 0.2
Total gas date: 4.60 Ma							
Plateau date (steps 5–9): 4.55 ± 0.15 Ma							

Note: Mineral separates and/or basalt or andesite groundmass concentrates were derived from 22 rock samples. Samples were crushed, ground, and sieved to 60–120 mesh size (250–125 μm). Mineral concentrates were passed through magnetic separator and heavy liquids and then hand-picked to >99% purity. Biotite and hornblende were treated in a dilute bath of ~10% HCl for 15 min, sanidine in a dilute bath of ~10% HF between 5 and 15 min. Phenocrysts and xenoliths were removed from groundmass concentrates using magnetic separator or hand-picking and final concentrates were washed in a dilute bath of ~10% HCl for ~15 minutes. All samples then were cleaned with reagent-grade acetone, alcohol, and deionized water, and air-dried in an oven at 95 °C. Between 11 and 350 mg of mineral and 200–1000 mg of groundmass concentrate were wrapped in aluminum or copper foil packages and encapsulated in silica vials along with neutron-fluence standards prior to irradiation. Standards for this experiment are hornblende MMhb-1 with percent K = 1.555, $^{40}Ar_R = 1.624 \times 10^{-9}$ mol/gm, and K-Ar age = 520.4 Ma (Samson and Alexander, 1987) and Fish Canyon Tuff (FCT) sanidine with an internally calibrated age of 27.84 Ma as measured against MMhb-1. For irradiation, an aluminum canister was loaded with six silica vials, each containing samples and standards similar to that described by Snee et al. (1988). Standards were placed between every two samples as well as at top and bottom of each silica vial. Samples were irradiated in one of six different irradiation packages in TRIGA reactor at U.S. Geological Survey in Denver, Colorado. Irradiation time was 20–40 h at 1 MW. Each irradiation package was rotated at 1 rpm during irradiation. All samples and standards were analyzed in Denver Argon Laboratory of U.S. Geological Survey using a Mass Analyzer Products 215 rare-gas mass spectrometer on a Faraday-cup collector or a VG Isotopes, Ltd, Model 1200B mass spectrometer fitted with an electron multiplier. For samples analyzed on VG Model 1200B mass spectrometer, computer program of Haugerud and Kunk (1988) was used to collect data. Each sample was heated in a double-vacuum, low-blank resistance furnace (similar to that described by Staudacher et al., 1978) for 10 or 20 min, in a series of 6–14 steps, to a maximum of 1650 °C, and analyzed using standard stepwise heating technique described by Snee (1982). Each standard was degassed to release argon in a single step at 1250 °C for MMhb-1, hornblende, or at 1350 °C for FCT sanidine. For every argon measurement, five isotopes of argon (^{40}Ar, ^{39}Ar, ^{38}Ar, ^{37}Ar, and ^{36}Ar) are measured; in some cases mass 44 (CO_2) was also monitored. Detection limit at time of these experiments was 2×10^{-17} mol of argon. Standard techniques were employed to produce $^{40}Ar/^{39}Ar$ age spectra, apparent K/Ca diagrams, and isochron diagrams as described by Snee (2002).

*Abundance of radiogenic ^{40}Ar and K-derived ^{39}Ar is measured in volts and calculated to five decimal places. Voltage may be converted to moles using 1.160×10^{-12} mol argon per volt signal. $^{40}Ar_R/^{39}Ar_K$ is calculated to three decimal places. All three are rounded to significant figures using analytical precision.

†$^{40}Ar_R/^{39}Ar_K$ has been corrected for mass discrimination. Mass discrimination was determined by calculating $^{40}Ar/^{36}Ar$ ratio of aliquots of atmospheric argon pipetted from fixed pipette on extraction line; ratio during these experiments was between 296.6 and 299.1, which was corrected to 295.5 to account for mass discrimination. $^{40}Ar_R/^{39}Ar_K$ was corrected for all interfering isotopes of argon including atmospheric argon. ^{37}Ar and ^{39}Ar, which are produced during irradiation, are radioactive and their abundances were corrected for radioactive decay. Abundances of interfering isotopes from K and Ca were calculated from reactor production ratios determined by irradiating and analyzing pure CaF_2 and K_2SO_4; K_2SO_4 was degassed in vacuum furnace prior to irradiation to release extraneous argon. Corrections for Cl-derived ^{36}Ar were determined using method of Roddick (1983). Production ratios for this experiment were determined $(^{40}Ar/^{39}Ar)_K$, $(^{38}Ar/^{39}Ar)_K$, $(^{37}Ar/^{39}Ar)_K$, $(^{36}Ar/^{37}Ar)_{Ca}$, $(^{39}Ar/^{37}Ar)_{Ca}$, and $(^{38}Ar/^{37}Ar)_{Ca}$; measured values are available upon request.

§To calculate apparent K/Ca ratios, divide $^{39}Ar_K/^{37}Ar_{Ca}$ by 2. Accuracy of apparent K/Ca ratios is dependent upon fast to thermal neutron ratios in particular reactor. U.S. Geological Survey TRIGA reactor correction factor has not been determined since Dalrymple et al. (1981). Because reactor fuel in USGS TRIGA has been changed since 1981, this ratio must be viewed as approximate, but is internally consistent for each sample and reveals within-sample variability.

**Apparent ages and associated errors were calculated from raw analytical data then rounded using associated analytical errors. Apparent ages of each fraction include error in J value (0.11%), which was calculated from reproducibility of splits of argon from several standards. Apparent ages were calculated using decay constants of Steiger and Jäger (1977). All apparent age errors are cited at 1 s. Uncertainties in calculations for apparent age of individual fractions were calculated using equations of Dalrymple et al. (1981) and critical value test of McIntyre (1963). Plateaus were determined according to method of Fleck et al. (1977).

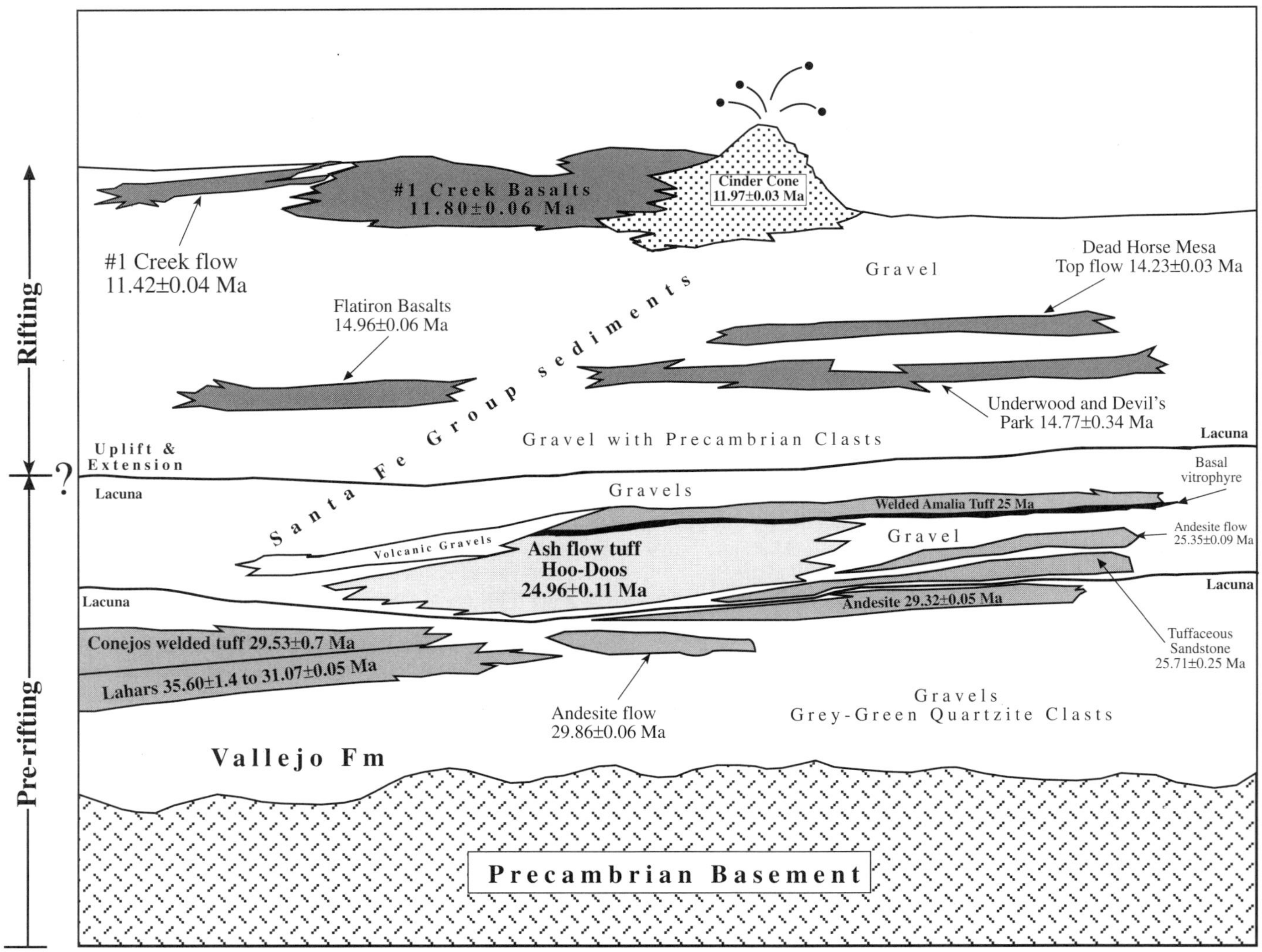

Figure 3. Diagrammatic section of volcanic and basin-fill rocks from the Underwood preserved volcanic sequence within the upper reaches of Costilla River drainage system at the Colorado–New Mexico border. Section includes both prerift and synrift volcanic and sedimentary units. $^{40}Ar/^{39}Ar$ ages determined for each unit are shown.

Culebra Range

The volcanic sequence of the Culebra Range spans an apparent age range from 31 to 12 Ma. At the base of the sequence, the Eocene Vallejo Formation unconformably overlies Proterozoic basement rocks (Fig. 3). Above the Vallejo Formation are boulders of welded tuff and rare laharic conglomerate, possibly related to volcanism in the San Juan volcanic field 100 km to the west. Pillmore et al. (1973) reported a K/Ar age of 35.6 ± 1.4 Ma from a hornblende separate from this unit, which represents one of the oldest Tertiary ages reported for this area. From this same unit, we obtained two $^{40}Ar/^{39}Ar$ ages from biotite and hornblende separates that yielded a preferred date of 29.53 ± 0.08 Ma (Fig. 4A) and a plateau date of 31.07 ± 0.05 Ma (Fig. 4B), respectively.

These laharic breccias are overlain by two andesite lava flows intercalated with locally derived conglomerates (Fig. 3). The upper and lower andesites yielded groundmass concentrate preferred dates of 29.32 ± 0.05 Ma (Fig. 4C) and 29.86 ± 0.06 Ma (Fig. 4D), respectively.

Distal deposits of the Amalia Tuff and associated volcaniclastic rocks erupted from the Questa caldera 40 km to the south conformably overlie locally derived basin-fill sediments above the andesites. A sanidine separate was obtained from reworked nonwelded tuffaceous sandstone from the lowermost unit in this section (Fig. 3). This sample yielded a plateau date of 25.71 ± 0.25 Ma (Fig. 4E). Overlying this tuffaceous sandstone is a single andesite flow that yielded a disturbed spectrum with a preferred date of 25.35 ± 0.09 Ma (Fig. 4F); this is a weighted-average date for the two oldest concordant steps comprising nearly 45% of the released $^{39}Ar_K$. We note that the two steps that define the apparent age of 25.4 Ma have relatively low K/Ca ratios, which could enhance the effect of incorporated excess ^{40}Ar.

Overlying these reworked tuff deposits and thin pebble to cobble gravel Santa Fe deposits is a nonwelded outflow of the

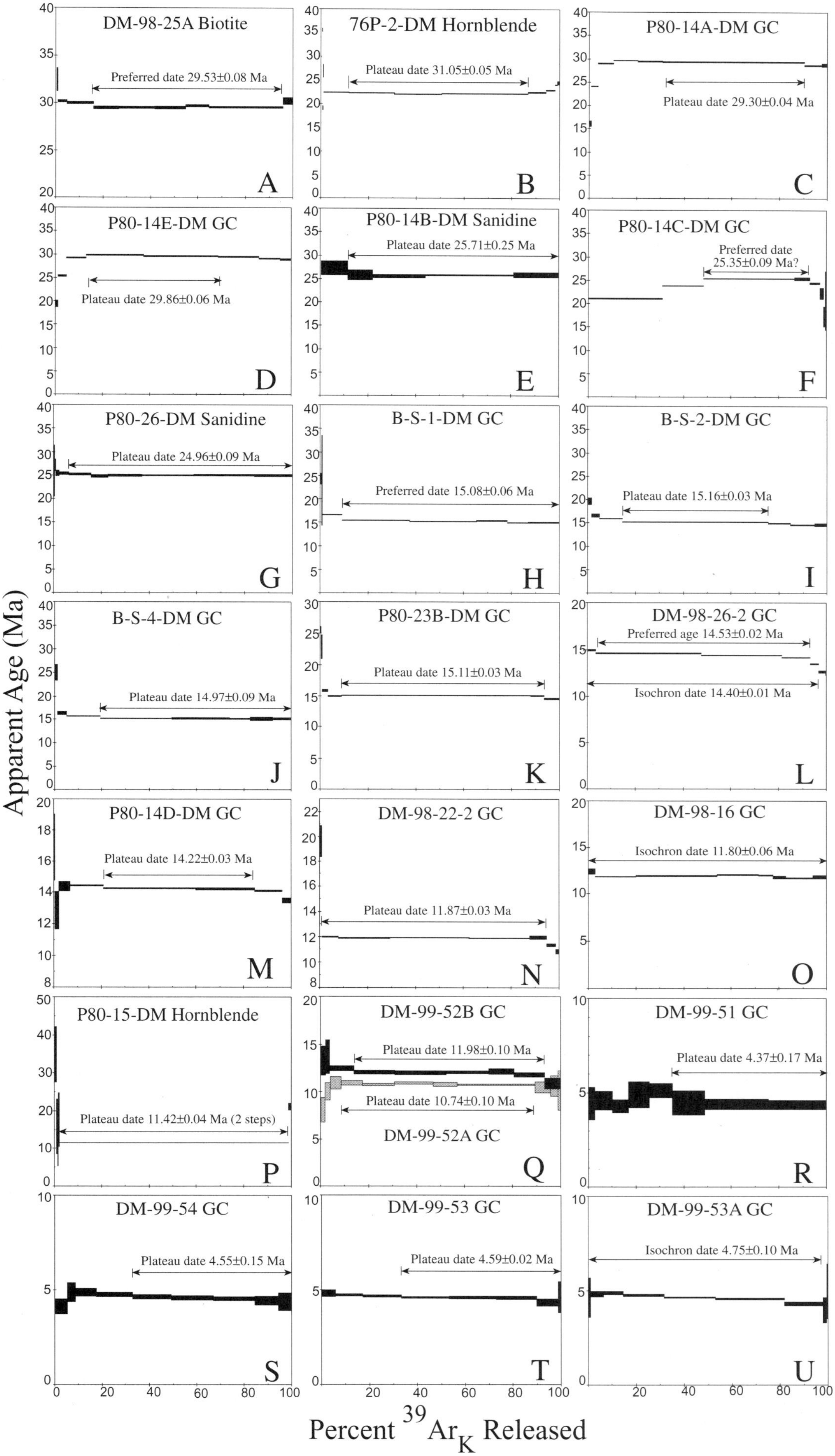

Figure 4. $^{40}Ar/^{39}Ar$ age spectra plots for various volcanic rocks contained within the preserved volcanic sequence and San Luis Valley from the northern Rio Grande Rift. Interpreted ages for each sample are indicated on individual age spectrum. GC = groundmass concentrate.

Amalia Tuff. Sanidines from this distal outflow sheet yield a plateau date of 24.96 ± 0.11 Ma (Fig. 4G). This nonwelded tuff grades vertically into a thin vitrophyre phase near the base of the welded Amalia Tuff (Fig. 3). Pillmore et al. (1973) and Pillmore and Laurie (1976) determined a K/Ar sanidine date of 23.40 ± 0.5 Ma for this welded tuff, implying a post-Amalia Tuff eruption from the Questa system. We view this K-Ar date as an inaccurate apparent age in light of the mineralogical and textural similarity to Amalia Tuff exposures elsewhere, and the virtual absence of any other regional welded tuff associated with the Questa complex.

Block faulting along the eastern rift margin and subsequent erosion of the bounding upper Proterozoic rocks fed detritus to local paleorift basins, resulting in thick accumulations of Santa Fe Group sediments. Broad erosional surfaces cut into these deposits were overlain by basaltic andesite lava flows from local eruptive centers, resulting in thick, but aerially restricted flow sequences as thick as 100 m. Three groundmass concentrates from this sequence were obtained for analyses, and yielded a preferred date of 15.08 ± 0.06 Ma (Fig. 4H) and two plateau dates of 15.16 ± 0.05 Ma (Fig. 4I) and 14.97 ± 0.09 Ma (Fig. 4J). A related flow, ~2 km to the east, yielded a plateau date of 15.11 ± 0.03 Ma (Fig. 4K). All four dates are statistically identical at the 95% confidence level, yielding a weighted mean date of 15.07 ± 0.07 Ma, a reasonable estimate for the extrusion age of the flows.

Stratigraphically higher basaltic lava flows overlie both the older 15 Ma volcanic sequence and Santa Fe Group sediments. Two samples from these flows yield a preferred date of 14.53 ± 0.02 Ma (Fig. 4L) and a plateau date of 14.22 ± 0.03 Ma (Fig. 4M), respectively. These dates are statistically different, and both age spectra show stepping down in apparent age, characteristic of the effect from recoiled ^{39}Ar (Turner and Cadogan, 1974; Hess and Lippolt, 1986). Because this recoil effect is less pronounced in Figure 4M, we accept it as the best estimate for the age of extrusion of the basalt. During post-15 Ma extensional deformation, these flows were tilted east ~20°–25°. Basin-fill deposition continued until the final volcanic episode occurred, with the eruption of basaltic andesite and dacite lavas.

The youngest volcanic rocks in the Culebra Range section consist of a small dissected cinder cone and related basaltic and dacitic lava flows (Fig. 3). The cinder cone yielded a plateau date of 11.87 ± 0.03 Ma (Fig. 4N). A thick basaltic lava flow associated with this cone yielded a preferred date of 11.95 ± 0.03 Ma and an associated isochron date of 11.80 ± 0.06 Ma (Fig. 4O). An overlying hornblende andesite lava flow and related dike at the top of the section yielded a hornblende plateau date of 11.42 ± 0.04 Ma (Fig. 4P).

San Luis Valley

The southern San Luis Valley is underlain by the northernmost exposures of Servilleta basalt of the Taos Plateau volcanic field (Fig. 2). At the latitude of the Colorado–New Mexico border, local eruptions of tholeiitic lavas are exposed on the banks of the modern Rio Grande River. Appelt (1998) obtained a plateau age of 4.6 Ma on Servilleta basalt flows ~25 km south of the state line but at approximately the same stratigraphic level as the uppermost lavas exposed east of the San Luis Hills. Servilleta basalt flows are preserved on San Pedro Mesa, one of several uplifted blocks along north-south–trending normal faults that have migrated westward into the modern rift basin. In places, these down-to-the-west faults have offset middle Pleistocene alluvium east of the Rio Grande, as well as the youngest rift-related basaltic eruptions at Mesita Cone (Thompson and Machette, 1989; Appelt, 1998). Late Pliocene to early Pleistocene faulting of the San Pedro Mesa lavas have displaced them from what was probably a similar stratigraphic position to the Servilleta lava flows preserved at river level.

Southeast of San Pedro Mesa, thick deposits (~200 m) of depositional basaltic andesite breccia deposited on Miocene and Pliocene sediments crop out on the western flanks of the Culebra Range. These breccias are lithologically similar to the youngest basaltic lavas of the Culebra Range sequence and are believed to be the westward margins of the deposits preserved higher in the range. Two samples from this deposit yielded statistically distinct plateau dates of 11.98 ± 0.10 Ma (Fig. 4Q) and 10.74 ± 0.10 Ma (Fig. 4Q). Both age spectra show good argon release patterns and were collected from the same unit. It is likely that this deposit represents tectonic mixing of distinctly different age lava flows during uplift of the Culebra sequence along active range-front faults.

Four samples from different locations along the San Pedro Mesa section yielded ages between 4.8 and 4.4 Ma. Two samples yielded plateau dates of 4.37 ± 0.17 Ma (Fig. 4R) and 4.55 ± 0.15 Ma (Fig. 4S). The first basalt sample (DM-99-51) was interbedded with Santa Fe gravels along Cuates Creek, and was collected east of Sanchez Reservoir at the base of the Culebra Range. The second basalt sample (DM-99-54) was interbedded with upper Santa Fe gravels and was collected ~32 km directly north on the east side of Highway 159. This Servilleta flow was subsequently rotated to the north and tilted 25°. Two other samples from a large flow on San Pedro Mesa yielded a plateau date of 4.75 ± 0.10 Ma (Fig. 4T) and an isochron date of 4.59 ± 0.02 Ma for an age spectrum that exhibits the effect of recoiled ^{39}Ar (Fig. 4U). These ages are broadly consistent with the 4.6 Ma ages reported by Appelt (1998) for Servilleta flows exposed on the valley floor.

DISCUSSION

Initiation of rifting

The onset of extension in the northern Rio Grande Rift was accompanied by the accumulation of basin-fill gravels of the Santa Fe Formation interbedded with basaltic lavas and ash-flow tuffs. These lavas were erupted onto broad alluvial plains and protobasins extending from southern New Mexico to central Colorado.

A number of geologic observations and temporal data provided by existing K-Ar and new $^{40}Ar/^{39}Ar$ age determinations suggest that initiation of extension in the northern Rio Grande Rift occurred ca. 27–25 Ma. These observations are supported by regional geologic mapping conducted during the 1970s and 1980s in addition to the new mapping conducted in the Culebra range for this study. Early observations were supported by K-Ar age determinations, which are less precise but broadly consistent with new $^{40}Ar/^{39}Ar$ age determinations reported in this study. In the following we summarize relevant data on the opening of the northern Rio Grande Rift, incorporating interpretation of $^{40}Ar/^{39}Ar$ age determinations and field observations reported herein.

Early age estimates for rift inception of ca. 27–26 Ma were proposed by Lipman and Mehnert (1975) based on observations of Hinsdale basalt flows interbedded with volcaniclastic sediments of the Los Pinos Formation and in the southeastern San Juan Mountains. In addition, Bingler (1968) reported a K-Ar age of 25.9 Ma for a local rhyolite lava flow interbedded with the Los Pinos Formation in the San Juan Mountains. Both of these observations are interpreted to reflect active deposition of volcaniclastic debris shed from the volcanic highlands of the San Juan volcanic field into actively subsiding basins to the east.

The Amalia Tuff, erupted from the Questa caldera in the Latir volcanic field Lipman et al. (1986), is an important regional marker in northern New Mexico. Lipman et al. (1986) obtained a K-Ar age for the Amalia Tuff of 26.5 Ma based on an average of 6 samples from various Amalia Tuff outcrops. However, our new $^{40}Ar/^{39}Ar$ age of 24.96 ± 0.11 Ma for the Amalia Tuff appears to serve as a more accurate determination for rift inception. Outcrops of the Amalia Tuff occur as far north as our mapped section in the southern Culebra Range, 40 km north of the Questa caldera. Additional exposures are preserved on the horst block at Brushy Mountain in the center of the Taos Plateau, and on the west side of the rift at Tres Piedras, where nonwelded tuffs overlie Precambrian granite of the Tusas Mountains. It is not known if faulting associated with extension of the rift was present during the deposition of the Amalia Tuff, but the resulting caldera was subsequently dissected by a range-bounding high-angle normal fault along the west margin of the Taos Range in the southern Sangre de Cristo Mountains.

The Amalia Tuff exposures in the Culebra Range section are preserved in a basin-fill sequence of Santa Fe Group sediments (Fig. 3). These exposures provide the only suggestion that the Amalia Tuff was locally erupted onto shallow subsiding basins related to the onset of extension. Other Amalia Tuff exposures are deposited directly on Tertiary or Precambrian rocks that currently are elevated relative to downdropped blocks of the modern rift system. This supports the interpretation that the paleogeography of the region beyond the Latir volcanic field was characterized by a broad erosional surface with only local depositional centers present, such as those of the Culebra Range and the southeastern San Juan Mountains.

The San Luis Hills, the surface expression of the same intrarift horst exposed at Brushy Mountain, are in the axial center of the modern rift. The horst is capped by basalt lava flows dated by the K-Ar method as ca. 26–25 Ma that overlie andesite lava flows dated by K-Ar as ca. 26 Ma (Thompson et al., 1991). These basaltic rocks are correlative with the Hinsdale Formation lavas reported by Lipman and Mehnert (1975), interbedded with Los Pinos Formation sediments in the southeastern San Juan Mountains. Unlike the region immediately to the west, however, the preserved section in the horst has remained a topographic high throughout rift history.

The genetic link to extension for these basalts is based largely on the sharp compositional contrast to underlying eroded intermediate composition volcanic rocks and the compositional similarity to basin-bound lava flows to the immediate west.

The sedimentary record northeast of the San Luis Hills provides additional limitations on the timing of extension. Near Fort Garland, the sedimentary record extends back to 30 Ma (A.Wallace, 2002, written commun.). His data suggest a time frame from which rifting began in the San Luis Valley. Wallace (unpublished data) has documented the depositional base of the Santa Fe Group and inferred the age to be ca. 25 Ma. A basalt flow dated as ca. 18.9 Ma interbedded with Santa Fe sediments at this locality was subsequently tilted during extension. Prior to this age the eastern margin of the rift still exhibited broad erosional basins and paleovalleys, as evidenced by gravels that were still being shed from more easterly source regions, suggesting that at 19 Ma the northern Rio Grande Rift was still a relatively shallow basin, and a major period of uplift occurred in the middle Miocene.

These observations of the late Cenozoic volcanic and sedimentary record of northern New Mexico and southern Colorado place new limitations on initiation of rifting and contain valuable information related to uplift and extension in the northern Rio Grande Rift. We present a simple approach to determining uplift rates for this part of the rift based on the $^{40}Ar/^{39}Ar$ age data we obtained along with their relative elevation data for preserved sections, and compare this to exhumation rates based on apatite fission track data north and south of our study area.

Uplift rates

Although numerous geologic mapping and petrologic studies have been carried out in the northern Rio Grande Rift, data pertaining to relative rates of uplift and extension are limited. Pazzaglia and Kelley (1998) and Kelley et al. (1992) determined the exhumation rates from apatite fission-track dating (AFT) for the southern Sangre de Cristo Mountains (Taos Range) to be ~280–140 m/m.y. and obtained similar uplift rates for the Wet Mountains 60 km to the north. Samples in these studies were collected from Proterozoic and Paleozoic basement rocks and upper to middle Tertiary volcanic and igneous rocks. Their results are varied, but for the most part indicate that a major period of uplift occurred in the middle Miocene (ca. 15 Ma). Kelley and Duncan (1984) obtained uplift rates for the Santa Fe region of 125 m/m.y. more than 200 km to the south, and 173 m/m.y. for the Taos Range, 100 km to the south. Overall, these uplift rates are broadly similar, and variations between the different locations may result from thermal resetting of apatite fission tracks due to younger plutonism, late Tertiary volcanism, and possible lingering thermal effects from Laramide thrust plates, or may reflect variable rates of uplift across the region.

Wallace (2002, written commun.) also determined that a major period of uplift in the region occurred in the middle Miocene based upon detailed mapping of prerift and synrift sedimentary units that are cut by rift-related faults. Based upon his mapping, several stratigraphic sections indicate that there is a dramatic change in clast lithologies from the base to the uppermost sections of the Santa Fe Formation. The lower sections contained lithologies similar to those far to the east (Mount Mestas and the Spanish Peaks igneous centers) mixed with local Paleozoic rocks. Later, clasts were derived exclusively from more proximal source areas (Proterozoic and Paleozoic basement rocks from the Culebra Range). This indicates that uplift to the east occurred sometime after 23 Ma, based upon the age range for Mount Mestas and the Spanish Peaks (25–21 Ma). However, $^{40}Ar/^{39}Ar$ age determinations (Wallace and Snee, 1995, personal commun.) of several key volcanic units within and below the Santa Fe Formation provide data on the timing of volcanism and extension. Prerift sediments and Oligocene volcanic rocks have been dated as 29.6 Ma. A 19.9 Ma $^{40}Ar/^{39}Ar$ date from a conformable basalt flow within the Santa Fe tilts 26° to the east, indicating that faulting and block rotation occurred after 19.9 Ma or at the earliest, the middle Miocene. The $^{40}Ar/^{39}Ar$ dates of these volcanic units and the sedimentary record from the area indicate that a major period of uplift occurred during the middle Miocene.

With the available data for uplift and extension in the northern Rio Grande Rift, we now put forth a very simple approach to exhumation rates based upon our $^{40}Ar/^{39}Ar$ age data from the southern Sangre de Cristo Mountains and the San Luis Valley. Figure 5 is a simplified cross section of the study area that extends from the preserved volcanic section of prerift and synrift volcanic rocks in the east to the San Luis Hills area in the west, and depicts the relative elevations of the volcanic rocks.

From field studies and age determinations on several volcanic rocks from the region, we have been able to determine two datum points with which to calculate exhumation rates. On a first-order scale, we determined an exhumation rate of 58 m/m.y. This rate was calculated using two reference points (D1 in Fig. 5) that contain 26–25 Ma volcanic rocks from the San Luis Hills at an elevation of 2450 m (~250 m above the valley bottom) and 25 Ma volcanic rocks from the crest of the Culebra Range at an elevation of 3900 m (~1450 m difference between the San Luis Hills volcanics). The second, more precise datum (D2 in Fig. 5) is based on the age and elevation differences of Servilleta basalt flows at San Pedro Mesa (elevation ~2600 m), dated as 4.7 Ma, and similar age Servilleta basalts located at the valley bottom represented by the Rio Grande River (elevation ~2200 m). This second datum gives an exhumation rate of 87 m/m.y. Recent $^{40}Ar/^{39}Ar$ dating of faulted lavas from the western part of the Ocate volcanic field (Fig. 1) by Olmstead (2000) indicated an exhumation rate of ~43 m/m.y. based on offset of 4.46 Ma basalt flows. This exhumation rate is similar to our rate of 58 m/m.y. for the D1 surface.

We acknowledge that we can't discriminate between uplifted flanks and downdropped rift valleys and margins. Our terminology (exhumation rates) already accommodates this uncer-

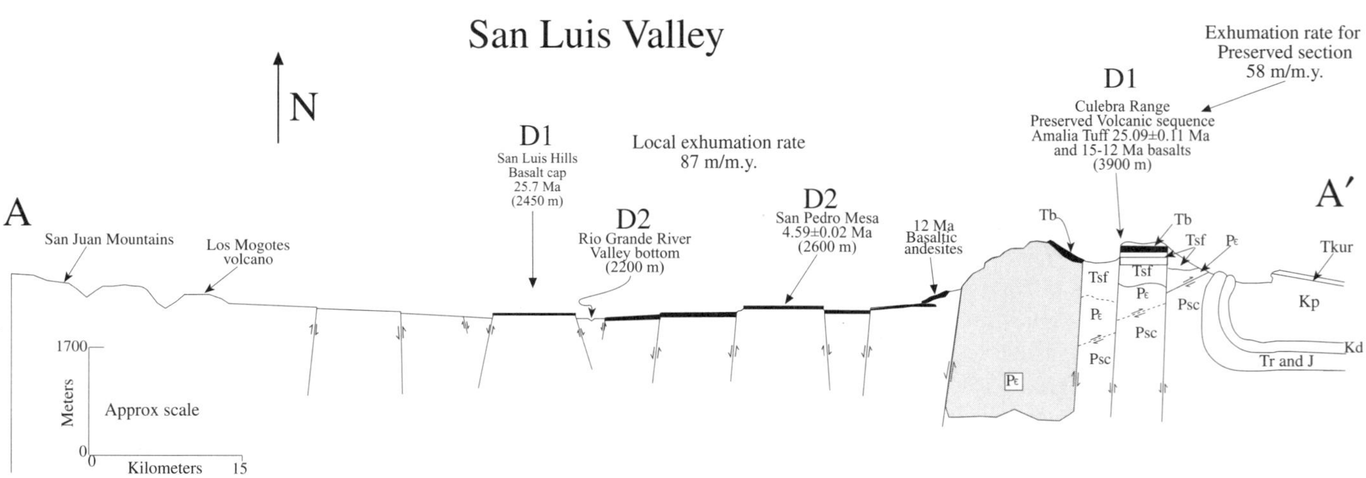

Figure 5. Cross section from San Juan volcanic field to west, across Rio Grande Valley, to the volcanic sequence preserved near the crest of the Culebra Range to the east. Vertical exaggeration is shown to show relief of the profile. Section showing volcanic sequence is diagrammatic and not to scale. Symbols: Tsf/v = Tertiary Santa Fe Group–Vallejo Formation rocks undifferentiated; AT = Amalia Tuff; Tb = Tertiary basalts; Tr and J = Triassic and Jurassic rocks; Kd = Dakota Group; Kp = Pierre, Niobrara, and Trinidad Formations undifferentiated; Tkur = Vermejo and Raton Formations, undifferentiated; Psc = Permian Sangre de Cristo Formation; P€ = Precambrian basement rocks.

tainty. The majority of the rifting may be regional downdropping of basins subsequent to uplift of the Colorado Plateau. Even with regional erosion rates of ~8–9 cm/k.y. (100 m/m.y.) (D. Schmidt, 2001, personal commun.), our modest exhumation rates will likely exceed erosion rates resulting in the present-day topography of the northern Rio Grande Rift. Although erosion rates of 80–90 m/m.y. are equal to or slightly higher than our estimated exhumation rates, the old preserved volcanic surfaces argue against abnormally high erosion rates in the region; erosion rates appear to be highly variable for the northern Rio Grande Rift and may have been highly variable during the 25 m.y. of rift development.

On the basis of age data, exhumation and uplift rates determined by Pazzaglia and Kelley (1998), Kelley and Duncan (1984), and Wallace (2002, written commun.), and our age data and field observations obtained from the San Luis Valley region, it appears that a major period of uplift and subsequent extension in the northern Rio Grande Rift occurred during the middle Miocene.

On the basis of age data; exhumation and uplift rates determined by Pazzaglia and Kelley (1998), Kelley and Duncan (1984), and Wallace (unpublished data); and our age data and field observations obtained from the San Luis Valley region, it appears that a major period of uplift and subsequent extension in the northern Rio Grande Rift occurred during the middle Miocene. Our research has major implications for northern Rio Grande Rift tectonic development and can be applied to other major rift zones. Determining exhumation and uplift rates for rift systems will help enhance our understanding of the volcanic and structural changes and extensional regime experiences through time.

REFERENCES CITED

Appelt, R.M., 1998, $^{40}Ar/^{39}Ar$ geochronology and volcanic evolution of the Taos plateau volcanic field, northern New Mexico and southern Colorado [M.S. thesis]: Socorro, New Mexico Institute of Mining and Technology, 207 p.

Bingler, E.C., 1968, Geology and mineral resources of Rio Arriba County: Bulletin—New Mexico Bureau of Mines and Mineral Resources, v. 91, 158 p.

Brister, B.S., and Gries, R.R., 1994, Tertiary stratigraphic and tectonic development of the Alamosa basin (northern San Luis Basin), Rio Grande rift, south-central Colorado, *in* Killer, G.R., and Cather, S.M., eds., Basins of the Rio Granderift: Structure, stratigraphy, and tectonic setting: Geological Society of America Special Paper 291, p. 39–58.

Cordell, L., 1978, Regional geophysical setting of the Rio Grande rift: Geological Society of America Bulletin, v. 89, p. 1073–1090.

Dalrymple, G.B., and Lanphere, M.A., 1969, Potassium-argon dating: Principles, techniques, and applications to geochronology: San Francisco, W.H. Freeman and Company, 258 p.

Dalrymple, G.B., Alexander, E.C., Lanphere, M.A., and Kraker, G.P., 1981, Irradiation of samples for $^{40}Ar/^{39}Ar$ dating using the Geological Survey TRIGA reactor: U.S. Geological Survey Professional Paper 1176, 55 p.

Fleck, R.J., Sutter, J.F., and Elliot, D.H., 1977, Interpretation of discordant $^{40}Ar/^{39}Ar$ age spectra of Mesozoic tholeiites from Antarctica: Geochimica et Cosmochimica Acta, v. 41, p. 15–32.

Gibson, S.A., Thompson, R.N., Leat, P.T., Dickin, A.P., Morrison, M.A., Hendry, G.L., and Mitchell, J.G., 1992, Asthenosphere-derived magmatism in the Rio Grande rift, western USA: Implications for continental break-up, *in* Storey B.C., Alabaster, T., and Pankhurst, R.J., eds., Magmatism and the causes of continental break-up: Geological Society [London] Special Publication 68, p. 61–89.

Haugerud, R.A., and Kunk, M.J., 1988, ArAr*, a computer program for reduction of $^{40}Ar/^{39}Ar$ data: U.S. Geolological Survey Open File Report 88-261, 68 p.

Hess, J.C., and Lippolt, H.J., 1986, Kinetics of Ar isotopes during neutron irradiation: ^{39}Ar loss from minerals as a source of error in $^{40}Ar/^{39}Ar$ dating: Chemical Geology (Isotope Geoscience Section), v. 59, no. 4, p. 223–236.

Keller, G.R., Cordell, L., Davis, G.H., Peeples, W.J., and White, G., 1984, A geophysical study of the San Luis basin: New Mexico Geological Society Guidebook, 35th Field Conference, Rio Grande rift: Northern New Mexico, p. 51–57.

Kelley, S.A., and Duncan, I.J., 1984, Tectonic history of the northern Rio Grande rift derived from apatite fission-track geochronology: New Mexico Geological Society Guidebook, 35th Field Conference, Rio Grande rift: Northern New Mexico, p. 67–73.

Kelley, S.A., Chapin, C.E., and Corrigan, J., 1992, Late Mesozoic to Cenozoic cooling histories of the flanks of the northern and central Rio Grande rift—Colorado and New Mexico: Bulletin—New Mexico Bureau of Mines and Mineral Resources, v. 145, 39 p.

Kleinkopf, D.M., Peterson, D.L., and Johnson, R.B., 1970, Reconnaissance geophysical studies of the Trinidad quadrangle, south-central Colorado: United States Geological Survey Professional Paper 700B, p. 78–85.

Leat, P.T., Thompson, R.N., Morrison, M.A., Hendry, G.L., and Dickin, A.P., 1988, Compositionally-diverse Miocene-recent rift-related magmatism in northwest Colorado: Partial melting, and mixing of mafic magmas from 3 different asthenospheric and lithospheric mantle sources:Journal of Petrology, Special Lithosphere Issue, p. 351–377.

Lipman, P.W., and Mehnert, H.H., 1975, Late Cenozoic basaltic volcanism and development of the Rio Grande depression in the southern Rocky Mountains, *in* Curtis, B.F., ed., Cenozoic history of the southern Rocky Mountains: Geological Society of America Memoir, v. 144, p. 119–154.

Lipman P.W., and Mehnert, H.H., 1979, The Taos plateau volcanic field, northern Rio Grande rift, New Mexico, *in* Reiker, R.E., ed., Rio Grande rift: Tectonics and magmatism: American Geophysical Union, p. 289–311.

Lipman, P.W., and Reed. J.C., Jr., 1989, Geologic map of the Latir volcanic field and adjacent areas, northern New Mexico: U.S. Geological Survey Miscellaneous Investigations, Map I-1907, scale 1:48 000, 1 sheet.

Lipman, P.W., Mehnert, H.H., and Naeser, C.W., 1986, Evolution of the Latir volcanic Field, northern New Mexico, and its relation to the Rio Grande rift, as indicated by K-Ar and fission-track dating: Journal of Geophysical Research, v. 91, p. 6329–6345.

McIntyre, D.B., 1963, Precision and resolution in geochronometry, *in* Albritton, C.C., ed., The fabric of geology: Reading, Massachusetts, Addison-Wesley, p. 112–134.

Miggins, D.P., Snee, L.W., Kunk, M.J., Pillmore, C.L., and Stern, C.R., 2000, The magmatic evolution of the Raton Basin using $^{40}Ar/^{39}Ar$ geochronology: Geological Society of America Abstracts with Programs, v. 32, no. 7, p. A159.

Olmsted, B.W., 2000, $^{40}Ar/^{39}Ar$ investigations of the Ocate Volcanic Field, north-central New Mexico [M.S. thesis]: Socorro, New Mexico Institute of Mining and Technology, 197 p.

O'Neill, J.M., and Mehnert, H.H., 1988a, Late Cenozoic physiographic evolution of the Ocate volcanic field, north-central New Mexico: U.S. Geological Survey Open File Report 80-928, 41 p.

Pazzaglia, F.J, and Kelley, S.A., 1998, Large-scale geomorphology and fission-track thermochronology in topographic and exhumation reconstructions of the southern Rocky Mountains: Rocky Mountain Geology, v. 33, no. 2, p. 229–257.

Penn, B.S., 1994, An investigation of the temporal and geochemical characteristics, and the petrogenetic origins of the Spanish Peaks intrusive rocks of south-central Colorado [Ph.D. thesis]: Golden, Colorado, School of Mines, 199 p.

Perry, F.V., Baldridge, W.S., and Depaolo, D.J, 1987, Role of asthenosphere and lithosphere in the genesis of late Cenozoic basaltic rocks from the Rio Grande rift and adjacent regions of the southwestern United States: Journal of Geophysical Research, v. 92, p. 9193–9213.

Personius, S.F., and Machette, M.N., 1984, Quaternary and Pliocene faulting in the Taos Plateau region, northern New Mexico, *in* Baldridge, W.S., Dickerson, P.W., Riecker, R.W., and Zidek, J., eds., Rio Grande rift: Northern New Mexico: New Mexico Geological Society Guidebook, 35th Field Conference, p. 83–90.

Pillmore, C.L., and Laurie, C.O., 1976, Second day road log from Raton to Underwood Lakes through the Raton coalfield via the York Canyon mine, Vermejo Park and Gold Creek, *in* Ewing R.C., and Kves, B.S., eds., Guidebook of Vermejo Park, Northeastern New Mexico: New Mexico Geological Society, 27th Field Conference, p. 25–47.

Pillmore, C.L., Obradovich, J.D., Landreth, J.O., and Pugh, L.E., 1973, Mid-Tertiary volcanism in the Sangre de Cristo Mountains of northern New Mexico: Geological Society of America Abstracts with Programs, v. 5, p. 502.

Roddick, J.C., 1983, Generalized numerical error analysis with applications to geochronology and thermodynamics: Geochimica et Cosmochimica Acta, v. 51, p. 2129–2135.

Samson, S.D., and Alexander, E.C., 1987, Calibration of the interlaboratory ^{40}Ar/^{39}Ar dating standard, MMhb-1: Chemical Geology (Isotope Geoscience Section), v. 66, p. 27–34.

Snee, L.W., 1982, Emplacement and cooling of the Pioneer Batholith, southwestern Montana [Ph.D. thesis]: Columbus, Ohio, Ohio State Univ., 320 p.

Snee, L.W., 2002, Argon thermochronology of mineral deposits—a review of analytical methods, formulations, and selected applications: U.S. Geological Survey Bulletin 2194 (in press).

Snee, L.W., Sutter, J.F., and Kelly, W.C., 1988, Thermochronology of economic mineral deposits: Dating the stages of mineralization at Panasqueira, Portugal, by high precision ^{40}Ar/^{39}Ar age-spectrum techniques on muscovite: Economic Geology, v. 83, p. 335–354.

Staudacher, T.H., Jessburger, E.K., Dorflinger, D., and Kiko, J., 1978, A refined ultrahigh-vacuum furnace for rare-gas analysis: Journal of Physics, E, Scientific Instruments, v. 11, p. 781–784.

Steiger, R.H., and Jäger, E., 1977, Subcommission of geochronology: Convention of the use of decay constants in geo- and cosmochronology: Earth and Planetary Science Letters, v. 36, p. 359–362.

Stormer, J.C., Jr., 1972a, Ages and nature of volcanic activity of the southern high plains, New Mexico and Colorado: Geological Society of America Bulletin, v. 83, p. 2443–2448.

Stormer, J.C., Jr., 1972b, Mineralogy and petrology of the Raton-Clayton volcanic field, northwestern New Mexico: Geological Society of America Bulletin, v. 83, p. 3299–3322.

Stroud, J.R., 1997, The geochronology of the Raton-Clayton volcanic field, with implications for volcanic history and landscape evolution [M.S. thesis]: Socorro, New Mexico Institute of Mining and Technology, 164 p.

Thompson, R.A., and Machette, M.N., 1989, Geologic map of the San Luis Hills area, Conejos and Costilla counties, Colorado: U.S. Geological Survey Miscellaneous Investigations Series Map, I-1906, 1:50 000 scale, 1 sheet.

Thompson, R.A., Dungan, M.A., and Lipman, P.W., 1986, Multiple differentiation processes in early-rift calc-alkaline volcanics, northern Rio Grande rift, New Mexico: Journal of Geophysical Research, v. 91, no. B6, p. 6046–6058.

Thompson, R.A., Johnson, C.M., and Mehnert, H.H., 1991, Oligocene basaltic volcanism of the northern Rio Grande rift: San Louis Hills, Colorado: Journal of Geophysical Research, v. 96, no. B8, p. 13577–13592.

Turner, G., and Cadogan, P.H., 1974, Possible effects of ^{39}Ar recoil in ^{40}Ar/^{39}Ar dating: Geochimica et Cosmochimica Acta, Supplement 5, Proceedings of the Fifth Lunar Science Conference, p. 1601–1615.

Tweto, O., 1979, The Rio Grande system in Colorado, *in* Riecker, R.C., ed., Rio Grande rift: Tectonics and magmatism, American Geophysical Union, p. 35–56.

Upson, J.E., 1941, The Vallejo Formation—new early Tertiary red-beds in south-central Colorado: American Journal of Science, v. 239, p. 577–589.

Wallace, A.R., 1995, Cenozoic rift-related sedimentation and faulting, northern Culebra Range, southern Colorado: New Mexico Geological Society Guidebook, 46th Field Conference, Geology of the Santa Fe region, p. 147–154.

York, D., 1969, Least squares fitting of a straight line with correlated errors: Earth and Planetary Science Letters, v. 5, p. 320–344.

Manuscript Accepted to the Society October 15, 2001

Printed in the U.S.A.

Geological Society of America
Special Paper 362
2002

Lithospheric mantle beneath Arabia: A Pan-African protolith modified by the Afar and older plumes, rather than a source for continental flood volcanism?

Joel Baker
Danish Lithosphere Centre, Øster Voldgade 10, 1350 Copenhagen K, Denmark, and Department of Geology, Royal Holloway University of London, Egham, Surrey TW20 0EX, UK
Gilles Chazot
Department of Geology, Royal Holloway University of London, Egham, Surrey TW20 0EX, UK and Laboratoire de Geologie, Universite Blaise Pascal, 5, rue Kessler, 63038 Clermont-Ferrand cedex, France
Martin A. Menzies
Matthew Thirlwall
Department of Geology, Royal Holloway University of London, Egham, Surrey TW20 0EX, UK

ABSTRACT

The geochemistry of Oligocene-Miocene flood volcanic rocks associated with breakup of the Africa-Arabian continent reveals possible source contributions from the shallow mantle lithosphere, the nature and origin of which can be delimited by the study of mantle xenoliths entrained by posterosional Pliocene-Quaternary volcanoes located on the uplifted margin. Trace element and isotopic studies of these xenoliths show that the Arabian lithospheric mantle contains several components including one formed during, or shortly after, accretion of the Arabian Pan-African shield ca. 700 Ma. Since that time influx of small volume melts has modified the lower lithosphere and, most recently, the Afar plume may have acted as a source for small volume metasomatizing melts. No evidence exists in the geochemistry of the spinel lherzolite xenoliths for Pan-African accretion of arc terranes (suprasubduction) or the presence of an Archean keel (>2500 Ma). The pre-Oligocene enriched component of the Arabian lithospheric mantle is too minor to have played a role in generating significant volumes of enriched melt that could have produced such characteristics in some Oligocene flood basalts with similar enriched isotopic characteristics. It seems that mantle plumes and asthenospheric-derived melts have repeatedly overprinted the lithospheric mantle in this region, rather than having been contaminated by melts derived from lithospheric mantle sources.

INTRODUCTION

Fingerprinting the chemical and isotopic composition of the lithospheric mantle underlying the Arabian Peninsula is pivotal to evaluating the potential contributions of different mantle sources to flood and intraplate volcanism associated with the volcanic rifted margins of the Red Sea. Studies of continental flood volcanism (e.g., Hergt et al., 1989; Lightfoot et al., 1990, 1993), coupled with theoretical considerations of the thermal and chemical aspects of melt generation in the lithospheric

Baker, J., Chazot, G., Menzies, M., and Thirlwall, M., 2002, Lithospheric mantle beneath Arabia: A Pan-African protolith modified by the Afar and older plumes, rather than a source for continental flood volcanism?, *in* Menzies, M.A., Klemperer, S.L., Ebinger, C.J., and Baker, J., eds., Volcanic Rifted Margins: Boulder, Colorado, Geological Society of America Special Paper 362, p. 65–80.

mantle (Gallagher and Hawkesworth, 1992; Saunders et al., 1992), have concluded that the lithospheric mantle might be a major source for continental flood volcanism. Such arguments have important implications for physical and thermal models of large igneous province formation, continental breakup, and, in particular, the causative relationships between rifting and volcanism. However, others have doubted the validity of models where large volumes of melt of basaltic composition can be generated within the lower lithosphere, essentially a dry and refractory rock (Arndt and Christensen, 1992; Menzies, 1992).

While general considerations of the physical, chemical, and isotopic characteristics of continental flood volcanism and the lithospheric mantle can shed some light on whether the lithospheric mantle may be a source of flood basalts (e.g., Gallagher and Hawkesworth, 1992; Arndt and Christensen, 1992), the opportunity exists in Yemen to directly compare the chemical and isotopic signatures of continental flood volcanism and spatially associated lithospheric mantle xenoliths. In this chapter, the chemical and isotopic signature of Arabian lithospheric mantle xenoliths is examined, and its bearing on the age, character, and development of Arabian lithospheric mantle along with its possible contribution to continental flood volcanism is discussed. The new data obtained on spinel facies mantle xenoliths collected from southern Yemen and the southern Red Sea are compared with data from mantle xenoliths in Saudi Arabia (Henjes-Kunst et al., 1990; Blusztajn et al., 1995) and Zabargad Island (Brueckner et al., 1988; Hamelin and Allègre, 1988).

SAMPLE DESCRIPTION

The xenoliths were collected from two Pliocene-Quaternary alkali basalt volcanoes in southern Yemen, (1) at Bir Ali on the south coast of Yemen, ~150 km east of Aden, and at Ataq in the Balhaf graben (Varne, 1970; Varne and Graham, 1971; Menzies and Murthy, 1980; Chazot et al., 1996a, 1996b, 1997), and (2) Kod Ali Island, a small volcanic diatreme close to the Eritrean coast in the southern Red Sea (Hutchinson and Gass, 1971).

Xenoliths from Bir Ali include anhydrous spinel lherzolites, pyroxenites, and meta–igneous gabbroic granulites. The Bir Ali lherzolites are equigranular with a slightly foliated texture and contain brown spinel. Xenoliths from the Ataq volcanic field comprise both anhydrous and hydrous spinel lherzolites that are discussed in more detail in the following paragraph.

Xenoliths from Kod Ali Island comprise anhydrous spinel lherzolites, clinopyroxenites, and gabbros. The Kod Ali lherzolites are protogranular, and composed of large olivines and subordinate orthopyroxene, clinopyroxene, and spinel. Pyroxenites from both Bir Ali and Kod Ali Island are dominated by Cr-diopside with minor amounts of olivine, orthopyroxene, and red-brown spinel.

The Ataq xenoliths are petrographically heterogeneous, including equigranular, protogranular, and porphyroclastic types. The hydrous peridotites contain abundant apple green Cr-diopside replacing primary enstatite and large amphibole ± apatite crystals. Most hydrous samples have undergone partial or complete breakdown and melting of amphibole ± clinopyroxene, leaving behind melt pockets several millimeters across. The melt pockets contain small newly formed olivine + clinopyroxene + spinel crystals in a base of silicate glass. One sample, JK2, is cut by a 1-cm-wide clinopyroxenite or websterite vein. Chazot et al. (1996b) described the petrography and mineral chemistry of the Ataq mantle xenoliths and commented on the possible origin of these silicate glass melt pockets through breakdown of the amphibole during entrainment and ascent of the xenoliths in their host magmas.

ANALYTICAL RESULTS

Chemical and Sr-Nd-Pb isotope data for minerals separated from the Ataq, Bir Ali, and Kod Ali xenoliths are presented in Tables 1 and 2. Table 1 also includes limited chemical data for some representative whole-rock samples. Details of analytical techniques and uncertainties are given in Appendix 1 and Chazot et al. (1996a, 1996b).

Elemental data

REE data. The mantle xenoliths can be divided into three groups on the basis of rare earth element (REE) data obtained on the mineral separates (Fig. 1).

Strongly light REE enriched. Minerals from the hydrous Ataq peridotites are all strongly light REE enriched (Ce/Yb_N = 10–185) (Fig. 1A), and exhibit middle REE enrichment relative to the heavy REEs (Gd/Yb_N = 2.7–16.8). All minerals from these peridotites have very high REE abundances, typically >10× chondritic values for the light REEs and middle REEs.

Moderate light REE enrichment. Clinopyroxene separates from the anhydrous Ataq peridotites have moderate light REE enrichment (Fig. 1B; Ce/Yb_N = 0.6–2.6; Ce/Sm_N = 2–3) compared with the hydrous Ataq peridotites. Moreover, the clinopyroxenes are also all marked by a distinctive depletion in the middle REE relative to the heavy REE (Gd/Yb_N = 0.55–0.90). Compared with the hydrous Ataq peridotites, clinopyroxenes from the anhydrous samples have much lower REE abundances, typically <30× chondritic values for the light REE and 1–5× chondritic values for the middle and heavy REEs.

Light REE depleted. Clinopyroxenes from the anhydrous Bir Ali and Kod Ali peridotites and pyroxenites are light REE depleted relative to the heavy REEs (Fig. 1C; Ce/Yb_N = 0.15–0.68), and have flat or slightly middle REE depleted profiles. REE abundances are always low, <10× chondritic values.

The limited REE data available for the whole-rock samples mirror those of the mineral data. An anhydrous Bir Ali lherzolite (BA8) and anhydrous Ataq lherzolite (JK4) are both light REE depleted, whereas the hydrous Ataq lherzolites (JK3 and JK8) are strongly and variably light REE enriched (Fig. 1D). Moreover, the anhydrous Bir Ali and Ataq lherzolite both have relatively flat middle and heavy REE chondrite-normalized

TABLE 1. CHEMICAL DATA FOR MINERALS AND REPRESENTATIVE WHOLE ROCKS FROM THE BIR ALI AND ATAQ (SOUTHERN YEMEN), AND KOD ALI ISLAND (SOUTHERN RED SEA) XENOLITH SUITES

Sample		Rb	Ba	Sr	U	Th	Pb	Zr	Nb	Hf	La	Ce	Nd	Sm	Eu	Gd	Dy	Er	Yb	Y
JK2 cpx*	1.†	2.10	4.55	N.D.	N.D.	1.71	N.D.	9.11	0.156	N.D.	N.D.	N.D.	N.D.	N.D.	N.D.	N.D.	N.D.	N.D.	N.D.	8.03
	2.	1.22	0.26	607	N.D.	1.21	N.D.	5.12	0.21	0.083	30.2	73.5	33.8	4.76	1.58	3.58	2.04	0.964	0.747	6.42
	4.	N.D.	N.D.	N.D.	0.310	N.D.	1.88	N.D.	N.D.	N.D.	20.9	56.4	26.3	3.74	1.36	2.15	1.45	0.672	0.628	N.D.
JK3 wr	3.	3.23	210.5	432.8	N.D.	2.31	1.18	2.68	0.26	N.D.	43.5	66.6	11.5	0.93	0.37	1.50	0.64	0.44	0.48	3.58
JK8 cpx	1.	1.01	25.3	1135	N.D.	1.37	N.D.	7.82	1.71	N.D.	N.D.	N.D.	N.D.	N.D.	N.D.	N.D.	N.D.	N.D.	N.D.	40.9
	2.	N.D.	0.60	1070	N.D.	0.83	N.D.	2.84	0.60	N.D.	31.9	99.3	78.8	19.3	6.05	18.0	12.6	3.86	2.28	32.3
	3.	N.D.	N.D.	N.D.	N.D.	N.D.	2.77	N.D.	N.D.	N.D.	26.4	80.0	64.8	15.1	5.02	13.9	8.95	3.00	1.84	N.D.
JK8 am	1.	14.0	530	N.D.	N.D.	1.30	N.D.	5.05	26.8	N.D.	N.D.	N.D.	N.D.	N.D.	N.D.	N.D.	N.D.	N.D.	N.D.	33.6
	2.	6.93	501	1030	N.D.	0.97	N.D.	5.06	22.8	2.16	29.9	90.1	68.7	15.4	4.90	14.8	9.91	2.94	1.37	26.4
	4.	N.D.	N.D.	N.D.	0.336	N.D.	2.87	N.D.	N.D.	N.D.	22.3	74.5	53.3	12.6	4.07	10.7	7.62	2.44	1.36	N.D.
JK8 ap	1.	0.89	N.D.	N.D.	N.D.	N.D.	N.D.	0.94	1.43	N.D.	N.D.	N.D.	N.D.	N.D.	N.D.	N.D.	N.D.	N.D.	N.D.	N.D.
	2.	N.D.	1160	33400	N.D.	293	127	0.382	N.D.	5.13	3770	6110	2180	342	91.0	194	72.0	20.0	4.21	276
	4.	N.D.	N.D.	N.D.	103.4	N.D.	54.9	N.D.	N.D.	N.D.	3913	6820	2313	317	82.4	197	84.9	21.7	9.37	N.D.
JK8 wr	3.	1.66	2.48	173.6	N.D.	0.30	0.99	2.79	4.48	N.D.	13.5	25.2	12.0	2.63	0.74	2.74	1.19	0.46	0.42	3.76
JK1 cpx	2.	N.D.	0.25	35.3	N.D.	0.59	N.D.	21.2	2.49	0.62	4.20	6.77	2.63	0.82	0.315	1.82	1.83	1.32	1.62	9.77
	3.	0.79	N.D.	64.4	N.D.	0.59	1.68	18.8	0.68	N.D.	4.79	10.8	3.04	1.22	0.46	1.63	1.94	1.26	1.28	11.9
	4.	0.535	N.D.	67.6	0.125	N.D.	0.241	N.D.	N.D.	N.D.	4.93	11.5	4.63	0.966	0.378	1.25	1.81	1.21	1.11	N.D.
JK4 cpx	2.	N.D.	0.036	45.9	N.D.	0.18	N.D.	5.52	0.068	0.112	1.38	2.20	0.964	0.297	0.170	0.496	0.916	1.00	1.13	5.05
	3.	0.38	N.D.	64.4	N.D.	0.59	1.29	5.42	0.30	N.D.	1.30	1.95	1.15	0.51	0.15	0.70	1.10	0.84	0.78	6.23
	4.	0.036	N.D.	74.2	0.129	N.D.	0.323	N.D.	N.D.	N.D.	0.969	1.59	0.709	0.203	0.074	0.388	0.810	0.652	0.708	N.D.
JK4 wr	3.	0.60	N.D.	6.23	N.D.	0.06	1.17	0.80	0.10	N.D.	0.20	0.31	0.39	0.26	0.10	0.26	0.25	0.19	0.27	1.30
JK10 cpx	4.	0.141	N.D.	224.4	0.596	N.D.	2.557	N.D.	N.D.	N.D.	8.25	7.35	0.530	2.13	0.170	0.665	1.00	0.734	0.778	N.D.
BA8 cpx	1.	0.055	0.33	25.1	N.D.	0.39	N.D.	21.9	0.092	N.D.	N.D.	N.D.	N.D.	N.D.	N.D.	N.D.	N.D.	N.D.	N.D.	17.6
	2.	N.D.	0.25	16.3	N.D.	N.D.	N.D.	33.8	0.101	1.15	0.776	2.83	3.79	1.67	0.708	2.58	3.10	1.81	1.92	16.2
	3.	0.29	N.D.	20.3	N.D.	0.32	0.45	23.0	0.14	N.D.	2.64	6.40	5.13	2.16	0.83	2.84	3.70	2.67	1.89	18.5
	4.	N.D.	N.D.	N.D.	0.090	N.D.	0.042	N.D.	N.D.	N.D.	1.77	4.33	3.38	1.26	0.531	1.92	2.80	1.78	1.61	N.D.
BA8 wr	3.	0.50	N.D.	5.04	N.D.	0.04	0.60	3.26	0.12	N.D.	0.19	0.51	0.67	0.33	0.15	0.49	0.53	0.37	0.42	0.94
KA cpx	4.	0.011	N.D.	69.9	N.D.	N.D.	0.081	N.D.	N.D.	N.D.	0.150	1.27	2.86	1.42	0.614	2.32	3.45	2.27	2.09	N.D.
K1 cpx	4.	0.0023	N.D.	1.10	N.D.	N.D.	0.011	N.D.	N.D.	N.D.	0.291	0.789	0.726	0.292	0.122	0.481	0.73	0.122	0.464	N.D.
K2 cpx	4.	0.049	N.D.	2.86	N.D.	N.D.	0.012	N.D.	N.D.	N.D.	0.308	0.921	0.907	0.373	0.158	0.628	0.99	0.647	0.612	N.D.
K3 cpx	4.	N.D.	N.D.	N.D.	N.D	N.D	0.014	N.D.	N.D.	N.D.	0.431	1.038	0.820	0.309	0.127	0.208	0.779	0.517	0.479	N.D.

* cpx = clinopyroxene; am = amphibole; ap = apatite; wr = whole rock.
† 1. ID ICP-MS data; 2. SIMS data; 3. ICP-MS data; 4. TIMS ID data.

TABLE 2. SR-ND-PB ISOTOPIC AND RB-SR, SM-ND, U-PB TRACE ELEMENT DATA FROM THE BIR ALI AND ATAQ (SOUTHERN YEMEN), AND KOD ALI ISLAND (SOUTHERN RED SEA) XENOLITH SUITES

Sample number	Mineral phase	Rb (ppm)	Sr (ppm)	$^{87}Sr/^{86}Sr$	Sm (ppm)	Nd (ppm)	$^{143}Nd/^{144}Nd$	U (ppm)	Pb (ppm)	$^{206}Pb/^{204}Pb$	$^{207}Pb/^{204}Pb$	$^{208}Pb/^{204}Pb$
Hydrous amphibole ± apatite-bearing peridotites from Ataq (southwestern Yemen):												
AT2	cpx	N.D.	N.D.	0.70364	N.D.	N.D.	0.512949	N.D.	N.D.	N.D.	N.D.	N.D.
AT3	cpx	N.D.	N.D.	0.70343	N.D.	N.D.	0.512905	N.D.	N.D.	N.D.	N.D.	N.D.
AT15	cpx	N.D.	N.D.	0.70344	N.D.	N.D.	0.512898	N.D.	N.D.	N.D.	N.D.	N.D.
	amph	N.D.	N.D.	0.70344	N.D.	N.D.	0.512898	N.D.	N.D.	N.D.	N.D.	N.D.
JK2	melt	N.D.	N.D.	0.70356	N.D.	N.D.	0.512902	N.D.	N.D.	18.887	15.571	38.834
	cpx	N.D.	N.D.	0.70355	3.741	26.25	0.512902	0.310	1.88	18.891	15.570	38.834
	vein cpx	N.D.	N.D.	0.70356	3.178	24.05	0.512912	0.305	1.92	18.723	15.568	38.657
	opx	N.D.	N.D.	0.70360	N.D.	N.D.	0.512897	N.D.	N.D.	N.D.	N.D.	N.D.
JK3	amph	N.D.	N.D.	0.70347	N.D.	N.D.	0.512937	N.D.	N.D.	19.004	15.591	39.046
JK7	cpx	N.D.	N.D.	0.70352	N.D.	N.D.	0.512927	N.D.	N.D.	18.954	15.592	38.921
JK8	cpx	N.D.	N.D.	0.70351	N.D.	N.D.	0.512921	N.D.	N.D.	N.D.	N.D.	N.D.
	amph	N.D.	N.D.	0.70354	12.62	53.29	0.512882	0.336	2.87	18.951	15.587	38.945
	ap	N.D.	N.D.	0.70355	316.7	2313	0.512917	103.4	54.9	18.955	15.588	38.950
Anhydrous peridotites from Ataq (southwestern Yemen):												
JK1	cpx	0.535	67.60	0.70401	0.9660	4.627	0.512553	0.125	0.241	18.326	15.415	38.410
	repeat	N.D.	N.D.	0.70397	N.D.	N.D.	0.512552	N.D.	N.D.	N.D.	N.D.	N.D.
JK4	cpx	0.036	74.19	0.70335	0.2031	0.7091	0.512898	N.D.	N.D.	N.D.	N.D.	N.D.
	repeat	N.D.	N.D.	0.70329	N.D.	N.D.	0.512890	N.D.	N.D.	N.D.	N.D.	N.D.
JK10	cpx	0.141	224.4	0.70355	0.5300	2.126	0.512982	0.596	2.557	18.853	15.567	38.827
	repeat	N.D.	N.D.	0.70349	N.D.	N.D.	0.512972	N.D.	N.D.	N.D.	N.D.	N.D.
Anhydrous peridotites from Bir Ali (southwestern Yemen):												
BA8	cpx	N.D.	N.D.	0.70291	1.257	3.379	0.513145	0.090	0.0422	17.887	15.499	37.749
Pyroxenite from Bir Ali (southwestern Yemen):												
BA4	cpx	N.D.	N.D.	0.70344	N.D.	N.D.	0.513115	N.D.	N.D.	17.321	15.499	37.181
Granulites from Bir Ali (southwestern Yemen):												
BA22	wr	N.D.	N.D.	0.70476	N.D.	N.D.	0.512754	N.D.	N.D.	N.D.	N.D.	N.D.
BA24	wr	N.D.	N.D.	0.70285	N.D.	N.D.	0.512728	N.D.	N.D.	N.D.	N.D.	N.D.
Anhydrous peridotites from Kod Ali island (southern Red Sea):												
KA	cpx	0.011	69.9	0.70219	1.422	2.859	0.513519	N.D.	0.0807	18.183	15.492	37.795
	repeat	N.D.	N.D.	0.70216	N.D.	N.D.	0.513512	N.D.	N.D.	N.D.	N.D.	N.D.
K117	cpx	N.D.	N.D.	0.70306	N.D.	N.D.	0.512968	N.D.	0.1207	18.274	15.616	38.250
Pyroxenites from Kod Ali island (southern Red Sea):												
K1	cpx	0.0023	1.10	0.70354	0.292	0.726	0.512966	N.D.	0.0106	18.709	15.618	38.671
K2	cpx	0.0490	2.86	0.70365	0.373	0.907	0.512965	N.D.	0.0119	18.669	15.598	38.585
K3	cpx	N.D.	N.D.	0.70346	0.309	0.820	0.512959	N.D.	0.0136	18.841	15.606	38.770

Note: All trace element data in this table were determined by thermal ionization mass spectrometric isotope dilution analysis. N.D. = no data.

* cpx = clinopyroxene; am = amphibole; opx = orthopyroxene; ap = apatite; wr = whole rock.

† Menzies and Murthy (1980) reported the following isotopic data for a hydrous Ataq peridotite: AT15 cpx = 0.70406, 0.51252; amph = 0.70408, 0.51301.

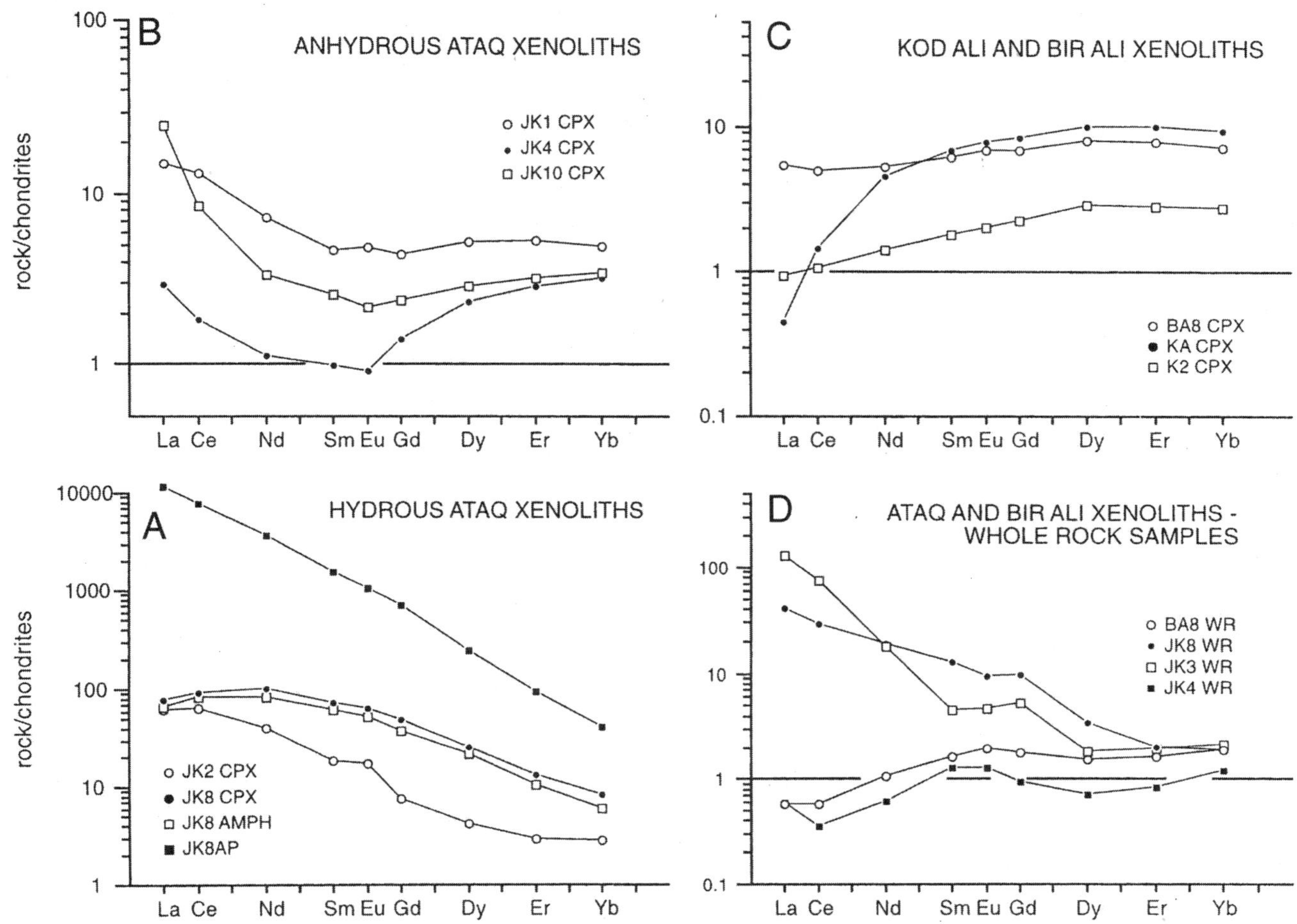

Figure 1. A–C: Chondrite-normalized multielement plots of rare earth element data for minerals separated from southern Red Sea and Yemen mantle xenoliths D: Representative data for four whole-rock samples. Normalization values for chondrites are from Nakamura (1974).

patterns compared with the hydrous Ataq lherzolites, which are middle REE enriched relative to the heavy REEs (Fig. 1D). The light REE depletion of the anhydrous Ataq lherzolite (JK4) reflects the fact that only relatively small amounts of moderately light REE enriched clinopyroxene are present in this sample. Both the anhydrous Ataq and Bir Ali peridotites have small upturns in their REE patterns at La.

Other trace elements. Primitive-mantle-normalized multielement patterns for the mineral separates and whole-rock samples display considerable variation (Fig. 2) and mimic the differences between the three groups of xenoliths identified on the basis of REE chemistry. Mineral separates from the hydrous Ataq peridotites (Fig. 2A) have extremely enriched incompatible trace element patterns, with the exception of Ti, Zr, and Nb (apart from Nb in the amphibole from sample JK8). Trace element concentrations for the large ion lithophile elements (LILE), Th, U, and light and middle REEs are typically >10× primitive mantle in the clinopyroxene and amphibole and >>100× primitive mantle in the apatite separate.

Clinopyroxenes from the anhydrous Ataq peridotites have multielement patterns that are one of two types (Fig. 2B): (1) very depleted incompatible trace element patterns (<10× or even <1180 primitive mantle) with marked enrichment in U, Pb, and Sr, and, to a lesser extent, the light REEs (La) relative to the other trace elements (JK4 and JK10); (2) moderately depleted incompatible trace element patterns (1–10× primitive mantle) with a relatively smooth profile (JK1). The anhydrous Ataq peridotites contain clinopyroxenes that are more depleted in the moderately incompatible trace elements than those from the anhydrous Bir Ali and Kod Ali peridotites (Fig. 2C). Clinopyroxenes from the anhydrous Bir Ali and Kod Ali peridotites have strongly depleted incompatible trace element patterns with trace element abundances typically <<10× primitive mantle. Clinopyroxene from the Bir Ali sample (BA8) has a large negative Nb anomaly, but lacks the negative Zr and Ti anomalies of minerals from the hydrous Ataq samples. This sample also exhibits minor Th, U, La, and Ce enrichment.

Primitive mantle-normalized multielement patterns of four whole-rock samples display the same general features as those of the mineral separates. The hydrous Ataq lherzolites (JK3 and JK8) are highly enriched in the incompatible trace elements with the exception of Nb and Zr. Nb depletion is not as marked in the

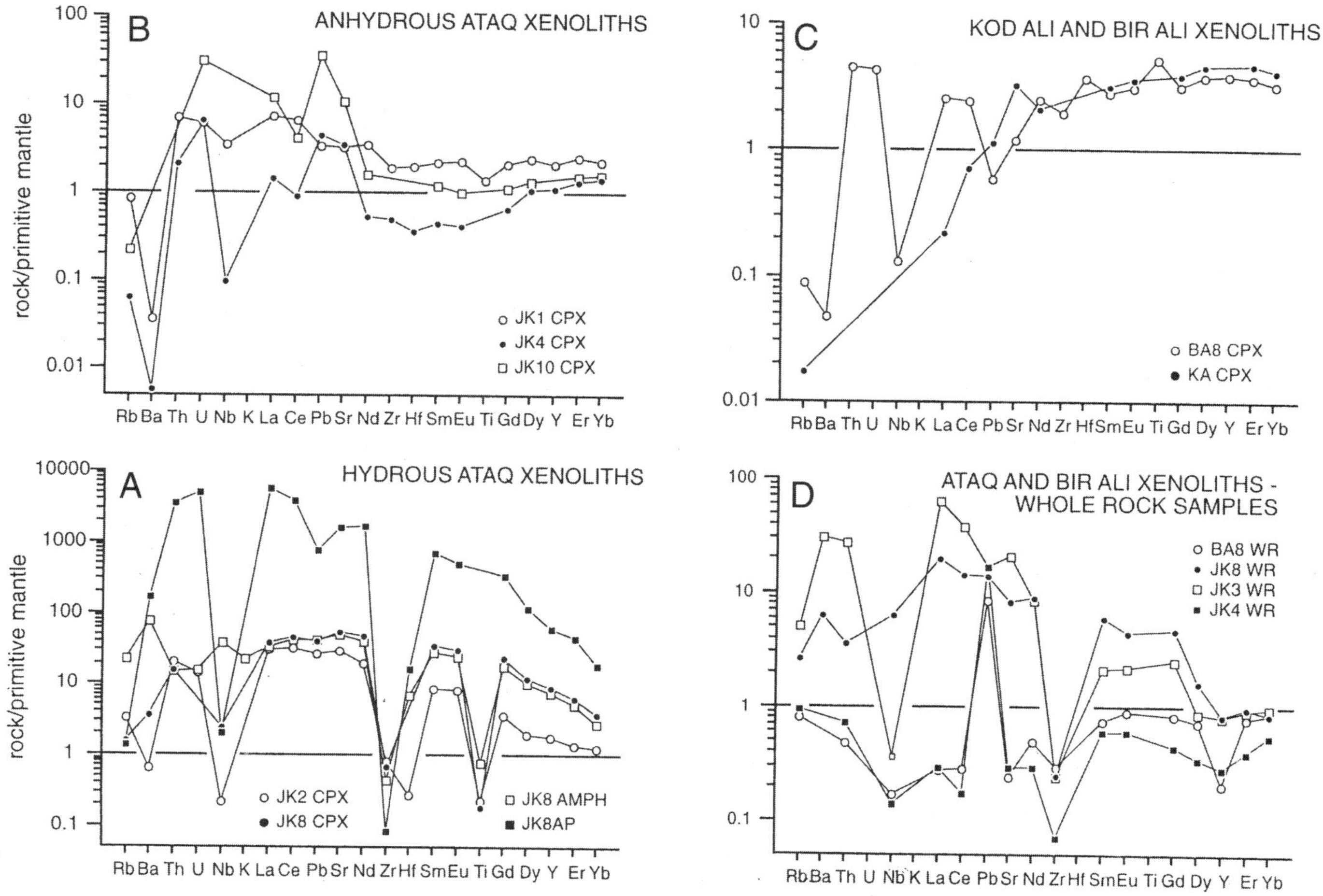

Figure 2. A–C: Primitive-mantle normalized multielement plots of trace element data for minerals separated from southern Red Sea and Yemen mantle xenoliths. D: Representative data for four whole-rock samples. Normalization values for chondrites are from Sun and McDonough (1989).

amphibole-rich peridotite (JK8) compared with the other hydrous Ataq peridotite (e.g., JK3). Anhydrous Bir Ali and Ataq lherzolites have depleted multielement patterns with small negative Nb and Zr anomalies.

Isotopic data

Sr-Nd isotope data. Nearly all the mantle minerals have Sr-Nd isotope ratios that are more depleted than bulk earth ($^{87}Sr/^{86}Sr$ = 0.7022–0.7036; $^{143}Nd/^{144}Nd$ = 0.5135–0.5129; Fig. 3A), although one sample (JK1) has a relatively enriched isotopic composition ($^{87}Sr/^{86}Sr$ = 0.7040; $^{143}Nd/^{144}Nd$ = 0.5125). In some respects, the Sr-Nd isotope data mirror the trace element characteristics of the xenoliths from southern Yemen and the Red Sea and allow us to further refine the subdivisions made on the basis of the trace element chemistry. Minerals from the hydrous Ataq xenoliths have a very restricted range in Sr-Nd isotope ratios ($^{87}Sr/^{86}Sr$ = 0.7034–0.7036; $^{143}Nd/^{144}Nd$ = 0.51288–0.51298; Fig. 3). No evidence for any intermineral Sr-Nd isotopic heterogeneity exists in the hydrous Ataq xenoliths despite significant differences in Rb/Sr and Sm/Nd ratios, e.g., JK8 clinopyroxene, amphibole, and apatite $^{147}Sm/^{144}Nd$ = 0.083–0.156.

Studies of ocean ridge volcanism in the Gulf of Aden (Schilling et al., 1992) and recent intraplate volcanism in Afar (Vidal et al., 1991; Deniel et al., 1994) have helped define the geochemical signature of the Afar plume. As noted by Baker et al. (1998), minerals from the hydrous Ataq xenoliths have Sr-Nd isotope ratios identical to that assigned to the Afar plume (Fig. 3A). Such isotopic characteristics are similar, but not identical, to many type 1B (i.e., clinopyroxenes from peridotites with $Ce/Yb_N > 1$) xenoliths from Saudi Arabia and Zabargad Island. The Yemen samples and the Afar plume have slightly higher $^{87}Sr/^{86}Sr$ ratios at the same $^{143}Nd/^{144}Nd$ ratios compared with the Saudi Arabian samples (Fig. 3B).

Anhydrous Ataq xenoliths contain minerals with the same Sr-Nd isotope ratios as minerals from the hydrous Ataq xenoliths (JK4 and JK10) or, in the case of JK1, the minerals have an enriched Sr-Nd isotopic composition (Table 1; Fig. 3). The different Sr-Nd isotope ratios of JK1 compared with JK4 and JK10

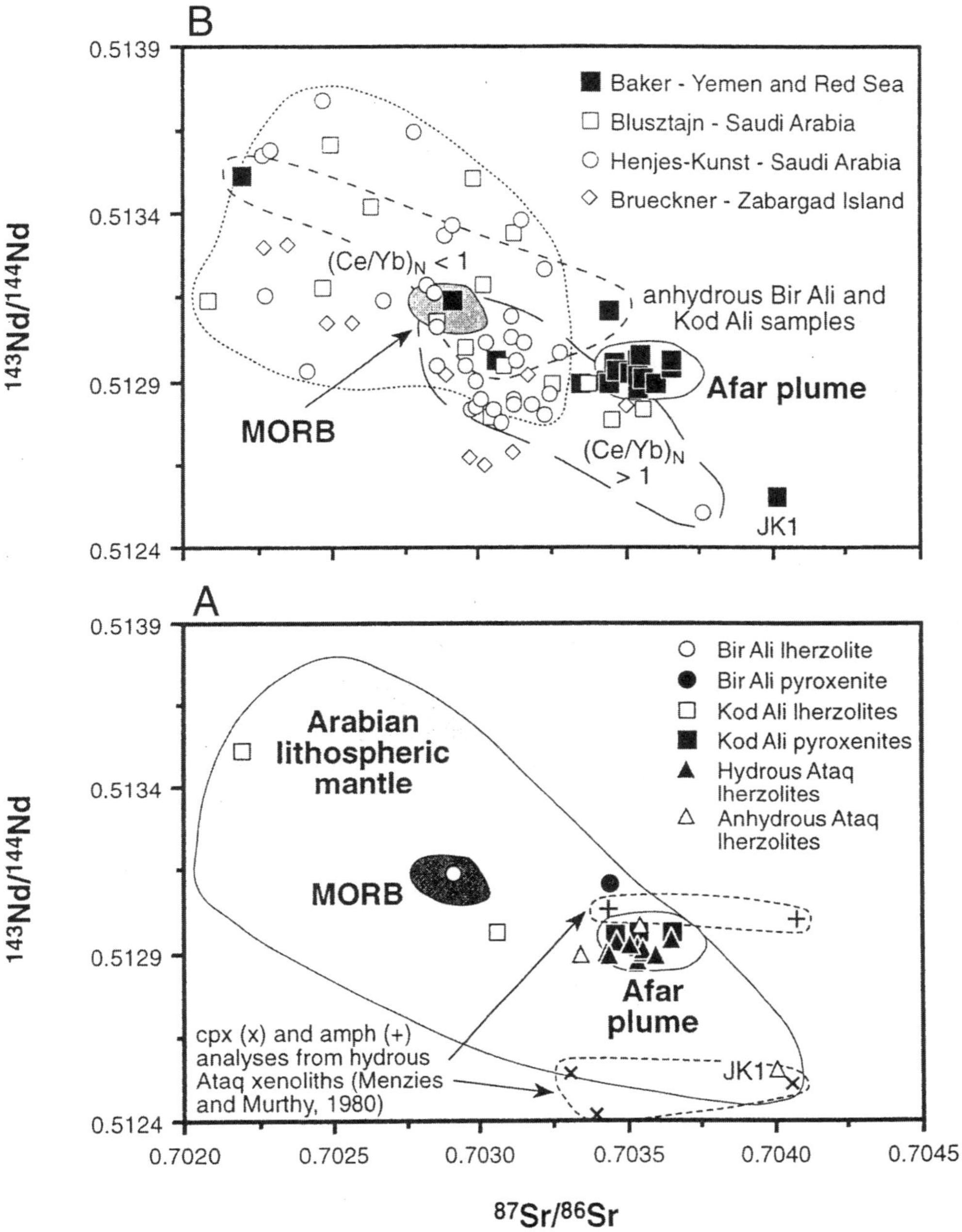

Figure 3. A: Sr-Nd isotopic composition of minerals separated from southern Red Sea and Yemen mantle xenoliths compared with isotopic compositions of mid-ocean ridge basalt (MORB), Afar plume, and previously published analyses of Arabian lithospheric mantle. See text for references and discussion. B: Sr-Nd isotopic composition of minerals separated from southern Red Sea and Yemen mantle xenoliths compared with previously published analyses of mantle rocks from Saudi Arabia and Zabargad Island. Data sources: MORB—Schilling et al. (1992); Afar plume—Vidal et al. (1991), Schilling et al. (1992), Deniel et al. (1994), and Baker et al. (1996a); Arabian lithospheric mantle—Brueckner et al. (1988), Henjes-Kunst et al. (1990), and Blusztajn et al. (1995).

are consistent with the subtle differences in the trace element chemistry of the clinopyroxenes outlined in the previous section. The isotopic composition of JK1 is similar to that of a single clinopyroxene from a xenolith brought to the surface in the Harrat As Shamah volcanic field, Jordan (Henjes-Kunst et al., 1990).

Minerals from the anhydrous Bir Ali and Kod Ali peridotites are characterized by generally higher $^{143}Nd/^{144}Nd$ and, to a lesser extent, lower $^{87}Sr/^{86}Sr$ ratios than those from the hydrous Ataq peridotites (Fig. 3A). Sr and Nd isotope ratios of some anhydrous samples overlap that of mid-ocean ridge basalts (MORB) from the Gulf of Aden. However, sample KA has much lower $^{87}Sr/^{86}Sr$ and higher $^{143}Nd/^{144}Nd$ ratios, respectively, than the local depleted-mantle MORB reservoir and is similar to clinopyroxenes from type 1A ($Ce/Yb_N < 1$) mantle xenoliths sampled elsewhere in Arabia (Fig. 3B). The high $^{143}Nd/^{144}Nd$ ratios of the Bir Ali and Kod Ali clinopyroxenes are in keeping with their light REE depleted character. Previously published Sr-Nd isotopic data for whole-rock Kod Ali peridotites are similar to that presented here, although Sr isotope ratios extend to marginally higher values ($^{87}Sr/^{86}Sr$ = 0.7037–0.7047; $^{143}Nd/^{144}Nd$ = 0.51294–0.51299; Menzies and Hawkesworth, 1987). Pyroxenites from Bir Ali and Kod Ali have Sr-Nd isotope ratios that are similar to those of the hydrous Ataq xenoliths and the Afar plume, although they, and in particular the Bir Ali pyroxenite, have slightly higher $^{143}Nd/^{144}Nd$ ratios.

The Sr-Nd isotope data helps us consolidate the subdivisions of the xenolith suite based on their elemental characteristics. Overall, the Sr-Nd isotope data for Ataq, Bir Ali, and Kod Ali show some overlap with data from Saudi Arabia and Zabargad Island, but the majority of the data trend from the MORB field toward Sr-Nd isotopic compositions similar to those of the Afar plume. The data define a tight linear anticorrelation on a Sr-Nd isotope diagram with a range in Sr and Nd isotopes similar to those in the data from Saudi Arabia and Zabargad Island.

Previous research on the Ataq and Kod Ali rocks was undertaken by Menzies and Murthy (1980) and Menzies and Hawkesworth (1987). Our results for the hydrous Ataq xenoliths contrast markedly with those of Menzies and Murthy (1980), who observed marked isotopic heterogeneity between coexisting clinopyroxene and amphibole, the clinopyroxenes always having relatively enriched Sr-Nd isotopic compositions (Fig. 3A). We failed to reproduce the low $^{143}Nd/^{144}Nd$ ratios reported therein for clinopyroxenes separated from hydrous Ataq xenoliths. Sr isotope ratios for clinopyroxene and amphibole reported by Menzies and Murthy (1980) range from 0.7033 to 0.7041, and are clearly at odds with those presented here, including reanalysis of minerals from one of the same samples (AT15) (see Table 2 footnotes). We attribute this discrepancy to the lack of acid leaching of samples and to the small size of the analyzed samples (0.01 g) in the older study. Menzies and Hawkesworth (1987) reported Sr-Nd isotopic data for two whole-rock peridotites from Kod Ali island. Although these peridotites have $^{143}Nd/^{144}Nd$ ratios within the range of that reported here for clinopyroxene from the Kod Ali samples, they have much higher $^{87}Sr/^{87}Sr$ ratios (0.7037–0.7047). It seems likely that the whole-rock samples were not acid leached prior to analysis.

Pb isotope data. Pb isotope ratios vary widely in the xenolith suite; e.g., $^{206}Pb/^{204}Pb$ = 17.3–19.0 (Fig. 4). Overall, the data define an array subparallel to the Northern Hemisphere Reference Line (NHRL) in the $^{208}Pb/^{204}Pb$ versus $^{206}Pb/^{204}Pb$ diagram, the majority of samples having Pb isotope ratios more radiogenic than MORB (Fig. 4). Other samples extend to more unradiogenic Pb isotope ratios that either overlap the MORB field or extend to more unradiogenic Pb isotope ratios.

In detail, minerals separated from the xenoliths define two broad groups in Pb isotopic space, and these groupings appear to link to the subdivisions based on trace element and Sr-Nd isotope ratios (Fig. 4).

Radiogenic Pb isotopes. Minerals from the hydrous Ataq peridotites have relatively radiogenic and uniform Pb isotope ratios ($^{206}Pb/^{204}Pb$ = ~18.7–19.0), despite large differences in U/Pb and Th/Pb ratios between the various samples. For example, no evidence exists for any intermineral Pb isotopic heterogeneity in sample JK8 despite substantial variation in U/Pb and Th/Pb ratios between the amphibole and apatite separates (e.g., $^{238}U/^{204}Pb$ [μ] = 7.49–137). Pb isotope ratios of the hydrous Ataq peridotites, like their Sr-Nd isotope ratios, are precisely the same as those assigned to the Afar plume and Oligocene flood basalts from Yemen, which were derived from this plume (Vidal et al., 1991; Schilling et al., 1992; Baker et al., 1996a, 1996b, 1998). Clinopyroxenes from two of the anhydrous Ataq peridotites (JK4 and JK10) also have Pb isotope ratios that are precisely the same as the hydrous Ataq peridotites and the Afar plume.

The Kod Ali pyroxenites also have similar Pb isotope ratios to these Ataq samples and the Afar plume, but have slightly elevated $^{207}Pb/^{204}Pb$ compared with the hydrous Ataq xenoliths at similar $^{206}Pb/^{204}Pb$ ratios. However, the Kod Ali pyroxenite clinopyroxenes overlap the Pb isotope ratios of the hydrous Ataq xenoliths in the $^{206}Pb/^{204}Pb$ versus $^{208}Pb/^{204}Pb$ plot (Fig. 4). Given the very low Pb contents of these samples (~0.01 ppm), and the ~2 standard deviation uncertainty in $^{207}Pb/^{204}Pb$ ratios illustrated in Figure 4 (as a result of mass fractionation), it is possible that the slightly higher $^{207}Pb/^{204}Pb$ ratios of the Kod Ali pyroxenites compared with these samples is not a real feature, but is actually a function of mass fractionation. Further Pb isotopic work is in progress to assess the true $^{207}Pb/^{204}Pb$ ratios of these samples.

Unradiogenic Pb isotopes. Bir Ali and Kod Ali anhydrous peridotites have $^{208}Pb/^{204}Pb$ and $^{206}Pb/^{204}Pb$ ratios close to those of MORB. One lherzolite from Kod Ali has a relatively high $^{207}Pb/^{204}Pb$ given its $^{206}Pb/^{204}Pb$ ($\Delta 7/4 = +14.4$) and lies considerably above the NHRL compared with the other samples. Clinopyroxenes from the anhydrous Bir Ali lherzolites and pyroxenites have relatively unradiogenic $^{206}Pb/^{204}Pb$ ratios (Fig. 4) and high $\Delta 7/4$ values distinct from MORB and Afar plume compositions.

Clinopyroxene from the anhydrous Ataq peridotite, JK1, has a distinctive Pb isotopic signature, in keeping with its unusually enriched Sr-Nd isotopic composition. This sample has less radiogenic Pb isotope ratios than the other Ataq samples ($^{206}Pb/^{204}Pb$ = 18.3), and is marked by a particularly low $^{207}Pb/^{204}Pb$ ratio given its $^{206}Pb/^{204}Pb$ ratio ($\Delta 7/4 = -6.3$). Clinopyroxene from JK1 is also marked by the highest $^{238}U/^{204}Pb$ ($\mu = 33.2$) ratio of any clinopyroxene from this study.

No published Pb isotope data are available for comparable mantle xenoliths from the rest of the Arabian peninsula. However, a small number of whole-rock analyses of peridotites from Zabargad Island (northern Red Sea; Hamelin and Allègre, 1988) are also shown in Figure 4. Although these peridotites have variable Pb isotope ratios, they have Pb isotope ratios similar to those of the southern Yemen and Red Sea xenoliths, but are slightly closer to the NHRL in the $^{206}Pb/^{204}Pb$ versus $^{208}Pb/^{204}Pb$ plot. No complementary trace element or Sr-Nd isotope data are available for these samples, and they are not considered further here.

DISCUSSION

The discussion that follows integrates the xenolith data presented herein with those previously published on xenoliths from Saudi Arabia and Jordan and outlines the current understanding

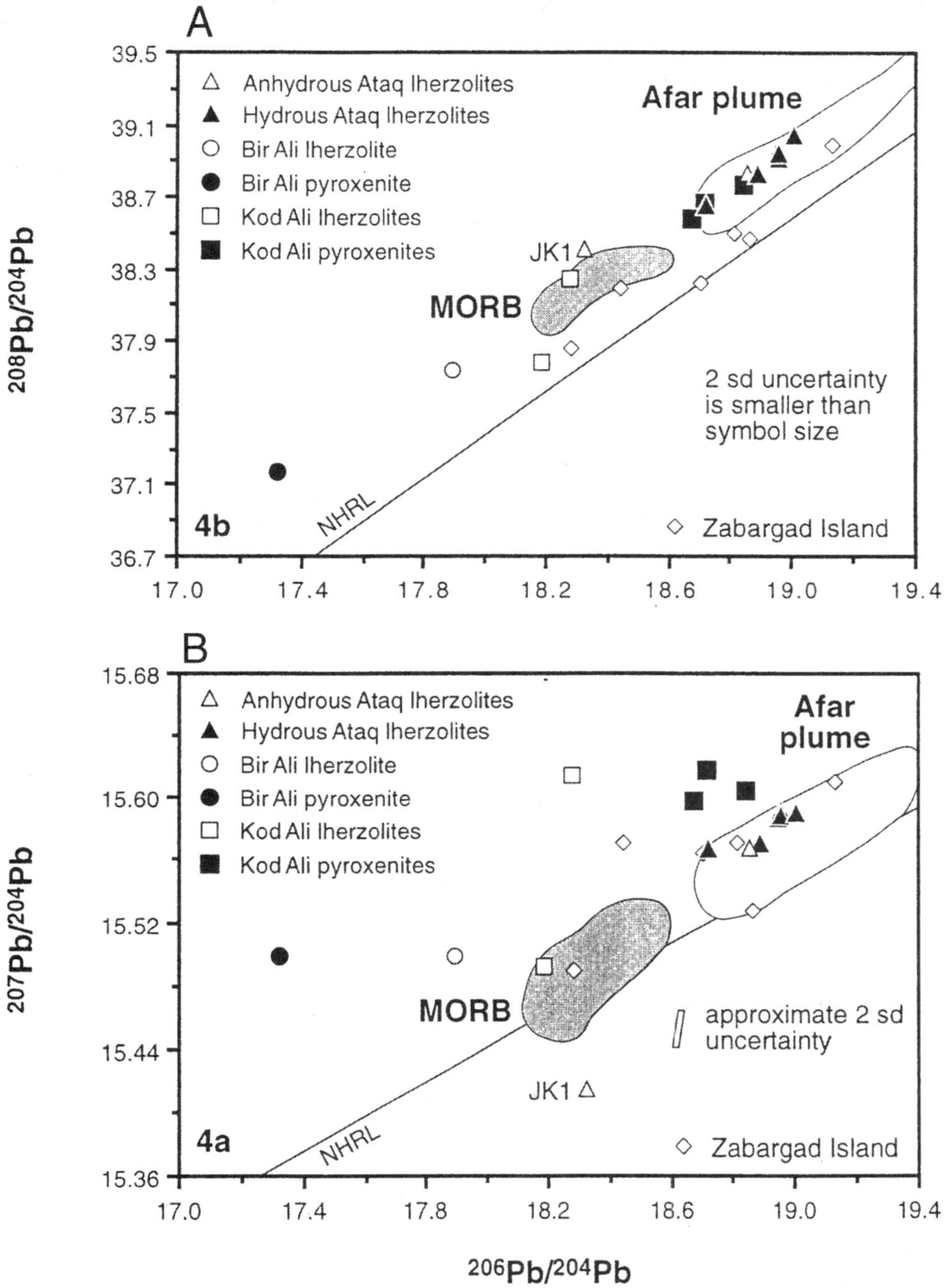

Figure 4. Pb isotopic systematics of minerals separated from southern Yemen and southern Red Sea mantle xenoliths compared with mid-ocean ridge basalt (MORB), Afar plume, and previously published analyses of Arabian lithospheric mantle. Data sources for MORB and Afar plume are same as Figure 3. Data for Zabargad Island are from Hamelin and Allègre (1988).

of the composition of the Arabian lithospheric mantle. Brueckner et al. (1988), Henjes-Kunst et al. (1990), and Blusztajn et al. (1995) proposed several closely related models for the formation and development of Arabian lithospheric mantle. We briefly review these here before proposing our preferred model, which integrates the southern Yemen and Red Sea xenolith data into these models. We stress that our model mostly reflects refinement and confirmation of some of the conclusions of these earlier studies. The possibility that the lithospheric mantle is a source for continental flood volcanism in Yemen (and Ethiopia) is explored.

Composition of the Arabian lithospheric mantle

The southern Yemen and Red Sea xenoliths have a wide range in elemental and isotopic compositions. The xenoliths are similar to those identified in other studies of Arabian lithospheric mantle and fall into several groups (Figs. 1, 2, 3, and 4).

Depleted components. These samples are marked by generally depleted trace element compositions, variably but often highly depleted Sr-Nd isotope ratios, and relatively unradiogenic Pb isotope ratios. Type 1A xenoliths from Saudi Arabia and Jordan have similar chemical and Sr-Nd isotopic characteristics and, together with the Yemen samples, are close to a ca. 700 Ma isochron that passes through depleted mantle (Fig. 5). As noted by Brueckner et al. (1988), Henjes-Kunst et al. (1990), and Blusztajn et al. (1995), the linear data array can be interpreted as a mixing line or a mantle isochron, or possibly a combination of both. Some samples have affinities to MORB in their elemental characteristics, their Sr-Nd isotopes, and/or their Pb isotopes. These may represent accreted mantle material added to the lithosphere recently (Pan-African), and maintain an identity similar to the asthenosphere depending on elemental (Rb/Sr, Sm/Nd, U-Th/Pb) fractionation during accretion and related melting events shortly thereafter.

Enriched components. Three samples from the Zabargad Island peridotite and one xenolith from Jordan have moderately enriched Sr-Nd isotopic compositions like JK1, an anhydrous xenolith from Ataq. These samples also plot close to the ca. 700 Ma reference isochron (Fig. 5); however, JK1 and the Jordanian sample are below this isochron, suggesting they are older than 700 Ma, or that they had an initial Nd isotope ratio lower than depleted mantle at their time of formation, which is probable if melts were derived from a plume or subduction zone. We stress that this enriched component, at least as preserved in the xenolith record, forms a very minor component (<2%) of the Arabian lithospheric mantle as sampled by the volcanic rocks. Also significant is that chemical enrichment of these xenoliths is not particularly marked; the two most isotopically extreme xenoliths have clinopyroxenes that only contain 3–5 ppm Nd, which is comparable to concentrations in clinopyroxenes from the isotopically depleted or type 1A xenoliths from the Arabian Peninsula.

Hydrous and anhydrous xenoliths from Ataq have Sr-Nd-Pb isotopic characteristics consistent with a petrogenetic link to mantle material like that associated with the Afar plume (Baker et al., 1998). The hydrous Ataq xenoliths are marked by clear evidence for modal metasomatism (Cr diopside + amphibole ± apatite) and strong chemical enrichment in all incompatible trace elements with the exception of the high field strength elements. In contrast, the anhydrous xenoliths exhibit no evidence for modal metasomatism, but contain clinopyroxene with enrichment in U, Pb, Sr, and, to a lesser extent, Th and the light REEs (La) relative to other trace elements. These metasomatized Ataq xenoliths have Sr-Nd isotopic characteristics similar to type 1B xenoliths from Saudi Arabia. Only limited and incomplete chemical data exist for type 1B xenoliths from Saudi Arabia (Blusztajn et al., 1995), but these exhibit some of the features observed in the Ataq xenoliths. Common features include enrichment in Sr and the light REEs ± middle REEs without concomitant enrichment in Zr and Ti. Some pyroxenites from Kod Ali have Sr-Nd-Pb isotope ratios similar to the Ataq peridotites and the Afar plume, but they are marked by chemical depletion.

General model for formation and modification of Arabian lithospheric mantle

Previous models. Brueckner et al. (1988) studied the peridotite massif exposed on Zabargad Island in the northern Red Sea, and proposed that these mantle rocks define a mixing array in a $^{147}Sm/^{144}Nd$ versus $^{143}Nd/^{144}Nd$ plot (Fig. 5) between a depleted mantle MORB-type component and a low $^{143}Nd/^{144}Nd$ component. The ca. 675 Ma age defined by the apparent isochron was interpreted to reflect the time at which a homogeneous mantle reservoir differentiated into two or more reservoirs, either within the mantle or lower crust, with the same $^{143}Nd/^{144}Nd$ ratios but different Sm/Nd ratios. With time this el-

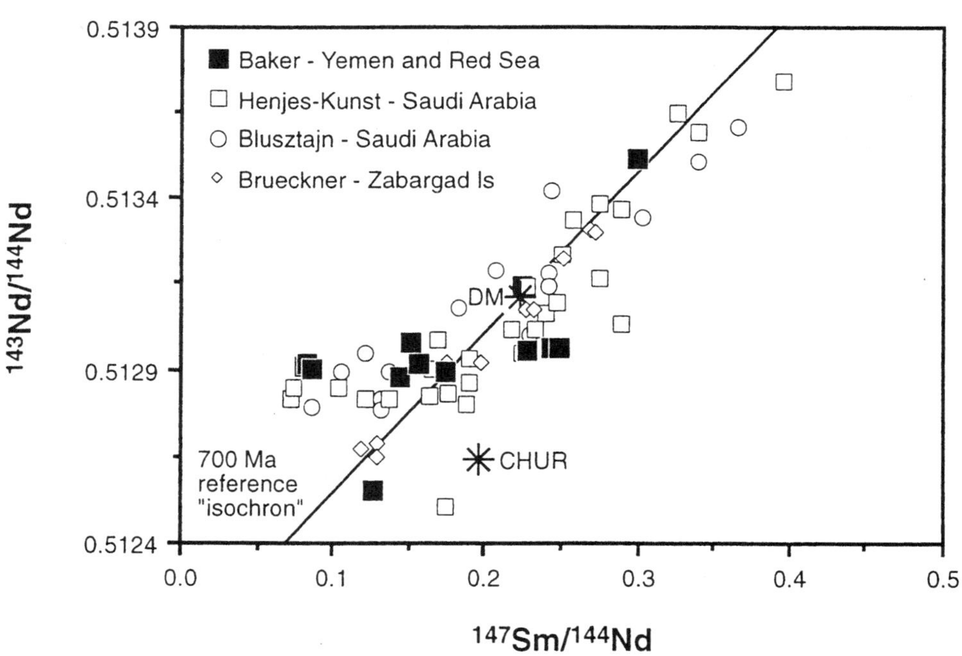

Figure 5. $^{147}Sm/^{144}Nd$-$^{143}Nd/^{144}Nd$ correlation exhibited by Arabian lithospheric mantle xenoliths. Much of the Arabian lithospheric mantle data plot close to a 700 Ma reference isochron for depleted mantle. However, most of Ataq xenoliths and many type 1B xenoliths from Saudi Arabia plot to left of this isochron, suggesting recent enrichment with component with $^{143}Nd/^{144}Nd$ = ~0.5128–0.5129. Data sources as in Figure 3. DM = present-day depleted mantle composition; CHUR = present-day chondritic uniform reservoir composition.

emental heterogeneity evolved into isotopic heterogeneity. During the early stages of Red Sea rifting these heterogeneous reservoirs underwent mixing. However, there would be no reason to invoke any later mixing process if the initial differentiation events ca. 675 Ma were on a small scale and were able to survive subsequent homogenization while residing at lithospheric mantle temperatures and pressures.

Henjes-Kunst et al. (1990) studied the Sr-Nd isotopic composition of spinel lherzolites entrained in Cenozoic volcanic fields erupted in Saudi Arabia and Jordan. These authors postulated that the mantle represented by the type 1A xenoliths beneath this region fractionated from a chondritic type source 1.3–1.9 Ga, and ca. 700 Ma a further differentiation event occurred at the time of Pan-African crustal growth. The type 1B xenoliths were interpreted to have various metasomatic overprints that were the result of subduction modification (producing high Sr/Nd ratios and low Ti contents) or modification by alkaline basalts with ocean island basalt (OIB)–like composition (high Ti contents) in latest Pan-African time. Old enrichment events by OIB-like melts may have resulted in formation of samples with relatively high Rb/Sr and low Sm/Nd ratios, leading to low $^{143}Nd/^{144}Nd$ and slightly elevated $^{87}Sr/^{86}Sr$ ratios compared with the other xenoliths. Mobilization of the various enriched components generated in late Pan-African time, during the early stages of Red Sea rifting (30 Ma), then led to the observed chemical and isotopic variability of the Arabian lithospheric mantle.

Blusztajn et al. (1995) presented further Sr-Nd isotope data and a suite of trace elements for clinopyroxenes separated from spinel lherzolites in Saudi Arabia. These authors concluded that the type 1A xenoliths represent long-term depleted lithospheric mantle that was generated from an already depleted (upper?) mantle source. Some of the type 1A xenoliths show evidence for incipient enrichment in Sr and the light REE. Blusztajn et al. (1995) stressed the difficulty in ruling out the possibility that the type 1A xenoliths were in fact representatives of modern MORB mantle. However, given the lack of any petrological and chemical evidence for preexisting garnet in these samples, they concluded that these xenoliths were not derived from the asthenosphere. Blusztajn et al. then considered the link between the type 1A xenoliths and Saudi Arabian crustal rocks by comparing Sm-Nd and Rb-Sr chondritic uniform reservoir (CHUR) model ages. Bluzstajn et al. failed to demonstrate any link between the lithospheric mantle and the crust on the basis of these ages.

Proposed model. Our preferred model to explain the chemical and isotopic variation of Arabian lithospheric mantle and, in particular, how it pertains to the origin of the components of the Arabian lithospheric mantle identified in our samples involves the following.

Depleted component. Formation of an Arabian lithospheric mantle protolith from an already depleted asthenospheric(?) mantle source occurred during accretion and growth of Pan-African continental crust ca. 700 Ma. The generally depleted character of the type 1A lithospheric mantle xenoliths is consistent with the chemically primitive and isotopically depleted nature of juvenile Neoproterozoic Pan-African crust (e.g., Duyverman et al., 1982). Similar depleted mantle Nd model ages for juvenile Pan-African crust and the type 1A xenoliths (Fig. 6) in this region support the cogenetic link between the Arabian lithospheric mantle protolith and growth of continental crust in Pan-African time. The chemically and isotopically depleted Bir Ali and Kod Ali lherzolites, and the Bir Ali pyroxenite, can be considered representative of original Pan-African lithospheric mantle, although sample BA8 may exhibit some evidence for recent incipient chemical enrichment (Fig. 2C).

We note that few xenoliths record Nd depleted mantle model ages that overlap those of older nucleii of continental crust that have been identified in Pan-African crustal basement (Fig. 6), suggesting that old cratonic Archean lithospheric mantle (1) did not survive major episodes of crustal growth (e.g., Pan-African accretion); (2) was subsequently removed by thermotectonic processes (e.g., Tertiary breakup and magmatism); (3) is not represented in the xenolith record due to biased sampling by intraplate magmas; or (4) is not represented in the biased sample set collected for this study, which focused on anhydrous and hydrous spinel lherzolites.

Enriched components. Sample JK1 has an enriched Sr-Nd isotopic composition that is suggestive of either enrichment by a component with an unusual isotopic composition or the relatively long term response to changes in Rb/Sr, Sm/Nd, and U/Pb ratios. Clinopyroxene from JK1 yields a depleted mantle Nd model age of 889 Ma, which can be considered a maximum estimate for the enrichment event that produced this sample. The Nd model age using a metasomatizing agent that had a bulk earth composition at the time of enrichment is somewhat younger (183 Ma). Clinopyroxene from JK1 has an unusually low Δ7/4 and a high μ value. Simple calculations show that this sample would have had a Pb isotopic composition that was at, or slightly above, the NHRL between 220 and 360 Ma (using $\mu = 0$ and 10 to back-calculate the NHRL Pb isotopic compositions). We conclude that sample JK1 reflects a relatively old enrichment event that occurred ca. 200–400 Ma, and involved enrichment by small degree partial melts with $\varepsilon Nd > 0$, but substantially less than depleted mantle. The rather smooth incompatible trace element pattern for clinopyroxene from JK1 suggests that the metasomatism was probably related to small degree melts from an asthenospheric or OIB source during intraplate volcanism rather than a subduction-driven metasomatic process.

The other anhydrous and hydrous Ataq peridotites must reflect more recent metasomatism than that responsible for the enriched component represented by JK1. Large variations in parent-daughter ratios such as Sm/Nd, U/Pb, and Th/Pb characterize minerals from these xenoliths, yet they are essentially isotopically homogeneous. Simple calculations suggest that the enrichment event that produced these samples can be no older than a few millions of years (Pb isotope ratios) or a few tens of million years (Nd isotope ratios). Although it might be argued that the Pb isotopic systematics could have been reset at mantle temperatures

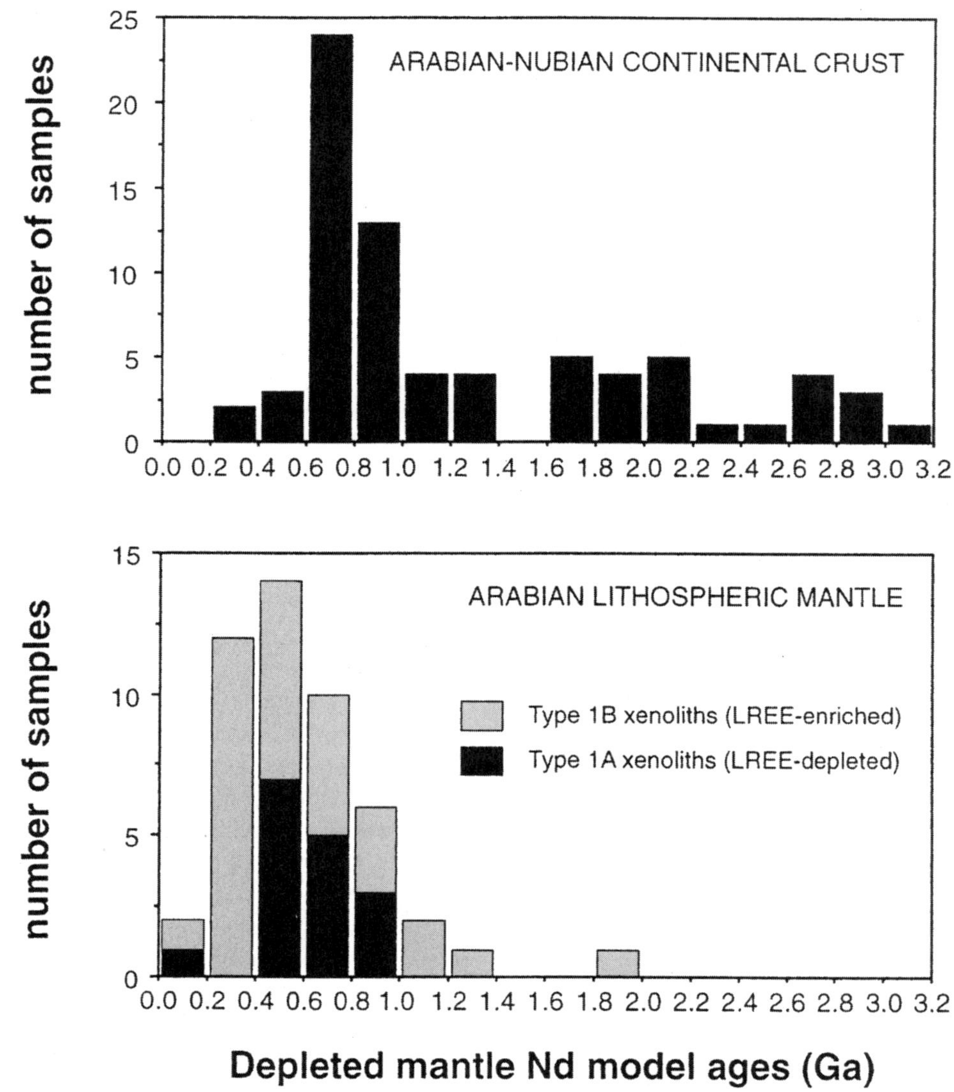

Figure 6. Histogram of depleted mantle Nd model ages for Arabian lithospheric mantle and continental crust from Sudan, Saudi Arabia, and Yemen compiled from literature. LREE—light rare earth element. Data sources for Arabian lithospheric mantle as in Figure 3. Data sources for Arabian continental crust are from Bokhari and Kramers (1981), Duyverman et al. (1982), Stern and Kröner (1993), Stacey and Hedge (1984), Kröner et al. (1987), Davidson and Wilson (1989), McGuire and Stern (1993), and Windley et al. (1996).

and pressures or during entrainment, it is unlikely that this could be the case for the Nd isotopic system. We concur with earlier studies of these rocks (Baker et al., 1998) that the evidence for recent enrichment is compelling and that the common Sr-Nd-Pb isotopic signature of the Afar plume, primary Oligocene flood basalts in this region and these xenoliths, reflects metasomatism of the depleted Arabian lithospheric mantle protolith by the Afar plume during or shortly after Oligocene flood volcanism in Yemen.

Differences between the hydrous and anhydrous (JK4 and JK10) Ataq xenoliths can be accounted for by invoking different metasomatizing agents. The hydrous Ataq xenoliths have been extensively modally metasomatized and overprinted by carbonatitic melts, which accounts for the extreme enrichment in most incompatible trace elements, but not the high field strength elements. Conversely, the anhydrous Ataq xenoliths, JK4 and JK10, only exhibit enrichment in fluid-mobile incompatible trace elements (U, very light REEs, Sr, and Pb). These samples underwent incipient cryptic metasomatism by fluids derived from a source with an isotopic signature similar to the carbonatitic melts that metasomatized the other xenoliths.

The similar Sr-Nd-Pb isotopic signature of the recently enriched Ataq xenoliths, other type 1B xenoliths from the Arabian lithospheric mantle, and the Afar plume is strong evidence that mantle plumes may have repeatedly played an important role in modifying Arabian lithospheric mantle since its formation ca. 700 Ma. The Afar plume may be a bigger feature beneath this region than hitherto realized, and may be responsible for more widespread recent metasomatism of Arabian lithospheric mantle than hitherto believed. This would require some heterogeneity of the Afar plume source, because subtle differences characterize the isotopic composition of the type 1B xenolith suite throughout Arabia. Alternatively, whereas some of this enrichment is demonstrably a relatively recent event (e.g., the modally metasomatized Ataq xenoliths), and was intimately associated with Oligocene flood volcanism and the Afar plume, it is also plausible that other parts of the Arabian lithospheric mantle represent older Proterozoic enrichment events (e.g., JK1 and other type 1B xenoliths). Perhaps enrichment of the Arabian lithospheric mantle by older mantle plumes with similar, but subtly different, isotopic compositions to the Afar plume produced the Arabian type 1B xenoliths (and JK1). However, the relatively homogeneous Nd isotopic composition of these xenoliths, which exhibit a range in Sm/Nd ratios, requires the enrichment to have occurred recently (<<200 Ma), or subsequent recent mixing of the lithospheric mantle to have produced the near horizontal trend in Figure 5.

It is interesting to compare the isotopic signature of the Arabian lithospheric mantle xenoliths from Yemen with those from Tanzania in the East Africa rift, because both areas are associated with Archean continental crust. Cohen et al. (1984) and Chesley et al. (1999) reported Sr-Nd-Pb-Os isotopic data for spinel and garnet facies mantle rocks from xenoliths entrained in young volcanic rocks in Tanzania close to the Tanzanian craton. The isotopic data clearly record the presence of Archean lithospheric mantle material below Tanzania (e.g., 2.8–2.0 Ga Re depletion ages for spinel; Nd isotope ratios as low as 0.5113), which contrasts with the lack of such material in association with the Archean crustal fragments preserved in Yemen. We suggest that the following three factors may account for the absence of preserved Archean lithospheric mantle beneath Yemen. (1) The Archean crustal fragments were detached from their lithospheric mantle roots during construction of Pan-African lithosphere 700 Ma or, more probably, (2) full-fledged continental rifting associated with the Afar plume beneath Yemen has successfully eroded the older lithospheric mantle record. (3) However, it is also possible that a Jurassic rifting event that affected Yemen and resulted in subsidence and subsequent deposition of a thick Jurassic to Cretaceous sequence of limestone and sandstone also served to remove any Archean lithospheric mantle.

Lower lithospheric contribution to continental flood or intraplate volcanism?

Extension of the lithosphere, heating by underlying thermal anomalies, or passage of melts produced at sublithospheric depths through the lithospheric mantle might all lead to a substantial contribution of the Arabian lithospheric mantle to Cenozoic flood and intraplate volcanism.

Rigorous evaluation of the possible contribution of Arabian lithospheric mantle requires assessment of three further problems. First, the primary chemical and isotopic signatures of flood and intraplate volcanism in this region need to be unequivocally identified. Second, if a common chemical and isotopic signature is shared by the lithospheric mantle xenoliths and volcanic rocks, does this reflect a lithospheric mantle contribution to the magmas or modification of the lithospheric mantle by the magmas? Third, can the xenoliths melt to produce sufficient volumes of magmas with appropriate major and trace characteristics of the continental volcanism?

Two important questions are relevant as to the role of the lithospheric mantle as a possible source for continental volcanism in this region. Some authors (e.g., Hart et al., 1989; Vidal et al., 1991; Deniel et al., 1994; Chazot and Bertrand, 1993) contend that flood basalts in Yemen and Ethiopia with $^{143}Nd/^{144}Nd$ = ~0.5125 are derived from the lithospheric mantle. The xenolith record can be used to assess whether this is plausible, and by what mechanism the lithospheric mantle might contribute to continental volcanism. Isotopically enriched xenoliths like JK1 are not a major component of the lithospheric mantle (<2%). Whole-scale melting of the lithospheric mantle would be likely to involve melt fractions of 0.1 or smaller. Partial melting of this scale, <<10% of the thickness of the lithospheric mantle (~70 km), could theoretically produce lavas to a maximum thickness of 140 m with the appropriately enriched Sr-Nd isotopic compositions. Furthermore, lithospheric contamination of sublithospheric-derived melts by small degree, highly enriched melts from samples like JK1, resulting in erupted lavas with chemical and isotopic signatures swamped by the lithospheric mantle, can be ruled out as a major process. Primary flood basalts from Yemen contain ≥10 ppm Nd (Baker et al., 1996a) and isotopically enriched xenoliths probably contain <0.25 ppm Nd (5% modal clinopyroxene). Small degree melting (0.1%) of the <10% of the 70 km of lithospheric mantle with the appropriate isotopic characteristics would only produce a 7-m-thick highly enriched melt with Nd = 250 ppm. Some 10% contamination of melts with $^{143}Nd/^{144}Nd \leq 0.5129$ and Nd = 10 ppm is required to produce a hybrid with $^{143}Nd/^{144}Nd \leq 0.5126$, limiting the thickness of any such hybrid produced to <70 m, which is insignificant compared with the total thickness of basalt erupted in this province.

However, one final point is of much greater significance when examining whether an enriched lithospheric mantle contribution to continental flood volcanism occurred in this region; the Sr-Nd isotopically enriched xenolith JK1 has an inappropriate composition in Pb isotopic space to generate the trends observed in the flood volcanic rocks that are attributed to an enriched lithospheric mantle contribution (unradiogenic Pb isotope ratios with high Δ7/4 and Δ8/4).

The second possible role for a lithospheric mantle source to continental volcanism in this region concerns the origin of the numerous intraplate volcanic fields located on the Arabian plate. Baker et al. (1997) concluded that such fields in Yemen are derived from the highly (chemically) enriched lithospheric mantle produced recently by metasomatism of the lithospheric mantle by the Afar plume. However, other intraplate fields are characterized by isotopic compositions subtly different from the Yemen intraplate volcanic fields (and Afar plume) and are also located beyond the presumed influence of this plume. Further consideration of the source of these other Arabian intraplate fields, possibly in Arabian lithospheric mantle enriched by older plumes, awaits detailed isotopic (in particular Hf-Pb) study of these volcanic rocks and their entrained type 1A and 1B xenoliths.

CONCLUSIONS

The Arabian lithospheric mantle is chemically and isotopically heterogeneous. Parts of the lithospheric mantle are clearly fertile and are enriched in volatiles and incompatible trace elements. Arabian lithospheric mantle largely has a depleted Sr-Nd isotopic signature that reflects stabilization of the lithospheric mantle from a depleted mantle source during and shortly after Pan-African crustal accretion. No unequivocal remnants of ancient pre-Pan-African lithospheric mantle that might have been accreted with older Archean cratonic fragments are present in the xenolith suites. Some peridotites record isotopic and/or incompatible trace element evidence for post-Pan-African enrichment events of different ages. The enrichment events appear to be the result of metasomatism by OIB or plume-like melts rather than Pan-African subduction processes. The most recent enrichment event characterizes xenoliths from Yemen that were enriched by melts and fluids from the Afar plume during or shortly after Oligocene flood volcanism. In many respects, these data complement, but further refine, previous models for development of the Arabian lithospheric mantle (Henjes-Kunst et al., 1990; Blusztajn et al., 1995).

Examination of the chemical and isotopic composition of the Arabian lithospheric mantle supports the view that isotopically enriched continental flood volcanism in this region was the result of crustal contamination (Baker, 1996; Baker et al., 1996a, 1997, 2000) rather than an enriched lithospheric mantle contribution. However, the lithospheric mantle may have been an important source for smaller volume postbreakup intraplate volcanism throughout the Arabian plate.

ACKNOWLEDGMENTS

We acknowledge the support of the Association of Commonwealth Universities, The Royal Society, British Petroleum, and the Natural Environment Research Council in the pursuance of this research.

APPENDIX 1. ANALYTICAL TECHNIQUES

Trace element data

Trace element data were determined by one or more of the following four techniques.

Isotope dilution inductively coupled plasma–mass spectrometric (ID ICP-MS) analysis. Trace element analyses were determined on acid-leached bulk mineral separates, taken into solution by attack with HF/HNO_3 and evaporation with HNO_3, HCl, and HNO_3, before final dissolution in 10% HNO_3. The solution was spiked with mixed Zr-Ba and U-Pb spikes: following measurement of isotope ratios at the Natural Environment Research Council ICP-MS facility, concentrations of Zr, Ba, U, and Pb were determined by isotope dilution (~4% error). Concentrations of Rb, Sr, Y, Nb, and Th (5%–8% error) were determined by ratio to Zr or U, with mass bias calibrated by international standards.

Ion microprobe (SIMS) analysis. SIMS analyses were performed with a Cameca IMS 4f ion microprobe at the University of Edinburgh, using an 8 nA primary beam of O^- and a beam size of 0–25 μm. Fuller details of the analytical technique are described by Chazot et al. (1996a, 1996b). Precision and accuracy of the ion microprobe data are better than ±15% for Hf, ±15% for Nb and Ba in the trace-element-depleted clinopyroxenes, and ±10% for all other elements.

Inductively coupled plasma–mass spectrometric (ICP-MS) analysis. ICP-MS analyses on a small number of samples were made on unleached mineral separates at Universite Blaise Pascal (France).

Thermal ionization mass spectrometric isotope dilution (TIMS ID) analysis. TIMS ID analyses for the rare earth elements (REE), Rb, Sr, U, and Pb, were carried out at the Royal Holloway University of London Radiogenic Isotope Laboratory (UK). Acid-leached mineral separates were taken into solution by standard acid-digestion techniques (HF/HNO_3 followed by evaporation of HNO_3, HCl, and HNO_3) and the resulting solution quantitatively split into fractions for TIMS ID analysis and Sr-Nd-Pb isotopic analysis. The TIMS ID fraction was spiked with a mixed REE spike, a mixed U-Pb spike, and a mixed Rb-Sr spike. U-Pb-Rb-Sr-REE were progressively separated using standard anion (REE-U-Pb) and cation (Rb-Sr) exchange techniques. Reproducibility of ID analyses is typically better than ±0.5%. Blanks for ID and isotopic analyses were typically less than: Rb, 0.5 ng; Sr, 1.0 ng; REE, La, 230 pg; Ce, 430 pg; Nd, 170 pg; Sm, 20 pg; Eu, 4 pg; Dy, 20 pg; Er, 12 pg; Yb, 12 pg; U, 0.5 ng; Pb, 0.5 ng. These blanks are negligible with the exception of Rb in samples JK4, JK10, BA8, KA, and K2 clinopyroxene, and U and Pb in the extremely trace-element-depleted samples (BA8, KA, K2), and trace element concentrations for all these samples were blank corrected.

In many cases, trace element analyses were determined on the same samples using multiple techniques. Trace element concentrations determined using the different techniques often exhibit considerable variation for individual elements, in particular: (1) low abundance elements in the depleted clinopyroxenes (e.g., Rb, Ba, Pb, Nb); (2) differences observed between in situ analyses and analyses of bulk mineral separates where the bulk mineral separates can clearly contain trace element rich inclusions or contaminants (e.g., note the high Ba content of JK8 clinopyroxene determined by ID ICP-MS compared with the SIMS value); (3) differences between leached (ID ICP-MS and TIMS ID) and unleached analyses (ICP-MS) of mineral separates (leached mineral separates almost without exception have higher trace element abundances); (4) relatively high Pb contents determined by ICP-MS, even in incompatible trace element depleted minerals that relate to Pb memory levels of the

ICP-MS technique. However, regardless of the analytical technique used, the trace element data clearly define the trace element differences between the minerals from the hydrous Ataq, anhydrous Ataq, and the anhydrous Bir Ali and Kod Ali xenoliths discussed herein.

Sr-Nd-Pb isotope data

Sr-Nd-Pb isotopic analyses were made on hand-picked, acid-leached mineral separates as previously described. Details of the mass spectrometry were described in Thirlwall (1991a, 1991b). Blank corrections were only necessary for the Pb isotopic analyses of samples BA8, KA, and K2, but are generally small compared with the large difference in Pb isotopic composition these samples have compared to the Ataq mineral separates. Reproducibilities of isotopic analyses monitored by multiple analyses of standards are better than: $^{87}Sr/^{86}Sr$, ±0.00002; $^{143}Nd/^{144}Nd$, ±0.000012; $^{206}Pb/^{204}Pb$, ±0.010; $^{207}Pb/^{204}Pb$, ±0.012; $^{208}Pb/^{204}Pb$, ±0.028. Reproducibility of Pb isotope ratios determined on low-level samples may be as much as 100% greater than these values.

REFERENCES CITED

Arndt, N.T., and Christensen, U.R., 1992, The role of lithospheric mantle in continental flood volcanism: Thermal and geochemical constraints: Journal of Geophysical Research, v. 97, p. 10967–10981.

Baker, J.A., 1996, Stratigraphy, geochronology and geochemistry of Cenozoic volcanism in western Yemen [Ph.D. thesis]: London, University of London, 386 p.

Baker, J.A., Snee, L.W., and Menzies, M.A., 1996a, A brief Oligocene period of flood volcanism in Yemen: Implications for the duration and rate of continental flood volcanism at the Afro-Arabian triple junction: Earth and Planetary Science Letters, v. 138, p. 39–55.

Baker, J.A., Thirlwall, M.F., and Menzies, M.A., 1996b, Sr-Nd-Pb isotopic and trace element evidence for crustal contamination of plume-derived flood basalts: Oligocene flood volcanism in western Yemen: Geochimica et Cosmochimica Acta, v. 60, p. 2559–2581.

Baker, J.A., Chazot, G., Menzies, M., and Thirlwall, M.F., 1998, Metasomatism of the shallow mantle beneath Yemen by the Afar plume: Implications for mantle plumes, flood volcanism, and intraplate volcanism: Geology, v. 26, p. 431–434.

Baker, J.A., Menzies, M.A., Thirlwall, M.F., and Macpherson, C.G., 1997, Petrogenesis of Quaternary intraplate volcanism, Sana'a, Yemen: Plume-lithosphere interaction and polybaric melt hybridisation: Journal of Petrology, v. 38, p. 1359–1390.

Baker, J.A., Macpherson, C.G., Menzies, M.A., Thirlwall, M.F., Al-Kadasi, M., and Mattey, D.P., 2000, Resolving crustal and mantle contributions to continental flood volcanism, Yemen: Constraints from mineral oxygen isotope data: Journal of Petrology, v. 41, p. 1805–1820.

Blusztajn, J., Hart, S.R., Shimuzu, N., and McGuire, A.V., 1995, Trace-element and isotopic characteristics of spinel peridotite xenoliths from Saudi Arabia: Chemical Geology, v. 123, p. 53–65.

Bokhari, F.Y., and Kramers, J.D., 1981, Island-arc character and late Precambrian age of volcanics at Wadi Shwas, Hijaz, Saudi Arabia: Geochemical and Sr and Nd isotopic evidence: Earth and Planetary Science Letters, v. 54, p. 409–422.

Brueckner, H.K., Zindler, A., Seyler, M., and Bonatti, E., 1988, Zabargad and the isotopic evolution of the sub–Red Sea mantle and crust: Tectonophysics, v. 150, p. 163–176.

Chazot, G., and Bertrand, H., 1993, Mantle sources and magma–continental crust interactions during early Red Sea–Aden rifting in southern Yemen: Elemental and Sr, Nd, Pb isotope evidence: Journal of Geophysical Research, v. 98, p. 1819–1835.

Chazot, G., Menzies, M.A., and Harte, B., 1996a, Determination of partition coefficients between apatite, clinopyroxene, amphibole, and melt in natural spinel lherzolites from Yemen: Implications for wet melting of the lithospheric mantle: Geochimica et Cosmochimica Acta, v. 60, p. 423–437.

Chazot, G., Menzies, M.A., and Harte, B., 1996b, Silicate glasses in spinel lherzolites from Yemen: Origin and chemical composition: Chemical Geology, v. 134, p. 159–179.

Chazot, G., Lowry, D., Menzies, M.A., and Mattey, D.P., 1997, Oxygen isotopic composition of megacrysts and minerals in hydrous and anhydrous mantle xenoliths from Yemen and Nunivak Island: Geochimica et Cosmochimica Acta, v. 61, p. 161–169.

Chesley, J.T., Rudnick, R.L., and Lee, C.-T., 1999, Re-Os systematics of mantle xenoliths from the East African rift: Age, structure, and history of the Tanzanian craton: Geochimica et Cosmochimica Acta, v. 63, p. 1203–1271.

Cohen, R.S., O'Nions, R.K., and Dawson, J.B., 1984, Isotope geochemistry of xenoliths from East Africa: Implications for development of mantle reservoirs and their interaction: Earth and Planetary Science Letters, v. 68, p. 209–220.

Davidson. J.P., and Wilson, I.R., 1989, Evolution of an alkali basalt–trachyte suite from Jebel Marra volcano, Sudan, through assimilation and fractional crystallization: Earth and Planetary Science Letters, v. 95, p. 141–160.

Deniel, C., Vidal, P., Coulon, C., Vellutini, P.-J., and Piguet, P., 1994, Temporal evolution of mantle sources during continental rifting: The volcanism of Djibouti (Afar): Journal of Geophysical Research, v. 99, p. 2853–2869.

Duyverman, H.J., Harris, N.B.W., and Hawkesworth, C.J., 1982, Crustal accretion in the Pan-African: Nd and Sr isotope evidence from the Arabian shield: Earth and Planetary Science Letters, v. 59, p. 315–326.

Gallagher, K., and Hawkesworth, C.J., 1992, Dehydration melting and the generation of continental flood basalts: Nature, v. 358, p. 57–59.

Hamelin, B., and Allègre, C.J., 1988, Lead isotope study of orogenic lherzolite massifs: Earth and Planetary Science Letters, v. 91, p. 117–131.

Hart, W.K., Woldegabriel, G., Walter, R.C., and Mertzman, S.A., 1989, Basaltic volcanism in Ethiopia: Constraints on continental rifting and mantle interactions: Journal of Geophysical Research, v. 94, p. 7731–7748.

Henjes-Kunst, F., Altherr, R., and Baumann, A., 1990, Evolution and composition of the lithospheric mantle underneath the western Arabian peninsula: Constraints from Sr-Nd isotope systematics of mantle xenoliths: Contributions to Mineralogy and Petrology, v. 105, p. 460–472.

Hergt, J., Chappell, B.W., McCulloch, M.T., MacDougall, I., and Chivas, A.R., 1989, Geochemical and isotopic constraints on the origin of the Jurassic dolerites of Tasmania: Journal of Petrology, v. 30, p. 841–883.

Hutchinson, R., and Gass, I.G., 1971, Mafic and ultramafic inclusions associated with undersaturated basalt on Kod Ali Island, southern Red Sea: Contributions to Mineralogy and Petrology, v. 31, p. 94–101.

Kröner, A., Stern, R.J., Dawoud, A.S., Compston, W., and Reischmann, T., 1987, The Pan-African continental margin in northeastern Africa: Evidence from a geochronological study of granulites at Sabaloka, Sudan: Earth and Planetary Science Letters, v. 85, p. 91–104.

Lightfoot, P.C., Hawkesworth, C.J., Devey, C.W., Rogers, N.W., and Van Calsteren, P.W., 1990, Source and differentiation of Deccan Trap lavas: Implications of geochemical and mineral chemical variations: Journal of Petrology, v. 31, p. 1165–1200.

Lightfoot, P.C., Hawkesworth, C.J., Hergt, J., Naldrett, A.J., Gorbachev, N.S., Fedorenko, V.A., and Doherty, W., 1993, Remobilisation of the continental lithosphere by a mantle plume: Major-, trace element and Sr-, Nd-, Pb-isotopic evidence from picritic and tholeiitic lavas of the Noril'sk District, Siberian Traps, Russia: Contributions to Mineralogy and Petrology, v. 114, p. 171–188.

McGuire, A.V., and Stern, R.J., 1993, Granulite xenoliths from western Saudi Arabia: The lower crust of the late Precambrian Arabian-Nubian shield: Contributions to Mineralogy and Petrology, v. 114, p. 395–408.

Menzies, M., 1992, The lower lithosphere as a major source for continental flood basalts: A re-appraisal, *in* Alabaster, T., and Pankhurst, R.J., eds., Magmatism and the causes of continental break-up: Geological Society [London] Special Publication 68, p. 31–39.

Menzies, M.A., and Hawkesworth, C.J., 1987, Upper mantle processes and composition, *in* Nixon, P.H., ed., Mantle xenoliths: Chichester, England, Wiley, p. 725–738.

Menzies, M.A., and Murthy, V.R., 1980, Nd and Sr isotope geochemistry of hydrous mantle nodules and their host alkali basalts: Implications for local heterogeneities in metasomatically veined mantle: Earth and Planetary Science Letters, v. 46, p. 323–334.

Nakamura, N., 1974, Determination of REE, Ba, Mg, Na and K in carbonaceous and ordinary chondrites: Geochimica et Cosmochimica Acta, v. 38, p. 757–775.

Saunders, A.D., Storey, M., Kent, R.W., and Norry, M.J., 1992, Consequences of plume-lithosphere interaction, *in* Alabaster, T., and Pankhurst, R.J., eds., Magmatism and the causes of continental break-up: Geological Society [London] Special Publication 68, p. 41–60.

Schilling, J.-G., Kingsley, R.H., Hanan, B., and McCully, B.L., 1992, Nd-Sr-Pb isotopic variations along the Gulf of Aden: Evidence for mantle plume–continental lithosphere interaction: Journal of Geophysical Research, v. 97, p. 10927–10966.

Stacey, J.S., and Hedge, C.E., 1984, Geochronologic and isotopic evidence for early Proterozoic crust in the eastern Arabian shield: Geology, v. 12, p. 310–313.

Stein, M., and Hofmann, A., 1993, Fossil plume head beneath the Arabian lithosphere: Earth and Planetary Science Letters, v. 114, p. 193–209.

Stern, R.J., and Kröner, A., 1993, Late Precambrian crustal evolution in NE Sudan: Isotopic and geochronologic constraints: Journal of Geology, v. 101, p. 555–574.

Sun, S.S., and McDonough, W.F., 1989, Chemical and isotopic systematics of oceanic basalts: Implications for mantle composition and processes, *in* Saunders, A.D., and Norry, M.J., eds., Magmatism in the ocean basins: Geological Society [London] Special Publication 42, p. 313–345.

Thirlwall, M.F., 1991a, Long-term reproducibility of multicollector Sr and Nd isotope ratio analyses: Chemical Geology (Isotope Geoscience Section), v. 94, p. 85–104.

Thirlwall, M.F., 1991b, High precision multicollector isotopic analyses of low levels of Nd as oxide: Chemical Geology (Isotope Geoscience Section), v. 94, p. 13–22.

Varne, R., 1970, Hornblende lherzolite and the upper mantle: Contributions to Mineralogy and Petrology, v. 27, p. 45–51.

Varne, R., and Graham, A.L., 1971, Rare earth abundances in hornblende and clinopyroxene of a hornblende lherzolite xenolith: Implications for upper mantle fractionation processes: Earth and Planetary Science Letters, v. 13, p. 11–18.

Vidal, P., Deniel, C., Vellutini, P.J., Piguet, P., Coulon, C., Vincent, J., and Audin, J., 1991, Changes of mantle source in the course of a rift evolution: Geophysical Research Letters, v. 18, p. 1913–1916.

Windley, B.F., Whitehouse, M.J., and Ba-Bttat, M.A.O., 1996, Early Precambrian gneiss terranes and Pan-African island arcs in Yemen: Crustal accretion of the eastern Arabian Shield: Geology, v. 24, p. 131–134.

Manuscript Accepted by the Society October 15, 2001

Printed in the U.S.A.

Geological Society of America
Special Paper 362
2002

Petrogenesis of the Late Cretaceous tholeiitic magmatism in the passive margins of northeastern Madagascar

Leone Melluso
Vincenzo Morra
Pietro Brotzu
Massimo D'Antonio
Lucia Bennio
Dipartimento di Scienze della Terra, Università di Napoli Federico II, Via Mezzocannone 8, 80134 Napoli, Italy

ABSTRACT

New chemical and Sr-Nd isotopic data on the Late Cretaceous mafic dike swarm intruding the Archean-Proterozoic crystalline basement in the Tamatave–Sainte Marie Island sector (northeast coast passive margin), and on lavas and dikes of the northeastern part of the Mahajanga sedimentary basin (passive margin after the opening of the Jurassic-Cretaceous Somali basin), allow better knowledge of the chemical variations observed in the northern part of the Madagascan igneous province. Two distinct basalt groups have been identified.

Group 1 basalts have low light to heavy rare earth element (REE) ratios [$(La/Yb)_n$ = 2.2–2.9], low Zr/Y and Nb/Y (4–6 and 0.2–0.4, respectively), low $(^{87}Sr/^{86}Sr)_{88}$ (0.7034–0.7042), and high to moderate $\varepsilon_{Nd}(88)$ (+5.1 to +1.5). Subgroup 1a comprises basalts with the same light to heavy REE ratios [$(La/Yb)_n$ = 2.7–3], Zr/Y and Nb/Y (4.5–5.8 and 0.2–0.3, respectively), and slightly high $(^{87}Sr/^{86}Sr)_{88}$ (0.7042–0.7048) at the same $\varepsilon_{Nd}(88)$ (+5.4 to +4.4) of the group 1 basalts. Group 2 basalts have high light to heavy REE ratios [$(La/Yb)_n$ = 5.3–7.8], high Zr/Y and Nb/Y (7–11 and 0.5–0.8, respectively), relatively high $(^{87}Sr/^{86}Sr)_{88}$ (0.7045–0.7057), and low $\varepsilon_{Nd}(88)$ (+3.8 to +1). The basalts of the groups 1, 1a, and 2 cannot be linked by closed-system magma-differentiation processes, and require distinct mantle sources. The major and trace element variations of the Tamatave dikes of the group 1-1a are compatible with moderate degrees of crystal fractionation (~60%) from the least (MgO = 7.3 wt%) to the most evolved compositions (MgO = 4.2 wt%), involving the separation of plagioclase, augite, pigeonite, and minor oxides, perhaps accompanied by crustal contamination or differences in the $^{87}Sr/^{86}Sr$ ratios. The mantle sources of the group 1-1a basalts seem to be located well within the spinel stability field, whereas a larger contribution of melts derived from garnet-bearing residual mantle is observed in the geochemistry and in the melting models of the group 2 basalts. The chemical and isotopic composition of both rock groups indicate their ultimate provenance from variably enriched lithospheric mantle sources; there is no clear evidence of a hotspot component like that found in the present-day lavas of the Marion–Prince Edward archipelago. The sources of this volcanism seem to be significantly similar to those of the Mahableshwar and Ambenali basalts of the later erupted Deccan Traps, located on formerly contiguous parts of the Gondwana lithosphere.

Melluso, L., Morra, V., Brotzu, P., D'Antonio, M., and Bennio, L., 2002, Petrogenesis of the Late Cretaceous tholeiitic magmatism in the passive margins of northeastern Madagascar, *in* Menzies, M.A., Klemperer, S.L., Ebinger, C.J., and Baker, J., eds., Volcanic Rifted Margins: Boulder, Colorado, Geological Society of America Special Paper 362, p. 81–95.

INTRODUCTION AND GEOLOGIC SETTING

Late Cretaceous flood basalt volcanism is recorded throughout Madagascar (Besairie and Collignon, 1972; Fig 1A). The major sedimentary basins of Mahajanga and Morondava, and the entire eastern coast, preserve significant amounts of the volcanic record in the form of lava flows, minor intrusive complexes, and dike swarms. There is also a volcanic record directly intruding or above the Precambrian basement (the Androy complex, the Ejeda-Bekily dike swarm, and scattered outcrops in the northwest), but these outcrops are localized in a few places and the extent of the province directly overlying the basement is unknown, due to subsequent widespread erosion and weathering. The volcanic activity throughout the island developed from 92 to 84 Ma, according to the most recent age determinations (e.g., Storey et al., 1995; Torsvik et al., 1998).

The Cretaceous volcanic rocks overlie the Precambrian basement and a Paleozoic-Cretaceous sedimentary sequence. The Precambrian basement of northeastern Madagascar is composed of Late Archean to Proterozoic metamorphic and

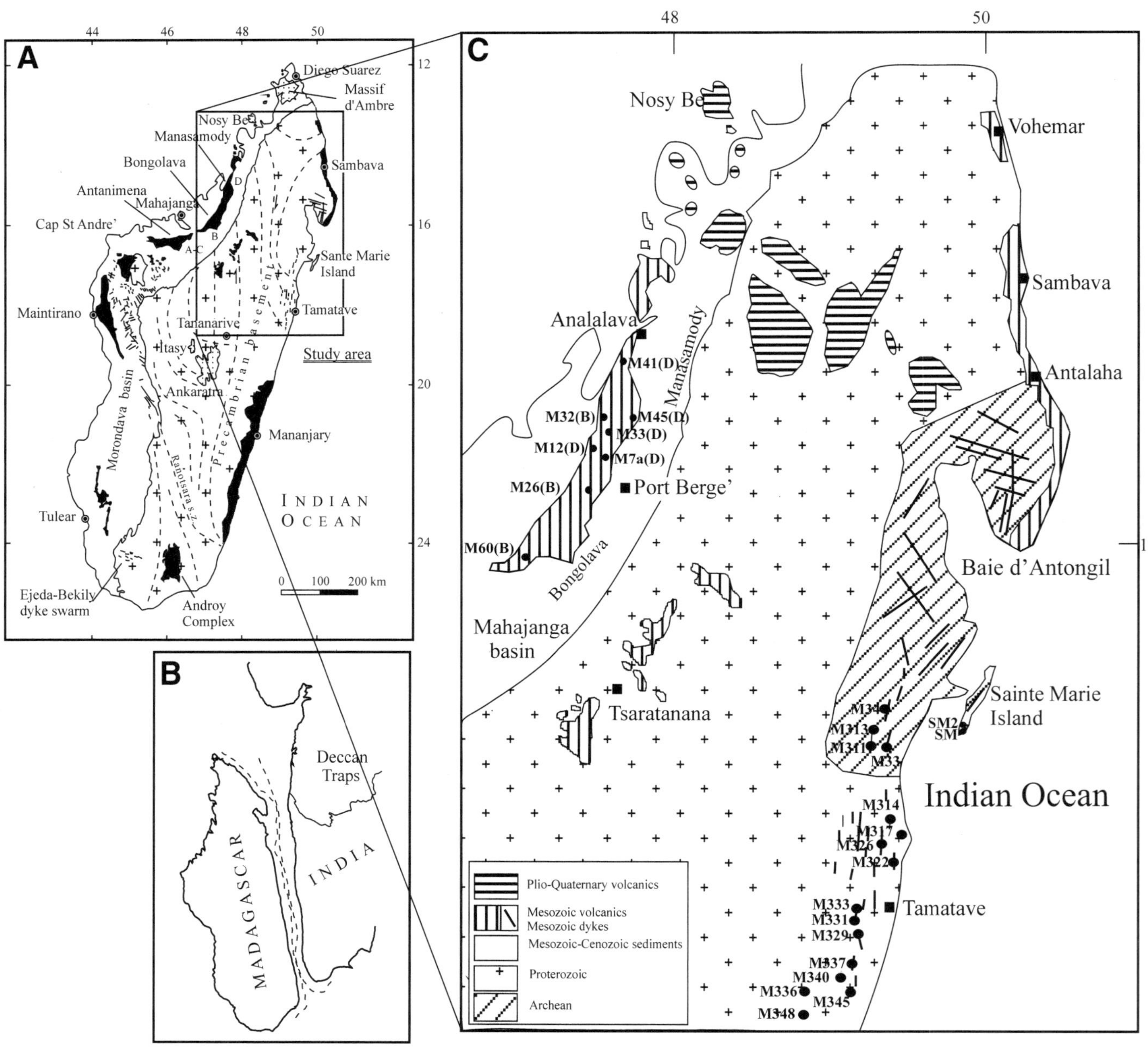

Figure 1. A, B, C: Geological sketch map of Madagascar. Areas in black in A indicate main outcrops of Cretaceous volcanic rocks. B shows relative paleoposition of India and Madagascar in predrift time (after Katz and Premoli, 1979). Location of samples considered in this study is shown in C.

granitoid rocks. At Sainte Marie Island (Fig. 1C), U-Pb and Sm-Nd model ages on tonalitic basement gneisses cluster ca. 3.2 Ga (Middle Archean; Tucker et al., 1999). Middle Proterozoic to Late Proterozoic ages (Pan-African; 800–520 Ma) are widespread in the metamorphic rocks of the basement (Tucker et al., 1999). Sedimentation in the Mahajanga (and Morondava) basin started from the late Paleozoic, as a part of the Karoo rift system of eastern Africa (Bosellini, 1989), and developed in a shallow marine to continental environment. A sedimentary basin is not well developed, or is much narrower, in the eastern coastal Madagascar (Besairie and Collignon, 1972).

When compared to major continental flood basalt provinces, like the Deccan Traps and the Paraná, a thorough knowledge of the geochemical characteristics and regional chemical variations of the Madagascar province is lacking. Storey et al. (1997) provided the most recent synthesis of data on this volcanic province: they used major, trace element, and Sr-Nd-Pb isotopic data, mostly from the eastern part of the province (Sambava, Tamatave, and Mananjary sectors). They argued that mid-ocean ridge basalt (MORB) and continental lithospheric mantle gave the most important chemical contribution to the melt budget, and that only a few basalts in the southeastern part of the province share chemical characteristics with the present-day products of the Marion hotspot.

The volcanic rocks of this province are thought to have been generated by the impingement of the Marion hotspot plume head below the Madagascar lithosphere (e.g., Storey et al., 1995). However, in the Androy complex, proximal to the inferred position of the Marion hotspot ca. 88 Ma, there is no evidence of a plume-related component in the geochemistry of the lavas. Furthermore, tholeiitic, alkaline, and strongly alkaline rocks from a dike swarm propagating away from the Androy complex (the Ejeda-Bekily swarm; Fig. 1A) have a strong lithospheric imprint (Storey et al., 1997; Dostal et al., 1992; Mahoney et al., in prep.).

In the northern part of the province, the mafic rocks are characterized by compositions with unradiogenic lead (e.g., $^{206}Pb/^{204}Pb$ = 16.7–17.5), unlike those of the southern part of the province, which trend toward more radiogenic lead isotopic compositions ($^{206}Pb/^{204}Pb$ to 18.6). This led Storey et al. (1997) to invoke major involvement of old continental lithospheric mantle in the source regions of the northeastern coast basalts. In addition, Mahoney et al. (1991) argued that parts of lithospheric mantle similar to those involved in the petrogenesis of the northern Madagascar basalts could have been entrained in the source of anomalous MORB of the 39–41°E sector of the Southwest Indian Ridge (cf. Mahoney et al., 1992; le Roex et al., 1989).

Two different magma series were found in the rocks of the central western Madagascar (the northern part of the Morondava basin): one tholeiitic, evolving from basalts to rhyodacites with coupled fractionation and crustal contamination, and the other transitional to alkaline, with a more restricted compositional range (Melluso et al., 2001). The primitive magmas of both series are distinctly poor in incompatible trace element contents with respect to magmas from other areas of the province, and some basalts carry even light rare earth element (REE) depletion, more characteristic of typical normal MORB. Therefore, no chemical correlations with the eastern coast basalts were shown to be possible.

A reconnaissance study of the basaltic samples of the Mahajanga basin was reported by Melluso et al. (1997), who recognized four different magma groups with significant chemical and Sr isotopic differences. The groups A and C, found southwest of Mahajanga (Fig. 1A), have low contents of high field strength elements (e.g., Nb < 8 ppm) and high $(^{87}Sr/^{86}Sr)_{88}$ (0.707–0.708) La/Nb (≈2) and Ba/Nb (>60), and strong negative Nb anomalies in mantle normalized diagrams. The group B and D basalts are found in the eastern Mahajanga basin (Fig. 1A) and have lower $(^{87}Sr/^{86}Sr)_{88}$ (0.7038–0.7039; group B, and 0.7049–0.7053, group D) and generally much higher incompatible element contents (see Melluso et al., 1997). Melluso et al. (1997) noted that the group D basalts represent the most incompatible element–enriched tholeiitic rocks of the province, and proposed that the chemical features of the basalts were derived from melting of heterogeneous lithospheric mantle.

Storey et al. (1995) reported whole-rock and phenocryst $^{40}Ar/^{39}Ar$ age determinations on Mahajanga, Sambava, and Tamatave rocks ranging from 90.1 ± 1.2 Ma (Sambava) to 85.9 ± 1.4 Ma (Tamatave). Torsvik et al. (1998) obtained a U-Pb age of 91.6 ± 0.3 Ma on zircons and baddeleyites separated from a Sambava gabbro.

This work aims to correlate the chemical composition of lava flows and dikes found in the eastern part of the Mahajanga basin (the basalts of groups B and D of Melluso et al., 1997) with the volcanic rocks erupted onto the northern part of the eastern passive margin (the Sambava and Tamatave transects). We will use new trace element and Nd isotopic data for the rocks of the Mahajanga basin, integrating those already published by Melluso et al. (1997), and a new chemical and Sr-Nd isotopic data set for the Tamatave–Sainte Marie Island dike swarm, which significantly widens the chemical and isotopic range provided by Storey et al. (1997) on the same swarm. The basalts of the A and C groups of the Mahajanga basin, which have a chemical and isotopic composition strongly indicative of a crustal contribution (cf. Melluso et al., 1997), belong to a different area of the Mahajanga basin (cf. Razafindrazaka et al., 1999). Therefore, we do not consider them further herein. An assessment of the role of lithospheric and sublithospheric mantle below the two passive margins of the northern part of the Madagascan province is also given.

Sampling sites

The location of the samples (lava flows and dikes) from the Mahajanga basin was given in Melluso et al. (1997). The lava flows gently dip seaward, and the few dikes sampled in the area between the lava flows and the crystalline basement have random orientation and the same chemistry as the nearby lava flows. There is some evidence that the group D basalts could be slightly younger, or partially overlapping in the upper sequence of the group B basalts, because no group D basalts were found in the Bongolava sector, whereas group B basalt was found at the end of the sequence in parts of the Manasamody area (Fig. 1C). The location of the Tamatave dikes was given in Storetvedt et al. (1992). The Tamatave dikes trend roughly north-northeast–south-southwest, mostly parallel to the eastern Madagascar coastline. The Sainte Marie Island is ~60 km long and maximum 6–7 km wide (Fig. 1C). The two dike samples come from its southern end. Information about the Sambava sector and on Tamatave dikes can be found in Storey et al. (1997).

ANALYTICAL TECHNIQUES

The new major and trace element analyses of the Tamatave dike samples were carried out at Napoli with a Philips PW1400 X-ray fluorescence spectrometer (CISAG, Napoli), according to methods and analytical uncertainties described in Melluso et al. (1997, 2001). Loss on ignition (LOI) was measured with standard gravimetric techniques after igniting powders at 1100 °C in a muffle furnace, and Na_2O was analyzed with atomic absorption spectrophotometry (Napoli) (Table 1). Mineral compositions on samples of the Mahajanga basin and Sainte Marie Island dikes were determined with a wavelength-dispersive–energy-dispersive spectrometry (WDS-EDS) equipped CAMECA SX50 microprobe (Consiglio Nazionale delle Ricerche–Centro di Studio per il Quaternario e l'Evoluzione Ambientale, Rome) utilizing silicates and oxides as standards (Table 2). REEs and additional trace element analyses were carried out with inductively coupled plasma–mass spectrometry (ICP-MS) at Centre des Recherches Petrographiques et Geochimiques, Nancy, France (Table 3).

For the isotopic determinations, 0.3 g of powder was strongly leached with warm 6N HCl for 30 min, then rinsed thoroughly in pure subboiling double-distilled water, and finally dissolved with high-purity HF-HNO_3-HCl mixtures. Sr and Nd were extracted by conventional ion-exchange chromatographic techniques. The total blank was ~6 ng Sr and 4 ng Nd. Measurements were obtained using a VG354 double-collector thermal ionization mass spectrometer (CISAG, Napoli) running in peak jumping mode, by normalizing the data to $^{86}Sr/^{88}Sr$ = 0.1194 and $^{146}Nd/^{144}Nd$ = 0.7219 for mass-fractionation effects. The quoted error is twice the standard deviation of the mean (2σ) and is $\pm 1 \times 10^{-5}$. Repeated analyses of NBS-987 standard yielded a mean value of $^{87}Sr/^{86}Sr$ = 0.71024 ± 0.00001 (N = 50) and the La Jolla Nd standard a mean value of $^{143}Nd/^{144}Nd$ = 0.511826 ± 0.000010 (N = 26). The Sr and Nd isotopic analyses of the sample SM (a Sainte Marie Island dike) were performed at the isotope laboratory of SOEST, University of Hawaii, with techniques and analytical uncertainties described in Melluso et al. (2001). The isotope data were age corrected to 88 Ma using the X-ray fluorescence (XRF) Rb and Sr and ICP-MS Sm and Nd elemental data on the whole rocks (Table 4).

CLASSIFICATION AND PETROGRAPHY

The igneous rocks the Tamatave–Sainte Marie Island and Sambava sectors and those of the eastern part of the Mahajanga basin are classified as basalts and (subordinately) basaltic andesites, according to the total alkali-silica (TAS) diagram (Le Bas et al., 1986). They belong to the tholeiitic series, and have hypersthene + quartz or olivine + hypersthene in their CIPW norms. Major and trace element analyses are reported in Table 1, and representative compositions of the major mineral phases are reported in Table 2.

The dikes of Tamatave and Sainte Marie Island are dolerites, with plagioclase, augite, and pigeonite (the latter often enclosed in augite), locally abundant Fe-Ti oxides, apatite ± sulfides, and, in coarser grained holocrystalline facies, interstitial quartz ± alkali feldspar. The Mahajanga basin basalts are petrographically olivine-free varieties, except for a few samples of group B, and have plagioclase and augite as the most abundant phases, with lower amounts of pigeonite ± orthopyroxene, Fe-Ti oxides, accessory apatite, and very fresh rhyolitic glass. Storey et al. (1997) gave petrographic descriptions of the Sambava rocks and described a plagioclase-phyric basalt (sample SAM92-33B) that has high Al_2O_3 (19.8 wt%) and MgO (10.5 wt%), and very low trace element contents (e.g., Zr and Y = 6 ppm).

MAGMATIC EVOLUTION AND GEOCHEMISTRY

Magmatic evolution of the Tamatave–Sainte Marie Island dikes

The new data on the Tamatave–Sainte Marie dikes allow a more detailed picture of magma-differentiation processes recorded in this dike suite. The MgO content of the samples ranges from 7.5 to 4.2 wt%; Mg_v [atomic Mg/(Mg + Fe^{2+}) with Fe_2O_3/FeO = 0.15] between 0.56 and 0.39. Therefore, there are no compositions approaching those of mantle-derived magmas. The dikes of Sainte Marie Island have among the lowest MgO contents, and therefore have the most evolved chemistry. CaO ranges from 11 to 8 wt%, Sc from 42 to 30 ppm, and Cr from 180 to 21 ppm, all decreasing with MgO; TiO_2 increases from 2 to 4 wt%, Fe_2O_{3t} from 12 to 17 wt%, P_2O_5 from 0.2 to 1.0 wt%, Zr from 110 to 400 ppm, Nb from 7 to 20 ppm, and Y from 22 to 65 ppm, all increasing with decreasing MgO (Fig. 2). Storey et al. (1997) reported Fe_2O_{3t} contents as high as 18 wt% (sample TAM92-30). Mass-balance calculations (Stormer and Nicholls, 1978) from the least differentiated dike M333 to the dike SM indicate 58% total fractionation of plagioclase An_{68} (41%), augite $Ca_{38}Mg_{47}Fe_{15}$ (38%), pigeonite $Ca_9Mg_{55}Fe_{36}$ (18%), and Fe-Ti oxides (3%) (ΣR^2 = 0.48). Similar results (56% total fractionation of 41% plagioclase, 41% augite, and 18% pigeonite; ΣR^2 = 0.19) were obtained for the transition from M333 to the sample

TABLE 1. MAJOR AND TRACE ELEMENT ANALYSES OF THE TAMATAVE–SAINTE MARIE ISLAND DOLERITES

Area	Sample	Group	SiO_2	TiO_2	Al_2O_3	Fe_2O_{3t}	MnO	MgO	CaO	Na_2O	K_2O	P_2O_5	LOI	Mg_V	Sc	V	Cr	Ni	Zn	Rb	Sr	Y	Zr	Nb	Ba	Nb/Y	Zr/Y	Zr/Nb
Tamatave	M333	1	49.94	2.15	13.18	12.98	0.22	7.35	11.20	2.43	0.33	0.24	1.27	0.56	38	386	180	83	103	7	264	23	117	7	99	0.29	5.0	17.6
Tamatave	M322	1	50.45	2.06	12.84	13.57	0.21	7.23	10.75	2.26	0.38	0.25	1.03	0.55	38	382	121	81	109	7	230	25	126	7	111	0.29	5.0	17.0
Tamatave	M336	1	51.07	2.38	12.37	13.33	0.24	7.16	10.70	1.92	0.46	0.38	1.52	0.55	36	395	126	63	122	9	310	33	202	11	160	0.34	6.1	17.7
Tamatave	M34	1	50.29	2.46	12.94	13.45	0.22	6.78	10.69	2.41	0.41	0.35	0.70	0.53	32	406	125	70	108	10	315	28	141	8	151	0.30	5.1	17.0
Tamatave	M314	1	50.21	2.48	12.24	14.90	0.24	6.60	10.31	2.39	0.37	0.26	0.94	0.50	34	404	130	69	129	5	220	29	160	8	109	0.28	5.6	19.7
Tamatave	M329	1	50.70	2.07	13.49	13.03	0.21	6.58	11.05	2.16	0.42	0.28	0.75	0.53	35	377	134	72	102	6	249	26	132	7	136	0.27	5.0	18.6
Tamatave	M33	1	50.17	2.57	12.87	13.83	0.22	6.52	10.68	2.47	0.34	0.33	0.61	0.51	35	436	104	67	101	5	322	28	145	13	133	0.44	5.1	11.6
Tamatave	M313	1	50.71	2.66	12.63	13.94	0.23	6.40	10.31	2.26	0.56	0.31	1.34	0.51	38	430	85	58	116	11	322	30	109	9	188	0.29	3.6	12.4
Tamatave	M317	1	50.26	2.53	12.36	14.97	0.23	6.29	10.11	2.56	0.42	0.26	0.64	0.49	38	413	121	66	124	7	220	30	166	10	115	0.33	5.5	16.7
Tamatave	M337	1	49.60	3.09	11.84	15.44	0.26	6.12	9.95	2.77	0.50	0.44	0.87	0.47	42	555	152	62	129	7	206	37	170	8	135	0.22	4.6	21.1
Tamatave	M340	1	52.66	2.35	12.92	13.36	0.24	5.26	8.71	3.14	0.81	0.55	1.50	0.47	31	344	65	39	127	18	372	38	247	14	201	0.39	6.6	17.0
Tamatave	M345	1	49.97	4.14	10.35	17.40	0.26	5.23	9.25	2.31	0.58	0.52	1.48	0.40	36	527	44	35	166	12	250	44	253	14	181	0.32	5.8	18.1
Tamatave	M311	1	51.68	2.85	13.05	14.80	0.23	4.70	9.18	2.47	0.60	0.43	2.11	0.42	27	382	59	36	131	11	312	39	230	11	248	0.29	5.9	20.0
Tamatave	M326	1	51.68	3.03	13.15	14.43	0.22	4.27	9.74	2.49	0.58	0.40	1.22	0.40	33	463	44	37	127	9	277	35	199	13	164	0.36	5.7	15.7
Mahajanga	M32	1 (B)	50.26	2.37	13.94	13.09	0.18	6.57	11.01	2.17	0.14	0.26	1.46	0.53	34	366	213	105	115	4.6	298	22	138	8	98	0.36	6.3	17.3
Sainte Marie	SM2	1a	49.78	3.73	12.20	16.63	0.21	5.17	8.10	2.60	0.76	0.82	0.51	0.41	33	379	21	21	149	14	253	57	316	19	172	0.33	5.5	16.6
Sainte Marie	SM	1a	51.16	3.26	12.85	15.58	0.21	4.51	9.00	2.17	0.64	0.61	1.06	0.39	29	459	43	35	144	19	271	49	282	18	161	0.36	5.8	16.0
Tamatave	M331	2	50.54	4.26	12.37	12.76	0.17	6.60	9.77	2.35	0.64	0.54	1.72	0.54	24	455	65	75	124	8	591	26	282	21	192	0.82	11.0	13.4
Tamatave	M348	2	48.50	4.67	12.10	14.55	0.19	6.10	10.20	2.59	0.64	0.46	0.20	0.49	30	552	71	67	123	10	554	24	207	17	181	0.70	8.7	12.5

Note: Mg_V = atomic Mg/(Mg + Fe^{2+}) with Fe_2O_3/F_3O = 0.15. The chemical analysis of sample M32 from the Mahajanga basin (group B of Melluso et al., 1997) is also given. See text for the subdivision into groups. LOI = weight loss on ignition.

TABLE 2. REPRESENTATIVE CHEMICAL ANALYSES OF MAJOR MINERAL PHASES IN THE MAHAJANGA BASIN AND SAINTE MARIE ISLAND DOLERITES

Group	Phase	SiO_2	TiO_2	Al_2O_3	FeO_t	MnO	MgO	CaO	Na_2O	K_2O	Total	Ca	Fe*	Mg	An	Ab	Or
1	augite	51.77	1.05	2.50	9.34	0.19	16.20	18.58	0.24		99.9	38.3	15.3	46.4			
1	augite	51.51	0.94	1.58	14.87	0.29	14.43	15.99	0.25		99.9	33.4	24.7	41.9			
1a	augite	51.10	0.98	1.63	15.86	0.39	15.02	14.49	0.22		99.7	30.1	26.4	43.5			
2	augite	50.46	1.34	2.24	11.78	0.32	16.30	16.78	0.29		99.5	34.3	19.3	46.4			
2	augite	51.16	1.08	1.50	13.70	0.44	15.02	16.30	0.20		99.4	33.8	22.9	43.3			
1	pigeonite	52.97	0.40	0.47	20.57	0.48	21.37	3.83	0.06		100.2	7.7	32.9	59.5			
1	pigeonite	51.95	0.62	0.83	21.58	0.44	19.08	5.16	0.08		99.7	10.6	35.2	54.3			
2	orthopyroxene	53.53	0.47	1.30	15.04	0.29	26.78	2.25	0.05		99.7	4.4	23.3	72.4			
1	plagioclase	50.97		30.89	0.69			13.65	3.63	0.09	99.9				67.2	32.3	0.5
1	plagioclase	52.57		29.51	0.87			12.12	4.44	0.15	99.7				59.6	39.5	0.9
2	plagioclase	52.23		29.83	0.83			12.46	4.21	0.23	99.8				61.2	37.4	1.3
2	plagioclase	51.55		30.16	0.79			13.19	3.8	0.15	99.6				65.2	34.0	0.9
2	plagioclase	55.16		27.78	0.82			10.23	5.51	0.37	99.9				49.6	48.3	2.1

Note: An, Ab, Or in mol%; Ca, Mg, Fe* (Fe + Mn) in mol%.

TABLE 3. RARE EARTH AND OTHER TRACE ELEMENT CONTENTS OF TAMATAVE AND MAHAJANGA BASIN SAMPLES

Sample group	Tamatave M333 1	Tamatave M329 1	Mahajanga M32 1,B	S. Marie SM 1a	Mahajanga M7a 2,D	Tamatave M331 2	Tamatave M348 2
La	12.16	10.24	10.64	20.40	30.38	27.44	20.17
Ce	28.42	24.70	26.07	53.14	71.24	66.29	48.69
Pr	4.30	3.39	3.75	7.70	9.27	8.95	6.78
Nd	20.33	17.35	18.19	35.46	43.69	41.89	33.07
Sm	5.86	4.71	4.91	9.65	9.81	10.41	7.68
Eu	1.97	1.46	1.85	2.88	3.32	3.02	2.56
Gd	5.78	4.76	5.75	9.21	9.76	8.73	7.04
Tb	0.88	0.74	0.85	1.54	1.33	1.14	0.96
Dy	5.53	4.43	5.21	9.35	7.67	6.42	5.17
Ho	1.07	0.97	0.97	2.04	1.39	1.10	0.97
Er	2.77	2.47	2.54	4.85	3.26	2.87	2.41
Tm	0.41	0.37	0.38	0.74	0.42	0.36	0.30
Yb	2.74	2.48	2.43	4.71	2.63	2.37	1.87
Lu	0.38	0.39	0.38	0.62	0.38	0.36	0.27
Hf	3.59	3.16	3.63		7.34	7.40	5.63
Ta	0.66	0.54	0.64		1.74	1.61	1.37
Pb	1.78	1.57	1.76		5.18	4.13	2.63
Th	1.11	0.95	0.88		2.86	2.84	1.90
U	0.22	0.19	0.22		0.62	0.56	0.40
Eu/Eu*	1.03	0.94	1.06	0.93	1.04	0.97	1.07
(La/Yb)n	2.99	2.78	2.95	2.92	7.79	7.81	7.28
Ce/Pb	16.0	15.7	14.8		13.7	16.0	18.5

Note: $Eu/Eu^* = Eu_n/(Gd_nSm_n)^{1/2}$.

TAM92-30 (Storey et al., 1997). The results of MELTS modeling (Ghiorso and Sack, 1995) indicate major plagioclase and Ca-rich clinopyroxene removal in the low-pressure fractionation of dike M333 toward low-Mg compositions. Therefore, crystal fractionation seems to account for the observed chemical variations of most Tamatave–Sainte Marie Island dikes. Relatively high amounts of gabbroic fractionation, starting from basalts similar to the most mafic rocks in this study, are a typical feature of tholeiitic liquid lines of descent before Fe-Ti oxide saturation (Cox, 1980; Cox and Hawkesworth, 1985).

The most Mg-rich Tamatave dikes overlap chemically to the Sambava rocks (Storey et al., 1997) and to the B group basalts of the Mahajanga basin (Melluso et al., 1997). However, a few Tamatave dikes (M348, M331) plot at higher Nb contents and Zr/Y and Nb/Y ratios (cf. Table 1; Fig. 2). These outliers have a similar chemical composition to the group D basalts of the Mahajanga basin (cf. Melluso et al., 1997). Most Tamatave dikes have systematically higher Zr/Nb ratios (17.1 ± 2.5) than those of the group D basalts of the Mahajanga basin (14.0 ± 1.1; Melluso et al., 1997), as well as lower Zr/Y (<7 vs. >9) and Nb/Y (0.2–0.4 vs. 0.5–0.8). Group D rocks of the Mahajanga basin have the lowest Sc and Y contents among the northern Madagascar basalts. Minor and trace element modeling indicates that the most Mg-rich Tamatave basalts can be parental magmas to the more differentiated Tamatave and Sainte Marie Island dikes (Fig. 2).

The chondrite-normalized REE diagrams (Fig. 3) of the samples from northeastern Madagascar show two markedly different patterns: one is characterized by low light to heavy REE fractionation [$(La/Yb)_n < 3$; Table 3)]. This pattern is present in the most Mg-rich Tamatave dikes, in the group B basalts of the Mahajanga basin (Melluso et al., 1997), and in the Sambava rocks (Storey et al., 1997). The Tamatave dikes (Storey et al., 1997) and the Sainte Marie dikes have identical $(La/Yb)_n$ ratios, and higher REE contents [$(Yb)_n > 20$; Fig. 3]. The other pattern is characterized by a stronger light to heavy REE fractionation [$(La/Yb)_n > 5.6$; Table 3], and is found in a few Tamatave dikes (M331, M348, TAM92-6; Storey et al., 1997) and in the group D basalts of the Mahajanga basin (Melluso et al., 1997). A large positive Eu anomaly is observed in the Tamatave dike TAM92-6 (Storey et al., 1997) (Eu/Eu* = 1.18; Table 3), possibly due to plagioclase accumulation.

Sr-Nd isotopic data

The Tamatave–Sainte Marie Island dikes have a large span in initial $^{87}Sr/^{86}Sr$ ratios and ε_{Nd}. The most Mg-rich samples (M333 and M329) have low $^{87}Sr/^{86}Sr$ (0.7038) and relatively high ε_{Nd} (+3.6 to +3.9). These values overlap with those observed in the Sambava dikes (0.7034–0.7042 and ε_{Nd} from +5.1 to +1.5, respectively; Storey et al., 1997), and are identical to those of the B group basalts of the Mahajanga basin ($^{87}Sr/^{86}Sr$

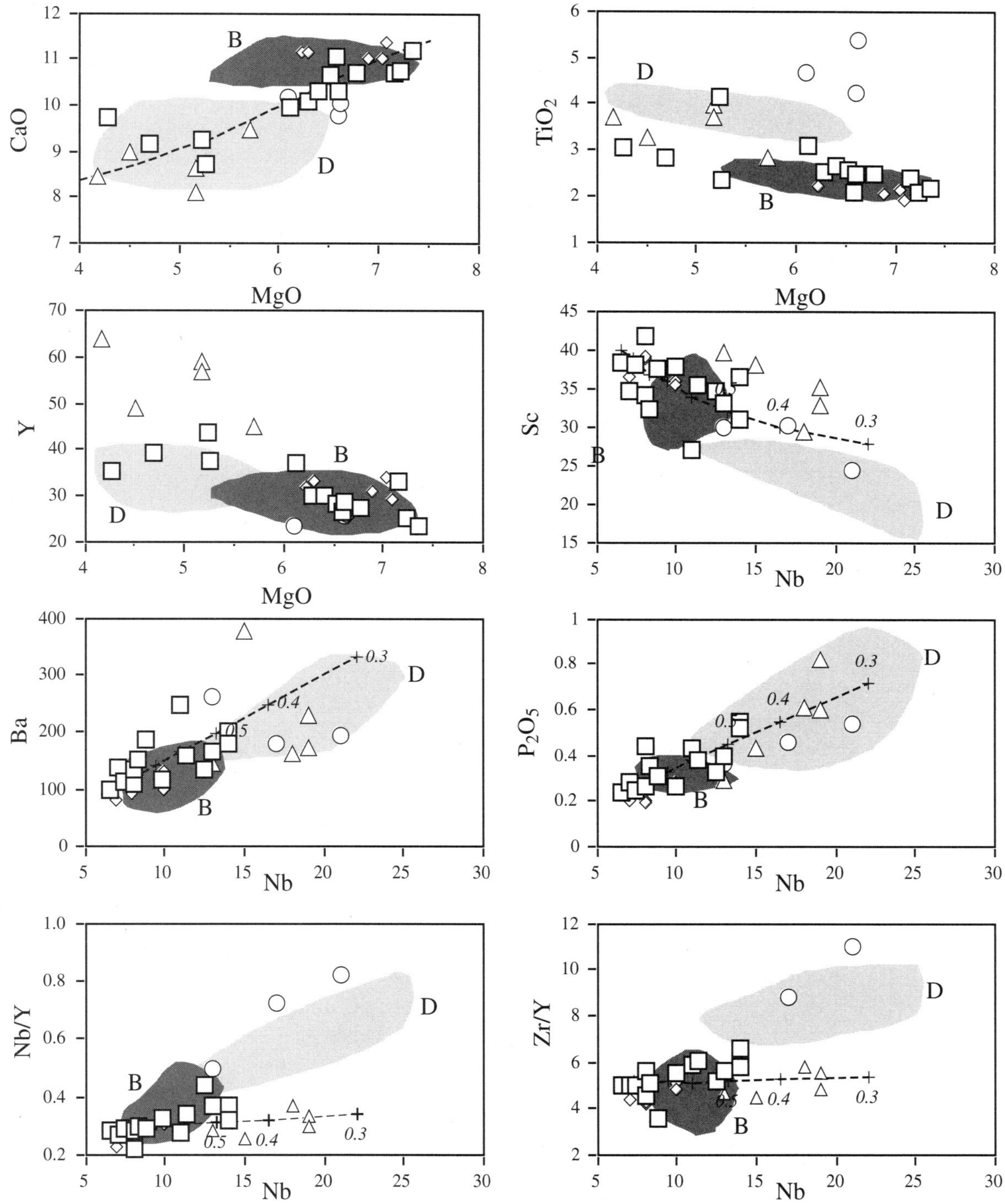

Figure 2. Representative major and trace element variations of rocks of this study. Squares, Tamatave dikes of this study (group 1); diamonds, Sambava rocks (group 1); circles, Tamatave dikes (group 2); triangles, Tamatave and Sainte Marie Island dikes (subgroup 1a); fields of basalts of group B and D of Mahajanga basin are also given (data from Melluso et al., 1997). Simulated fractionation curves of Tamatave–Sainte Marie Island dike swarm use Rayleigh fractionation equation starting from M333 and following partition coefficients: D_{Nb} and $D_{Ba} = 0$; $D_{Sc} = 1.3$; D_P and $D_{Zr} = 0.1$; $D_Y = 0.14$. Residual liquid fractions (*f*) of 0.5, 0.4, and 0.3 are indicated on fractionation curves. Fractionation trend obtained by MELTS modeling, starting from M333 (P = 1 kbar; f_{O_2} set to QFM buffer), is shown in CaO-MgO diagram.

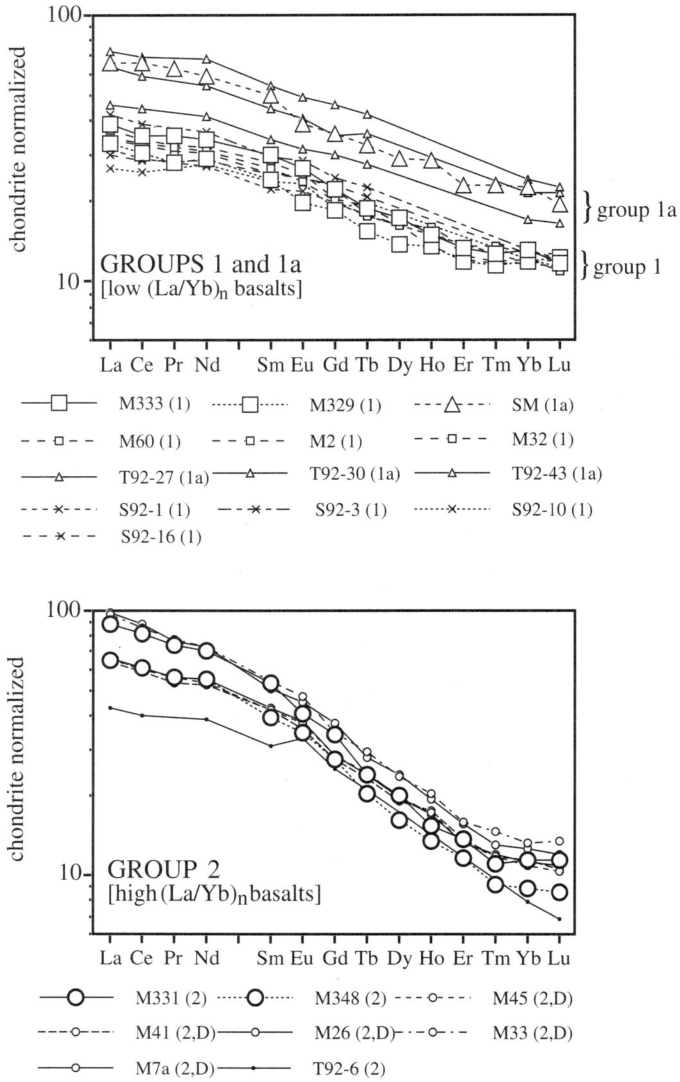

Figure 3. Chondrite-normalized rare earth element diagrams for samples of northeastern Madagascar. Normalization value is recommended chondrite of Boynton (1984). Other data are from Melluso et al. (1997) and Storey et al. (1997). Samples are divided taking into account their $(La/Yb)_n$ ratios (see text), and therefore distinguish group 1-1a and group 2 basalts (see text).

= 0.7038–0.7039 and ε_{Nd} from +3.2 to +3.7, respectively; data from Melluso et al. [1997] and this paper; Fig. 4; Table 4). All these samples belong to the basalt types with low $(La/Yb)_n$. The more evolved dike of Sainte Marie Island and the Tamatave dikes with low $(La/Yb)_n$ have slightly higher $^{87}Sr/^{86}Sr$ (0.7043–0.7048), at similar ε_{Nd} (+5.4 to +4.4) than the Mg-rich Tamatave dikes.

The Tamatave dike M331 [with high $(La/Yb)_n$ and Nb/Y] has $^{87}Sr/^{86}Sr$ = 0.7055 and ε_{Nd} = +1.7. These values are almost identical to those measured in the lava flows and dikes of the group D basalts of the Mahajanga basin ($^{87}Sr/^{86}Sr$ = 0.7050–07053, and ε_{Nd} = +1.8 to +1; Melluso et al. [1997]; Table 4), and to sample TAM92-6 (Storey et al., 1997) (Fig. 4). The other sample with high $(La/Yb)_n$ (M348) has lower $^{87}Sr/^{86}Sr$ (0.7045) and higher ε_{Nd} (+3.8).

Mafic magma types

Summarizing the previous sections, we can distinguish two main groups of basalts (and a subgroup) in northeastern Madagascar.

Group 1. These basalts have low Zr/Y, Nb/Y (4–6 and 0.2–0.4, respectively), low light to heavy REE fractionation [$(La/Yb)_n$ = 2.2–2.9], low $^{87}Sr/^{86}Sr$ (0.7034–0.7042), and high ε_{Nd} (+1.5 to +5.1). This group includes the group B basalts of the Mahajanga basin, the Sambava basalts (Storey et al., 1997), and the most Mg-rich Tamatave dikes. The subgroup 1a is represented by evolved basalts with low light to heavy REE fractionation [$(La/Yb)_n$ = 2.7–3.0], Zr/Y and Nb/Y (4.5–5.8 and 0.2–0.3, respectively), and slightly high $^{87}Sr/^{86}Sr$ (0.7042–0.7048) at ε_{Nd} = +4.4 to +5.4. This subgroup includes Tamatave dikes (Storey et al., 1997), and dikes from Sainte Marie Island.

Group 2. These basalts have high Zr/Y and Nb/Y (7–11 and 0.5–0.8, respectively), marked light to heavy REE fractionation [$(La/Yb)_n$ = 5.3–7.8], high $^{87}Sr/^{86}Sr$ (0.7045–0.7057), and low ε_{Nd} (+3.8 to +1). This group includes the group D basalts of the Mahajanga basin, and a few dikes of the Tamatave area (M331, M348; TAM92-6, Storey et al., 1997).

The data are given in Figures 3, 4, and 5 according to this subdivision. These chemical criteria allow chemical correlation between the basalts of Sambava and Tamatave and those found in the eastern part of the Mahajanga basin.

DISCUSSION

Relationships between group 1-1a and group 2 basalts

Two different basalt groups are present in the northeastern part of the Madagascar igneous province. There is sufficient chemical and isotopic evidence to deduce that the basalts in groups 1 and 2 are not comagmatic. For example, the much higher Nb at given Y and Zr contents of group 2 basalts, as well as their significantly stronger REE fractionation, higher $^{87}Sr/^{86}Sr$, and lower ε_{Nd} (Figs. 2, 3, 4, and 5; Tables 3 and 4) cannot be produced by closed-system crystal fractionation of plagioclase and pyroxene. Group 2 basalts have higher $^{87}Sr/^{86}Sr$ and lower ε_{Nd} ($^{143}Nd/^{144}Nd$) than group 1 basalts, but identical La/Nb and Ba/Nb ratios (cf. Fig. 6). This argues against derivation of group 2 basalts from combined fractionation and crustal contamination starting with group 1 basalts.

Relationships between group 1 and subgroup 1a basalts

Subgroup 1a basalts are chemically more evolved than group 1 basalts, have higher contents of incompatible elements,

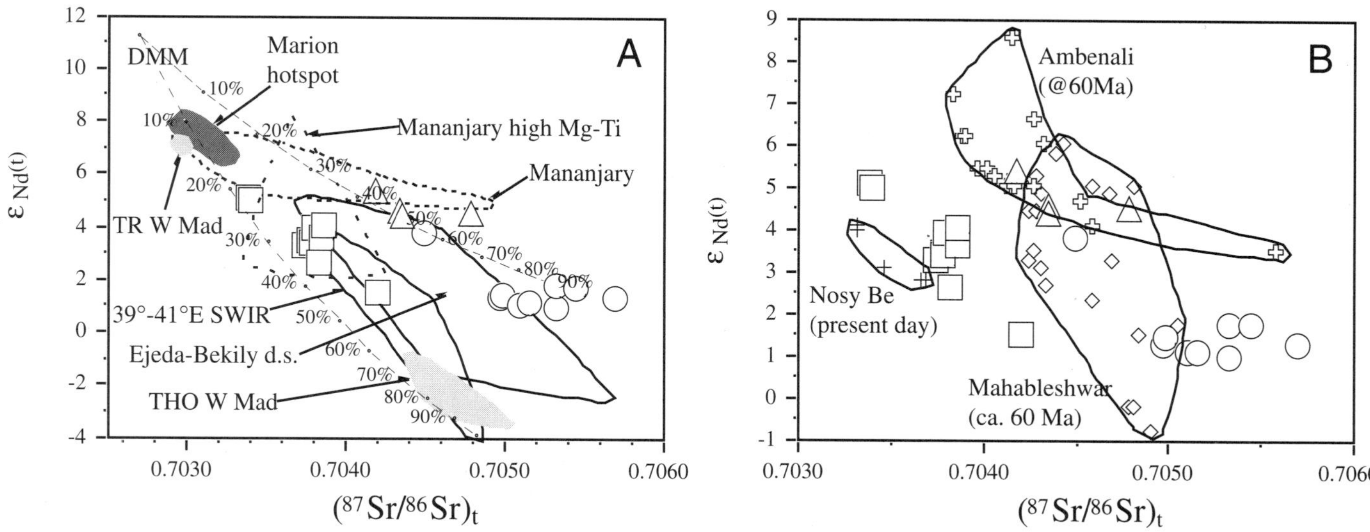

Figure 4. A: $(^{87}Sr/^{86}Sr)_t$–$\varepsilon_{Nd(t)}$ diagram for basalts of northeastern Madagascar (Mad). Squares, group 1 basalts; triangles, subgroup 1a basalts; circles, group 2 basalts. Also plotted are Marion hotspot (present-day values), field of basalts from 39°–41°E sector of Southwest Indian Ridge (SWIR; Mahoney et al., 1992; recalculated at 88 Ma), Mg-rich samples from central western sector of province (Melluso et al, 2001), and basalts from southeastern coast (Mananjary sector; Storey et al., 1997). Two mixing curves are calculated between mid-ocean ridge basalt (MORB) source and isotopically most extreme sample of southwest Indian Ridge and between MORB source and an average isotopic composition of group 2 basalts. $^{87}Sr/^{86}Sr_{MORB} = 0.7027$; $\varepsilon Nd_{MORB} = +11.3$; Sr and Nd elemental data of sources from Reiners et al. (2000). B: Same diagram with fields of Ambenali and Mahableshwar basalts of Deccan Traps (data from Cox and Hawkesworth, 1985; Lightfoot and Hawkesworth, 1988; Lightfoot et al., 1990). Present-day isotopic values of late Cenozoic mafic alkaline rocks of Nosy Be archipelago (northern Madagascar) are also plotted (data from Melluso and Morra, 2000).

and higher $^{87}Sr/^{86}Sr$ at similar ε_{Nd}. However, the group 1-1a basalts share an identical range in $(La/Yb)_n$ and other elemental ratios, and there are no major changes in the La/Nb or Nb/U ratios with increasing degree of differentiation. Only the $^{87}Sr/^{86}Sr$ ratios and the degree of chemical evolution discriminate between group 1 and 1a basalts. The slightly higher $^{87}Sr/^{86}Sr$ ratio of the subgroup 1a basalts could be the result of small-scale isotopic heterogeneity, crustal contamination, or moderate interaction with seawater (cf. Fig. 6).

The mantle-normalized incompatible element diagrams of the group 1 and 2 basalts (Fig. 5) show similar features: the marked peak at Pb, small negative Nb (Ta) troughs with respect

TABLE 4. $^{87}Sr/^{86}Sr$, $^{143}Nd/^{144}Nd$ ISOTOPIC RATIOS AND $\varepsilon_{Nd}(88)$ OF THE TAMATAVE AND MAHAJANGA SAMPLES

	Sample	Group	$^{87}Sr/^{86}Sr$ measured	$^{87}Sr/^{86}Sr$ at 88Ma	$^{143}Nd/^{144}Nd$ measured	$^{143}Nd/^{144}Nd$ at 88 Ma	ε_{Nd}	ε_{Nd} at 88 Ma	Nd model age Ga
Tamatave	M333	1	0.70391	0.70385	0.51281	0.51271	3.4	3.6	0.67
Tamatave	M329	1	0.70393	0.70379	0.51282	0.51273	3.6	3.9	0.56
Mahajanga	M60 (B)	1	0.70379	0.70377	0.51280	0.51270	3.2	3.5	0.64
Mahajanga	M32 (B)	1	0.70384	0.70378	0.51280	0.51271	3.2	4.5	0.59
Mahajanga	M2 (B)	1	0.70391	0.70386	0.51283	0.51273	3.7	4	0.59
S. Marie	SM	1a	0.70459	0.70434	0.51285	0.51276	4.2	4.6	0.48
Tamatave	M331	2	0.70549	0.70545	0.51270	0.51261	1.2	1.7	0.67
Tamatave	M348	2	0.70456	0.70449	0.51280	0.51272	3.2	3.8	0.46
Mahajanga	M41 (D)	2	0.70508	0.70498	0.51269	0.51260	1	1.5	0.76
Mahajanga	M45 (D)	2	0.70507	0.70497	0.51268	0.51259	0.8	1.3	0.74
Mahajanga	M12 (D)	2	0.70519	0.70510					
Mahajanga	M26 (D)	2	0.70521	0.70515	0.51267	0.51258	0.6	1.1	0.75
Mahajanga	M33 (D)	2	0.70543	0.70532	0.51266	0.51258	0.4	1	0.71
Mahajanga	M7a (D)	2	0.70541	0.70532	0.51270	0.51262	1.2	1.8	0.87

Note: The $^{87}Sr/^{86}Sr$ data of the Mahajanga samples are from Melluso et al. (1997). Calculation of model ages assumes $^{143}Nd/^{144}Nd_{DM} = 0.5131$ and $^{147}Sm/^{144}Nd_{DM} = 0.24$. See text for the subdivision into groups.

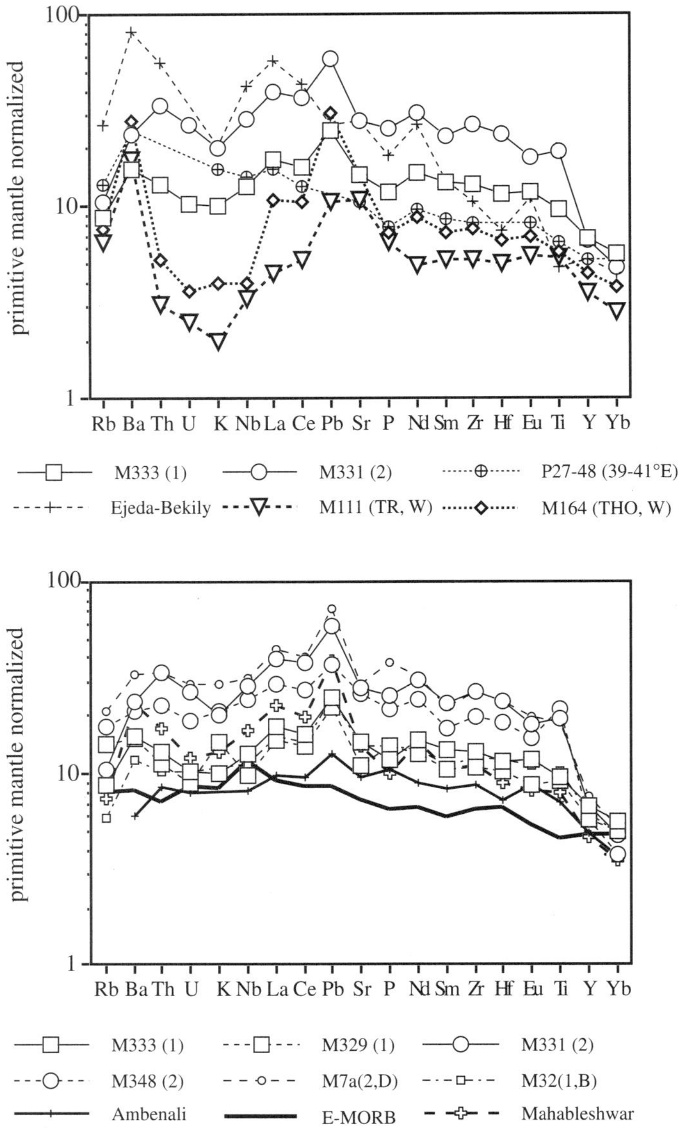

Figure 5. Primitive mantle-normalized incompatible element patterns (primitive mantle values after Sun and McDonough, 1989) of rocks of northeastern Madagascar. Two patterns of basalts of groups B (group 1) and D (group 2) of Mahajanga basin (data from this work and Melluso et al., 1997), two basalts from western coast (data from Melluso et al., 2001), two basalt samples of Ambenali and Mahableshwar formations of Deccan Traps (L. Melluso, unpublished data), average enriched mid-ocean ridge basalt (E-MORB) (Sun and McDonough, 1989), and anomalous MORB from 39–41°E Southwest Indian Ridge (P14-27; data from le Roex et al., 1989) are plotted for comparison.

to La, no or positive Ti peaks, and the overall concave-downward patterns at different levels of enrichment. Rb, Ba, and Th are slightly enriched relative to U and K. The analyses provided by Storey et al. (1997) have identical patterns, with an anomalously high Pb content in TAM92-6 (Pb = 7.9 ppm; ~100 times primitive mantle; Ce/Pb = 4), maybe due to cumulus feldspar (cf. the positive Eu anomaly of this sample), or to interaction with lower crust or lithospheric mantle, given its low $^{206}Pb/^{204}Pb$ (16.718). TAM92-6 has anomalously high Nb/U (87) and low Zr/Y (5.6) ratios (Storey et al., 1997).

The ranges of Ce/Pb and Nb/U of the northeastern Madagascar basalts tend, on the average, to be slightly lower than worldwide MORB–ocean island basalt (OIB) averages (Ce/Pb = 25±5 and Nb/U = 47±10; Hofmann et al., 1986). They are markedly higher than the values observed in Madagascan basement rocks (Nb/U = 9 and Ce/Pb = 3; Melluso et al., 2001) (Nb/U = 2.2–4.7 and Ce/Pb = 1.6–4; Kröner et al., 1999), or in the rhyodacites found on top of the Cretaceous volcanic sequence in central western Madagascar (Nb/U = 3 and Ce/Pb = 4.3; Melluso et al., 2001). Therefore, there is no clear evidence of major crustal input in the northeastern Madagascan basalts, though crustal contamination processes cannot be ruled out.

The chemical features, and the relatively high $^{87}Sr/^{86}Sr$ and low ε_{Nd} of the rocks indicate that the source of the northeastern Madagascar basalts cannot be a mixture of MORB-like asthenosphere and the present-day source of the Marion hotspot magmas, because of the need for additional low ε_{Nd}, high $^{87}Sr/^{86}Sr$ components. With the lack of evidence for major crustal contamination processes, which usually increase $^{87}Sr/^{86}Sr$, Ba/Nb, and La/Nb, and decrease ε_{Nd}, Nb/U, and Ce/Pb of the magmas (cf. Fig. 6), it is possible that these enriched components are located in the mantle lithosphere.

Modeling of the mantle sources of the northeastern Madagascar province

Because none of the compositions observed in northeastern Madagascar can be considered as a mantle-derived primary magma, it is likely that multiple saturation of Cr-spinel, olivine, clinopyroxene, and plagioclase occurred before the emplacement of all the magmas to shallow crustal levels. Therefore, we have modeled the extent of partial melting utilizing ratios of elements that are not modified by low to moderate fractionation of the phases mentioned in the preceding. In the Lu/Hf versus Sm/Nd diagram (Fig. 7), group 1-1a basalts appear to have formed by small-degree melts (3%–5%) of sources composed by mixed incompatible element–depleted and –enriched mantle in the spinel stability field. In contrast, group 2 basalts were derived from similar, or slightly smaller, degrees of partial melting (~3%) with a higher contribution from residual garnet-bearing mantle (Fig. 7). Derivation of the group 2 basalts from melts generated within the garnet peridotite stability field is also supported by the unusually high Zr/Y ratio of the latter.

Modeling of the Sr-Nd composition of group 1-1a and 2 basalts was performed assuming mixing between MORB mantle (a depleted baseline end member) and two different enriched mantle components, both assumed to be lithospheric. One is represented by Sr-Nd isotopic composition similar to that of group

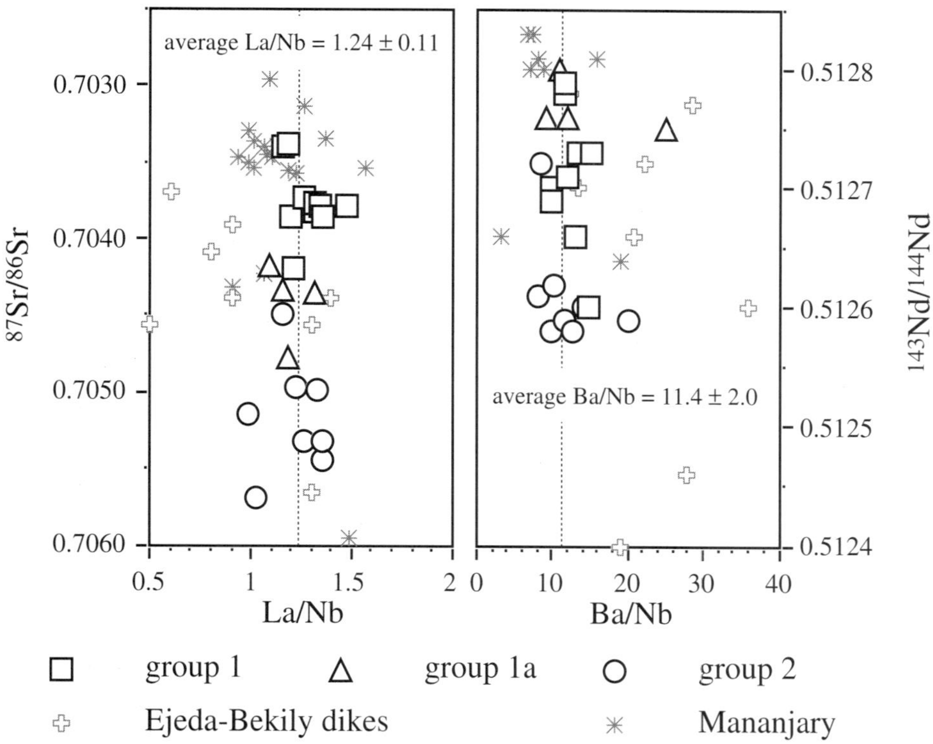

Figure 6. La/Nb vs. $^{87}Sr/^{86}Sr$ and Ba/Nb vs. $^{143}Nd/^{144}Nd$ for northeastern Madagascar basalts; data on Mananjary sector and Ejeda-Bekily dikes are from Storey et al. (1997), Mahoney et al. (1991), and Dostal et al. (1992). Average La/Nb and Ba/Nb (excluding two outliers) for northeastern Madagascan samples utilized in this study are also reported.

2 basalts, and the other by the most extreme Sr-Nd isotopic composition of the 39–41°E Southwest Indian Ridge (Mahoney et al., 1992). The two mixing curves (Fig. 4A) indicate that the compositional range of group 1-1a basalts could be accounted for by 20%–50% components added to a depleted mantle.

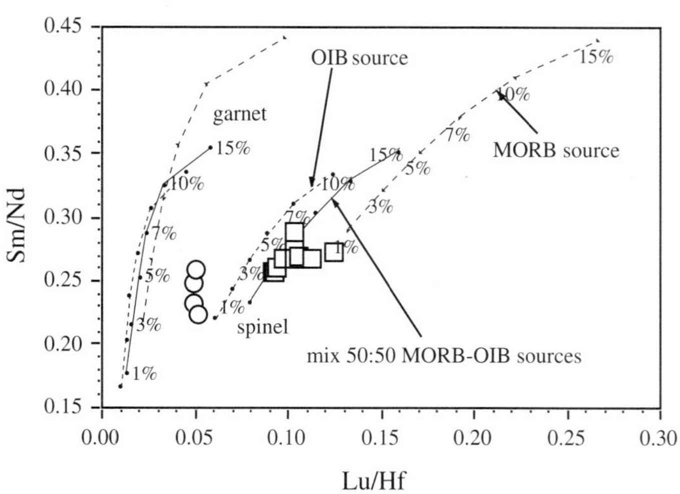

Figure 7. Lu/Hf vs. Sm/Nd diagram with results of nonmodal fractional melting model. Ocean island (OIB) and mid-ocean ridge basalt (MORB) source end members are from Reiners et al. (2000). Source chosen is 50:50 mix of MORB and OIB source end members (partial melting trend is continuous line). Partition coefficients are taken from Johnson (1998). Symbols as in Figure 6.

Regional correlations and source processes

The basalts from Mananjary (southeastern coast) mostly plot in the highest part of the trend depicted by group 1 basalts (Fig. 4A), thus confirming their derivation from different sources (see also Fig. 6). A few scattered samples from the southern coast plot close to the field of subgroup 1a. The high-Mg-Ti basalts of the Mananjary sector (Storey et al., 1997) do not display a coherent trend, plotting throughout the area covered by the other Madagascan rocks. Some similarities between the rocks of this study and alkali basalts and basanites of the Ejeda-Bekily dike swarm (Mahoney et al., 1991) are observed in the Sr-Nd isotopic space (Fig. 4A) and in the patterns and ratios between incompatible elements (Figs. 5 and 6). However, the Ejeda-Bekily alkaline dikes have some very high Ba/Nb values (Fig. 6), and negative Ti anomalies, features not observed in the northeastern Madagascan basalts. The group 1 basalts trend toward the Southwest Indian Ridge basalts, whereas farther south, the basalts of the Mananjary sector trend toward components similar to those of the group 2 basalts.

Marked differences are observed between the northern Madagascar basalts and those that crop out in the central western sector of the province (cf. Melluso et al., 2001). The depletion of most incompatible elements in the Mg-rich basalts of central western Madagascar and their isotopic composition make the transitional basalts the products of melting of sources similar to normal MORB. In contrast, the tholeiitic basalts may be the product of slightly higher degrees of partial melting of a source enriched with very low amounts of crustal material (Melluso et al., 2001). The transitional basalts of central western Madagascar plot close to the present-day Marion field, whereas the tholeiitic basalts plot within the field of the anomalous basalts from the Southwest Indian Ridge. The central western Madagascar basalts are thus chemically and isotopically very different from the group 1-1a and group 2 basalts. The group 1 basalts plot very close to the (present-day) isotopic composition of the mafic alkaline volcanic rocks of the Nosy Be archipelago (Figs. 1C and 4B; Melluso and Morra, 2000). This could indicate a similar mantle source involved in the petrogenesis of both volcanic provinces.

The isotopic compositions of the basalts of the Southwest Indian Ridge at 39–41°E, when recalculated back to 88 Ma, are on an extension of the trend of group 1 basalts in the $^{87}Sr/^{86}Sr–\varepsilon_{Nd}$ diagram (Fig. 4A). These anomalous MORBs have Pb isotopic ratios (e.g., $^{206}Pb/^{204}Pb$ = 16.8–17.4) and Ba/Nb (Ba/Nb = 9–22; Mahoney et al., 1992; le Roex et al., 1989) in the range of group 1-1a basalts.

The data of this work provide evidence that the two passive margins of northern Madagascar were flooded by magmas coming from similar, though heterogeneous, mantle sources, and that there is no clear evidence that crustal contamination played a major role during the differentiation of these magmas. The basalts found in the Mahajanga basin could have been generated below the eastern coastal Madagascar, and then traveled toward west and north. This possibility cannot be dismissed, but, to date, it is difficult to demonstrate that northern Madagascar was covered by a flood basalt sequence, except for a few inliers located close to the border of the Mahajanga basin (Fig. 1C). Moreover, there is no trace of Mesozoic dike swarms crosscutting the Precambrian basement between the eastern coast and the Mahajanga basin, whereas dikes crosscutting the early Mesozoic sedimentary rocks of this basin have been observed and sampled (Melluso et al., 1997). The relative abundance of dikes in the Tamatave area with a group 2 affinity is minor. Consequently, the hypothesis for these rocks being feeders for the lava flows of the Mahajanga basin seems to be difficult to demonstrate. Therefore, we conclude that a relatively local origin for the Mahajanga basin basalts is more likely.

The presence of a core of an Archean craton, located beneath the area covered by the Tamatave and Sainte Marie Island dikes, has been described (cf. Windley et al., 1994). In $^{207}Pb^{*}/^{235}U$ versus $^{206}Pb^{*}/^{238}U$ diagrams, the upper intersection of isotope compositions of zircon populations with the Concordia curve recently confirmed this (Tucker et al., 1999). The age of the enrichment events in the mantle sources of basalts is often related to age and type of lithosphere through which the basalts were erupted (e.g., Hawkesworth et al., 1986; Menzies, 1989). Worldwide evidence suggests that the Archean lithosphere is chemically more refractory and incompatible element–depleted than post-Archean lithosphere (cf. Menzies, 1989; Sweeney and Watkeys, 1990; McDonough, 1990). Taking all this into account, we find that there is no clear evidence of an Archean signature in the chemical composition and in the Nd model ages of the samples from this study. This is particularly evident for the dikes of Sainte Marie Island, which crosscut Archean gneisses (Tucker et al., 1999). Therefore, it is inferred that the sources of this volcanism could have been last modified during the Pan-African orogeny, and that the eventual older (Archean) lithospheric mantle, if still present beneath the area, did not contribute significant amounts to the melt budget.

The postulated position of the Marion hotspot 88 Ma is just south of the southern coast of Madagascar, not too far from the Androy complex, i.e., ~1000 km south of the Tamatave area (Fig. 1). The volume of the erupted volcanics in the Madagascar province is known to be ~200 m thick in the Mahajanga basin (Besairie and Collignon, 1972; Razafindrazaka et al., 1999), with the exception of the Androy complex. Therefore, there is no need to invoke very high temperatures in the mantle (>1400 °C), because they could favor higher initial pressures of melting and increase melt production (cf. McKenzie and Bickle, 1988), even though thick, cold lithosphere may act as a barrier inhibiting the upward propagation of sublithospheric melt columns. The lithosphere of the eastern coast was thinned shortly before the opening of the Mascarene basin, and possibly later than the lithosphere of the Mahajanga basin. This may have begun in the late Paleozoic–early Mesozoic as it was a part of the Karoo rift system. Moreover, there is evidence of uplift of the Mahajanga basin shortly before the outpouring of the volcanic rocks (Razafindrazaka et al., 1999). The possibility of rift-related decompression melting of the mantle beneath northern Madagascar, possibly favored by increase of the ambient temperature due to a thermal anomaly, seems to be a plausible working model.

Metasomatic processes in the Madagascan lithosphere could have enriched it in small amounts of more fusible (±volatile bearing) phases, which later participated in the Cretaceous melting event. The correlation between the deeper depth of initiation of melting and the enriched chemical and isotopic features of group 2 basalts could indicate that the deeper parts of the Madagascan lithosphere were more enriched than the spinel-bearing mantle, and therefore that the Madagascan lithosphere could be chemically and isotopically stratified (Fig. 8).

Relationships between the northeastern Madagascar basalts and the Ambenali and Mahableshwar basalts of the Deccan Traps

Mahoney et al. (1991) remarked on the general chemical similarities between the southwestern coast basalts with the Ambenali and Mahableshwar basalts of the Deccan Traps, views reiterated by Melluso et al. (1997), in a discussion of the geochemistry of the group B and D basalts of the Mahajanga basin. The two Deccan formations are the most widespread in the southern, uppermost part of the Western Ghats (e.g., Cox and Hawkesworth, 1985; Devey and Lightfoot, 1986). The northeastern margin of Madagascar was located close to the west coast of India and to the Western Ghats, according to paleogeographic reconstructions (Katz and Premoli, 1979; Fig. 1B).

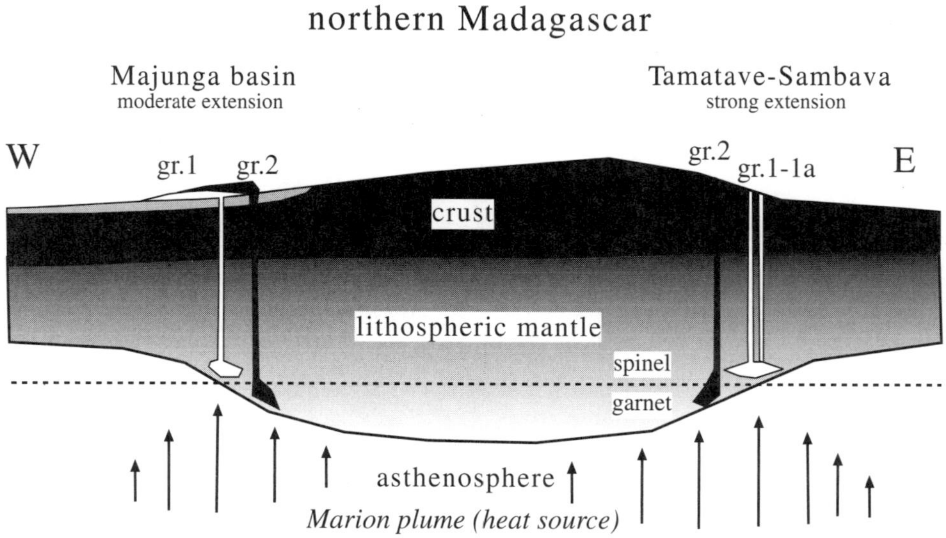

Figure 8. Schematic section of lithosphere of northern Madagascar (not to scale).

Therefore, one would expect similarities if the sources are located within, or are influenced by, the lithospheric mantle. It was argued that the chemical and isotopic characteristics of the Mahableshwar basalts could have been derived from a chemical contribution from the Indian lithosphere (Cox and Hawkesworth, 1985; Lightfoot et al., 1990). The geochemical features of the Ambenali basalts are currently interpreted to have been derived from asthenospheric mantle similar to the source of transitional MORB (Lightfoot et al., 1990; Mahoney et al., 1982). The Ambenali and Mahableshwar basalts share many features with group 1-1a basalts. For example, the $(La/Yb)_n$ of the Ambenali ranges from 2.25 to 3.15 [average $(La/Yb)_n = 2.59 \pm 0.28$] and the $(La/Yb)_n$ of the Mahableshwar basalts ranges from 1.76 to 4.53 [average $(La/Yb)_n = 3.34 \pm 0.64$] (Cox and Hawkesworth, 1985; Lightfoot and Hawkesworth, 1988; Lightfoot et al., 1990; Widdowson et al., 2000). Similar Zr/Y ratios are reported for the northeastern Madagascar basalts and the two basalt formations of the Western Ghats (e.g., Zr/Y = 3.4–4.9, average 3.98 ± 0.3, for the Ambenali, and Zr/Y = 3.6–6.6, average 5.03 ± 0.65, for the Mahableshwar; data sources as in the preceding and Melluso et al., 1997) (Fig. 9). The similarities in the trace element patterns between the two Deccan formations and group 1-1a basalts, with marked Pb peaks and overall concave downward patterns, are seen in Figure 5. However, differences are observed between the Mahableshwar and group 2 basalts, such as the lower Zr/Nb ratios of the former (Zr/Nb = 11.2 ± 1.7, average from Melluso et al., 1997; Zr/Nb = 9.7, average from Widdowson et al., 2000) and the much higher Zr/Y of the latter. The Ambenali and, particularly, the Mahableshwar basalts, have Sr-Nd isotopic ranges that encompass those of subgroup 1a and group 2 basalts (Fig. 4) but differ from the group 1 basalts. Furthermore, the range of present-day $^{206}Pb/^{204}Pb$ for the Ambenali and Mahableshwar basalts overlap the range of the northeastern Madagascar basalts (Storey et al., 1997). Chemical and isotopic arguments therefore suggest that some of the components involved in the petrogenesis of the northern Madagascar, Mahableshwar, and Ambenali basalts could be similar, because they were conjugate volcanic rifted margins.

CONCLUSIONS

This study of the basalt samples recovered from the two passive margins of northern Madagascar (the eastern coastal margin and the Mahajanga basin) shows that they belong to two chemical groups. Group 1-1a has a mildly enriched incompatible element signature, relatively low $^{87}Sr/^{86}Sr$ (<0.7048) and high εNd (>+2.6), and was derived from spinel-bearing mantle sources. This group is widespread in both margins, and is likely the most voluminous magma type found in northern Madagascar. Group 2 is decidedly more incompatible element–enriched, and carries a distinctive Sr-Nd isotopic composition ($^{87}Sr/^{86}Sr$ > 0.7045 and εNd < +3.8). The basalts of this group are encountered as lava flows and dikes in the northeastern Mahajanga basin and in a few Tamatave dikes, and were derived from garnet-bearing mantle, evidence of a deeper melt column than that of group 1-1a basalts.

The mantle sources of these volcanic rocks carry a chemical and isotopic signature different from that observed in the basalts erupted in the southern (Storey et al., 1997), western, and central western areas (Melluso et al., 1997, 2001), and do not have any resemblance to the present-day composition of the Marion hotspot basalts. Therefore, the sublithospheric component in the petrogenesis of the Madagascar igneous province is more likely depleted (MORB like) mantle, and not a plume-like enriched component. Passive margin formation seems to focus (or channelize) melt production, and this is the case on the east-

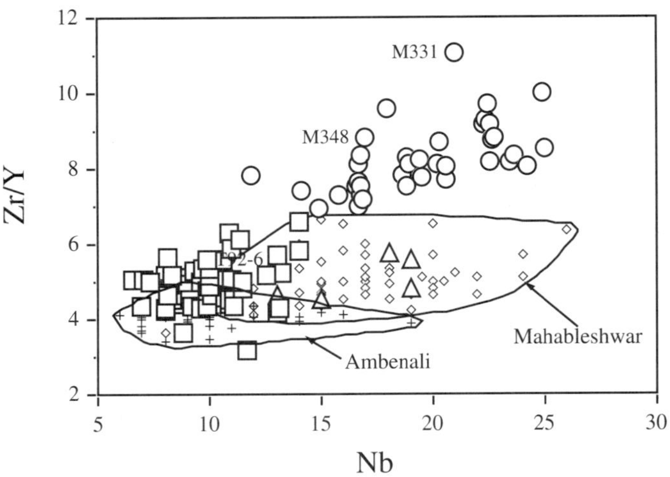

Figure 9. Diagram of Zr/Y vs. Nb (ppm) for samples of northwestern Madagascar and for Ambenali and Mahableshwar basalts of Deccan Traps. Data sources: Melluso et al. (1997), Storey et al. (1997), this paper, Lightfoot et al. (1990), and Lightfoot and Hawkesworth (1988).

ern coast of Madagascar. The Mahajanga basin may have been proximal to hotter than normal upwelling mantle.

Work is needed to link the northern Madagascar basalts to those of the southeastern outcrops, and to identify the characteristics of the sources of southernmost Madagascar, which are known to involve both mantle and crustal parts of the lithosphere (Mahoney et al., 1991; Storey et al., 1997).

ACKNOWLEDGMENTS

We thank Karsten Storetvedt for making available dike samples from the Tamatave area. We also thank Antonio Canzanella, Vincenzo Monetti, Rino Ricci, Marcello Serracino, and Khalil Spencer for their help in the X-ray fluorescence setting, the wet chemical work, drafting Figure 1, the microprobe analyses, and isotopic determinations of the SM sample, respectively. Early versions of the manuscript benefited from critical reading of, and discussions with, John Mahoney, and from helpful comments of David Peate, Dougal Jerram, and Martin Menzies. This research was partially supported by Ministero dell'Università e Ricerca Scientifica (PRIN 1998-2000 to Brotzu).

REFERENCES CITED

Besairie, H., and Collignon, M., 1972, Geologie de Madagascar—Les terraines sedimentaires: Annales Geologiques Madagascar, v. 35, 553 p.

Bosellini, A., 1989, The continental margins of Somalia: Their structural evolution and sequence stratigraphy: Memorie di Scienze Geologiche, v. 41, p. 373–458.

Boynton, W.V., 1984, Cosmochemistry of the rare earth elements: Meteorite studies, *in* Henderson, P., ed., Rare earth element geochemistry: Amsterdam, Elsevier, p. 63–114.

Cox, K.G., 1980, A model for flood basalt vulcanism: Journal of Petrology, v. 21, p. 629–650.

Cox, K.G., and Hawkesworth, C.J., 1985, Geochemical stratigraphy of the Deccan Traps at Mahabaleshwar, Western Ghats, India, with implications for open-system processes: Journal of Petrology, v. 26, p. 355–377.

Devey, C.W., and Lightfoot, P.C., 1986, Volcanological and tectonic control of stratigraphy and structure in the western Deccan traps: Bulletin of Volcanology, v. 48, p. 195–207.

Dostal, J., Dupuy, C., Nicollet, C., and Cantagrel, J.M., 1992, Geochemistry and petrogenesis of upper Cretaceous basaltic rocks from southern Madagascar: Chemical Geology, v. 97, p. 199–218.

Ghiorso, M.S., and Sack, R.O., 1995, Chemical mass transfer in magmatic processes. 4. A revised and internally consistent thermodynamic model for the interpolation and extrapolation of liquid-solid equilibria in magmatic systems at elevated temperatures and pressures: Contributions to Mineralogy and Petrology, v. 119, p. 197–212.

Hawkesworth, C.J., Mantovani, M.S.M., Taylor, P.N., and Palacz Z., 1986, Evidence from Paraná of south Brazil for a continental contribution to Dupal basalts: Nature, v. 322, p. 356–359.

Hofmann, A.W., Jochum, K.P., Seufert, M., and White, W.M., 1986, Nb and Pb in oceanic basalts: New constraints on mantle evolution: Earth and Planetary Science Letters, v. 79, p. 33–45.

Johnson, K.T.M., 1998, Experimental determination of partition coefficients for rare earth and high-field-strength elements between clinopyroxene, garnet, and basaltic melt at high pressures: Contributions to Mineralogy and Petrology, v. 133, p. 60–68.

Katz, M.B., and Premoli, C., 1979, India and Madagascar in Gondwanaland based on matching Precambrian lineaments: Nature, v. 279, p. 312–315.

Kröner, A., Windley, B.F., Jaeckel, P., Brewer, T.S., and Razakamanana T., 1999, New zircon ages and regional significance for the evolution of the Pan-African orogen in Madagascar: Journal of the Geological Society of London, v. 156, p. 1125–1135.

Le Bas, M.J., Le Maitre, R.W., Streckeisen, A., and Zanettin, P., 1986, A chemical classification of volcanic rocks based on the total alkali-silica diagram: Journal of Petrology, v. 27, p. 745–750.

le Roex, A.P., Dick, H.J.B., and Fisher, R.L., 1989, Petrology and geochemistry of MORB from 25°E to 46°E along the Southwest Indian Ridge: Evidence for contrasting styles of mantle enrichment: Journal of Petrology, v. 30, p. 947–986.

Lightfoot, P.C., and Hawkesworth, C.J., 1988, Origin of Deccan Trap lavas: Evidence from combined trace element and Sr-Nd- and Pb-isotope studies: Earth and Planetary Science Letters, v. 91, p. 89–104.

Lightfoot, P.C., Hawkesworth, C.J., Devey, C.W., Rogers, N.W., and Van Calsteren, P.W.C., 1990, Source and differentiation of Deccan Trap lavas: Implications of geochemical and mineral chemical variations: Journal of Petrology, v. 31, p. 1165–1200.

Mahoney, J.J., Nicollet, C., and Dupuy, C., 1991, Madagascar basalts: Tracking oceanic and continental sources: Earth and Planetary Science Letters, v. 104, p. 350–363.

Mahoney, J.J., le Roex, A.P., Peng, Z.X., Fisher, R.L., and Natland, J.H., 1992, Southwestern limits of Indian Ocean Ridge mantle and the origin of low $^{206}Pb/^{204}Pb$ mid-ocean ridge basalt: Isotope systematics of the central southwest Indian Ridge (17°–50°E): Journal of Geophysical Research, v. 97, p. 19771–19790.

Mahoney J.J., Macdougall, J.D., Lugmair, G.W., Murali, A.V., Sankar Das, M., and Gopalan, K., 1982, Origin of the Deccan Trap flows at Mahabaleshwar inferred from Nd and Sr isotopic and chemical evidence: Earth and Planetary Science Letters, v. 60, p. 47–60.

McDonough, W.F., 1990, Constraints on the composition of the continental lithospheric mantle: Earth and Planetary Science Letters, v. 101, p. 1–18.

McKenzie, D.P., and Bickle, M.J., 1988, The volume and composition of melt generated by extension of the lithosphere: Journal of Petrology, v. 29, p. 625–679.

Melluso L., and Morra V., 2000, Petrogenesis of late Cenozoic mafic alkaline rocks of the Nosy Be archipelago (northern Madagascar): Relationships with the Comorean magmatism: Journal of Volcanology and Geothermal Research, v. 56, p. 129–142.

Melluso L., Morra V., Brotzu, P., and Mahoney, J.J., 2001, The Cretaceous igneous province of Madagascar: Geochemistry and petrogenesis of lavas and dykes from the central-western sector: Journal of Petrology, v. 42, p. 1249–1278.

Melluso, L., Morra, V., Brotzu, P., Razafiniparany, A., Ratrimo, V., and Razafimahatratra, D., 1997, Geochemistry and Sr-isotopic composition of the Cretaceous flood basalt sequence of northern Madagascar: Petrogenetic and geodynamic implications: Journal of African Earth Sciences, v. 24, p. 371–390.

Menzies, M.A., 1989, Cratonic, circumcratonic and oceanic mantle domains beneath the western United States: Journal of Geophysical Research, v. 94, p. 7899–7915.

Razafindrazaka, Y., Randriamananjara, T., Piqué, A., Thouin, C., Laville, E., Malod, J., and Rehault, J.P., 1999, Extension et sedimentation au Paleozöique terminal et au Mesozoique dans le bassin de Mahajanga (Nord-Ouest de Madagascar): Journal of African Earth Sciences, v. 28, p. 949–959.

Reiners, P.W., Hammond, P.E., McKenna, J.M., and Duncan, R.A., 2000, Young basalts of the central Washington Cascades, flux melting of the mantle, and trace element signatures of primary arc magmas: Contributions to Mineralogy and Petrology, v. 138, p. 249–264.

Storetvedt, M., Mitchell, J.G., Abranches, M.C., Maaloe, S., and Robin, G., 1992, The coast-parallel dykes of east Madagascar: Age of intrusion, remagnetization and tectonic aspects: Journal of African Earth Sciences, v. 15, p. 237–249.

Storey, M., Mahoney, J.J., and Saunders, A.D., 1997, Cretaceous basalts in Madagascar and the transition between plume and continental lithosphere mantle sources, *in* Mahoney, J.J., and Coffin, M.F., eds., Large igneous provinces: Continental, oceanic and planetary flood volcanism: American Geophysical Union Geophysical Monograph 100, p. 95–122.

Storey, M., Mahoney, J.J., Saunders, A.D., Duncan, R.A., Kelley, S.P., and Coffin, M.F., 1995, Timing of hot spot–related vulcanism and the breakup of Madagascar and India: Science, v. 267, p. 852–855.

Stormer, J.C., Jr., and Nicholls, J., 1978, XLFrac: A program for interactive testing of magmatic differentiation models: Computers and Geosciences, v. 4, p. 143–159.

Sun, S.-S., and McDonough, W.F., 1989, Chemical and isotopic systematics of oceanic basalts: Implications for mantle composition and processes, *in* Saunders, A.D., and Norry, M.J., eds., Magmatism in the ocean basins: Geological Society [London] Special Publication 42, p. 313–345.

Sweeney, R.J., and Watkeys, M.K., 1990, A possible link between Mesozoic lithospheric architecture and Gondwana flood basalts: Journal of African Earth Sciences, v. 10, p. 707–716.

Torsvik, T.H., Tucker, R.D., Ashwal, L.D., Eide, E.A., Rakotosolofo, N.A., and de Wit, M.J., 1998, Late Cretaceous magmatism of Madagascar: Paleomagnetic evidence for a stationary hotspot: Earth and Planetary Science Letters, v. 164, p. 221–232.

Tucker, R.D., Ashwal, L.D., Handke, M.J., Hamilton, M.A., Le Grange, M., and Rambeloson, R.A., 1999, U-Pb geochronology and isotope geochemistry of the Archean and Proterozoic rocks of north-central Madagascar: Journal of Geology, v. 107, p. 135–153.

Wedepohl, K.H., 1995, The composition of the continental crust: Geochimica et Cosmochimica Acta, v. 59, p. 1217–1232.

Widdowson, M., Pringle, M.S., and Fernandez, O.A., 2000, A post K-T boundary (early Paleocene) age for Deccan-type feeder dykes, Goa, India: Journal of Petrology, v. 41, p. 1177–1194.

Windley, B.F., Razafiniparany, A., Razakamanana, T., and Ackermand, D., 1994, Tectonic framework of the Precambrian of Madagascar and its Gondwana connections: A review and reappraisal: Geologische Rundschau, v. 83, p. 642–659.

Manuscript Accepted by the Society October 15, 2001

Printed in the U.S.A.

Geological Society of America
Special Paper 362
2002

Silicic volcanism: An undervalued component of large igneous provinces and volcanic rifted margins

Scott E. Bryan*
Department of Earth Sciences, University of Queensland, St. Lucia, Brisbane, Queensland 4072, Australia
Teal R. Riley*
British Antarctic Survey, Natural Environment Research Council, High Cross, Madingley Road, Cambridge CB3 0ET, UK
Dougal A. Jerram*
Department of Geological Sciences, University of Durham, South Road, Durham DH1 3LE, UK
Christopher J. Stephens*
Resource Service Group Pty Ltd., 35 Ventnor Avenue, West Perth, Western Australia 6005, Australia
Philip T. Leat*
British Antarctic Survey, Natural Environment Research Council, High Cross, Madingley Road, Cambridge CB3 0ET, UK

ABSTRACT

Silicic volcanic rocks are associated with most, if not all, continental flood basalt provinces and volcanic rifted margins, where they can form substantial parts of the eruptive stratigraphy and have eruptive volumes >10^4 km^3. Poor preservation of silicic volcanic rocks following kilometer-scale uplift and denudation of the volcanic rifted margins, however, can result in only deeper level structural features being exposed (i.e., dike swarms, major intrusions, and deeply subsided intracaldera fills; e.g., North Atlantic igneous province). The role of silicic magmatism in the evolution of a large igneous province and rifted margin may therefore be largely overlooked.

There are silicic-dominated igneous provinces that have extrusive volumes comparable to those of mafic large igneous provinces (>10^6 km^3), but that have low proportions of basalt expressed at the surface. Some silicic large igneous provinces are associated with intraplate magmatism and continental breakup (e.g., Jurassic Chon Aike province of South America, Early Cretaceous eastern Australian margin), whereas others are tectonically and geochemically associated with backarc environments (e.g., Sierra Madre Occidental). Silicic volcanic rocks formed in these two environments are similar in terms of total eruptive volumes, dominant lithologies, and rhyolite geochemistry, but show fundamental differences in tectonic setting and basalt geochemistry.

Large-volume ignimbrites are the dominant silicic volcanic rock type of continental flood basalt and silicic large igneous provinces. Individual silicic eruptive units can have thicknesses, areal extents, and volumes that are comparable to, or exceed,

*E-mails: sbryan@earth.uq.edu.au; t.riley@bas.ac.uk; D.A.Jerram@durham.ac.uk; chris.stephens@rsg.com.au.; p.leat@bas.ac.uk.

Bryan, S.E., Riley, T.R., Jerram, D.A., Stephens, C.J., and Leat, P.T., 2002, Silicic volcanism: An undervalued component of large igneous provinces and volcanic rifted margins, *in* Menzies, M.A., Klemperer, S.L., Ebinger, C.J., and Baker, J., eds., Volcanic Rifted Margins: Boulder, Colorado, Geological Society of America Special Paper 362, p. 97–118.

interbedded flood basalt lavas. Caldera complexes, with diameters typically 10–30 km, represent the main eruptive sources for the large volumes of silicic magma, and may range from regional sag structures to complex volcanic-tectonic collapse structures controlled by tectonic stresses and preexisting crustal architecture.

The largest volume silicic igneous provinces occur along accreted continental margins, whereas continental flood basalt provinces have been emplaced on or adjacent to Archean cratons. Large-volume silicic igneous provinces ultimately reflect large-scale crustal melting processes in response to lithospheric extension and high thermal (± mass) input from underlying hot mantle. Partial melting of hydrous, mafic-intermediate composition (amphibolite) crust is critical in generating the large volumes of predominantly I-type silicic igneous melt. In these cases, subduction as much as hundreds of millions of years prior to the emplacement of the silicic igneous province seems crucial in producing a hydrous lower crustal source receptive to melting.

INTRODUCTION

Large igneous provinces are the most significant accumulations of mafic material at the Earth's surface after basaltic and associated intrusive rock emplaced at mid-ocean ridges (Coffin and Eldholm, 1992, 1994). Continental flood basalts, synonymous with mafic large igneous provinces, typically represent short-lived (<5 m.y.), high-rate (0.1 to >1 km^3/yr), large-volume ($\geq 1 \times 10^6$ km^3) eruptive events, mostly related to breakup of continents (White and McKenzie, 1989; Coffin and Eldhom, 1994; Fig. 1). Most research has focused on the geochronology and basaltic geochemistry of continental flood basalt provinces (e.g., papers in Storey et al., 1992; Mahoney and Coffin, 1997). Recent studies have also focused on the flow morphology and internal structures of flood basalt lavas; these studies indicate that the mafic magmas were emplaced as inflated compound pahoehoe lava fields via prolonged, episodic eruptions (e.g., Self et al., 1996, 1997; Thordarson and Self, 1998). However, the stratigraphy, magma generation, and eruption and emplacement processes of any associated silicic volcanic rocks have received less attention because of a sampling bias toward the basalt lavas and/or poor exposure (e.g., Paraná).

Silicic volcanic rocks have long been recognized as being associated with continental flood basalt provinces (e.g., Paraná-Etendeka and North Atlantic igneous provinces). However, the silicic volcanic rocks are generally regarded as being (1) small volume (10^4 km^3); (2) emplaced late in the eruptive sequence/history as their ascent was hindered by their higher viscosity; (3) proximal to the main locus of melt generation; and (4) formed by secondary crustal melting resulting from local heating by basaltic intrusions (e.g., White and McKenzie, 1989; White, 1992).

It is increasingly recognized that some large igneous provinces related to continental breakup are dominated by silicic volcanic rocks, with basalt forming a minor or nonexistent part. Two such silicic large igneous provinces include the Early Cretaceous volcanic rifted margin of eastern Australia (Bryan et al., 1997, 2000), and the Jurassic Chon Aike province of South America and Antarctic Peninsula (Pankhurst et al., 1998, 2000; Riley and Leat, 1999). The petrogenesis of these silicic-dominated large igneous provinces is more complex than typical basaltic large igneous provinces because of their wider variety of volcanic and intrusive compositions. Considerable debate has centered on the relative roles of fractional crystallization, assimilation-fractional crystallization, partial melting, and magma mixing in the generation of the silicic magmas (e.g., see discussions in Pankhurst et al., 1998; Ewart et al., 1998a; Wark, 1991). The calc-alkaline chemistry of the rhyolites also provides ambiguity when interpreting the tectonic setting of magmatism, and as a consequence, the tectonic setting of some large silicic igneous provinces may have previously been wrongly interpreted (e.g., eastern Australia; see Ewart et al., 1992; Bryan et al., 1997).

We propose the term "silicic large igneous province" to describe those volcanic-plutonic provinces with the following characteristics: (1) extrusive volumes of >10^5 km^3; (2) compositions are >75 vol% dacite-rhyolite; and (3) rhyodacite-rhyolite compositions near the hydrous granite minimum. We point out that silicic large igneous provinces characterize both intraplate environments (being related to continental breakup), and active convergent margins undergoing extension (being related to backarc or arc rifting; e.g., Sierra Madre Occidental, Mexico; Taupo volcanic zone, New Zealand). The latter examples show a close spatial-temporal relationship to subduction-related tectonism and igneous activity (e.g., Ward, 1995). Silicic igneous provinces from extended intraplate and continental margin settings show similarities in eruptive volumes ($\geq 10^5$ km^3), dominant lithologies (ignimbrite), and composition (intermediate-silicic calc-alkaline I-type magmas), making tectonic discrimination between them difficult.

In this chapter, we review continental flood basalt provinces that have a significant silicic volcanic component, and large igneous provinces that are silicic dominated (Whitsunday volcanic province, eastern Australia; Chon Aike province, South America–Antarctic Peninsula; Fig. 1). We make the following points.

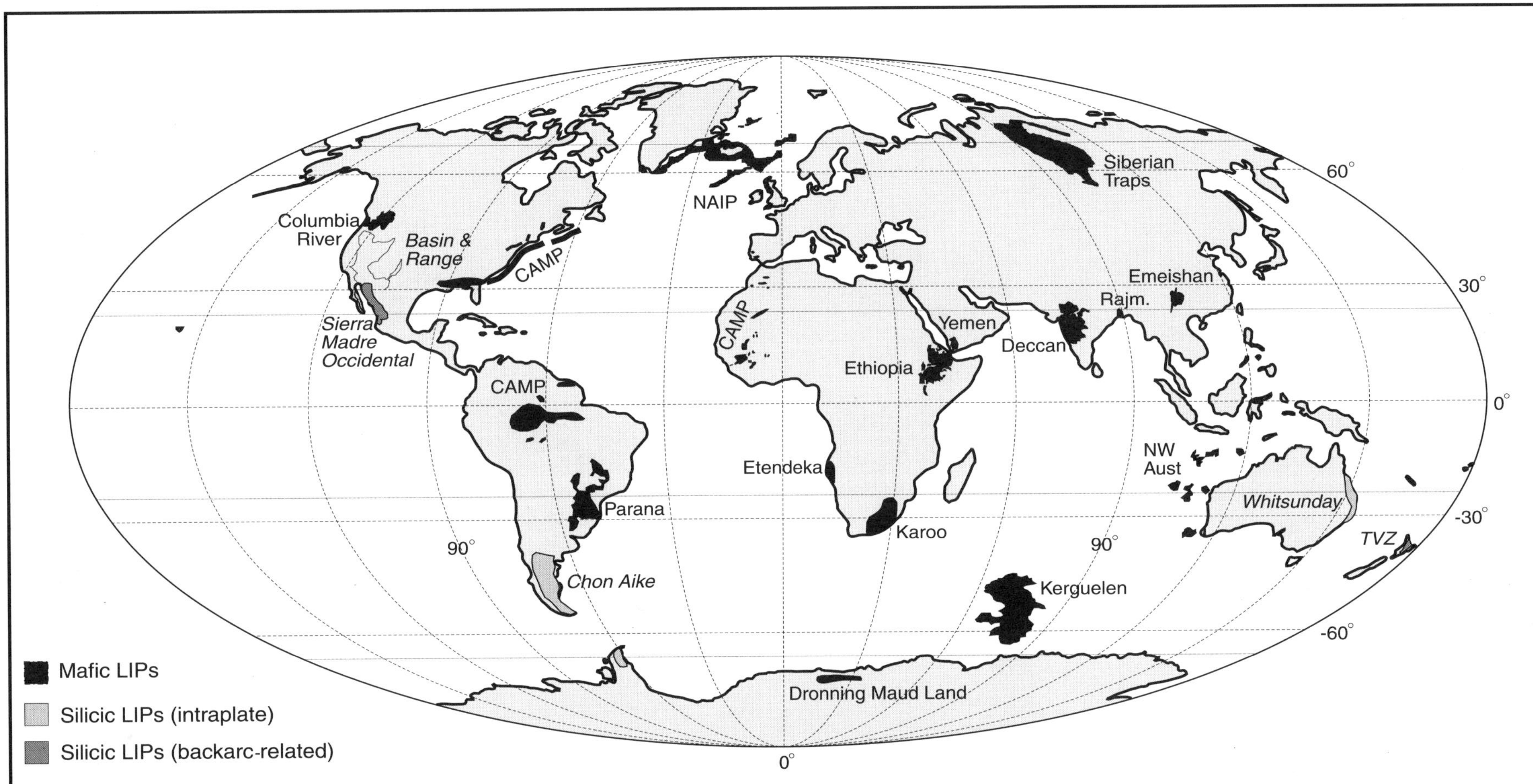

Figure 1. Distribution of large igneous provinces (LIPs) discussed in this chapter; silicic LIPs are in italics. NAIP, North Atlantic igneous province; CAMP, Central Atlantic magmatic province; Rajm. Rajmahal basalts; TVZ, Taupo volcanic zone; NW Aust, Northwest Australian oceanic plateaus: Cuvier, Roo Rise, Scott, Wallaby, and Naturaliste. Modified from Coffin and Eldholm (1994).

1. Silicic magmatism is an integral part of continental flood basalt provinces and can be of large volume (>10^4 km^3).

2. Large silicic igneous provinces exist with extrusive volumes comparable to those defined for mafic large igneous provinces, but are not associated with large extrusive volumes of mafic magma.

3. Silicic large igneous provinces (with ≤10% mafic igneous rocks) and continental flood basalts (with ≤10% silicic igneous rocks) represent end members, and there is an absence of large igneous provinces with subequal proportions of mafic and silicic igneous rocks.

4. Some silicic large igneous provinces are associated with, or may be transitional to, intraplate magmatism (e.g., Whitsunday volcanic province), whereas others are tectonically and geochemically associated with backarc processes (e.g., Sierra Madre Occidental).

Although large-volume silicic magmatism ultimately reflects continental extension, hot mantle upwelling, and intrusion of basaltic magma into the crust, a hydrous lower crust is also considered critical for the generation of large volumes of rhyolite magma.

CONTINENTAL FLOOD BASALT PROVINCES WITH SILICIC VOLCANISM

All continental flood basalt provinces of Jurassic to Cretaceous age associated with the breakup of Gondwana have varying amounts of silicic igneous rocks associated with them, as do the Tertiary North Atlantic, Yemen-Ethiopia, and Columbia River provinces (Table 1). Silicic volcanic rocks, previously unrecognized or considered, have also been intersected in the latest drilling program of the Kerguelen oceanic plateau (Ocean Drilling Program [ODP] Leg 183; Frey et al., 2000; Moore et al., 2000), consistent with occurrences of silicic volcanic rocks (lavas, tuffs, and/or bentonites) in the synrift sequences along the Indian and western Australian margins (Kent et al., 1997;

TABLE 1. SUMMARY OF MAFIC LARGE IGNEOUS PROVINCES AND THE ERUPTIVE AGES AND CHARACTER OF ASSOCIATED SILICIC VOLCANISM

Province	Age of flood basalts (Ma)	Age of silicic magmatism (Ma)	Silicic igneous lithologies	References
Emeishan (China)	?253–250	251, 250	Bentonite/tuff, dacitic/trachytic to rhyolitic lavas; granite, syenite	Chung et al. (1998); Chung and Jahn (1995)
Siberian Traps	250	Not known	Not known	Renne et al. (1995); Sharma (1997)
Central Atlantic Magmatic Province	205–191 (peak at 200)	201–162; 165, 159	Trachyte, alkali rhyolite ignimbrite and breccia, dacitic fallout tuffs, rhyolite and high-K rhyolite; high-K calc-alkaline granite, granite, quartz syenite, syenite	Marzoli et al. (1999); Hames et al. (2000); Heatherington et al. (1999); Heatherington and Mueller (1991); Eby et al. (1992)
Karoo	183–184	179	Lava-like dacitic to rhyolite ignimbrite, lava, rhyolite dikes; syenite, granite, granophyre	Encarnación et al. (1996); Cleverly et al. (1984)
Ferrar	183–184	186	Pyroclastic fallout deposits (tuff), ?ignimbrite	Encarnación et al. (1996); Faure and Hill (1973); Elliot (1992, 2000)
Paraná-Etendeka	132	132	Latite to quartz latite rheoignimbrite, rhyolite dikes, minor rhyolitic lava breccia (Messum); granite, syenite	Ewart et al. (1998a); Renne et al. (1996)
Western Australia–India–Kerguelen	130–100	213, 145–135, 116–113	Bentonite and/or silicic tuffs, pre-breakup alkali rhyolite and/or trachyte (northwestern Australia), rhyolite lavas (India)	Frey et al. (1996); Colwell et al. (1994); Moore et al. (2000)
Deccan-Seychelles	66–65	62	Trachyte tuff, rhyolite lavas and tuff; syenite, granite	Hofmann et al. (2000); Devey and Stephens (1992); Sethna and Battiwalla (1977); Mahoney (1988)
North Atlantic igneous province	~62–58, 56–53	63–61, 56–52	Rhyolite ignimbrite, tuff, lava, dykes; granites	Hamilton et al. (1998); Sinton et al. (1998); Saunders et al. (1997); Bell & Emeleus (1988)
Ethiopia-Yemen	31–26	35–32, 29–28, 19–6	Rhyolitic ignimbrite, pyroclastic surge and fallout deposits, lava, dikes; A-type granites	Baker et al. (1996a, 1996b); Ukstins et al. (2000); Kampunzu and Mohr (1991)
Columbia River–Snake River Plain–Yellowstone	17.5–6 (peak between 17–15)	16.5–0.07	Rhyolitic ignimbrite, lava, pyroclastic fallout and surge deposits, dikes	Cummings et al. (2000); Chesley and Ruiz (1998); Leeman (1982); Hooper (1997)

Note: Large igneous provinces exhibit the following features: (1) they are dominated by basalt, with minor silicic igneous rocks; (2) silicic volcanism can predate, be coincident with, or postdate the main phase of basalt volcanism; and (3) the silicic rocks are dry, high-temperature eruptives.

Colwell et al., 1994). Only the Siberian Traps have no documented silicic igneous rocks, although the temporally related Emeishan province of China contains dacitic to rhyolitic lavas, tuffs, and associated intrusives (Table 1; Chung et al., 1998). The Paraná-Etendeka province is probably the best known of these provinces for containing silicic volcanic rocks, which have been the focus of several studies (e.g., Milner, 1988; Milner et al., 1992, 1995; Garland et al., 1995; Ewart et al., 1998a). Silicic volcanic rocks only account for a small percentage (typically <5% volume) of continental flood basalt provinces, however, and very large volumes of mafic volcanic rocks (~10^6 km^3) emplaced rapidly (a few million years) are characteristic. In the following section, we examine the role of silicic volcanism in some of the best documented continental flood basalt provinces.

Karoo (Lebombo-Mwenezi)-Ferrar

The Karoo continental flood basalt province (Fig. 1) comprises an extensive series of predominantly mafic rocks and less voluminous silicic rocks that erupted prior to the fragmentation of Gondwana. Mafic magmatism peaked at 183 ± 1 Ma and was followed by smaller volumes of mafic and silicic volcanism to 179 Ma (Duncan et al., 1997). The top of the Karoo volcanic sequence is not preserved. It is important to note that although the basaltic volcanic rocks of the Karoo continental flood basalt are distributed extensively over the southern African continent, the silicic volcanic rocks are concentrated around the continental margins and sites of continental rifting (Duncan et al., 1984), mostly in the Lebombo-Mwenezi region of southern Africa (Fig. 2A). Of the total preserved thickness of the Lebombo and Mwenezi sequences, ~30% is composed of silicic volcanic rocks (Duncan et al., 1984). The Lebombo rhyolites are typically sheetlike, having features common to both lavas and ignimbrites, but have been interpreted as high-temperature (lava-like) ignimbrites (Cleverley, 1979). Individual rhyolite units extend for as much as 60 km along strike, and reach maximum thicknesses of 200 m. The rhyolites have an estimated maximum thickness of 5 km in the Lebombo monocline and an estimated volume of 35 000 km^3 (Cleverly et al., 1984), accounting for <2% of the total volume of the preserved Karoo province. The rhyolites yield ages of 178–180 Ma (Rb-Sr, Allsopp et al., 1984; $^{40}Ar/^{39}Ar$, Duncan et al., 1997), which are younger than the main phase of mafic volcanism at 182–183 Ma (Encarnación et al., 1996; Duncan et al., 1997), although they occur interbedded with basalt lavas.

In the synchronous Antarctic Ferrar province, silicic explosive volcanism mainly preceded, but was also contemporaneous with the eruption of the Ferrar Group tholeiitic basalts, both of which were emplaced into a volcanic-tectonic rift system (Elliot, 1992). The Hanson Formation of the central Transantarctic mountains comprises ~240 m of silicic tuffs, tuffaceous sandstones, and subordinate quartzose sandstones, and is overlain by extrusive basaltic rocks of the Ferrar Group (Elliot, 1992, 2000). A Rb/Sr whole-rock isochron on four tuffs yielded an age of 186 ± 9 Ma (Faure and Hill, 1973), but may be affected by widespread zeolitization (Elliot, 2000). However, we note that this age for the Hanson Formation overlaps with widespread silicic explosive volcanism of the Chon Aike province and the component Brennecke and Mount Poster Formations of the Antarctic Peninsula (Pankhurst et al., 2000; see following). The predominantly sedimentary nature of the Hanson Formation may indicate that it is a distal equivalent of the volcanic-dominated Mount Poster and Brennecke Formations.

The silicic volcanic rocks of the Lebombo monocline are primarily rhyolitic in composition, although a number of samples have lower silica contents, being classified as dacites and rhyodacites (Duncan et al., 1984). Most rhyolites have relatively uniform Sr and Pb initial isotopic ratios similar to those of the flood basalts, suggesting derivation from a similar source (Cleverly et al., 1984). However, the lowermost rhyolites interbedded with basalts (Mkutshane Beds) have strong enrichments in radiogenic Sr and Pb, and have been interpreted to be either crustal melts or strongly contaminated basaltic magmas (Cleverly et al., 1984). Strong enrichment of the incompatible elements (e.g., Nb, Zr, Y, Ba, Rb, Sr), primitive initial Sr and Nd isotopic ratios, and modeling suggest the bulk of the rhyolites were generated as a result of ~10% partial melting of newly underplated Karoo basaltic magma (gabbro sills or cumulate) following crustal thinning and decompressional melting (Cleverly et al., 1984; Harris and Erlank, 1992; Cox, 1993). However, Bristow et al. (1984) concluded that melting of mafic lower crustal material (e.g., mafic granulite) could also be a plausible source for the rhyolites, and consistent with the model of Cleverly et al. (1984). The low ^{18}O values for the Lebombo rhyolites further require large-scale interaction of the source with meteoric water at high temperatures (Harris and Erlank, 1992; Harris, 1995). The widespread basalts and/or dolerites (Lesotho-type of Marsh and Eales, 1984) have calc-alkaline or subduction-related geochemical signatures (Cox, 1983) interpreted to reflect contamination by subcontinental lithospheric mantle (e.g., Sweeney et al., 1994), and the Rooi Rand dike swarm is partly mid-ocean ridge basalt–like (Watkeys et al., 2000).

Paraná-Etendeka

The Paraná-Etendeka province (Fig. 1) was erupted just prior to the opening of the South Atlantic ca. 130 Ma (Renne et al., 1992, 1996; Turner et al., 1994). The province is dominated by mafic lavas; silicic volcanic rocks account for only 3% of the preserved total erupted volume, equivalent to ~20 000 km^3 (Harris and Milner, 1997). However, the silicic volcanic rocks crop out over an estimated combined area of 170 000 km^2, and have proved critical in correlating the eruptive stratigraphies on either side of the South Atlantic Ocean (Milner et al., 1995). More silicic rocks occur as intrusions in central complexes associated with the volcanism (e.g., Messum and Brandberg igneous centers; Martin et al., 1960; Milner and Ewart, 1989; Schmitt et al., 2000; Fig. 2B). The Etendeka region in particu-

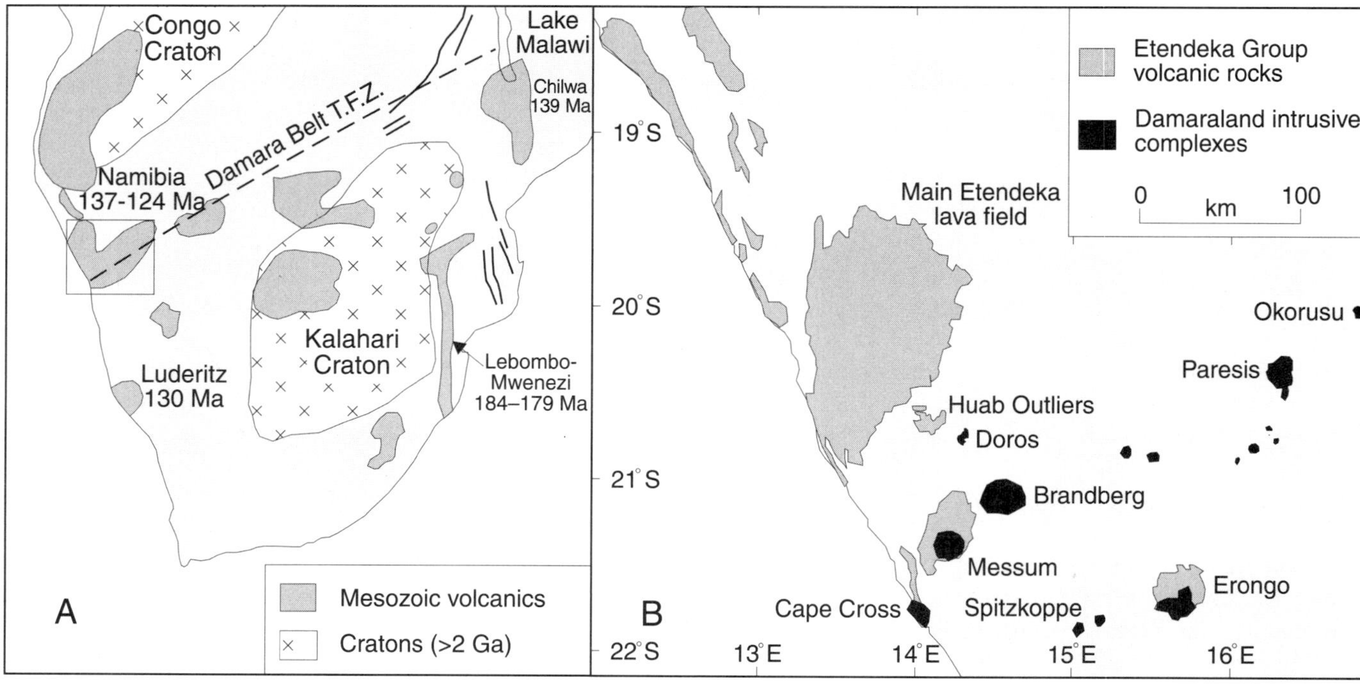

Figure 2. A: Distribution of Mesozoic volcanic rocks in southern Africa and structural setting of Early Cretaceous magmatism of the Etendeka province at the western end of a major continental-scale transfer fracture zone (T.F.Z.) separating >2 Ga cratons. Modified from Kampunzu and Popoff (1991). B: Map of northwestern Namibia showing distribution of Etendeka Group volcanic rocks and Damaraland subvolcanic complexes, which define a linear belt trending east-northeast up to 350 km inland. The Messum complex was an eruptive center for some quartz latite units (Milner and Ewart, 1989; Ewart et al. 1998a), and recent studies have also identified the Doros center as a feeder for a lower sequence of flood basalt lavas (Jerram et al., 1999a). After Milner et al. (1992).

lar has proved critical, with some eruptive centers being identified for both mafic (e.g., Doros igneous center, Jerram et al., 1999a) and silicic units (e.g., Messum, Milner et al., 1995; Fig. 2B). Although the silicic components make up only a small volume of the province overall, the dimensions of individual silicic eruptive units are as substantial as those mapped for flood basaltic lavas (see Jerram, this volume). Individual silicic units occur as extensive, flat-lying, low-aspect-ratio sheets with average thicknesses of 70–250 m, strike lengths of >340 km, areal extents of >25 000 km^2, and eruptive volumes to 6340 km^3 (Milner et al., 1992, 1995; Ewart et al., 1998a). Silicic volcanic rocks occur interbedded with basalt lavas related to the main pulse of flood volcanism, but also characterize the upper parts of the preserved stratigraphy (Fig. 3). An earlier phase of rhyolitic volcanism preceding the eruption of the voluminous quartz latites was identified at Messum (Messum core rhyolites of Ewart et al., 1998b).

The silicic volcanic rocks are referred to as rhyolite or rhyodacite in the Paraná (e.g., Garland et al., 1995), but the terms "quartz latite" and "latite" are used in the Etendeka (e.g., Milner, 1988; Milner et al., 1992). In the International Union of Geological Sciences total alkalis-silica classification (Le Maitre, 1989), these rocks plot continuously across the dacite-rhyolite-trachyte fields, and have a distinctive composition (e.g., high Fe, alkalies; low Al in relation to silica content; Ewart et al., 1998a). Thermometry indicates temperatures >1000 °C for the quartz latites, which is reflected in their lava-like, fluidal, and weakly phenocrystic character.

The quartz latite melts have been interpreted in terms of large-scale assimilation-fractional crystallization (AFC) processes, involving high degrees of lower and upper crustal melting, with thermal and material input from hybridized (low Ti-Zr) basaltic magmas (Ewart et al., 1998a, 1998b). The crustal end member has been interpreted to be dry granulitic lower crust (Harris et al., 1990; Harris and Milner, 1997), or the Middle Proterozoic restite source of the Damaran granites; some shallower crustal input is likely (Ewart et al., 1998a). Two chemically different rhyolites have been identified in the southern Uruguay portion of the province and are interpreted as either the products of extensive fractionation and crustal assimilation, or melts of preexisting mafic lower crust with subsequent extreme fractionation (Kirstein et al., 2000).

Deccan Traps

The main phase of mafic volcanism in the Deccan Traps (Fig. 1) occurred close to the Cretaceous-Tertiary boundary ca. 66 Ma (Duncan and Pyle, 1988; Hofmann et al., 2000). Silicic

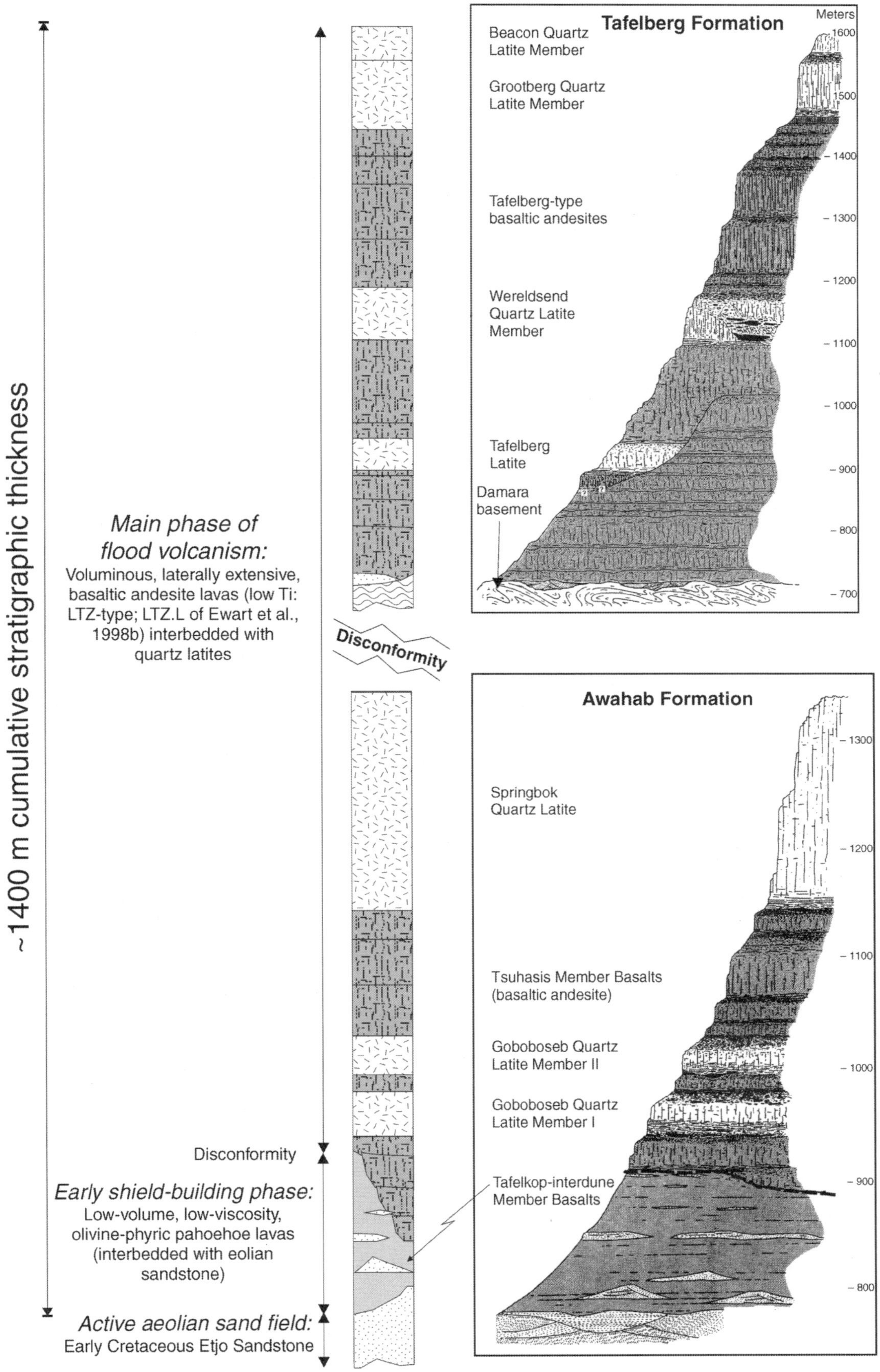

Figure 3. Stratigraphic subdivisions through the Etendeka Group, northwestern Namibia, showing the interbedded nature and thicknesses of quartz latite units within main phase of flood basalt volcanism. Adapted from Milner (1988) and Jerram et al. (1999a, Fig. 5). Meters are above sea level.

volcanic rocks are known from the Deccan Traps, especially around Bombay (Sethna and Battiwala, 1977; Lightfoot et al., 1987) within the Narmada lineament (Mahoney, 1988), and from drill holes to the north of the Seychelles (Devey and Stephens, 1992, and references therein). Rhyolites from the Deccan province have an estimated volume of ~500 km^3 (Lightfoot et al., 1987), accounting for <1% of the total eruptive volume. Near Bombay, the rhyolites occur mainly as laterally extensive lavas, cropping out for as much as 20 km along strike; individual units are 20–100 m thick (Lightfoot et al., 1987). The rhyolites yield a younger age than the mafic rocks (61.5 ± 1.9 Ma; Rb/Sr whole-rock isochron age; Lightfoot et al., 1987). However, dacitic tuffs within the main phase of the Deccan flood basalts indicate that the contribution of more evolved volcanism may be greater than previously recognized (Widdowson, 1997; Widdowson et al., 1997).

The silicic volcanic rocks are peralkaline in character, as observed for silicic volcanics from other continental flood basalts (e.g., Ethiopia-Yemen), and trace element signatures indicate significant fractional crystallization in the evolution of the rhyolites (Lightfoot et al., 1987). Low initial $^{87}Sr/^{86}Sr$ ratios and Pb isotopic compositions similar to the associated flood basalts indicate derivation from a primitive mafic source, as observed in the Karoo rhyolites (e.g., Cleverly et al., 1984). The rhyolites were similarly interpreted to have been generated by partial melting of underplated basaltic magma (~10%–15%) with superimposed fractional crystallization (Lightfoot et al., 1987).

North Atlantic igneous province

The North Atlantic igneous province is related to the Tertiary opening of the North Atlantic Ocean and the separation of Greenland from northwestern Europe, and North America from Greenland (Fig. 1). Predominantly mafic magmatism occurred in two main phases: phase 1 (ca. 62–58 Ma) produced the terrestrial basaltic lava sequences; and phase 2 (56–53 Ma), linked to the breakup of the northeast Atlantic, produced the bulk of the seaward-dipping reflector sequences along the continental margins and the main series of basalts in central East Greenland (Saunders et al., 1997). Persistent volcanism that is commonly attributed to the Iceland plume has since occurred for the past ~60 m.y. in the central region (Fig. 1). The area covered by flood basaltic volcanism, both onshore and offshore, is ~1.3×10^6 km^2, and the eruptive volume is ~1.8×10^6 km^3 (Coffin and Eldholm, 1994; Eldholm and Grue, 1994), although White et al. (1987) suggested a total igneous volume, including erupted material and additions to the deeper crust, of between 5×10^6 and 1×10^7 km^3. A substantial portion of the province is offshore, and whereas the composition of the offshore volcanic sequences is generally regarded as basaltic, an unknown but perhaps significant proportion comprises silicic lavas and tuffs (Sinton et al., 1998).

Volcanism in western Britain was concentrated around the central intrusive complexes that comprise a wide variety of igneous rocks, from peridotite to granite (Saunders et al., 1997). Silicic igneous rocks are an important part of the province, largely occurring within, and typically emplaced early in the evolution of, the intrusive centers (Bell and Emeleus, 1988). Preservation of pyroclastic material is restricted mostly to the remains of caldera complexes (e.g., Rum, Mull, Arran), and thin, sanidine-bearing tuffs occurring throughout the lava fields (Bell and Emeleus, 1988; Emeleus et al., 1996; Bell et al., 1996). Two significant silicic volcanic units also occur in the Antrim Lava Group of Northern Ireland: the Tardree rhyolite complex and the Donalds Hill welded ignimbrite that occur interbedded with, and underlie, the basaltic lava pile, respectively (Meighan et al., 1984; Bell and Emeleus, 1988; Mitchell et al., 1999).

Andesitic to rhyolitic lavas and pyroclastic rocks were emplaced contemporaneously with basaltic lavas at the southeast Greenland margin at 62–61 Ma (Larsen et al., 1995; Sinton et al., 1998). The $^{40}Ar/^{39}Ar$ ages of 62.8 ± 0.6 and 62.4 ± 0.6 Ma for sanidine-bearing tuffs intercalated with the Eigg Lava Formation in western Britain (Pearson et al., 1996) are among the oldest ages for the entire North Atlantic igneous province (Saunders et al., 1997). Several granite intrusions in the British Tertiary igneous province (Arran, Skye) were also emplaced during phase 1 of Saunders et al. (1997). Silicic magmatism continued during phase 2; granites (e.g., Skye) and silicic volcanics were emplaced between ca. 56 and 52 Ma (Hamilton et al., 1998; Saunders et al., 1997). The Gardiner alkaline complex in East Greenland has been interpreted as the source for alkaline tuffs that occur in East Greenland (dated as 53.8 ± 0.3 Ma), Europe, and in North Atlantic cores (Heister et al., 2001). Silicic lavas erupted ca. 55 Ma along the Voring Plateau offshore Norway (Eldholm et al., 1987) and offshore Scotland (Sinton et al., 1998) were followed shortly thereafter by basaltic volcanism. These offshore localities from Europe show a stratigraphic succession from crust-derived intermediate-silicic magmatism to basaltic magmatism (Sinton et al., 1998).

The felsic volcanic rocks range from peraluminous to peralkaline compositions, although peralkaline compositions are absent from the major granite bodies (Meighan et al., 1992). Fractional crystallization has been an important process in their development; however, it is widely accepted that most silicic magmas represent the end products of fractionated, crustally contaminated basaltic melts, the proportion of the crustal component varying from ~10% to ~67% (Dickin and Exley, 1981; Meighan et al., 1992). The rhyolitic pitchstones of Mull, however, contain only small amounts (5%–10%) of basaltic magma, with both lower and upper crustal sources important in their generation (Preston et al., 1998a, 1998b).

Yemen-Ethiopia

Flood volcanism in Yemen-Ethiopia (Fig. 1) occurred between 31 and 26 Ma, with the peak of ages being between 31 and 29 Ma, and was related to the rifting of the Red Sea and the Gulf of Aden (Baker et al., 1996a; Hofmann et al., 1997). The province has an extrusive volume of >350 000 km^3 (Mohr,

1983), dominated by basaltic lavas, but includes a significant proportion of silicic volcanic rocks (Table 1). Rhyolitic ignimbrite, other pyroclastic rocks, and lavas occur toward the middle and top of the flood basalt lava pile in Ethiopia, and ages overlap the peak eruptive ages for the flood basalts (30.6–28.2 ± 0.1 Ma; Hofmann et al., 1997). Rhyolites and associated tuffs from the southern Red Sea Hills, Sudan, also have eruptive ages close to the main pulse of flood volcanism (29.9 ± 0.3 Ma, Kenea et al., 2001). In Yemen, rhyolitic pyroclastic rocks and lesser lava also predominate in the upper parts of the lava succession. The oldest rhyolitic units throughout Yemen are ca. 29.3–29 Ma, and bimodal volcanism continued until 26.5 Ma (Baker et al., 1996a). A period of widespread prerift basaltic and subordinate rhyolitic volcanism also occurred in southern Ethiopia between 45 and 33 Ma, prior to the better known flood basalt volcanic event of the Ethiopian Traps and Yemen (Ebinger et al., 1993, 2000; George et al., 1998). Silicic explosive eruptions, including voluminous rhyolitic ignimbrites, occurred ca. 40, 37, and 33 Ma during this earlier eruptive phase (Levitte et al., 1974; Davidson and Rex, 1980; Ebinger et al., 1993). Silicic volcanism (trachytes to peralkaline rhyolites and phonolites) has continued until recently during development of the Ethiopian rift, several centers being focused along the structural margins of the rift (Woldegabriel et al., 1990; Chernet et al., 1998; Ebinger et al., 2000). Younger (26–14 Ma) silicic magmatism also occurs along the Red Sea rift margin, and is dominated by trachytic to rhyolitic dikes and A-type granites that intrude basement and the flood basalt sequences (Capaldi et al., 1987; Chazot and Bertrand, 1995). The silicic igneous rocks show strong evidence for combined crustal assimilation and fractional crystallization of parental basaltic magmas (e.g., Chazot and Bertrand, 1995; Baker et al., 1996b), whereas crustal assimilation (to 30%) has significantly modified the primary chemistry of the flood basalts (Baker et al., 1996b).

In Yemen, ~50% of the volcanic sequences (~300 m thickness) comprise silicic pyroclastic rocks and, less commonly, rhyolitic lavas (Baker et al., 1996a, 1996b; Ukstins et al., 2000). Within the lower, almost entirely basaltic lava sections are interbedded distal facies of silicic explosive eruptions (fine-grained, 5-m-thick silicic fallout deposits), indicating penecontemporaneous effusive basaltic flood volcanism and explosive silicic volcanism (i.e., multiple vents; Ukstins et al., 2000). Silicic fallout deposits, welded ignimbrite, pyroclastic surge deposits, lavas, breccias, and sedimentary intervals reflecting reworking of silicic pyroclastic material characterize the upper parts of the volcanic stratigraphy in Yemen (Baker et al., 1996a). Individual eruptive units have thicknesses of 90 m, dispersal areas to 30 000 km^2, and eruptive volumes of ~600 km^3 (Ukstins et al., 2000). In the Sana'a area of Yemen, megabreccia containing a tuffaceous matrix and with blocks of dimensions to 70 × 15 × 20 m reflect a caldera-forming eruption at 29 Ma (Ukstins et al., 2000). The silicic volcanics show an up-sequence petrographic variation, tending to more evolved compositions in the upper units (Ukstins et al., 2000).

Summary of continental flood basalt provinces with silicic volcanism

Most, if not all, continental flood basalt provinces contain a significant contribution from silicic volcanism. The following points summarize the important aspects of silicic volcanism in the continental flood basalt provinces.

Low volume, olivine-rich, basalts-picrites represent the early mafic eruptions of continental flood basalts (e.g., North Atlantic igneous province, Etendeka; Jerram et al., 1999a), with slightly more evolved and/or contaminated mafic lavas of much larger volume and varying volumes of silicic volcanism characterizing the main phase of flood volcanism (Fig. 3).

Within the main pulse of flood basaltic volcanism, silicic volcanism can occur throughout and/or very early in the eruptive sequence (e.g., North Atlantic igneous province, Bell and Emeleus, 1988; Sinton et al., 1998; Mitchell et al., 1999; Ferrar, Elliot, 2000). In the Etendeka, Karoo and Yemen examples, silicic volcanics increase in abundance up the volcanic pile (Fig. 3). However, because the eruptive stratigraphy in each case is incompletely preserved, the proportion of silicic volcanism may be underestimated, and their occurrence late in the evolution of a province may be an artefact of preservation.

The interbedded silicic volcanic rocks, combined with palynology and crosscutting relationships of (silicic) intrusives, can be important in obtaining higher precision geochronology for continental flood basalts (e.g., Baker et al., 1996a; Emeleus et al., 1996; Bell et al., 1996; Hamilton et al., 1998).

The processes of rhyolite generation are dominated by crustal melting and assimilation by flood basaltic magmas with superimposed fractional crystallization (AFC; Etendeka, North Atlantic igneous province, Yemen), whereas remelting of mafic cumulate or lower crust is considered important for the Karoo and Deccan rhyolites. Crustal melting typically involved refractory, dry Archean to Proterozoic basement material, and the larger volume rhyolites all require some basaltic input into the magmas, in terms of both heat transfer and material addition. Crustal melting and consequent contamination of the flood basalt magmas is commonplace in the continental flood basalts, although contamination by melts from the subcontinental lithospheric mantle appears to be more important in the Karoo continental flood basalt (Ellam et al., 1992).

The dimensions of some silicic eruptive units, which are as voluminous, if not more so, than the largest recorded flood basaltic lavas, could have significant implications in terms of the environmental impact of continental flood basalt provinces. For example, the Goboboseb and Springbok quartz latite units erupted from Messum (Paraná-Etendeka province) have combined eruptive volumes of >8000 km^3 (Milner et al., 1995). Placed in perspective of crustal thickness, the Springbok quartz latite unit alone would have a magma sphere diameter of 23 km; this is only a minimum estimate of the magma chamber dimension, which is unlikely to have been totally evacuated (Ewart et al., 1998a).

SILICIC LARGE IGNEOUS PROVINCES

Silicic large igneous provinces have extrusive volumes comparable to mafic large igneous provinces, but have minor amounts of mafic magma expressed at the surface. Large volume silicic igneous provinces of Mesozoic to Cenozoic age[1] are recognized from two tectonic settings: those associated with continental breakup (e.g., Whitsunday volcanic province, eastern Australia; Chon Aike province, South America–Antarctic Peninsula), as well as active continental convergent margins undergoing extension (e.g., Sierra Madre Occidental, Mexico; Fig. 1). Despite showing lithological and geochemical similarities, the tectonics for the formation of these silicic igneous provinces are fundamentally different and therefore discussed separately here. In this chapter we also discuss the Taupo volcanic zone of New Zealand in the context of a modern example of a silicic large igneous province developing along an active continental margin. Although the Taupo volcanic zone does not fit our definition of a silicic large igneous province (i.e., eruptive volume), it shows many attributes that would have characterized the other silicic large igneous provinces and are emphasized here.

Associated with continental plate breakup

Two silicic large igneous provinces related to continental plate breakup have been recognized in the rock record: the Early Cretaceous Whitsunday volcanic province of eastern Australia (Ewart et al., 1992; Bryan et al., 1997, 2000), and the Jurassic Chon Aike province of South America and the Antarctic Peninsula (Pankhurst et al., 1998, 2000; Riley and Leat, 1999; Feráud et al., 1999; Riley et al., 2001). These provinces differ from well-known continental flood basalts by being overwhelmingly dominated (>80% volume) by intermediate to silicic magma compositions, with basalt being a minor component. These silicic large igneous provinces are (1) volumetrically dominated by ignimbrite; (2) active over prolonged periods (to 40 m.y.); and (3) spatially and temporally related to other mafic large igneous provinces and plate breakup (Bryan et al., 2000). The volcanic rocks typically show calc-alkaline affinities that resemble modern destructive plate margin volcanic rocks rather than bimodal or alkalic volcanism associated with continental flood basalts and continental rifts.

Whitsunday volcanic province. The Early Cretaceous Whitsunday volcanic province is part of a silicic-dominated pyroclastic volcanic belt that extended along the northeast Australian coast (Fig. 4) and was >900 km in strike length, >1 km thick,

[1] Recent studies are indicating the existence of another silicic large igneous province of Permo-Carboniferous age (~320–280 Ma) in the New England fold belt of Eastern Australia (e.g., Holcombe et al., 1997; Allen et al., 1998). The dimensions of this large igneous province remain poorly constrained at present, but silicic igneous rocks define a belt >1900 km long and ≥300 km wide (areal extent of >570 000 km^2), with volcanic sequences up to 1 km thick. The Proterozoic (1600–1585 Ma) Gawler Range volcanics (>25 000 km^2) of South Australia (e.g., Fanning et al., 1988; Giles, 1988; Creaser and White, 1991) may also be the remnants of a silicic large igneous province.

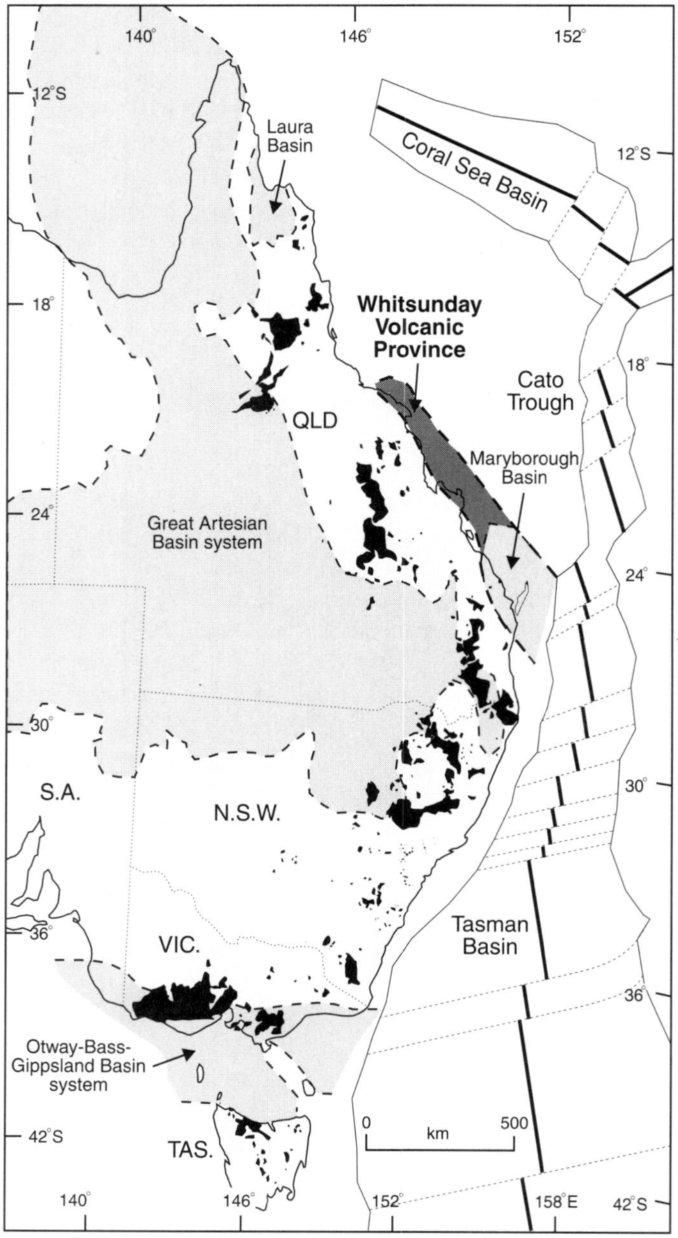

Figure 4. Location of the silicic-dominated Whitsunday volcanic province (132–95 Ma) and Early Cretaceous sedimentary basins of eastern Australia that contain >1.4 × 10^6 km^3 of coeval large igneous province-derived volcanogenic sediment (Bryan et al., 1997). Large igneous province magmatism was followed by: (1) kilometer-scale uplift of eastern margin of Australia beginning ca. 100–95 Ma (e.g., O'Sullivan et al., 1995, 1999); (2) seafloor spreading in Tasman Basin–Cato Trough–Coral Sea Basin occurring between 84 and 56 Ma (e.g., Veevers et al., 1991); and (3) intraplate alkaline volcanism (80–0 Ma, shown in black) that was partly synchronous with seafloor spreading, and that defines a broken belt 4400 km long along the highlands of eastern Australia. Intraplate alkaline volcanism occurred within 500 km of coastline, and has an extrusive volume of >20 000 km^3 (Johnson, 1989). QLD, Queensland; N.S.W., New South Wales; VIC., Victoria, TAS., Tasmania; S.A., South Australia.

and had an extrusive volume >10^5 km^3 (Clarke et al., 1971; Bryan et al., 1997). The remainder of the silicic large igneous province is interpreted to have been eroded and/or rifted from the Australian continent, now occurring on submerged continental ridges and marginal plateaus following continental breakup and seafloor spreading in the Late Cretaceous and Tertiary. The original extent of the volcanic belt is thought to have been >2500 km along the present eastern Australian plate margin, and igneous rocks of Early Cretaceous age are widespread elsewhere in eastern Gondwana, occurring in New Zealand, Lord Howe Rise, and Marie Byrd Land (see references in Bryan et al., 2000). Most of the eruptive products of silicic volcanism, however, are preserved as huge volumes of coeval volcanogenic sediment in adjacent sedimentary basins of eastern Australia (Fig. 4), where the volume of the volcanogenic sediment alone (>1.4×10^6 km^3; Bryan et al., 1997) exceeds that of several mafic continental flood basalts. Such substantial volumes of coeval volcanogenic sediment are not characteristic of other large igneous provinces, and voluminous pyroclastic eruptions were an important factor in generating fine-grained volcanic material that was rapidly delivered into these sedimentary basin systems (Bryan et al., 1997, 2000).

The main period of volcanic activity occurred between 120 and 105 Ma (Ewart et al., 1992). Lithologically, the volcanic sequences are volumetrically dominated by welded dacitic-rhyolitic lithic-rich ignimbrite, and some interpreted intracaldera ignimbrite units are as thick as 1 km (Clarke et al., 1971; Ewart et al., 1992; Bryan et al., 2000). Coarse lithic lag breccias containing clasts to 6 m diameter (Ewart et al., 1992) commonly cap the ignimbrites in proximal sections and record the onset of caldera collapse. The volcanic sequences record a multiple vent, but caldera-dominated, low-relief volcanic region (Bryan et al., 2000). Volcanism appears to have evolved from an early explosive phase dominated by intermediate compositions, to a later, bimodal effusive-explosive phase characterized by rhyolitic ignimbrites and lavas and primitive basaltic lavas and/or intrusives (Bryan et al., 2000). The ignimbrite-dominated sequences are intruded by gabbro and/or dolerite to rhyolite dikes (to 50 m width), sills, comagmatic granite (Ewart et al., 1992), and rarely in the intracaldera sequences, by welded pyroclastic dikes, interpreted to be vents for some of the ignimbrite-forming eruptions (Bryan et al., 2000).

Chemically, the suite ranges continuously from basalt to high-silica rhyolite, with calc-alkalic to high-K affinities (Ewart et al., 1992). The range of compositions is interpreted as being generated by two-component magma mixing and fractional crystallization superimposed to produce the rhyolites. The two magma components are (1) a volumetrically dominant partial melt of relatively young, nonradiogenic calc-alkaline crust; and (2) a within-plate tholeiitic basalt of enriched-MORB affinity (Ewart et al., 1992; Stephens et al., 1995).

Chon Aike province. The Chon Aike province of Patagonia (Fig. 5) extends from the Atlantic Coast to the Chilean side of the Andes (Pankhurst et al., 1998) and is correlated with the Jurassic silicic volcanic rocks of the Antarctic Peninsula (Riley and Leat, 1999). In eastern Patagonia, the volcanic rocks are predominantly flat lying and undeformed where they overlie crystalline basement rocks of Precambrian to earliest Jurassic age and Lower Jurassic rift-related sedimentary rocks. In contrast, silicic volcanic rocks of the Andean Cordillera form relatively narrow outcrops, which are locally deformed, tilted, and strongly affected by hydrothermal alteration.

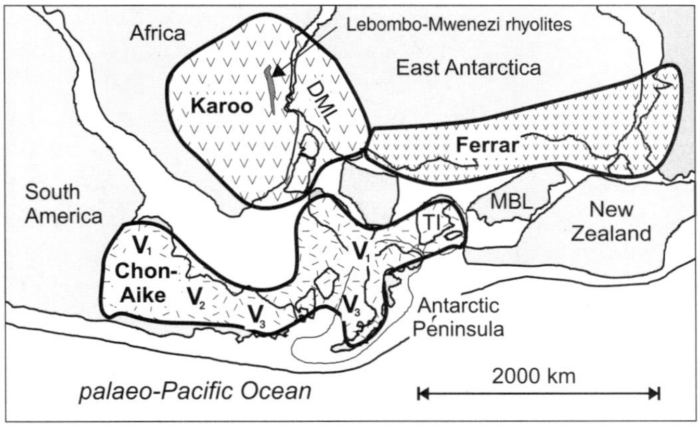

Figure 5. Reconstruction of western Gondwana prior to breakup showing the spatial relationship of the Jurassic Karoo and Ferrar continental flood basalt provinces to the silicic Chon Aike province. Note marginal setting and general migration of volcanic phases (V_1 to V_3) within Chon Aike province away from the locus of basaltic flood volcanism. DML, Dronning Maud Land; MBL, Marie Byrd Land; TI, Thurston Island. Modified after Pankhurst et al. (1998, 2000).

The province is dominated by phenocryst-poor ignimbrites, sourced from multiple caldera centers (e.g., Aragón et al., 1996; Riley and Leat, 1999), and vary in degree of welding from high-grade rheomorphic ignimbrites with parataxitic textures, to the volumetrically dominant, nonwelded, lithic-rich ignimbrites. Volumetrically minor rhyolite lavas, fallout deposits, debris-flow deposits, and epiclastic deposits are interbedded with the ignimbrites. The province is chemically bimodal, but is dominated by rhyolite, with only rare intermediate (basaltic andesite and/or andesite) compositions.

TABLE 2. SUMMARY OF AGES AND RELATIVE ERUPTIVE VOLUMES FOR THE THREE ERUPTIVE PHASES IDENTIFIED FOR THE CHON AIKE PROVINCE

Eruptive phase	Formations	Age (Ma)	Volume (km^3)
V1	Marifil (northeastern Patagonia) Brennecke, Mount Poster (southern Antarctic Peninsula)	188–178	45 000
V2	Chon Aike (Argentina) Mapple (Antarctic Peninsula)	172–162	130 000
V3	El Quemado (western Argentina) Ibañez (Chile)	157-153	55 000

Note: Data after Pankhurst et al. (1998, 2000).

The eruptive ages of silicic volcanic rocks from Patagonia and the Antarctic Peninsula have been defined by U-Pb ion microprobe dating (Pankhurst et al., 2000) and $^{40}Ar/^{39}Ar$ geochronology (Féraud et al., 1999). The age data indicate that volcanism continued for as long as 30 m.y., from the Early Jurassic to Late Jurassic, but occurred in three main phases (Table 2). The first phase of volcanism (188–178 Ma; V_1 of Pankhurst et al., 2000; Fig. 5) has a peak eruptive age of 184 ± 2 Ma, and brackets the peak of flood basalt volcanism of the Karoo and Ferrar provinces at 183 ± 1 Ma (e.g., Encarnación et al., 1996). The Tobífera Formation of southernmost Patagonia, however, yielded ages of 178 and 171 Ma (Pankhurst et al., 2000) and may partly span the interval between V_1 and V_2 (172–162 Ma). The final episode, V_3 (157–153 Ma), is confined to the Andean volcanic outcrops of Argentina and Chile (Table 2), although small granite bodies of this age occur both in western Patagonia and in the western Antarctic Peninsula, and may be subvolcanic equivalents. The three eruptive phases define an age progression of magmatism from an intraplate to continental margin setting, away from the locus of flood basalt volcanism in the Karoo province (Pankhurst et al., 2000; Fig. 5). Concomitant with the age progression is a change in rhyolite composition from intraplate (higher Nb, Zr; V_1) to calc-alkaline (lower Nb, Zr; V_2 and V_3) geochemical signatures. The Middle Jurassic silicic volcanic rocks of the Mapple Formation (Antarctic Peninsula) and those of the Chon Aike Formation (South America) are thought to have been generated as a result of anatexis of Grenvillian age hydrous mafic lower crust of andesitic composition, linked to pre-Middle Jurassic subduction, and superimposed fractional crystallization (Pankhurst and Rapela, 1995; Riley et al., 2001).

Associated with active continental convergent margins

These silicic igneous provinces are generally related to continental extension preceding backarc basin formation. They have extrusive volumes of the order of 10^4–10^5 km^3 and volcanic activity can be prolonged for 10–20 m.y. The silicic magmas commonly possess a more hydrous mineralogy (e.g., Ewart, 1979), and I-type calc-alkaline magma compositions.

Sierra Madre Occidental. The mid-Tertiary Sierra Madre Occidental province is the largest silicic volcanic province in North America (Fig. 1), comprising more than 350 calderas (Swanson and McDowell, 1984) in a volcanic-plutonic belt ~1200 km long and 200 km wide in northern Mexico (Ward, 1995), and with an estimated extrusive volume of as much as 250 000 km^3 (Cameron et al., 1980). Magmatism mostly occurred over a relatively short period (27–34 Ma, McDowell and Clabaugh, 1979; 28–31 Ma, Nieto-Samaniego et al., 1999). However, the province was contiguous with widespread, predominantly explosive silicic volcanism (ignimbrite flare-up) in the Basin and Range province of the western United States (Fig. 1), where >35 000 km^3 of dacitic to rhyolitic ignimbrite was emplaced between 31 and 20 Ma (Best and Christiansen, 1991; Lipman, 1984; Ward, 1991). The Sierra Madre Occidental formed at a time when andesitic (suprasubduction zone) magmatism was rare and during the period of greatest divergence of the Pacific and North American plates, resulting in extension of the continental margin (Ward, 1991). The province was emplaced through basement that varied from Precambrian ("Grenville," ≥1 Ga) to Phanerozoic "exotic" terranes accreted to the continental margin in the Mesozoic (Albrecht and Goldstein, 2000).

The Sierra Madre Occidental province is dominated by rhyolitic ignimbrites, with rare mafic to intermediate rocks, ranging in composition from 50 to 76 wt% SiO_2 (e.g., Wark, 1991; Albrecht and Goldstein, 2000). The volcanics have medium- to high-K calc-alkaline compositions; however, spatial differences exist in trace element (i.e., low and high Nb, Zr, Th, Rb suites) and isotopic compositions that reflect the age and composition of the basement through which the volcanics were erupted (Albrecht and Goldstein, 2000). Initial $^{87}Sr/^{86}Sr$ ratios range from 0.7044 to ~0.710, and show no systematic variation with rock composition (Wark, 1991; Albrecht and Goldstein, 2000). There is a divergence of opinion about the origin of the voluminous rhyolites. Some workers interpreted the rhyolites to have formed by lower crustal melting (e.g., Elston, 1984; Ruiz et al., 1988). Widespread partial melting of lower crust (due to heating by emplaced basalt) is also generally considered as the mechanism to account for the voluminous mid-Tertiary silicic volcanism of the western United States (e.g., Lipman and Glazner, 1991). In contrast, fractional crystallization of precursor, more mafic magmas (basalt-andesite) to produce the rhyolites has also been proposed (e.g., Cameron et al., 1980; Cameron and Hanson, 1982; Wark, 1991). Volume considerations, however, tend to argue against the derivation of the voluminous rhyolites by fractional crystallization alone (see also Pankhurst and Rapela, 1995), and Cameron et al. (1980) suggested that 80% of the mass of a basaltic andesite magma had to have been removed by fractional crystallization to produce the rhyolites. The relationship between volcanic compositions, isotopic ratios, and the age of basement reflects the strong effect of continental crust on the chemistry of the silicic magmas, and crustal contributions to basaltic magmas are 20%–70% (Albrecht and Goldstein, 2000).

Taupo volcanic zone. The Taupo volcanic zone (Fig. 1) is an exceptionally active area of young volcanism (≥90% is rhyolitic), heat flow, and tectonism accompanying rapid extension of continental crust (Wilson et al., 1984, 1995; Stern, 1985; Houghton et al., 1995). The Taupo volcanic zone is ~300 km long (200 km of which is on land) to 60 km wide, and comprises volcanic deposits ~2 km thick. It represents the youngest (1.6 Ma) and most southward expression of backarc to arc-rifting in the Taupo-Hikurangi arc-trench system (Cole, 1990; Wilson et al., 1995). Precursor explosive silicic volcanism of Miocene-Pliocene age occurs in the Coromandel volcanic zone to the northwest.

Bimodal but volumetrically dominant rhyolite volcanism characterizes the Taupo volcanic zone (Cole, 1990; Wilson et al., 1995). Rhyolite (15 000 km^3 bulk volume) is erupted mostly during caldera-forming ignimbrite eruptions from the central zone; arc-related andesite is an order of magnitude less abundant, and basalt and dacite are relatively minor in volume (<100 km^3 each; Wilson et al., 1995). At least 34 caldera-forming eruptions have occurred over the past 1.6 m.y.; several eruptions produced ignimbrites >300 km^3, and the 340 ka Whakamaru group ignimbrites have a volume of >1000 km^3 (Houghton et al., 1995; Brown et al., 1998). Basalts have been erupted from several locations within the Taupo volcanic zone (Cole, 1990), and all erupted examples represent small, monogenetic events 2–3 orders of magnitude smaller than the largest rhyolite eruptive events (Wilson et al., 1995). The Taupo volcanic zone basalts have geochemical signatures similar to other backarc basin basalts from oceanic-arc systems, with upwelling mantle involved in "unzipping" a segment of continental lithosphere (Gamble et al., 1993).

Generation of the rhyolites has been attributed to (1) fractionation of mantle-derived melts (e.g., Blattner and Reid, 1982; Conrad et al., 1988); (2) crustal melting of either basement greywacke-argillite (e.g., Ewart and Stipp, 1968; Cole, 1981) or igneous basement (Cole, 1990; Graham et al., 1992) in response to thermal input from the mantle; or (3) a combination of assimilation and fractional crystallization (e.g., McCulloch et al., 1994). The predominantly metaluminous character of the rhyolites, and Pb isotope studies (e.g., Graham et al., 1992; Brown et al., 1998), however, do not indicate substantial partial melting of, or significant contamination by, an aluminous radiogenic crustal component. Seismic studies in conjunction with petrogenetic models suggest that there is a substantial volume of mafic cumulate (or restite) material flooring the Taupo volcanic zone at >15 km depth (e.g., Hochstein et al., 1993; McCulloch et al., 1994; Wilson et al., 1995). Many rhyolitic ignimbrites in the Taupo volcanic zone display evidence for fractionation within a large magma system from a less evolved primary magma, but these primary magmas are complex with distinct trace element characteristics inherited from a heterogeneous source area, variable degrees of partial melting, and/or magma mixing (Brown et al., 1998).

DISCUSSION

The preceding outline illustrates that silicic volcanic rocks occur within most, if not all, the continental flood basalt provinces. Although silicic volcanics compose a small overall portion of continental flood basalts, they are locally dominant (e.g., Lebombo) or form considerable parts of the eruptive stratigraphy (e.g., 50% in the Etendeka province and Yemen; Milner et al., 1992; Ukstins et al., 2000). In contrast, some large igneous provinces have very large volumes of silicic igneous rock without associated large volumes of mafic igneous rock (e.g., Whitsunday volcanic province, Chon Aike province). Silicic large igneous provinces occur in both intraplate and convergent margin settings, with the latter examples related to backarc processes. The outline also illustrates that in terms of proportions of mafic to silicic igneous rock, no spectrum of large igneous provinces exists. Mafic and silicic large igneous provinces represent end members, each comprising ≤10% of silicic or mafic igneous rock, respectively, and large igneous provinces with subequal proportions of mafic to silicic igneous rocks are absent from the geologic record.

Relative eruptive age of silicic volcanism in continental flood basalt provinces

Despite previous assumptions that silicic volcanism postdates flood basalt volcanism, there is no consistent pattern among the continental flood basalt provinces. Several continental flood basalt provinces show multiple pulses of silicic volcanism that are apparent when data from the Karoo, Ferrar, and Chon Aike provinces are combined. In the Ferrar province, silicic volcanism mainly preceded the eruption of tholeiitic basalts, but also accompanied, at low intensity, Ferrar mafic magmatism (Elliot, 1992). The latest geochronology from the Chon Aike province (Pankhurst et al., 2000) indicates that the first phase of silicic volcanism (V_1; 188–178 Ma; Table 2), although spatially separate, predated and temporally overlapped the peak phase of Ferrar-Karoo magmatism at 183 Ma (e.g., Encarnación et al., 1996; Duncan et al., 1997), possibly correlating with silicic volcanic and volcaniclastic rocks (Hanson Formation; Elliot, 2000) underlying the Ferrar basalts. A younger phase of silicic volcanism, the 178 Ma Lebombo rhyolites, occurred in the Karoo province, postdating both the Karoo flood basalts and the MORB-like Rooi Rand dike swarm (Watkeys et al., 2000).

In the North Atlantic igneous province, silicic magmatism occurred during both phases of breakup of the northeast Atlantic. Silicic volcanic rocks are not only interbedded with the flood basalt lavas, but also represent some of the oldest volcanic products of this event (e.g., Mitchell et al., 1999; Sinton et al., 1998). Silicic pyroclastic activity appears to have occurred throughout the evolution of the Mull igneous center (Bell and Emeleus, 1988). In the Etendeka province, latites and quartz latites occur interbedded with basalt lavas during the main phase of flood basalt volcanism (Fig. 3), whereas an earlier phase of rhyolitic volcanism is recorded in the Messum Igneous Complex (Ewart et al., 1998b). Repeated episodes of silicic volcanism occurred prior to and during development of the Ethiopian Rift and Traps, although the silicic volcanics tend to occur toward the middle and top of the flood basalt lava piles (e.g., Ebinger et al., 1993; Baker et al., 1996a; Hofmann et al., 1997). Results from recent ODP drilling of the Kerguelen Plateau suggest that plateau growth by flood basalt volcanism between 110 and 115 Ma concluded with explosive subaerial silicic volcanism (Frey et al., 2000; Moore et al., 2000). However, the earliest eruptive history of the plateau remains undefined, and older silicic ash layers (bentonites) have been recovered from the margins of other submerged oceanic plateaus off the northwest Australian margin (Colwell et al., 1994; Von Rad and Thurow, 1992).

Thus, although most continental flood basalts were erupted during a time period of ~1 m.y., there is no consistent pattern to the eruption of silicic magmas, which may either precede (e.g., Ferrar), be coincident with (e.g., Paraná-Etendeka), or postdate (Karoo-Lebombo) the peak of basaltic flood volcanism.

Eruptive volume, areal extent, and nature of silicic volcanism associated with continental flood basalt provinces

The preservation of silicic volcanic rocks associated with continental flood basalts and along volcanic rifted margins can be limited by regional uplift (typically 1–2 km) and exhumation following emplacement, which results in significant erosion of the volcanic pile. As a consequence of these events, the silicic volcanic rocks are unlikely to be preserved where they occur predominantly in the upper parts of the stratigraphy. Only the deeper level features may be preserved, such as silicic dike swarms, granite intrusions, ring complexes, and deeply subsided caldera collapse structures (e.g., Arran and Rum igneous centers, North Atlantic igneous province; Bell and Emeleus, 1988).

A common assumption has been that the silicic volcanic rocks are small in volume (when compared to the associated flood basalts) and thus areal extent. The areal extent and eruptive volume of individual quartz latite sheets from the Paraná-Etendeka province are comparable to, and often exceed, those of flood basalt lavas (Milner et al., 1995; Jerram, this volume). As further work is undertaken on continental flood basalt provinces, the products of other individual silicic eruptions are being recognized as having large areal extents (10^4 km^2), thickness, and eruptive volumes (e.g., Mitchell et al., 1999; Ukstins et al., 2000).

The silicic volcanic component of continental flood basalt and silicic large igneous provinces is dominantly ignimbrite (often from caldera-forming eruptions); thin, widespread silicic ash-grade beds (tuffs, bentonites) are also significant (cf. McPhie et al., 2000). The lava-like character of the quartz latites of the Parana-Etendeka province is a reflection of their high eruptive temperatures and low magma viscosities (i.e., magma composition), and they have been interpreted as rheomorphic ignimbrites (Milner et al., 1992). Preservation of highly vesicular pumice and cuspate shard textures, typical of lower grade, calc-alkaline ignimbrites, should not be expected from eruptions of such magmas, and the occurrence of spatter and globule textures is more likely (see Hay et al., 1979; Milner et al., 1992). Coeval rhyolitic volcanics in southern Uruguay exhibit lower eruptive temperatures (850–950 °C) and have well-developed ignimbrite textures (Kirstein et al., 2000).

It has also been assumed that because of their higher viscosity, the ascent and consequent eruption of silicic magmas will be hindered, and therefore they will be emplaced late in the eruptive history of continental flood basalt provinces (e.g., White, 1992). This assumption is clearly invalidated by the quartz latites of the Paraná-Etendeka province that have viscosities approaching andesite (10^5 poise, Milner et al., 1992). Evidence such as globule textures further supports a low magma viscosity (Milner, 1988; Milner et al., 1992). Age data and stratigraphic information from the continental flood basalt provinces illustrate that silicic magmas are not always emplaced late in the eruptive history (e.g., Etendeka, Fig. 3; Ferrar, North Atlantic igneous province). The simultaneous existence of mafic and silicic magmas is also evident in the Messum Igneous Complex (Ewart et al., 1998b).

Such massive outpourings of silicic volcanism have the potential to dramatically alter and affect the environment into which they erupt (Jerram, this volume). In the Etendeka region, for example, the early phase of volcanism was characterized by the dynamic interaction between the flood basalt lavas and an active eolian sand field (Jerram et al., 1999b, 2000). The first occurrence of large-volume silicic eruptive units in the Etendeka (Goboboseb Member) marks the point in the stratigraphy where no further record of aeolian deposition occurs (Jerram et al., 1999b, 2000; Fig. 3).

The explosive eruptive style has been important in producing widespread silicic volcanic deposits such as ignimbrite sheets, Plinian fallout deposits, and distal ash layers occurring interbedded with the flood basalt lavas (e.g., Yemen; Ukstins et al., 2000; Ferrar province, Elliot, 1992; North Atlantic igneous province, Bell and Emeleus, 1988; Heister et al., 2001; northwest Australian margin, von Rad and Thurow, 1992). Widespread silicic ash layers may represent important marker horizons in an otherwise monotonous basaltic volcanic stratigraphy. The pyroclastic mode of fragmentation and dispersal also promotes large volumes of fine-grained volcanogenic sediment to be reworked into sedimentary basins, and the depositional record of any sedimentary basins adjacent to large igneous provinces can provide insightful information on the eruptive history of these provinces (e.g., eastern Australia, Bryan et al., 1997).

Eruptive centers for silicic volcanism in mafic and silicic large igneous provinces

Recognizing eruptive vent systems for flood basalt lavas has been extremely difficult due to a combination of burial by younger lavas, lack of exposure, lack of proximal-distal variation in lava facies, and because near-vent features such as spatter cones and ramparts can be easily eroded. Eruptive vents for flood basalts are generally considered to be fissure systems fed by dike swarms, following the identification of a linear feeder dike system at least 200 km wide and 450 km long for the Roza Member of the Columbia River Basalts (Swanson et al., 1975). Recent findings have also identified a shield volcanic feature as a source for some flood basalts, associated with the Doros layered gabbro complex in northwest Namibia (Fig. 2B; Jerram et al., 1999a; Jerram and Robbe, 2001; Marsh et al., 2001).

Geochemistry has been used to correlate silicic volcanic rocks with their eruptive sources. The Messum Igneous Complex has been identified as the eruptive center for the Goboboseb and Springbok quartz latite units in the Etendeka, based on the occurrence of chemically and mineralogically equivalent quartz monzonite plugs and a laccolith immediately peripheral to Messum (Milner and Ewart, 1989). In the North Atlantic igneous province, mineral compositions have been used to link alkaline tuffs in East Greenland, the North Sea, and Denmark with the Gardiner alkaline complex in East Greenland (Heister et al., 2001).

Caldera complexes represent the dominant eruptive source for silicic volcanic rocks associated with continental flood basalt and silicic large igneous provinces. Within the silicic large igneous provinces, caldera complexes are part of a multiple vent volcanic system that includes numerous extracaldera (and intracaldera) effusive vents (e.g., monogenetic scoria and/or spatter cones, tuff rings and/or cones or maars for mafic volcanic rocks, and lava domes or flow dome complexes for the more silicic volcanic compositions; e.g., Bryan et al., 2000). Thick dike swarms, often spatially associated with proximal lithofacies, are probably feeders to the effusive vents. Such multiple vent configuration is best illustrated by the interstratification of proximal and/or near-vent and distal volcanic lithofacies.

Evidence for calderas as source vents for the large-volume ignimbrites includes coarse lithic lag breccia facies and megabreccias within the ignimbrite eruptive units, and ponding relationships. Calderas, however, have proved difficult to identify, especially in the silicic large igneous provinces because of the scale of volcanism ($\geq 10^5$ km^2), burial, exhumation, faulting, and later deformation.

Caldera dimensions, although often difficult to define, appear to range between 10 and 30 km in diameter in the continental flood basalt provinces and silicic large igneous provinces. The Messum Igneous Complex is roughly circular in plan and ~18 km in diameter, whereas the neighboring Brandberg igneous center is ~30 km in diameter (Fig. 2B). Caldera dimensions in the Whitsunday volcanic province are probably 10–20 km, based on ponding relationships and distribution of proximal volcanic lithofacies such as lithic lag breccias. The Mount Poster Formation of the Antarctic Peninsula (V_1) is believed to form an intracaldera succession of monotonous welded ignimbrites, to 1 km in thickness, that define a caldera structure 30–40 km in diameter (Riley et al., 2001).

An important point is that caldera complexes may not be classical circular collapse structures (e.g., Valles-type), but involve collapse along volcanic-tectonic structures controlled by the preexisting crustal architecture and extension orientation. The vent regions and hence collapse structures may be more akin to graben-type structures (e.g., Moore and Kokelaar, 1998). Several phases of collapse and resurgence (e.g., Messum, Brandberg, Rhum) provide further complication to the eruptive history and vent structure of silicic eruptive centers (Ewart et al., 1998b; Emeleus et al., 1985). The Messum Igneous Complex is interpreted as a caldera structure superimposed on a large regional sag structure resulting from massive magma withdrawal (>8000 km^3) during quartz latite eruptions (Milner and Ewart, 1989; Ewart et al., 1998b). Despite the proximity of the Messum and Brandberg igneous complexes in the Etendeka province, these two enormous silicic plumbing systems appear to have acted independently (A. Ewart, 2000, personal commun.).

Available evidence indicates that silicic volcanic rocks (or their eruptive centers) are not necessarily proximal to the main locus of (mafic) melt generation, which is exacerbated by their wide dispersal from pyroclastic transport processes (buoyant Plinian-type columns and pyroclastic density currents). Multiple eruptive centers or caldera complexes occur within the continental flood basalt provinces (e.g., Paraná-Etendeka, North Atlantic igneous province) and silicic large igneous provinces (Whitsunday volcanic province, Chon Aike province). Many modern continental silicic calderas are sited at lineament intersections (e.g., Valles, Long Valley) and the same crustal anisotropy or structure is probably the most important control on caldera location in large igneous provinces. The igneous centers and/or calderas of the Etendeka province show a clear alignment defining an east-northeast–trending zone up to 350 km long from Cape Cross at the coast to Paresis and Okorusu in central Namibia (Schmitt et al., 2000; Fig. 2B); their locations are controlled by basement lineaments (e.g., the Omaruru and Autseib lineaments and the Ugab terrane, Clemson et al., 1999). Recent studies of silicic explosive volcanism associated with basaltic flood volcanism in Ethiopia-Yemen suggest that caldera complexes may also have migrated with time (Ukstins et al., 2000).

Generation of large-volume silicic igneous provinces

Although rhyolites can occur in a variety of tectonic settings, both oceanic and continental, large-volume (>10^4 km^3) silicic volcanism is restricted to continental margin settings, and to a lesser extent, continental interiors when associated with continental flood basalts. The silicic volcanic rocks associated with the continental flood basalt provinces listed in Table 1 are widely believed to be the end result of varying amounts of assimilation of partial melts of either anhydrous granulitic lower crust or mafic underplate at high temperatures by basaltic magmas, followed by extended fractional crystallization. For silicic large igneous provinces where the volume of silicic magma generated is at least an order of magnitude bigger, partial melting of lower crust is essential, the most suitable source materials being hydrated, calc-alkaline, and high-K calc-alkaline andesites and basaltic andesites and/or amphibolites (e.g., Roberts and Clemens, 1993). Basement to the Whitsunday volcanic province and Chon Aike province, and in part the Sierra Madre Occidental, comprises Paleozoic-Mesozoic volcanic and sedimentary rocks accreted and/or deposited along the continental margin. The involvement of Mesozoic to Paleozoic crust in magma genesis is supported by Nd model T_{DM} ages for the Whitsunday volcanic province (see Ewart et al., 1992), whereas mid-Late Proterozoic (Grenvillian) model ages are indicated for the crustal source in the eastern (interior) part of the Chon Aike province (Pankhurst and Rapela, 1995; Riley et al., 2001). These older depleted model ages may reflect either that of the sedimentary provenance or formation of the crust (Pankhurst et al.,

TABLE 3. SUMMARY OF THE IMPORTANT CRUSTAL PRECONDITIONS, MAGMATIC PROCESSES, AND ERUPTED PRODUCTS THAT LEAD TO THE DEVELOPMENT OF MAFIC AND SILICIC LARGE IGNEOUS PROVINCES

	Mafic large igneous province	Silicic large igneous province
Crustal setting	Craton interior	Accreted orogenic margin
Crustal composition and age	Refractory Archean–Proterozoic, dry mafic and/or silicic, brittle crust.	Fertile Proterozoic–Phanerozoic, hydrous crust with a large I-type (calc-alkaline) meta-igneous component.
Driving processes	Thermal and mass transfer into crust caused by hot mantle upwelling, and lithospheric extension	
Nature of crust and/or magma interaction	Crust with low preexisting geothermal gradient, melts to produce low volume, high temperature (dry) ternary granite minimum magma.	Widespread partial melting of crust (~20%) to produce large volumes of hydrous, ternary granite minimum magma.
Thermal and mass transfer characteristics	Crust-penetrating structures readily transfer mafic melt to surface. Mafic magma can be thermally and chemically insulated from crust by chilling along reservoir margins limiting further crustal melting.	Density/buoyancy filter caused by silicic melt zone, and lack of well-defined crust-penetrating structures, suppresses rise/transfer of mafic magma. Containment of mafic melt promotes further increase in temperature and degree of crustal partial melting.
Magmatic processes and geochemical signature	Magma processes dominated by FC/AFC producing large volumes of variably contaminated within-plate basalt. Volumetrically minor silicic magma generated by AFC/PM. Melting of mafic underplate may occur.	Magma processes dominated by mixing and AFC producing large volume, volatile-rich rhyolitic-rhyodacitic melt with calc-alkaline signature and highly contaminated mafic-intermediate magmas.
Eruption characteristics	Effusive, flood basalt lava-dominated volcanism. Variable proportions of silicic pyroclastic rocks and lesser lavas from calderas, central igneous complexes ± fissures.	Explosive silicic-dominated volcanism erupted from multiple caldera complexes with minor mafic-intermediate lavas. Highly variable, upper crustal structure/rheology controls character of upper crustal magma reservoirs and eruptive centers (plutons, calderas, rifts).

1998). Nevertheless, the long history of subduction and intrusion of hydrous melts into the lower crust along the proto-Pacific margin is considered crucial for the generation of the large-volume rhyolites of the Chon Aike province (Riley et al., 2001). This difference in lower crustal materials between mafic and silicic large igneous provinces (i.e., the presence of anhydrous or hydrous crust) led Stephens et al. (1995) to coin the term "wet" large igneous province to describe silicic large igneous provinces such as the Whitsunday volcanic province and Chon Aike province.

Table 3 illustrates the key processes that lead to the development of silicic large igneous provinces or continental flood basalt provinces with associated large-volume rhyolites. The presence of a fertile crustal source appears to be the main difference between silicic and mafic large igneous province formation. Large degrees of crustal partial melting, essential to produce the large volumes of rhyolitic magma, are controlled by the water content and composition of the crust and the large thermal input from the mantle. Although the thermal budget for mafic and silicic large igneous provinces is considered the same, hydrous crustal material will be more receptive to melting, and will begin to melt at lower temperatures. In contrast, melting of a refractory dry crust will be limited by prior depletions in minimum melt components and preexisting low geothermal gradient. Subsequent melting events will not only require higher temperatures to remelt, but produce less silicic (rhyodacitic) compositions.

Ancient and active convergent margins tend to be characterized by a fertile, hydrous lower crust that can readily melt. Long-lived subduction promotes the development of a hydrated lower crust and lithospheric mantle that can extend for several hundred kilometers from the active margin (e.g., Karoo, western United States; Fitton et al., 1988; Davis et al., 1993), particularly if significant lateral accretion has occurred over time. Previous subduction episodes may also have been important in the development of low-Ti source regions for some continental flood basalts (e.g., Hawkesworth et al., 1988). Heating and partial melting of a hydrous, mafic crust will generate intermediate to silicic composition melts (55%–75% SiO_2; Rapp and Watson, 1995). The silicic melts can act as a density barrier, preventing the mafic magmas from reaching the surface (cf. Huppert and Sparks, 1988), as will a lack of deep, crust-penetrating structures that can transfer mafic magma to the surface.

In contrast, continental flood basalts are emplaced on or adjacent to Archean cratons (Anderson, 1999), where the crust is relatively old (Proterozoic-Archean) and refractory, and any lower crustal melting would occur only at very high temperatures. Extensive mafic dike swarms (e.g., Central Atlantic magmatic province) imply a brittle crust with deep-penetrating structures that can channel mafic melt to the surface. In the cases of continental flood basalt provinces that have significant volumes of silicic volcanism, crustal melting and assimilation are generated by achieving such high temperatures at the base of the crust, caused by sustained thermal and material input of mafic magma. The Paraná-Etendeka rhyolites, for example, are anhydrous and had an eruption temperature in excess of 1050 °C (Harris and Milner, 1997), consistent with partial melting and assimilation of anhydrous crustal material at very high temperatures.

Distinguishing between intraplate and backarc-related silicic large igneous provinces

Large-volume silicic igneous provinces are not unique to continental breakup, but can characterize any continental region undergoing extension that is underlain by hot mantle and has a hydrous lower crust receptive to melting. This can occur in continental intraplate and subduction-related tectonic environments. The similarities in eruptive volumes, rhyolite geochemistry, and the ignimbrite-dominated character of the silicic igneous provinces related to plate breakup and those from extensional convergent continental margin settings makes distinguishing the two difficult.

Understanding the regional tectonic framework is a critical factor in determining the origin of such provinces. Detailed regional time-space analysis for western North America, for example, has shown that the Sierra Madre Occidental formed when suprasubduction zone andesitic volcanism was rare, and divergence between the Pacific and North American plates was greatest, and followed a period of rapid subduction, andesitic volcanism, trench-normal contraction, and widespread deformation during the Laramide orogeny (Ward, 1991, 1995). In contrast, some silicic large igneous provinces tend not to show such an intimate space-time relationship to convergent margin tectonism, and like mafic large igneous provinces, have been emplaced into intraplate regions remote from plate margin–related tectonics and commonly undergoing rifting.

The Whitsunday volcanic province and Chon Aike province were emplaced into intraplate environments, and show strong evidence for being spatially and temporally related to continental plate breakup (Fig. 4). They are characterized by large volume (>10^5 km^3), overwhelmingly rhyodacite-rhyolite compositions approaching the hydrous granite minimum that strongly suggest melting of a hydrous crust as the predominant source. However, associated mafic magma compositions show evidence of an intraplate signature that is important in defining a fundamentally intraplate (versus active convergent margin) tectonic setting.

CONCLUSIONS

Flood basalt lavas are now thought to have been emplaced as inflated pahoehoe-like lavas (e.g., Self et al., 1996; Thordarson and Self, 1998) that are relatively slow advancing (as low as 0.2–1.4 m/s; Keszthelyi and Self, 1998), from low eruptive rate (e.g., ~4000 m^3/s average total eruption rate for Roza Member; Self et al., 1997) and long-lived eruptions (years to tens of years; Thordarson and Self, 1998). In contrast, large-volume ignimbrites, which represent the dominant silicic volcanic rock type of continental flood basalts and silicic large igneous provinces, are emplaced from short-lived (hours to weeks), high rate (10^6 m^3/s, >2×10^8 kg/s; e.g., Carey and Sigurdsson, 1989), highly explosive eruptions. In many continental flood basalt provinces, individual silicic eruptive units commonly have thicknesses, areal extents, and volumes that are comparable to, and exceed those of, flood basalt lavas. Regional uplift of 1–2 km is not a prerequisite nor indicated to generate either basaltic or silicic volcanic rocks with large areal extent or run-out distances. However, regional surface uplift and exhumation following emplacement of the large igneous province (typically 1–2 km) can result in significant erosion of the volcanic pile, and silicic volcanics are unlikely to be preserved, especially if they occur predominantly in the upper parts of the stratigraphy. Only the deeper level features may be preserved, such as silicic dike swarms, granites, and the lower levels of caldera collapse structures (e.g., North Atlantic igneous province, Bell and Emeleus, 1988).

Recent detailed studies of continental flood basalt provinces are revealing complex spatial-temporal relationships between silicic and mafic volcanism. In general, the silicic magmas are erupted from caldera-type centers, different from the fissure-type feeder dike systems commonly invoked for flood basalt lavas. However, the simultaneous existence of mafic and silicic magmas in some eruptive centers (e.g., Messum Igneous Complex, Ewart et al., 1998b), suggests that (large) calderas could be the site of both basaltic lava and rhyolitic ignimbrite-forming eruptions.

It is also important to point out that the plume model was developed in part because of the requirement for continental flood basalts that large volumes of (mafic) magma be produced over short periods (~1 m.y.). In contrast, silicic large igneous provinces comprise similar volumes of magma produced over a much longer duration (e.g., up to 40 m.y.). Therefore, the generation of silicic large igneous provinces requires melting of hot mantle and sustained mantle upwelling rather than the transient impact of a large plume head as in plume models commonly applied to mafic large igneous provinces (e.g., Campbell and Griffiths, 1990; White and McKenzie, 1989).

The generation of large volumes of crustal melt in fundamentally intraplate settings requires high heat influx to the crust transported by mantle-derived mafic magmas. This amount of mantle melting requires that the mantle be hotter than normal but within the range of mantle temperatures expected for the generation of continental flood basalts. The key to generating large volumes of silicic magma preserved in silicic large igneous provinces is the prior formation of hydrous, highly fusible mafic lower crust. Partial melting (~20%–25%) of this fusible crust will generate melts of intermediate to silicic composition (Rapp and Watson, 1995; Roberts and Clemens, 1993). We conclude that earlier subduction was crucial in developing a hydrous, fusible lower crust (i.e., mafic to intermediate, transitional to high-K calc-alkaline metaigneous source), and consistent with their development along or near continental margins.

ACKNOWLEDGMENTS

We thank Tony Ewart, Stuart Brown, Simon Milner, and Henry Emeleus for helpful discussions and comments on an earlier version of the manuscript; Greg McHone for useful information on the Central Atlantic magmatic province; and Martin Menzies and Joel Baker for constructive reviews of the manuscript. Bryan acknowledges financial support from the Australian Research Council Large Grant A10027005 to Rod Holcombe and Chris Fielding.

REFERENCES CITED

Albrecht, A., and Goldstein, S.L., 2000, Effects of basement composition and age on silicic magmas across an accreted terrane–Precambrian crust boundary, Sierra Madre Occidental, Mexico: Journal of South American Earth Sciences, v. 13, p. 255–273.

Allen, C.M., Williams, I.S., Stephens, C.J., and Fielding, C.R., 1998, Granite genesis and basin formation in an extensional setting: The magmatic history of the northernmost New England Orogen: Australian Journal of Earth Sciences, v. 45, p. 875–888.

Allsopp, H.L., Manton, W.I., Bristow, J.W., and Erlank, A.J., 1984, Rb-Sr geochronology of Karoo felsic volcanics, *in* Erlank, A.J., ed., Petrogenesis of the volcanic rocks of the Karoo Province: Geological Society of South Africa Special Publication 13, p. 273–280.

Anderson, D.L., 1999, A theory of the earth: Hutton and Humpty Dumpty and Holmes, *in* Craig, G.Y., and Hull, J.H., eds., James Hutton—Present and future: Geological Society [London] Special Publication 150, p. 13–35.

Aragón, E., Rodriguez, A.M.I., and Benialgo, A., 1996, A calderas field at the Marifil Formation, new volcanogenic interpretation, Norpatagonian massif, Argentina: Journal of South American Earth Sciences, v. 9, p. 321–328.

Baker, J., Snee, L., and Menzies, M., 1996a, A brief Oligocene period of flood volcanism in Yemen: Implications for the duration and rate of continental flood volcanism at the Afro-Arabian triple junction: Earth and Planetary Science Letters, v. 138, p. 39–55.

Baker, J.A., Thirlwall, M.F., and Menzies, M.A., 1996b, Sr-Nd-Pb isotopic and trace element evidence for crustal contamination of plume-derived flood basalts: Oligocene flood volcanism in western Yemen. Geochimica et Cosmochimica Acta, v. 60, p. 2559–2581.

Bell, B.R., and Emeleus, C.H., 1988, A review of silicic pyroclastic rocks of the British Tertiary Volcanic Province, *in* Morton, A.C., and Parson, L.M., eds., Early Tertiary volcanism and the opening of the NE Atlantic: Geological Society [London] Special Publication 39, p. 365–379.

Bell, B.R., Williamson, I.T., Head, F.E., and Jolley, D.W., 1996, On the origin of a reddened interflow bed within the Palaeocene lava field of north Skye: Scottish Journal of Geology, v. 32, p. 117–126.

Best, M.G., and Christiansen, E.H., 1991, Limited extension during peak Tertiary volcanism, Great Basin of Nevada and Utah: Journal of Geophysical Research, v. 96, p. 13509–13528.

Blattner, P., and Reid, F.W., 1982, The origin of lavas and ignimbrites of the Taupo Volcanic Zone, New Zealand in light of oxygen isotope data: Geochimica et Cosmochimica Acta, v. 46, p. 1417–1429.

Bristow, J.W., Allsopp, H.L., Erlank, A.J., Marsh, J.S., and Armstrong, R.A., 1984, Strontium isotope characterization of Karoo volcanic rocks, *in* Erlank, A.J., ed., Petrogenesis of the volcanic rocks of the Karoo Province: Geological Society of South Africa Special Publication 13, p. 295–329.

Brown, S.J.A., Wilson, C.J.N., Cole, J.W., and Wooden, J., 1998, The 340 ka Whakamaru-group ignimbrites, Taupo Volcanic Zone, New Zealand: Geochemical evidence for non-sequential tapping of a zoned silicic magmatic system: Journal of Volcanology and Geothermal Research, v. 84, p. 1–37.

Bryan, S.E., Constantine, A.E., Stephens, C.J., Ewart, A., Schön, R.W., and Parianos, J., 1997, Early Cretaceous volcano-sedimentary successions along the eastern Australian continental margin: Implications for the break-up of eastern Gondwana: Earth and Planetary Science Letters, v. 153, p. 85–102.

Bryan, S.E., Ewart, A., Stephens, C.J., Parianos, J., and Downes, P.J., 2000, The Whitsunday Volcanic Province, central Queensland, Australia: Lithological and stratigraphic investigations of a silicic-dominated large igneous province: Journal of Volcanology and Geothermal Research, v. 99, p. 55–78.

Cameron, K.L., and Hanson, G.N., 1982, Rare earth element evidence concerning the origin of voluminous mid-Tertiary rhyolitic ignimbrites and related volcanic rocks, SMO, Chihuahua, Mexico: Geochimica et Cosmochimica Acta, v. 46, p. 1489–1503.

Cameron, M., Bagby, W.C., and Cameron, K.L., 1980, Petrogenesis of voluminous mid-Tertiary ignimbrites of the Sierra Madre Occidental: Contributions to Mineralogy and Petrology, v. 74, p. 271–284.

Campbell, I.H., and Griffiths, R.W., 1990, Implications of mantle plume structure for the evolution of flood basalts: Earth and Planetary Science Letters, v. 99, p. 79–93.

Capaldi, G., Chisea, S., Manetti, P., Ortsi, G., and Poli, G., 1987, Tertiary anorogenic granites of the western border of the Yemen Plateau: Lithos, v. 20, p. 433–444.

Carey, S., and Sigurdsson, H., 1989, The intensity of plinian eruptions: Bulletin of Volcanology, v. 51, p. 28–40.

Chazot, G., and Bertrand, H., 1995, Genesis of silicic magmas during Tertiary continental rifting in Yemen: Lithos, v. 36, p. 69–83.

Chernet, T., Hart, W.K., Aronson, J.L., and Walter, R.C., 1998, New age constraints on the timing of volcanism and tectonism in the northern Main Ethiopian Rift–southern Afar transition zone (Ethiopia): Journal of Volcanology and Geothermal Research, v. 80, p. 267–280.

Chesley, J.T., and Ruiz, J., 1998, Crust-mantle interaction in large igneous provinces: Implications from the Re-Os isotope systematics of the Columbia River flood basalts: Earth and Planetary Science Letters, v. 154, p. 1–11.

Chung, S.-L., and Jahn B., 1995, Plume-lithosphere interaction in generation of the Emeishan flood basalts at the Permian-Triassic boundary: Geology, v. 23, p. 889–892.

Chung, S.-L., Jahn, B., Genyao, W., Lo, C.-H., and Bolin, C., 1998, The Emeishan flood basalt in SW China: A mantle plume initiation model and its connection with continental break-up and mass extinction at the Permo-Triassic boundary, *in* Flower, M.F.J., ed., Mantle dynamics and plate interactions in East Asia: American Geophysical Union, Geodynamics, v. 27, p. 47–58.

Clarke, D.E., Paine, A.G.L., and Jensen, A.R., 1971, Geology of the Proserpine 1:250 000 sheet area, Queensland: Bureau of Mineral Resources Geology and Geophysics Report 144, 98 p.

Clemson, J., Cartwright, J., and Swart, R., 1999, The Namib Rift: A rift system of possible Karoo age, offshore Namibia, *in* Cameron, N., Bate, R., and Clure, V., eds., The oil and gas habitats of the South Atlantic: Geological Society [London] Special Publication 153, p. 381–402.

Cleverly, R.W., 1979, The volcanic geology of the Lebombo monocline in Swaziland: Transactions of the Geological Society of South Africa, v. 82, p. 227–230.

Cleverly, R.W., Betton, P.J., and Bristow, J.W., 1984, Geochemistry and petrogenesis of the Lebombo rhyolites, *in* Petrogenesis of the volcanic rocks of the Karoo Province: Geological Society of South Africa Special Publication 13, p. 171–195.

Coffin, M.F., and Eldholm, O., 1992, Volcanism and continental break-up: A global compilation of large igneous provinces, *in* Storey, B.C., Alabaster, T., and Pankhurst, R.J., eds., Magmatism and the causes of continental break-up: Geological Society [London] Special Publication 68, p. 17–30.

Coffin, M.F., and Eldholm, O., 1994, Large igneous provinces: Crustal structure, dimensions, and external consequences: Reviews of Geophysics, v. 32, p. 1–36.

Cole, J.W., 1981, Genesis of lavas of the Taupo Volcanic Zone, North Island, New Zealand: Journal of Volcanology and Geothermal Research, 10, 317–337.

Cole, J.W., 1990, Structural control and origin of volcanism in the Taupo Volcanic Zone, New Zealand: Bulletin of Volcanology, v. 52, p. 445–459.

Colwell, J.B., Symonds, P.A., and Crawford, A.J., 1994, The nature of the Wallaby (Cuvier) Plateau and other igneous provinces of the west Australian margin: Australian Geological Survey Organisation, AGSO Journal of Australian Geology and Geophysics, v. 15/1, p. 137–156.

Conrad, W.K., Nicholls, I.A., and Wall, V.J., 1988, Water-saturated and -undersaturated melting of metaluminous crustal compositions at 10 kb: Evidence for the origin of silicic magmas in the Taupo Volcanic Zone, New Zealand and other occurrences: Journal of Petrology, v. 29, p. 765–803.

Cox, K.G., 1983, The Karoo province of southern Africa: Origin of trace element enrichment patterns, *in* Hawkesworth, C.J., and Norry, M.J., eds., Continental basalts and mantle xenoliths: Nantwich, UK, Shiva, p. 139–157.

Cox, K.G., 1993, Continental magmatic underplating, *in* Cox, K.G., McKenzie, D.P., and White, R.S., eds., Melting and melt movement in the earth: Oxford, Oxford University Press, p. 155–166.

Creaser, R.A., and White, A.J.R., 1991, Yardea Dacite: Large volume, high temperature felsic volcanism from the Middle Proterozoic of South Australia: Geology, v. 19, p. 48–51.

Cummings, M.L., Evans, J.G., Ferns, M.L., and Lees, K.R., 2000, Stratigraphic and structural evolution of the middle Miocene synvolcanic Oregon-Idaho graben: Geological Society of America Bulletin, v. 112, p. 668–682.

Davidson, A., and Rex, D.C., 1980, Age of volcanism and rifting in southwestern Ethiopia: Nature, v. 283, p. 657–658.

Davis, J.M., Elston, W.E., and Hawkesworth, C.J., 1993, Basic and intermediate volcanism of the Mogollon-Datil volcanic field: Implications for mid-Tertiary tectonic transitions in southwestern New Mexico, USA, *in* Prichard, H.M., Alabaster, T., Harris, N.B.W., and Neary, C.R., eds., Magmatic processes and plate tectonics: Geological Society [London] Special Publication 76, p. 469–488.

Devey, C.W., and Stephens, W.E., 1992, Deccan-related magmatism west of the Seychelles-India rift, *in* Storey, B.C., Alabaster, T., and Pankhurst, R.J., eds., Magmatism and the causes of continental break-up: Geological Society [London] Special Publication 68, p. 271–291.

Dickin, A.P., and Exley, R.A., 1981, Isotopic and geochemical evidence for magma mixing in the petrogenesis of the Coire Uagneich Granophyre, Isle of Skye, N.W. Scotland: Contributions to Mineralogy and Petrology, v. 76, p. 98–108.

Duncan, A.R., Erlank, A.J., and Marsh, J.S., 1984, Regional geochemistry of the Karoo Igneous Province: Geological Society of South Africa Special Publication 13, p. 355–388.

Duncan, R.A., and Pyle, D.G., 1988, Rapid eruption of the Deccan flood basalts at the Cretaceous/Tertiary boundary: Nature, v. 333, p. 841–843.

Duncan, R.A., Hooper, P.R., Rehacek, J., Marsh, J.S., and Duncan, A.R., 1997, The timing and duration of the Karoo igneous event, southern Gondwana: Journal of Geophysical Research, v. 102, no. B8, p. 18127–18138.

Ebinger, C.J., Yemane, T., Woldegabriel, G., Aronson, J.L., and Walter, R.C., 1993, Late Eocene-Recent volcanism and faulting in the southern main Ethiopian rift. Journal of the Geological Society of London, v. 150, p. 99–108.

Ebinger, C.J., Yemane, T., Harding, D.J., Tesfaye, S., Kelley, S., and Rex, D.C., 2000, Rift deflection, migration and propagation: Linkage of the Ethiopian and eastern rifts, Africa: Geological Society of America Bulletin, v. 112, p. 163–176.

Eby, G.N., Krueger, H.W., and Creasy, J.W., 1992, Geology, geochronology, and geochemistry of the White Mountain batholith, New Hampshire: Geological Society of America Special Paper 268, p. 379–397.

Eldholm, O., and Grue, K., 1994, North Atlantic volcanic margins: Dimensions and production rates: Journal of Geophysical Research, v. 99, p. 2955–2968.

Eldholm, O., Thiede, J., and Taylor, E., 1987, Proceedings of the Ocean Drilling Program, Initial Results, Leg 104: College Station, Texas, Ocean Drilling Program, 783 p.

Ellam, R.M., Carlson, R.W., and Shirey, S.B., 1992, Evidence from Re-Os isotopes for plume-lithosphere mixing in Karoo flood basalt genesis: Nature, v. 359, p. 718–721.

Elliot, D.H., 1992, Jurassic magmatism and tectonism associated with Gondwanaland break-up: An Antarctic perspective, *in* Storey, B.C., Alabaster, T., and Pankhurst, R.J., eds., Magmatism and the causes of continental break-up: Geological Society [London] Special Publication 68, p. 165–184.

Elliot, D.H., 2000, Stratigraphy of Jurassic pyroclastic rocks in the Transantarctic Mountains: Journal of African Earth Sciences, v. 77, p. 77–89.

Elston, W., 1984, Subduction of young oceanic lithosphere and extensional orogeny in southwestern North America during mid-Tertiary time: Tectonics, v. 3, p. 229–250.

Emeleus, C.H., Wadsworth, W.J., and Smith, N.J., 1985, The early igneous and tectonic history of the Rhum Tertiary Volcanic Centre: Geological Magazine, v. 122, p. 451–457.

Emeleus, C.H., Allwright, E.A., Kerr, A.C., and Williamson, I.T., 1996, Red tuffs in the Palaeocene lava successions of the Inner Hebrides: Scottish Journal of Geology, v. 32, p. 83–89.

Encarnación, J., Fleming, T.H., Elliot, D.H., and Eales, H.V., 1996, Synchronous emplacement of Ferrar and Karoo dolerites and the early breakup of Gondwana: Geology, v. 24, p. 535–538.

Ewart, A., 1979, A review of the mineralogy and chemistry of Tertiary-Recent dacitic, latitic, rhyolitic, and related salic volcanic rocks, *in* Barker, F., ed., Trondhjemites, dacites, and related rocks: Amsterdam, Elsevier, p. 13–121.

Ewart, A., and Stipp, J.J., 1968, Petrogenesis of the volcanic rocks of the central North Island, New Zealand, as indicated by a study of Sr^{87}/Sr^{86} ratios, and Sr, Rb, K, U and Th abundances: Geochimica et Cosmochimica Acta, v. 32, p. 699–735.

Ewart, A., Schön, R.W., and Chappell, B.W., 1992, The Cretaceous volcanic-plutonic province of the central Queensland (Australia) coast: A rift related "calc-alkaline" province: Transactions of the Royal Society of Edinburgh, Earth Sciences, v. 83, p. 327–345.

Ewart, A., Milner, S.C., Armstrong, R.A., and Duncan, A.R., 1998a, Etendeka volcanism of the Goboboseb Mountains and Messum Igneous Complex, Namibia. 2. Voluminous quartz latite volcanism of the Awahab magma system: Journal of Petrology, v. 39, no. 2, p. 227–253.

Ewart, A., Milner, S.C., Armstrong, R.A., and Duncan, A.R., 1998b, Etendeka volcanism of the Goboboseb Mountains and Messum Igneous Complex, Namibia. 1. Geochemical evidence of Early Cretaceous Tristan plume melts and the role of crustal contamination in the Paraná-Etendeka CFB: Journal of Petrology, v. 39, no. 2, p. 191–225.

Fanning, C.M., Flint, R.B., Parker, A.J., Ludwig, K.R., and Blisset, A.H., 1988, Refined Proterozoic evolution of the Gawler Craton, South Australia, through U-Pb zircon geochronology: Precambrian Research, v. 40/41, p. 363–386.

Faure, G., and Hill, R.L., 1973, Age of the Falla Formation (Triassic), Queen Alexandra Range: Antarctic Journal of the United States, v. 8, no. 5, p. 264–266.

Féraud, G., Alric, V., Fornari, M., Bertrand, H., and Haller M., 1999, $^{40}Ar/^{39}Ar$ dating of the Jurassic volcanic province of Patagonia: Migrating magmatism related to Gondwana break-up and subduction: Earth and Planetary Science Letters, v. 172, p. 83–96.

Fitton, J.G., James, D., Kempton, P.D., Ormerod, D.S., and Leeman, W.P., 1988, Role of lithospheric mantle in the generation of Late Cenozoic basic magmas in the western U.S., *in* Menzies, M.A., and Cox, K.G., eds., Oceanic and continental lithosphere: Similarities and differences: Journal of Petrology Special Volume, p. 331–349.

Frey, F.A., McNaughton, N.J., Nelson, D.R., deLaeter, J.R., and Duncan, R.A., 1996, Petrogenesis of the Bunbury Basalt, Western Australia: Interaction between the Kerguelen plume and Gondwana lithosphere?: Earth and Planetary Science Letters, v. 144, p. 163–183.

Frey, F.A., Nicolaysen K., Weis, D., Wallace, P.J., and Leg 183 Shipboard Scientific Party, 2000, Origin and evolution of a submarine large igneous province: The Kerguelen Plateau and Broken Ridge, southern Indian Ocean: Penrose 2000 Volcanic Rifted Margins, Royal Holloway Campus, University of London, Department of Geology, Abstracts, p. 25.

Gamble, J.A., Smith, I.E.M., McCulloch, M.T., Graham, I.J., and Kokelaar, B.P., 1993, The geochemistry and petrogenesis of basalts from the Taupo Volcanic Zone and Kermadec Island Arc, S.W. Pacific: Journal of Volcanology and Geothermal Research, v. 54, p. 265–290.

Garland, F., Hawkesworth, C.J., and Mantovani, S.M., 1995, Description and petrogenesis of the Paraná rhyolites, southern Brazil: Journal of Petrology, v. 36, p. 1193–1227.

George, R., Rogers, N., and Kelley, S., 1998, Earliest magmatism in Ethiopia: Evidence for two mantle plumes in one flood basalt province: Geology, v. 26, p. 923–926.

Giles, C.W., 1988, Petrogenesis of the Proterozoic Gawler Range Volcanics, South Australia: Precambrian Research, v. 40/41, p. 407–427.

Graham, I.J., Gulson, B.L., Hedenquist, J.W., and Mizon, K., 1992, Petrogenesis of Late Cenozoic volcanic rocks from the Taupo Volcanic Zone, New Zealand in the light of new lead isotope data: Geochimica et Cosmochimica Acta, v. 56, p. 2797–2819.

Hames, W.E., Renne, P.R., and Ruppel, C., 2000, New evidence for geologically instantaneous emplacement of earliest Jurassic central Atlantic magmatic province basalts on the North American margin: Geology, v. 28, p. 859–862.

Hamilton, M.A., Pearson, D.G., Thompson, R.N., Kelley, S.P., and Emeleus, C.H., 1998, Rapid eruption of Skye lavas inferred from precise U-Pb and Ar-Ar dating of the Rum and Cuillin plutonic complexes: Nature, v. 394, p. 260–263.

Harris, C., 1995, The oxygen isotope geochemistry of the Karoo and Etendeka volcanic provinces of southern Africa: South African Journal of Geology, v. 98, p. 126–139.

Harris, C., and Erlank, A.J., 1992, The production of large-volume, low ^{18}O rhyolites during the rifting of Africa and Antarctica: The Lebombo Monocline, southern Africa: Geochimica et Cosmochimica Acta, v. 56, p. 3561–3570.

Harris, C., and Milner, S., 1997, Crustal origin for the Paraná rhyolites: Discussion of "Description and petrogenesis of the Paraná rhyolites, southern Brazil" by Garland et al. (1995): Journal of Petrology, v. 38, p. 299–302.

Harris, C., Whittingham, A.M., Milner, S.C., and Armstrong, R.A., 1990, Oxygen isotope geochemistry of the Karoo and Etendeka volcanic provinces of southern Africa: South African Journal of Geology, v. 98, p. 126–139.

Hawkesworth, C., Mantovani, M., and Peate, D., 1988, Lithosphere remobilisation during Paraná CFB magmatism, *in* Menzies, M.A., and Cox, K.G., eds., Oceanic and continental lithosphere: Similarities and differences: Journal of Petrology Special Volume, p. 205–223.

Hay, R.L., Hildreth, W., and Lambe, R.N., 1979, Globule ignimbrite of Mount Suswa, Kenya, *in* Chapin, C.E., and Elston, W.E., eds., Ash flow tuffs: Geological Society of America Special Paper 180, p. 167–175.

Heatherington, A.L., and Mueller, P.A., 1991, Geochemical evidence for Triassic rifting in southwestern Florida: Tectonophysics, v. 188, p. 291–302.

Heatherington, A.L., Mueller, P.A., and Nutman, A.P., 1999, A Jurassic granite from southern Georgia, U.S.A.: Silicic, extension-related magmatism along the southeastern coastal plain: Journal of Geology, v. 107, p. 375–384.

Heister, L.E., O'Day, P.A., Brooks, C.K., Neuhoff, P.S., and Bird, D.K., 2001, Pyroclastic deposits within the East Greenland tertiary flood basalts: Journal of the Geological Society of London, v. 158, p. 269–284.

Hochstein, M.P., Smith, I.E.M., Regenauer-Leib, K., and Ehara, S., 1993, Geochemistry and heat transfer processes in Quaternary rhyolitic systems of the Taupo Volcanic Zone, New Zealand: Tectonophysics, v. 223, p. 13–235.

Hofmann, C., Féraud, G., and Courtillot, V., 2000, ^{40}Ar/^{39}Ar dating of mineral separates and whole rocks from the western Ghats lava pile: Further constraints on duration and age of the Deccan Traps: Earth and Planetary Science Letters, v. 180, p. 13–27.

Hofmann, C., Courtillot, V., Féraud, G., Rochette, P., Yirgu, G., Ketefo, E., and Pik, R., 1997, Timing of the Ethiopian flood basalt event and implications for plume birth and global change: Nature, v. 389, p. 838–841.

Holcombe, R.J., Stephens, C.J., Fielding C.R., Gust, D.A., Little, T.A., Sliwa, R., McPhie, J., and Ewart, A., 1997, Tectonic evolution of the northern New England fold belt: Carboniferous to Early Permian transition from accretion to extension: Geological Society of Australia Special Publication 19, p. 66–79.

Hooper, P.R., 1997, The Columbia River Flood Basalt Province: Current status, *in* Mahoney, J.J., and Coffin, M.F., eds., Large igneous provinces: Continental, oceanic, and planetary flood volcanism: American Geophysical Union Geophysical Monograph 100, p.1–27.

Houghton, B.F., Wilson, C.J.N., McWilliams, M.O., Lanphere, M.A., Weaver, S.D., Briggs, R.M., and Pringle, M.S., 1995, Chronology and dynamics of a large silicic magmatic system: Central Taupo Volcanic Zone, New Zealand: Geology, v. 23, p. 13–16.

Huppert, H.E., and Sparks, R.S.J., 1988, The generation of granitic magmas by intrusion of basalt into continental crust: Journal of Petrology, v. 29, p. 599–624.

Jerram, D.A., and Robbe, O., 2001, Building a 3-D geologic model of a Flood Basalt: An example from the Etendeka, NW Namibia: Electronic Geoscience, v. 6, no. 1.

Jerram, D.A., Mountney, N., Holzförster, F., and Stollhofen, H., 1999a, Internal stratigraphic relationships in the Etendeka Group in the Huab Basin, NW Namibia: Understanding the onset of flood volcanism: Journal of Geodynamics, v. 28, p. 393–418.

Jerram, D.A., Mountney, N., and Stollhofen, H., 1999b, Facies architecture of the Etjo Sandstone Formation and its interaction with the basal Etendeka flood basalts of NW Namibia: Implications for offshore analogues, *in* Cameron, N., Bate, R., and Clure, V., eds., The oil and gas habitats of the South Atlantic: Geological Society [London] Special Publication 153, p. 367–380.

Jerram, D.A., Mountney, N., Howell, J., Long, D., and Stollhofen, H., 2000, Death of a sand sea: An active erg systematically buried by the Etendeka flood basalts of NW Namibia: Journal of the Geological Society of London, v. 157, p. 513–516.

Johnson, R.W., editor, 1989, Intraplate volcanism in Eastern Australia and New Zealand: Sydney, Cambridge University Press, p. 1–408.

Kampunzu, A.B., and Mohr, P., 1991, Magmatic evolution and petrogenesis in the East African Rift system, *in* Kampunzu, A.B., and Lubala, R.T., eds., Magmatism in extensional structural settings: The Phanerozoic African Plate: Berlin, Springer-Verlag, p. 85–136.

Kampunzu, A.B., and Popoff, M., 1991, Distribution of the main Phanerozoic African rifts and associated magmatism: Introductory notes, *in* Kampunzu, A.B., and Lubala, R.T., eds., Magmatism in extensional structural settings: Berlin, Springer-Verlag, p. 2–10.

Kenea, N.H., Ebinger, C.J., and Rex, D.C., 2001, Late Oligocene volcanism and extension in the southern Red Sea Hills, Sudan: Journal of the Geological Society of London, v. 158, p. 285–294.

Kent, W., Saunders, A.D., Kempton, P.D., and Ghose, N.C., 1997, Rajmahal Basalts, eastern India: Mantle sources and melt distribution at a volcanic rifted margin, *in* Mahoney, J.J, and Coffin M.F., eds., Large igneous provinces: Continental, oceanic, and planetary flood volcanism: American Geophysical Union Geophysical Monograph 100, p. 145–182.

Keszthelyi, L., and Self, S., 1998, Some physical requirements for the emplacement of long basaltic lava flows: Journal of Geophysical Research, v. 103, no. B11, p. 27447–27464.

Kirstein, L.A., Peate, D.W., Hawkesworth, C.J., Turner, S.P., Harris, C., and Mantovani, M., 2000, Early Cretaceous basaltic and rhyolitic magmatism in southern Uruguay associated with the opening of the South Atlantic: Journal of Petrology, v. 41, p. 1413–1438.

Larsen, H.C., Saunders, A.D., and Clift, P.D., 1995, Proceedings of the Ocean Drilling Program, Initial Results, Leg 152: College Station, Texas, Ocean Drilling Program, 975 p.

Le Maitre, R.W., editor, 1989, A classification of igneous rocks and glossary of terms, *in* Recommendations of the International Union of Geological Sciences Subcommission on the systematics of igneous rocks: Oxford, Blackwell, 193 p.

Leeman, W.P., 1982, Development of the Snake River Plain–Yellowstone Plateau Province, Idaho and Wyoming: An overview and petrologic model, *in* Bonnichsen, B., and Breckenridge, R.M., eds., Cenozoic geology of Idaho: Idaho Bureau of Mines and Geology Bulletin, v. 26, p. 155–177.

Levitte, D., Columbia, J., and Mohr, P., 1974, Reconnaisance geology of the Amaro Horst, southern Ethiopian rift: Geological Society of America Bulletin, v. 85, p. 417–422.

Lightfoot, P.C., Hawkesworth, C.J., and Sethna, S.F., 1987, Petrogenesis of rhyolites and trachytes from the Deccan Trap: Sr, Nd and Pb isotope and trace element evidence: Contributions to Mineralogy and Petrology, v. 95, p. 44–54.

Lipman, P.W., 1984, The roots of ash flow calderas in western North America: Windows into the tops of granite batholiths: Journal of Geophysical Research, v. 89, p. 8801–8841.

Lipman, P.W., and Glazner, A.F., 1991, Introduction to Middle Tertiary cordillerian volcanism: Magma sources and relations to regional tectonics: Journal of Geophysical Research, v. 96, p. 13193–13199.

McCulloch, M.T., Kyser, T.K., Woodhead, J.D., and Kinsley, L., 1994, Pb-Sr-Nd-O isotopic constraints on the origin of rhyolites from the Taupo Volcanic Zone of New Zealand: Evidence for assimilation followed by fractionation from basalt: Contributions to Mineralogy and Petrology, v. 115, p. 303–312.

McDowell, F.W., and Clabaugh, S.E., 1979, Ignimbrites of the Sierra Madre Occidental and their relation to the tectonic history of western Mexico: Geological Society of America Special Paper 180, p. 113–124.

McPhie, J., Allen, S.R., and Simpson, C., 2000, Extensive felsic lavas in intraplate volcanic provinces: Geological Society of Australia, Abstracts, v. 59, p. 328.

Mahoney, J.J., 1988, Deccan traps, *in* Macdougall, J.D., ed., Continental flood basalts: Dordrecht, Netherlands, Kluwer Academic Publishers, p. 151–194.

Mahoney, J.J., and Coffin, M.F., 1997, Large igneous provinces: Continental, oceanic, and planetary flood volcanism: American Geophysical Union Geophysical Monograph 100, 438 p.

Marsh, J.S., and Eales, H.V., 1984, The chemistry and petrogenesis of igneous rocks of the Karoo central area, southern Africa, *in* Erlank, A.J., ed., Petrogenesis of the volcanic rocks of the Karoo Province: Geological Society of South Africa Special Publication 13, p. 27–67.

Marsh, J.S., Ewart, A., Milner, S.C., Duncan, A.R., and Miller, R.McG., 2001, The Etendeka Igneous Province: Magma types and their stratigraphic distribution with implications for the evolution of the Paraná-Etendeka Flood Basalt Province: Bulletin of Volcanology, v. 62, p. 464–486.

Martin, H., Mathias, M, and Simpson, E.S.W., 1960, The Damaraland sub-volcanic ring complexes in south west Africa, *in* Sorgenfrei, T., ed., Report of the International Geological Congress 21 Session 13: Copenhagen, Norden, p. 156–174.

Marzoli, A., Renne, P.R., Piccirillo, E.M., Ernesto, M., Bellieni, G., and De Min, A., 1999, Extensive 200-million-year-old continental flood basalts of the Central Atlantic Magmatic Province: Science, v. 284, p. 616–618.

Meighan, I.G., Fallick, A.E., and McCormick A.G., 1992, Anorogenic granite magma genesis: New isotopic data for the southern sector of the British Tertiary Igneous Province: Transactions of the Royal Society of Edinburgh, Earth Sciences, v. 83, p. 227–233.

Meighan, I.G., Gibson, D., and Hood, D.N., 1984, Some aspects of Tertiary acid magmatism in NE Ireland, *in* Walsh, J.N., ed., The origin of the igneous rocks of the Tertiary igneous province: Mineralogical Magazine, v. 48, p. 351–363.

Milner, S.C., 1988, The geology and geochemistry of the Etendeka Formation quartz latites, Namibia [Ph.D. thesis]: Cape Town, South Africa, University of Cape Town, 263 p.

Milner, S.C., and Ewart, A., 1989, The geology of the Goboboseb Mountain volcanics and their relationship to the Messum Complex, Namibia: Communications of the Geological Survey of Namibia, v. 5, p. 31–40.

Milner, S.C., Duncan, A.R., and Ewart, A., 1992, Quartz latite rheoignimbrite flows of the Etendeka Formation, north-western Namibia: Bulletin of Volcanology, v. 54, p. 200–219.

Milner, S.C., Duncan, A.R., Whittingham, A.M., and Ewart, A., 1995, Trans-Atlantic correlation of eruptive sequences and individual silicic volcanic units within the Paraná-Etendeka igneous province: Journal of Volcanology and Geothermal Research, v. 69, p. 137–157.

Mitchell, W.I., Cooper, M.R., Hards, V.L., and Meighan, I.G., 1999, An occurrence of silicic volcanic rocks in the early Palaeogene Antrim Lava Group of Northern Ireland: Scottish Journal of Geology, v. 35, p. 179–185.

Mohr, P., 1983, Ethiopian flood basalt province: Nature, v. 303, p. 577–584.

Moore, C.L., Wallace, P., and ODP Leg 183 Shipboard Scientific Party, 2000, Products of explosive felsic volcanism within the basalt lava pile of the Kerguelen Plateau large igneous province, southern Indian Ocean: Geological Society of Australia, Abstracts, v. 59, p. 352.

Moore, I., and Kokelaar, P., 1998, Tectonically controlled piecemeal caldera collapse: A case study of Glencoe volcano, Scotland: Geological Society of America Bulletin, v. 110, p. 1448–1466.

Nieto-Samaniego, Á.F., Ferrari, L., Alaniz-Alvarez, S.A., Labarthe-Hernández, G., and Rosas-Elguera, J., 1999, Variation of Cenozoic extension and volcanism across the southern Sierra Madre Occidental volcanic province, Mexico: Geological Society of America Bulletin, v. 111, p. 347–363.

O'Sullivan, P.B., Kohn, B.P., and Cranfield, L., 1999, Fission track constraints on the Mesozoic to Recent thermotectonic history of the northern New England Orogen, southeastern Queensland: New England Orogen Conference Abstracts, Armidale, Australia, p. 285–293.

O'Sullivan, P.B., Kohn, B.P., Foster, D.A., and Gleadow, A.J.W., 1995, Fission track data from the Bathurst Batholith: Evidence for rapid Mid-Cretaceous uplift and erosion within the eastern highlands of Australia: Australian Journal of Earth Sciences, v. 42, p. 597–607.

Pankhurst, R.J., and Rapela, C.R., 1995, Production of Jurassic rhyolite by anatexis of the lower crust of Patagonia: Earth and Planetary Science Letters, v. 134, p. 23–36.

Pankhurst, R.J., Leat, P.T., Sruoga, P., Rapela, C.W., Márquez, M., Storey, B.C., and Riley, T.R., 1998, The Chon Aike silicic igneous province of Patagonia and related rocks in Antarctica: A silicic large igneous province: Journal of Volcanology and Geothermal Research, v. 81, p. 113–136.

Pankhurst, R.J., Riley, T.R., Fanning, C.M., and Kelley, S.R., 2000, Episodic silicic volcanism along the proto-Pacific margin of Patagonia and the Antarctic Peninsula: Plume and subduction influences associated with the break-up of Gondwana: Journal of Petrology, v. 41, p. 605–625.

Pearson D.G., Emeleus, C.H., and Kelley, S.P., 1996, Precise $^{40}Ar/^{39}Ar$ ages for the initiation of igneous activity in the Small Isles, Inner Hebrides and implications for the timing of magmatism in the British Tertiary Igneous Province: Journal of the Geological Society of London, v. 153, p. 815–818.

Preston, R.J., Bell, B.R., and Rogers, G., 1998a, The Loch Scridain xenolithic sill complex, Isle of Mull, Scotland: Fractional crystallisation, assimilation, magma-mixing and crustal anatexis in subvolcanic conduits: Journal of Petrology, v. 39, p. 519–550.

Preston, R.J., Hole, M.J., Still, J., and Patton, H., 1998b, The mineral chemistry and petrology of Tertiary pitchstones from Scotland: Transactions of the Royal Society of Edinburgh, Earth Sciences, v. 89, p. 95–111.

Rapp, R.P., and Watson, E.B., 1995, Dehydration melting of metabasalt at 8–32 kbar: Implications for continental growth and crust-mantle recycling: Journal of Petrology, v. 36, p. 891–932.

Renne, P.R., Glen, J.M., Milner, S.C., and Duncan, A.R., 1996, Age of Etendeka flood volcanism and associated intrusions in southwestern Africa: Geology, v. 24, p. 659–662.

Renne, P.R., Zichao, Z., Richards, M.A., Black, M.T., and Basu, A., 1995, Synchrony and causal relations between Permian-Triassic boundary crises and Siberian flood volcanism: Science, v. 269, p. 1413–1415.

Renne, P.R., Ernesto, M., Pacca, I.G., Coe, R.S., Glen, J.M., Prévot, M., and Richards, M.A., 1992, The age of the Paraná flood volcanism, rifting of Gondwanaland, and the Jurassic-Cretaceous boundary: Science, v. 258, p. 975–979.

Riley, T.R., and Leat, P.T., 1999, Large volume silicic volcanism along the proto-Pacific margin of Gondwana: Lithological and stratigraphical investigations from the Antarctic Peninsula: Geological Magazine, v. 136, p. 1–16.

Riley, T.R., Leat, P.T., Pankhurst, R.J., and Harris, C., 2001, Origins of large volume rhyolitic volcanism in the Antarctic Peninsula and Patagonia by crustal melting: Journal of Petrology, v. 42, no. 6, p. 1043–1065.

Roberts, M.P., and Clemens, J.D., 1993, Origin of high-potassium, calc-alkaline, I-type granitoids: Geology, v. 21, p. 825–828.

Ruiz, J., Patchett, P.J., and Arculus, R.J., 1988, Nd-Sr isotope composition of lower crustal xenoliths: Evidence for the origin of mid-Tertiary felsic volcanics in Mexico: Contributions to Mineralogy and Petrology, v. 99, p. 36–43.

Saunders, A.D., Fitton, J.G., Kerr, A.C., Norry, M.J., and Kent, R.W., 1997, The North Atlantic Igneous Province, *in* Mahoney, J.J., and Coffin, M.F., eds., Large igneous provinces: Continental, oceanic, and planetary flood volcanism: American Geophysical Union Geophysical Monograph 100, p. 45–93.

Schmitt, A.K., Emmermann, R., Trumbull, R.B., Bühn B., and Henjes-Kunst, F., 2000, Petrogenesis and $^{40}Ar/^{39}Ar$ geochronology of the Brandberg Complex, Namibia: Evidence for a major mantle contribution in metaluminous and peralkaline granites: Journal of Petrology, v. 41, p. 1207–1239.

Self, S., Thordarson, T., and Keszthelyi, L., 1997, Emplacement of continental flood basalt lava flows, *in* Mahoney, J.J, and Coffin M.F., eds., Large igneous provinces: Continental, oceanic, and planetary flood volcanism: American Geophysical Union Geophysical Monograph 100, p. 381–410.

Self, S., Thordarson, T., Keszthelyi, L., Walker, G.P.L., Hon, K., Murphy, M.T., Long, P., and Finnemore, S., 1996, A new model for the emplacement of Columbia River basalts as large, inflated pahoehoe lava flow fields: Geophysical Research Letters, v. 23, p. 2689–2692.

Sethna, S.F., and Battiwala, H.K., 1977, Chemical classification of the intermediate and acid rocks (Deccan Trap) of Salsette Island, Bombay: Journal of the Geological Society of India, v. 18, p. 323–330.

Sharma, M., 1997, Siberian traps, *in* Mahoney, J.J., and Coffin, M.F., eds., Large igneous provinces: Continental, oceanic, and planetary flood volcanism: American Geophysical Union Geophysical Monograph 100, p. 273–295.

Sinton, C.W., Hitchen, K., and Duncan, R.A., 1998, ^{40}Ar-^{39}Ar geochronology of silicic and basic volcanic rocks on the margins of the North Atlantic: Geological Magazine, v. 135, p. 161–170.

Stephens, C.J., Ewart, A., Bryan, S., and Schön, R.W., 1995, Rift-related, large-volume silicic volcanism associated with Lower Cretaceous continental breakup, eastern Australia: Boulder, Colorado, International Union of Geodesy and Geophysics 21 General Assembly, Abstracts, p. A443.

Stern, T.A., 1985, A back-arc basin formed within continental lithosphere: The central volcanic region of New Zealand: Tectonophysics, v. 112, p. 385–409.

Storey, B.C., Alabaster, T., and Pankhurst, R.J., editors, 1992, Magmatism and the causes of continental break-up: Geological Society [London] Special Publication 68, 404 p.

Swanson, E.R., and McDowell, F.W., 1984, Calderas of the Sierra Madre Occidental volcanic field western Mexico: Journal of Geophysical Research, v. 89, p. 8787–8799.

Swanson, D.A., Wright, T.H., and Helz, R.T., 1975, Linear vent systems and estimated rates of magma production and eruption for the Yakima basalt on the Columbia Plateau: American Journal of Science, v. 275, p. 877–905.

Sweeney, R.J., Duncan, A.R., and Erlank, A.J., 1994, Geochemistry and petrogenesis of central Lebombo basalts of the Karoo igneous province: Journal of Petrology, v. 35, p. 95–125.

Thordarson, T., and Self, S., 1998, The Rosa Member, Columbia River Basalt Group: A gigantic pahoehoe lava flow field formed by endogenous processes?: Journal of Geophysical Research, v. 103, no. B11, p. 27411–27445.

Turner, S., Regelous, M., Kelley, S., Hawkesworth, C., and Mantovani, M., 1994, Magmatism and continental break-up in the South Atlantic: High precision ^{40}Ar-^{39}Ar geochronology: Earth and Planetary Science Letters, v. 121, p. 333–348.

Ukstins, I.A., Baker, J., Al-Kadasi, M., Al-Subbary, A., Menzies, M., and Peate, D., 2000, Voluminous explosive silicic eruptions during Oligocene flood volcanism in Yemen: Royal Holloway, University of London, Department of Geology, Penrose 2000 Volcanic Rifted Margins, Abstracts, p. 85.

Veevers, J.J., Powell, C.McA., and Roots, S.R., 1991, Review of seafloor spreading around Australia. 1. Synthesis of the patterns of spreading: Australian Journal of Earth Sciences, v. 38, p. 373–389.

Von Rad, U., and Thurow, J., 1992, Bentonitic clays as indicators of Early Neocomian post-breakup volcanism off northwest Australia, *in* Proceedings of the Ocean Drilling Program, Scientific Results, Leg 122: College Station, Texas, Ocean Drilling Program, p. 213–232.

Ward, P.L., 1991, On plate tectonics and the geologic evolution of southwestern North America: Journal of Geophysical Research, v. 96, p. 12479–12496.

Ward, P.L., 1995, Subduction cycles under western North America during the Mesozoic and Cenozoic eras, *in* Miller, D.M., Busby C., eds., Jurassic magmatism and tectonics of the North American Cordillera: Geological Society of America Special Paper 299, p. 1–45.

Wark, D.A., 1991, Oligocene ash flow volcanism, northern Sierra Madre Occidental: Role of mafic and intermediate-composition magmas in rhyolite genesis: Journal of Geophysical Research, v. 96, p. 13389–13411.

Watkeys, M.K., Meth, D.L., and Harmer, R.E., 2000, The Lebombo and the Rooi Rand dyke swarm, southern Africa: A volcanic margin without a plume?: Royal Holloway, University of London, Department of Geology, Penrose 2000 Volcanic Rifted Margins, Abstracts, p. 86.

White, R.S., 1992, Magmatism during and after continental break-up, *in* Storey, B.C., Alabaster, T., and Pankhurst, R.J., eds., Magmatism and the causes of continental break-up: Geological Society [London] Special Publication 68, p. 1–16.

White, R.S., and McKenzie, D.P., 1989, Magmatism at rift zones: The generation of volcanic continental margins and flood basalts: Journal of Geophysical Research, v. 94, p. 7685–7729.

White, R.S., Spence, G.D., Fowler, S.R., McKenzie, D.P., Westbrook, G.K., and Bowen, A.N., 1987, Magmatism at rifted continental margins: Nature, v. 330, p. 439–444.

Widdowson, M., 1997, Tertiary palaeosurfaces of the SW Deccan, Western India: Implications for passive margin uplift, *in* Widdowson, M., ed., Palaeosurfaces: Recognition, reconstruction and palaeoenvironmental interpretation: Geological Society [London] Special Publication 120, p. 221–248.

Widdowson, M., Walsh, J.N., and Subbarao, K.V., 1997, The geochemistry of Indian bole horizons: Palaeoenvironmental implications of Deccan intravolcanic palaeosurfaces, *in* Widdowson, M., ed., Palaeosurfaces recognition, reconstruction, and palaeoenvironmental interpretation: Geological Society [London] Special Publication 120, p. 269–282.

Wilson, C.J.N., Houghton, B.F., McWilliams, M.O., Lanphere, M.A., Weaver, S.D., and Briggs, R.D., 1995, Volcanic and structural evolution of Taupo Volcanic Zone, New Zealand: A review: Journal of Volcanology and Geothermal Research, v. 68, p. 1–28.

Wilson, C.J.N., Rogan, A.M., Smith, I.E.M., Northey, D.J., Nairn, I.A., and Houghton, B.F., 1984, Caldera volcanoes of the Taupo Volcanic Zone, New Zealand: Journal of Geophysical Research, v. 89, p. 8463–8484.

Woldegabriel, G., Aronson, J.L., and Walter, R.C., 1990, Geology, geochronology, and rift basin development in the central sector of the Main Ethiopian Rift: Geological Society of America Bulletin, v. 102, p. 439–458.

MANUSCRIPT ACCEPTED BY THE SOCIETY OCTOBER 15, 2001

Printed in the U.S.A.

Geological Society of America
Special Paper 362
2002

Volcanology and facies architecture of flood basalts

Dougal A. Jerram*
Department of Geological Sciences, University of Durham, South Road, Durham DH1 3LE, UK

ABSTRACT

In terms of their detailed volcanology and facies architecture, continental flood basalts and associated volcanic rifted margins reveal important information to help our understanding of their evolution. Mafic volcanism, which makes up the majority of preserved material, is characterized by flows 2–3 m to several tens of meters thick, with ponded flows and occasional massive flow events of ~100 m thick. Although most of the flows are emplaced by the same mechanism as passive inflated sheets, a variety of different facies associations are dependent on flow volumes and to some extent flow composition. The largest silicic volcanic events in continental flood basalts are larger in volume than the largest recorded mafic events, and they are potentially more catastrophic if erupted as ignimbrite flows. The architecture of continental flood basalts and associated volcanic rifted margins is recorded by facies types and facies associations. Facies types, such as tabular-classic flows, braided-compound flows, or hyaloclastites, represent genetically related building blocks of the volcanic stratigraphy. Facies associations, such as downlap, onlap, and disconformities, relate how the volcanic facies are stacked together. Many of the facies associations occur on an intermediate to large basin-wide scale and may only be revealed by detailed field work, photogrammetry, and three-dimensional geological models.

INTRODUCTION

Flood basalt provinces are important because they provide information on the timing of major mantle events. They are associated with continental breakup, and they may be associated with global extinction events (e.g., Campbell et al., 1992; Hofman et al., 1997; Courtillot et al., 1999; Wignall, 2001). Such large-scale volcanic events generate complex basin fills and cover large areas of the planet. However, the majority of studies into continental flood basalt provinces have concentrated on geochemistry (e.g., Hawkesworth et al., 1988; Ewart et al., 1998), geochronology (e.g., Renne et al., 1996; Hamilton et al., 1998), and relationship with mantle plumes (e.g., White and McKenzie, 1989). Only a few recent studies have concentrated on the detailed facies architecture of flood basalts and their physical volcanology (Self et al., 1997; Jerram et al., 1999a).

Continental flood basalts consist of large accumulations of lava flows (e.g., the Paraná basin, South America, and the North Atlantic igneous province, which comprises the British Tertiary and Greenland igneous provinces). Such large accumulations do not have a simple, layer-cake stratigraphy, but they contain complex internal and external architectures (Jerram et al., 1999a, 1999b; Planke et al., 2000). Such architectures are governed by the volume of individual eruption events, the location and abundance of volcanic centers, and the evolution of these centers through time.

Some recent advances have improved our understanding of the volcanic processes occurring in the formation of flood basalts (e.g., Self et al., 1997). Coupled with more detailed studies of two-dimensional and three-dimensional facies relationships in flood basalts, this enables us to look at the facies architecture in a more pragmatic way to help understand their onset

*E-mail: D.A.Jerram@dur.ac.uk.

Jerram, D.A., 2002, Volcanology and facies architecture of flood basalts, *in* Menzies, M.A., Klemperer, S.L., Ebinger, C.J., and Baker, J., eds., Volcanic Rifted Margins: Boulder, Colorado, Geological Society of America Special Paper 362, p. 119–132.

and evolution through time. The intent of this contribution is not to provide a complete and comprehensive review, but rather to (1) outline the facies architecture of continental flood basalts, (2) summarize our current understanding of the physical volcanology of flood basalt provinces, (3) discuss how the styles of volcanology relate to the broader facies architecture of flood basalt provinces, and (4) outline some of the future goals toward understanding the development of flood basalts.

HOW DIFFERENT ARE FLOOD BASALTS FROM RECENT VOLCANIC EVENTS?

What makes the effusion of continental flood basalts different from present-day volcanism? Continental flood basalts, volcanic rifted margins, and oceanic plateaus represent the largest episodes of volcanism on the planet (Mahoney and Coffin, 1997). However, are the types and scales of individual volcanic events different than what has been observed on Earth in the recent past? In essence, is the difference between flood volcanism and normal volcanism (that not associated with continental flood basalt) a factor of (1) protracted and persistent eruption of material over a large period of time producing large volumes of volcanic material, or (2) a shorter period and/or periods of high-flux volcanism? Table 1 briefly summarizes and compares some recent and historic observations of modern volcanism with those observed from some of the better constrained continental flood basalts. In general, continental flood basalts appear to be a large mixture of different scales of volcanic event. Occasionally, very large individual events occur, resulting in massive volume outputs for single eruption events (e.g., the Goboboseb Member, Paraná-Etendeka, and the Roza Member, Columbia River; see Fig. 1). Often, the majority of flood volcanism has occurred over a short period of time (~1 m.y.) (e.g., Renne et al., 1996; Hofman et al., 1997), and is interpreted to originate from both fissure and center fed eruptions.

On a broader scale the association of continental flood basalts with continental breakup and mantle plumes (e.g., Storey, 1995; Hawkesworth et al., 1999) may be significant in determining the volcanological character of continental flood basalts. Although plume-related magmatism occurs today, most of this is through oceanic crust, and we currently have few examples of present-day plumes under the continents (e.g., Yellowstone, African Rift). It is not entirely clear that all continental flood basalts are associated with plumes (Hawkesworth et al., 1999). The major difference between normal and continental flood basalt volcanism seems to be the scale and time duration of activity.

Continental flood basalts show a persistence of eruptions through a short space of time, some with scales similar to that observed in recent history, punctuated with much larger volume individual events. Given this general description of continental flood basalts, what are the scales and geometries of units that stack up to form the internal architecture of the continental flood basalt province? Before examining the facies architecture of continental flood basalts it is important to understand the phys-

TABLE 1. SCALES AND VARIATIONS IN MAFIC VOLCANISM: A GENERAL COMPARISON BETWEEN CONTINENTAL FLOOD BASALTS AND HOLOCENE FLOWS

	Scale of flows	Thickness	Top of the lava surface	Internal layers	External relationships
Modern	Hawaii 51 km long over 300 days in 1859, Iceland (Laki) 65 km 1783–1784 15–20 km^3, Thjorsa flow 140 km 8500 years old	Individual lobes 1–5 m, ponded flows several meters to tens of meters	Variations of the order of meters to tens of meters, usually ± 5 m. Large variations are caused by lava flowing over preexisting topography.	Flows have a core/upper crust relationship. Cores are less vesicular in nature with simple platy jointing. Crusts have higher vesicle contents and have more irregular jointing patterns. Cores commonly make up >50% of the flow thickness.	Larger variations may be present, e.g., rifts, slopes.
Ancient	Columbia River Basalts (CRB) individual flows to 600 km (100 km common) Roza Member 1300 km^3	CRB 16–50 m, Etendeka 1–60 m, ponded flows to 100 m. Deccan 20 m.	Variations similar to modern flows of the order of meters to tens of meters usually ± 5 m	A similar core/crust relationship, e.g., flows in the Etendeka made of ~60% core and ~40% crust which have different properties. In CRB cores commonly compose 44%–75% of the flow thickness (Self et al., 1997).	Volcanic disconformities (Etendeka) disconformities of 300–400 m over 70 km

ical volcanology and emplacement models for continental flood basalts (discussed in the following).

PHYSICAL VOLCANOLOGY AND EMPLACEMENT

Groups of lava flows, in a way similar to parasequences in sedimentology (Van Wagoner et al., 1988), provide the fundamental building blocks of the volcanic sequence. Therefore, an understanding of lava flow emplacement and the physical volcanology of flood basalts is very important when considering the facies and three-dimensional stratigraphic architecture of continental flood basalts. In general, many continental flood basalts are bimodal in character (e.g., Milner et al., 1995; Menzies et al., 1997) with the mafic (basalts, basaltic andesites) and silicic (rhyolites, ignimbrites) volcanism. These two broad groups are discussed separately.

Mafic volcanism

The most abundant type of volcanism preserved in continental flood basalts is mafic; flows range in composition from 45 to 55 wt% SiO_2. Much of the early work that addressed the problems of emplacement of mafic units in continental flood basalts concentrated on the Columbia River Basalts (e.g., Swanson et al., 1975), invoking rapid emplacement mechanisms. One of the main reasons for this was the observation that flows in the Columbia River Basalts that traveled many tens to hundreds of kilometers showed little crystallization, and presumably limited synemplacement cooling, over their length. Therefore, the flows must have traveled rapidly, possibly turbulently, before any significant cooling occurred. One potential problem with this interpretation is that turbulent flow would result in a more rapid cooling. Very rapid eruption rates (>0.6 km^3/hr/linear km of fissure, fissure or dike widths >3 m) were suggested to reduce the overall flow time and reduce the impact of turbulent cooling until 100 km or more (Shaw and Swanson, 1970). In addition, a transition from turbulent to laminar flow toward the end of eruptive activity was used to explain the preservation of pahoehoe flow features in localized areas (Reidel and Tolan, 1992).

Following the work of Hon et al. (1994), Self et al. (1996a) presented a new model in which the Columbia River Basalts were emplaced through an inflation process. In this model, flows initially start as thin lobes of slow-moving pahoehoe lava breakouts. With continued injection of lava, the flow inflates, an insulating crust developing on top and new breakouts occurring at the front of the flow (see Self et al., 1996a, their Fig. 1). If lava production is maintained and flow continues, it will inflate to a thickness of several meters and possibly, in the case of flood basalts, several tens of meters thick. The introduction of the inflation model was followed by more detailed observations for the Columbia River Basalts (Self et al., 1997), and the suggestion that this model could be applied to continental flood basalts in general (Self et al., 1998). Keszthelyi and Self (1998) argued, on the basis of cooling models, that flows emplaced rapidly should be morphologically similar to channel-fed aa flows, while those emplaced under insulated conditions should be similar to tube-fed or sheet-

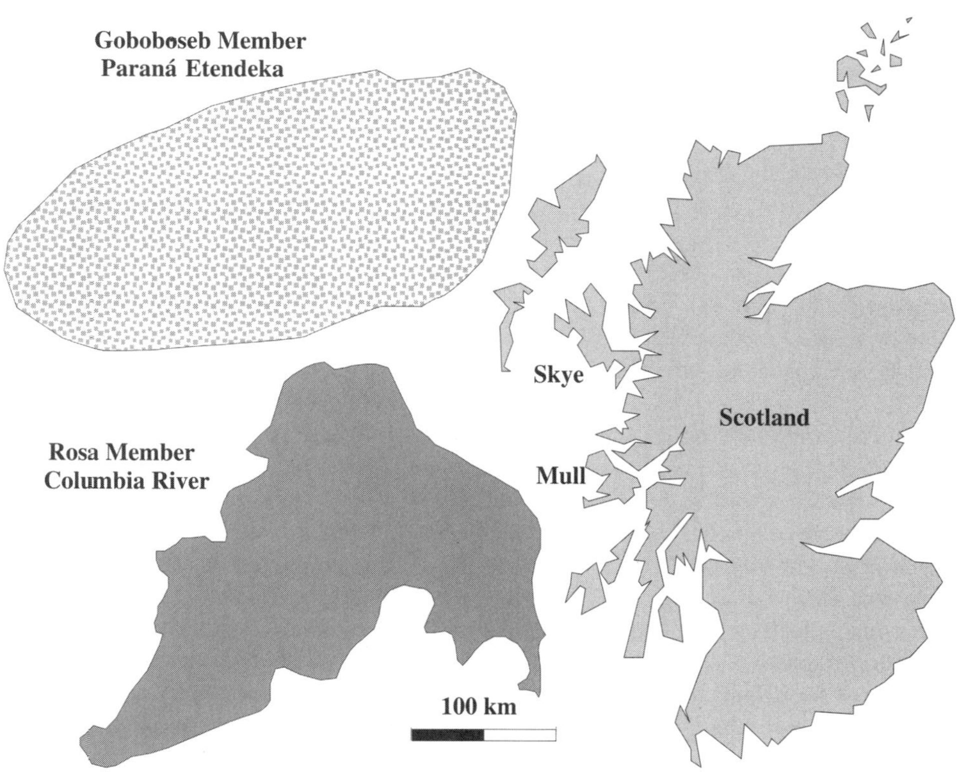

Figure 1. Comparison of areal extent and volume of Rosa Member basalt from Columbia River Basalts (~1300 km^3, Self et al., 1997) with Goboboseb Member silicic volcanic unit form Paraná-Etendeka (~2320 km^3, Milner et al., 1995) (map of Scotland given for comparison of scale).

like inflated pahoehoe flows. An inflation model is clearly favored because detailed field observations of volcanological phenomena are inconsistent with rapid emplacement, such as: pahoehoe ropes, pahoehoe lobes, inflation tumuli, and pipe vesicles (see Self et al., 1997; their Table 2). The best defined example from the Columbia River Basalts is the 1300 km^3 Roza Member (Thordarson and Self, 1998).

Workers have been recording similar volcanological observations for some of the classic larger continental flood basalts (e.g., British Tertiary [North Atlantic igneous province], Kent et al., 1997; Paraná-Etendeka, Jerram et al., 2000). Figure 2 provides some selected examples of preserved volcanic features from continental flood basalts (pahoehoe ropes; Fig. 2, A and B), lava tube and/or pipe (Fig. 2C), inflation lobes and/or tumuli (Fig. 2, D and E). Other features can include sediment-filled inflation fissures, feeder dikes, ponded flows, volcanic centers, and sheet flows (for good documented examples see Self et al., 1997; Jerram et al., 1999b, 2000; Jerram and Stollhofen, 2002).

If the inflation model, also known as the swell hypothesis (Self et al., 1998), can be applied to all types of mafic flow in continental flood basalts, then it provides a basis for estimating eruption rates (e.g., Self et al., 1997). If, however, this hypothesis only applies to a limited number of flow types in continental flood basalts, then further work is required to distinguish flow types directly related to the inflation emplacement mechanism from others that form by other processes; e.g., aa flows. The approach of comparison between Earth volcanism, recent and continental flood basalts, with other planetary flows can be very useful because here full flow morphologies and top surface structures are often visible for very large eruptions (Keszthelyi et al., 2000).

Silicic volcanism

Silicic volcanism plays an important role in the development of continental flood basalts and associated margins, yet its significance has been overlooked (see also Bryan et al., this volume). Important factors that are poorly defined with respect to silicic volcanism in continental flood basalts are the overall volume (in terms of both erupted material and gases), abundance of igneous centers, and the general frequency and abundance of silicic eruptive events.

Unlike mafic volcanism, which can be fed from both volcanic centers and fissure systems, silicic volcanism is entirely associated with igneous centers. Examples such as the British Tertiary and the Etendeka contain a number of exposed igneous centers (e.g., Skye and Mull [British Tertiary], Messum and Erongo [Etendeka]), and exposure of these complexes is due, in part, to their erosion level. In many other examples (such as Siberia, and Paraná) the level of erosion and/or exposure makes it difficult to assess the amount and distribution of silicic igneous centers. Therefore, our current record of the abundance of silicic igneous centers is limited in many of the continental flood basalts.

When we examine the preserved eruptive sequences of continental flood basalts it is clear that in a few of the continental flood basalts there have been large eruptions of silicic material (e.g., Ethiopia-Yemen, Paraná-Etendeka). Some of the most significant of these eruptions are the large-volume flows found in the Paraná-Etendeka (Milner et al., 1992; Ewart et al., 1998). Milner et al. (1995) estimated the volume of three silicic units in the Paraná-Etendeka system to be 2320 km^3 (Goboboseb–PAVunit A), 6340 km^3 (Springbok–PAVunit B), and 3775 km^3 (Grootberg–PAVunit E/F), respectively. These silicic flow units are bigger than the largest recorded mafic flows (e.g., Roza Member, 1300 km^3 [Self et al., 1997]). The eruptive and emplacement mechanisms for these large flows are poorly understood. If they erupt as lava flows, they may have a limited environmental effect on a global scale. However, if these units represent rheo-ignimbrites (Milner et al., 1992), an ignimbrite flow of such enormous volume would have a large eruptive cloud and would have a far greater impact on the global environment. The significance of these large silicic flows is discussed further in the next section.

Estimating the general frequency and abundance of much smaller silicic eruptive events poses another problem. Occasional small rhyolites and ignimbrites have been found associated with continental flood basalts (e.g., Mitchell et al., 1999), and other pyroclastic evidence is present as remnants in caldera structures preserved in volcanic centers (e.g., Bell and Emeleus, 1988; Donaldson et al., 2001). However, unless silicic eruptions are of significant volume and/or welded, their preservation potential is limited. This is because loose pyroclastic material is easy to erode. A close inspection of many of the commonly considered mafic sequences (e.g., Skye main lava sequence) reveals evidence for silicic activity throughout the sequence. This evidence is only preserved as thin ash horizons, mixed up in boles, between the mafic flows (Bell and Emeleus, 1988; C. Emeleus, 2001, personal commun.). Due to the nature and scope of many of the previous studies on continental flood basalts, much of this type of data is yet to be collected and/or fully assessed.

Silicic volcanic components could potentially display a characteristic similar to mafic flows, i.e., eruptive events similar in scale and style to modern silicic volcanism with occasional massive volcanic events forming extensive thick silicic sheets. It is possible that the role of small- to moderate-scale silicic volcanism in continental flood basalts may be underestimated due to poor preservation of pyroclastic material.

Rates of eruption, scales of flows, and environmental impact

There has been much discussion on the possible links between flood volcanism and mass-extinction events. Wignall (2001) provided a comprehensive review of the debate about large igneous provinces and mass extinction. There is at least a partial correlation with continental flood basalts and mass-

Figure 2. Examples of physical volcanological features preserved in flood basalts. A, B: Molds of ropey pahoehoe textures preserved on underside of sediments from Etendeka, northwestern Namibia. C: Tube and/or pipe feature, British Tertiary (North Atlantic igneous province, NAIP), northwestern Scotland. D: Inflated tumuli or lobe, British Tertiary (NAIP), northwestern Scotland. E: Cliff section highlighting two lobes of same flow, British Tertiary (NAIP), northwestern Scotland (sheep for scale).

extinction events, and it is possible for flood volcanism to act as a causal mechanism to trigger dramatic changes in the global environment (Wignall, 2001). Key to understanding and assessing the potential impact of flood volcanism on the environment is a thorough understanding of the rates of eruption, scales of events, and a full record of the number of flows in a continental flood basalt. Many recent advances have been made in determining the geochronology of flood basalt eruptions. This has been mainly due to an increase in the accuracy of dating techniques (e.g., Hamilton et al., 1998). However, there is a limit to the level at which we can further define eruption rates using such techniques, with realistic resolutions commonly limited to 1 m.y. In addition, a full stratigraphic coverage of radiometric dates does not yet exist for many provinces.

There has been much debate about the emplacement style and rate of mafic flood volcanism (e.g., Self et al., 1997, and references therein). Self et al. (1998) discussed the environmental impact of flood basalts in terms of gas emissions from large-scale mafic flood volcanism. Estimates of eruption rates ~4000 m^3/s for the Roza Member, Columbia River Basalts (Self et al., 1998), are similar to the fastest rates recorded for the largest recorded modern lava flow, e.g., Laki, Iceland (Thordarson and Self, 1993). Using this eruption rate and estimates of SO_2 gas emissions from Laki (Thordarson et al., 1996), Self et al. (1998) postulated that an eruption on the scale of the Roza Member could release >12 000 Mt of SO_2 over 10 yr. The Roza Member is by far the best mapped, large-scale, individual mafic flow event. However, it is not clear that it is representative of continental flood basalts in general, and more examples of the geometry of individual flow units are required to provide further examples. The volatile contents (with respect to SO_2) of different continental flood basalt flows from different provinces are not entirely clear.

Another important environmental issue arises when one considers the role of explosive volcanism in the formation of flood basalt provinces. Explosive volcanism represents the most extreme of volcanic hazards. The effect of SO_2 released in recent explosive eruptions may act as a guide to the potential environmental effects of explosive volcanic activity that occurs in continental flood basalts. The 1991 eruption of Mount Pinatubo, for example, erupted 17 megatons of SO_2 into the atmosphere with an observed Earth surface cooling of 0.5–0.6 °C (Self et al., 1996b). Pinatubo erupted 5 km^3 of dacitic magma and was the second-largest eruption of the twentieth century (Self et al., 1996b). One of the largest explosive eruptions known in the recent geological past is the Toba eruption, ca. 73 ka (Wignall, 2001). The eruption is estimated to have produced at least 2800 km^3 and to have efficiently erupted gases and ash into the stratosphere for 9–14 days (Rose and Chesner, 1990). An estimated 2200–4400 megatons of sulfate aerosols were ejected into the stratosphere, and may have perturbed the climate by –5 °C (see discussion in Wignall, 2001).

If there is a significant component of small to moderate explosive eruptions occurring at the same time as the better preserved mafic components, then models of the environmental impact of continental flood basalts must take this into account. This is important because explosive eruptions would be far more efficient at transporting material into the atmosphere. The presence of large-volume silicic eruptions, such as the Goboboseb Member in the Paraná-Etendeka, poses an altogether different problem. The potential environmental impact of such large-volume (2320–6340 km^3) flows, if erupted as ignimbrites (Milner et al., 1992), could be catastrophic. Several of these flows have been recorded in the Etendeka, occurring in very close stratigraphic proximity, and are indistinguishable by radiometric dates (Renne et al., 1996).

It must be the goal of studies into the environmental impact of continental flood basalts to assess, as accurately as possible, the likely inputs from the whole of the continental flood basalt volcanic system into the atmosphere. Is the sustained slow release of sulfur from mafic eruptions more influential on climate than a sudden large-volume release, as during a large-volume explosive eruption? What are the consequences if both models are combined? Ideally, given better descriptions of the volumes and relations of facies within the volcanic stratigraphy, complex models could be generated that account for all of the volcanic inputs for each continental flood basalt.

FACIES ARCHITECTURE OF FLOOD BASALT PROVINCES

In this section the subject of the internal and external architecture of flood basalt provinces is addressed. Internal architecture relates to facies variations within the volcanic stratigraphy making up the continental flood basalt. External architecture relates to the relation of the continental flood basalt and the surrounding geology, i.e., the basin that hosts it. Issues discussed are the stratigraphic stacking patterns expected in continental flood basalts, the characteristic facies, and how continental flood basalts are manifested in three dimensions, in terms of their individual emplacement units and facies stacking relationships.

Some current issues that provide a greater understanding of the facies architecture in continental flood basalts are introduced, and the concepts of facies types, facies associations, and stacking patterns in continental flood basalts are discussed. The production of three-dimensional geological models of flood basalt provinces is discussed using the preliminary models of Jerram and Robbe (2001) from the Etendeka as an example.

Why look at the facies architecture of continental flood basalts?

One area of increasing interest is the internal and external architecture of flood basalt provinces in an offshore setting. Much of this is being fueled by industry exploration and exploitation of hydrocarbons in areas such as rift basins along continental margins with significant flood basalt cover (e.g., the

Kudu gas province, where hydrocarbon-bearing sediments are interbedded with flood basalts offshore southern Namibia; Jerram et al., 1999a). However, significant flood basalt cover makes it difficult to identify potential source/reservoir rocks and trap structures, which is known as the subbasalt imaging problem. Figure 3 highlights the problem of basalt cover masking a sedimentary basin beneath the North Atlantic igneous province. In such examples, a better understanding of the properties of the flood basalt cover will aid imaging techniques and methodologies aimed at the subbasalt imaging problem.

Significant sequences of basalt have been imaged along several volcanic rifted margins (e.g., Planke et al., 2000). It has recently been illustrated that on a seismic scale, volcanic rifted margins contain a variety of different seismic facies. These include landward flows, lava deltas, inner flows, inner and outer seaward-dipping reflectors, and outer highs (see Planke et al., 2000). In continental flood basalts it is often possible to get direct access to the stratigraphy and therefore examine the facies in more detail than on the seismic scale. If we can characterize the facies and facies architecture of continental flood basalts and, where possible, relate detailed observations to the larger scale basin-wide variations, we may be able to gain a better understanding into the evolution of continental flood basalts.

Concept of facies types and facies associations in continental flood basalts

A common way of characterizing large-scale sedimentary packages on a basin-wide scale is to identify and break down the facies components within the stratigraphy. These are then placed into conceptual models where facies relationships, stacking patterns, and architecture can be identified (e.g., the Book Cliffs, Utah; Van Wagoner et al., 1990). Can flood basalt components be identified in terms of different facies? It follows logically that different styles of volcanism will produce different packages, or facies, of volcanic units with different internal and external geometrical relationships. Would the concept of stacking patterns be of relevance in flood basalt provinces? In sedimentology, stacking pattern relationships are caused by the dynamic interrelationship between sea-level changes, subsidence, and sediment supply (Jervey, 1988). In the flood basalt volcanic system, supply rate and subsidence are important, as is the periodicity of volcanic eruptions (e.g., Pyle, 1998). An issue is whether volcanic episodes follow a distinctive pattern of periodicity during the development of the flood basalt province, and if so, whether this can be recognized in the internal stratigraphic architecture of continental flood basalts. The potential relationship of geo-

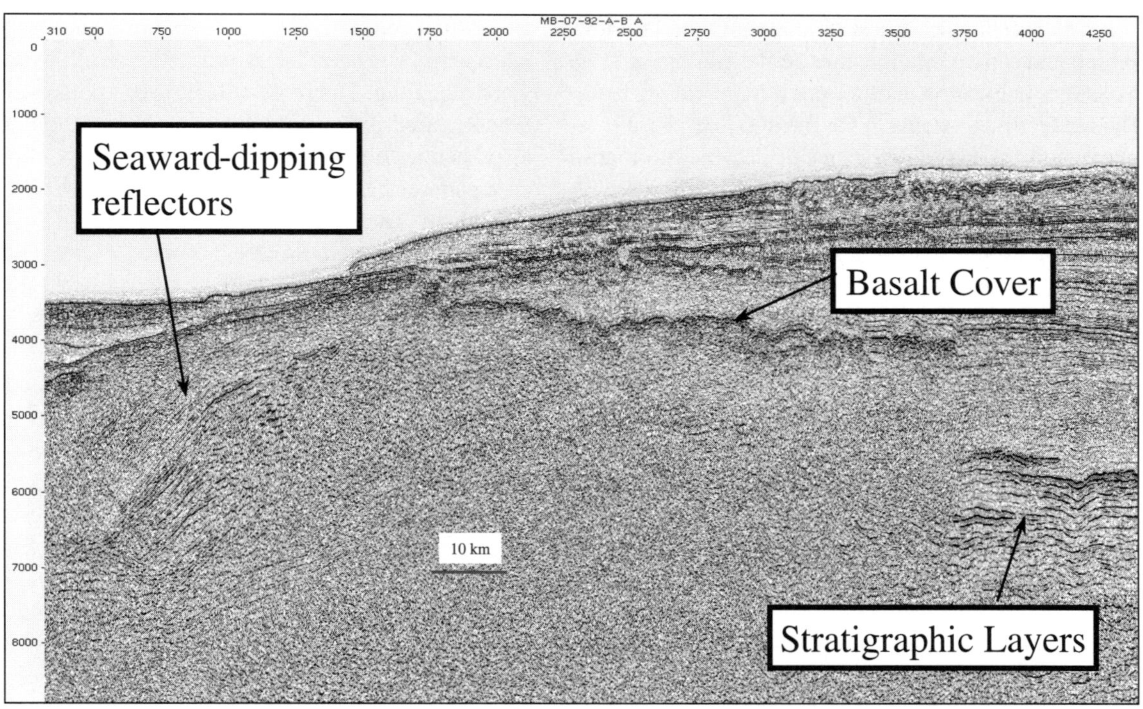

Figure 3. Seismic line MB-07-92 off Norway highlighting problem of subbasalt imaging (MB-07-92 NPD is owned by Norwegian Petroleum Directorate [NPD]; image courtesy of Elf Norge).

chemistry to flood basalt facies types is important, e.g., whether a certain geochemical group of lavas (such as high MgO) produce a certain facies type.

Stacking patterns in flood basalts may be complex because of the interrelationships between volcanic style, eruptive volume, duration, and periodicity. Defining stacking patterns, however, is of vital importance in being able to relate geometrical patterns in well-exposed flood basalts with what we may expect to find in offshore settings. Jerram et al. (1999b) mapped the stratigraphic architecture of the basal units in the Etendeka province and identified a paleovolcanic feature, the Doros Volcano, on the basis of stacking pattern relationships. This was later confirmed by three-dimensional modeling (Jerram and Robbe, 2001) and linked with the Doros igneous center on geochemical grounds (Marsh et al., 2001).

Continental flood basalt facies types are introduced in Table 2, with a description of the facies and an indication where known examples exist. Table 3 highlights some of the facies associations that have been documented from continental flood basalts. These facies and facies associations are based on personal observations from the Etendeka, Karoo, Columbia River, and British Tertiary provinces, and from published observations and personal communications based on a variety of continental flood basalts. The facies and facies associations outlined in Tables 2 and 3 are not a definitive list. However, it is hoped that they will provide a useful framework for future detailed architectural studies in continental flood basalts. Figure 4 highlights two examples of facies and facies associations. The example from the British Tertiary shows a braided-compound facies followed by tabular-classic facies (Fig. 4 A and B). Figure 4 (C and D) shows subtle downlap and toplap relationships in the Ethiopian Traps relating to changes in location of eruptions possibly along feeder fissures. The feeder dike systems in continental flood basalts are known to persist to the very later stages of province formation (Widdowson et al., 2000). However, an interesting facies association is that shield volcanoes occur in the early and late stages of provinces, marking the start up and shut down of the continental flood basalt system (Table 3).

A detailed example of the internal architecture of the Etendeka flood basalts in the Huab Basin, northwestern Namibia, was described by Jerram et al. (1999b) (mapped over ~70 × 40 km). Figure 5 shows a facies interpretation of the stratigraphic section through the Etendeka data set (adapted from Fig. 9 in Jerram et al., 1999b). Both tabular-classic and compound-braided lava facies are observed. The compound-braided lava facies define a shield volcano and associated flow fields, and contain sediment interlayers (isolated barchan dunes that were blowing over the building volcanic pile; see Jerram et al., 2000). There is a burial-onlap volcanic disconformity between this and the overlying tabular-classic flow facies package 1: a truncation-onlap volcanic disconformity occurs between tabular-classic flow facies packages 1 and 2.

In many cases different facies overlap in origin; e.g., tabular-classic flows may originally start from compound flows but reach their thick tubular nature by the process of inflation, whereas compound-braided facies are constructed from the base up as a series of anastomosing small flow lobes. The two facies would have markedly different seismic characteristics, and such facies descriptions are important for tackling the subbasalt imaging problem. Here the possibility of describing a continental flood basalt and associated volcanic margin in terms of facies architecture is most useful.

Three-dimensional modeling of continental flood basalts

A solution to the subbasalt imaging problem can be aided by the development of detailed three-dimensional models of flood basalt provinces (e.g., Jerram and Robbe, 2001). Such three-dimensional models can help predict the scales and types of heterogeneities within the flood basalt cover that are causing problems with remote imaging techniques, e.g., seismic imagery. The models are constructed from key surfaces e.g., lava flow tops, volcanic disconformities, and are used as templates for synthetic forward modeling of seismic data. Work is under way addressing the subbasalt imaging problem in the northeastern Atlantic margin (e.g., European Union 5th Framework Project, SIMBA; NNE5-1999-20101).

Jerram and Robbe (2001) developed a three-dimensional model of the Etendeka flood basalts in the Huab Basin, northwestern Namibia, based on the detailed data set of Jerram et al. (1999b). An example of the construction of this geological model is shown in Figure 6. Three-dimensional basin models such as that of Jerram and Robbe (2001) can be used to reconstruct the evolution of the volcanic pile and can also be populated with facies and facies properties to allow for detailed forward modeling. The scale of facies variations observed can also be compared directly with seismic data. Photogrammetry has also been used successfully to measure three-dimensional outcrop patterns in areas of good exposure with limited access (e.g., Greenland, Dueholm and Pedersen, 1992). Such information can be incorporated directly into three-dimensional models.

FUTURE AIMS AND OBJECTIVES

This study has summarized the current understanding of volcanology and applied the concept of facies architecture to continental flood basalts. The following list is of some of the future aims and objectives (many of which are being worked on by different groups) that will improve our understanding of the volcanology and facies in continental flood basalts.

More data should be collected on the dimensions of individual eruption events preserved in continental flood basalts. Data on individual eruptive units will provide better limitations on eruption rates and consequently gas output estimates.

A better estimation of the role of silicic volcanism in continental flood basalts is required. More information is needed on the minor ash layers and other deposits between the better preserved mafic components: this will improve the amount and

TABLE 2. FACIES TYPES FOUND IN CONTINENTAL FLOOD BASALTS

Facies type	Schematic appearance
Tabular-classic flow facies Tabular laterally extensive thick flows (~50 m) several kilometers to tens of kilometers in lateral extent with some examples traveling hundreds of kilometers. The flows, where erupted in wet environments, have classic, well-developed columnar jointing patterns (Lyle, 2000). In arid environments e.g., Etendeka, columnar joints are poorly developed. Examples: Columbia River Basalts, where flows were erupted into arid environments the columnar joints are not very well developed or absent. Examples include Karoo, Paraná-Etendeka.	**Tabular-classic facies** 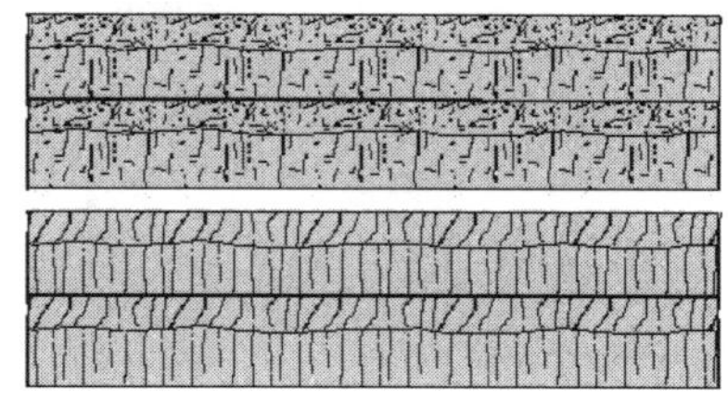
Compound-braided flow facies Thin anastamosing pahoehoe flow sheets and lobes up to several meters in thickness. Often associated with early low volume, low viscosity eruptions early in the formation of continental flood basalts (CFBs). Examples: British Tertiary (NAIP), Etendeka, Greenland (NAIP), Columbia River Basalts.	**Compound-braided facies** 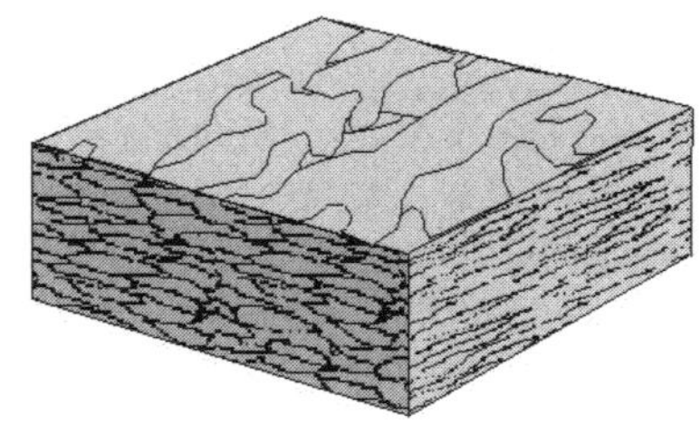
Dipping hyaloclastites Dipping prograding foresets, several meters to tens of thousands of meters thick, of volcaniclastic hyaloclasties. These signify eruption into lakes and seawater. Examples: commonly found in Greenland (NAIP) (e.g. Pedersen et al., 1998) with some examples in the British Tertiary (NAIP).	**Hyaloclastites** 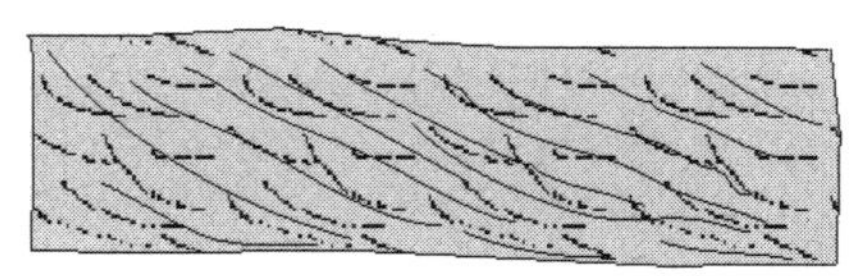
Ponded flows Ponded units are quite common in CFBs where eruptions fill preexisting topography. Units can be >100 m thick, and may internally differentiate during cooling and crystallization. Examples: Etendeka, British Tertiary (e.g., Preshal More, Skye) (Williamson and Bell, 1994).	**Ponded flows** 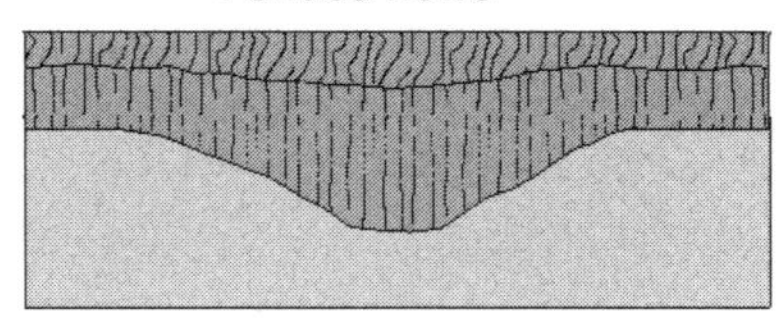
Sill facies Large sills and sill complexes tend to intrude around the base of CFB where the lava pile is in contact with the sediments it erupted onto. Sills often have a classic step like geometry (Francis, 1982) on a large scale making them "bowl" as in 3-D. Examples are found in the Paraná-Etendeka, Karoo-Ferrar, Greenland (NAIP)–British Tertiary (NAIP).	**Sill facies** 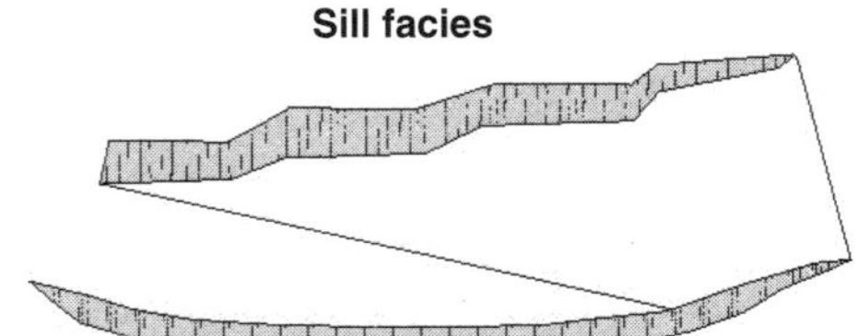
Sheeted dikes Often associated with igneous centers in CFBs. Concentrations of thin dikes cutting up through preexisting lava stratigraphy. Examples: Karoo, Paraná-Etendeka, British Tertiary (NAIP), Greenland (NAIP).	**Sheeted dike** 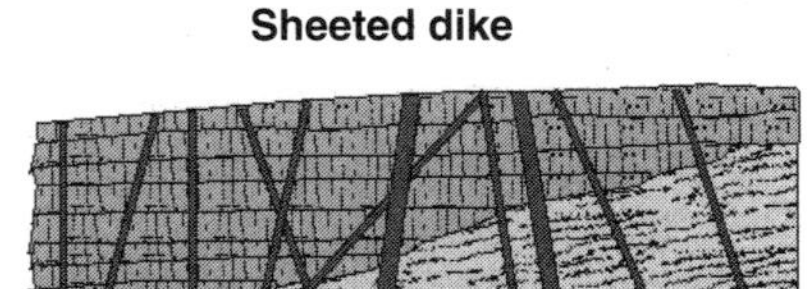

Note: Early volcaniclastic and/or flood lahars from the basal Karoo (Skilling, 2001) may be another facies type.

TABLE 3. FACIES ASSOCIATIONS FOUND IN CONTINENTAL FLOOD BASALTS

Facies association	Schematic appearance
Low-angle downlap and/or toplap Packages of lavas from different eruption sites, possibly along fissure. Each stacking pattern building up from a different direction. These may highlight significant eruption events. Examples: Ethiopian Traps (see Fig. 4), Deccan Traps, NAIP.	**Downlap/toplap** 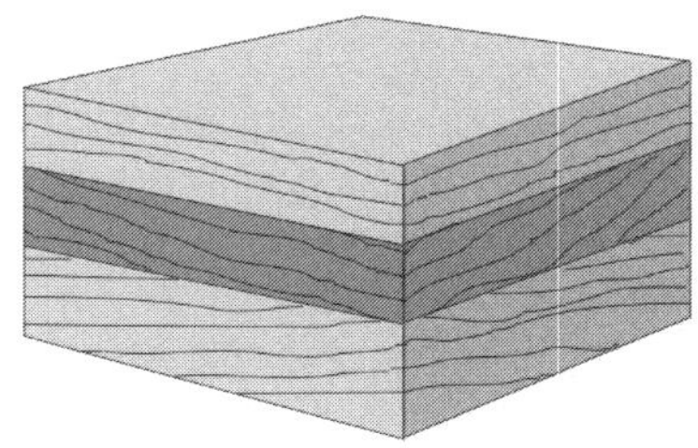
Volcanic disconformity Onlapping relationships between batches of tabular-classic flow facies resulting in disconformable relationships. These represent flows from different eruptive centers onlapping previous flows that have been eroded. Often very difficult to map, because the scale of the disconformities can be >50 km, and the two flow type facies are identical. Example: Etendeka (Jerram et al., 1999b). In many cases on a broader scale these disconformities must exist based on the distribution of different geochemical magma types: e.g., Paraná-Etendeka (Peate, 1997), Yemen (Menzies et al., 1997), Karoo (Marsh et al., 1997).	**Volcanic disconformity** 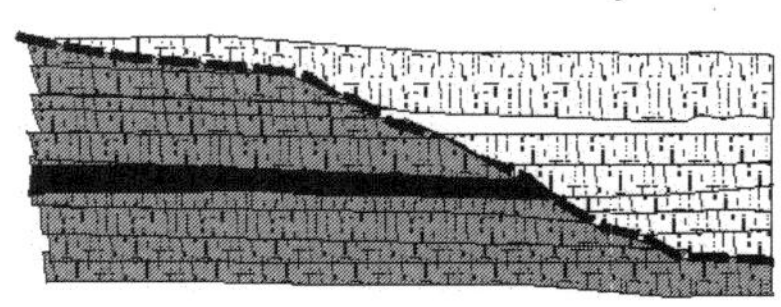
Onlap and/or burial—disconformity Onlapping relationships between batches of tabular-classic flow facies and compound-braided flow facies, representing shield volcanoes, resulting in disconformable relationships. Examples: Etendeka, Greenland (NAIP).	**Onlap and/or burial—disconformity** 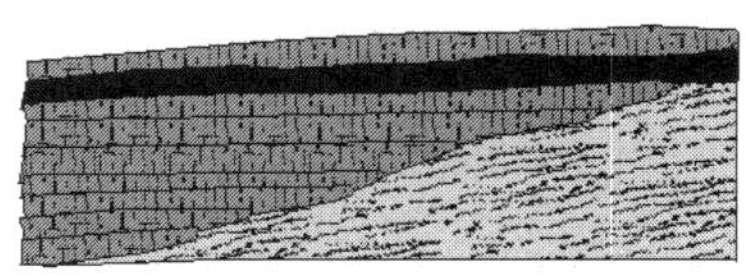
Shield volcanoes Usually associated with compound-braided flow facies, representing shield volcanoes preserved in the continental flood basalts (CFB). These tend to be restricted toward the base and the tops of the CFB as the flood volcanism starts up and shuts down. Examples: Etendeka, Greenland (NAIP), Ethiopian Traps.	**Shield volcano** 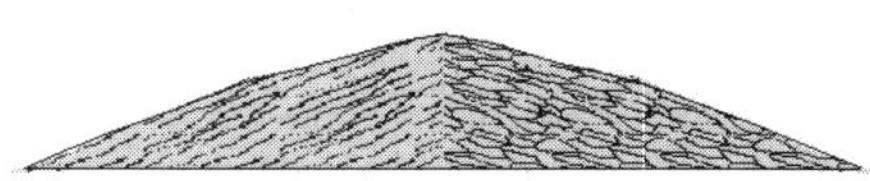
Sediment interlayers Sediments interbedded with volcanics. These are found mainly toward the base of the CFB system where there is some overlap between the active volcanic and active sedimentary systems. Examples: Etendeka, Greenland (NAIP), Ethiopian Traps, Deccan Traps.	**Sediment interlayer** 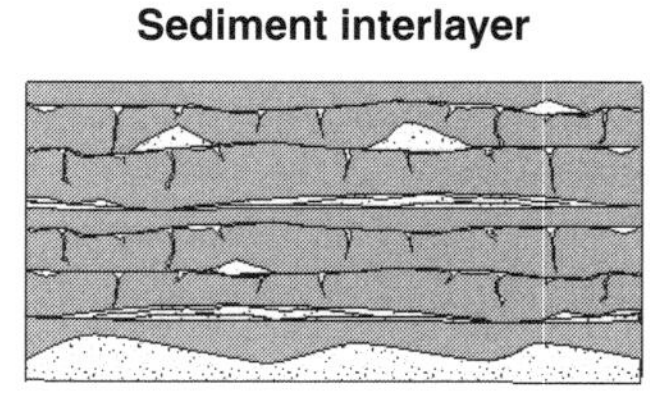

Note: Other examples include syn-volcanic rifting folded/faulted disconformities—Etendeka, Ethiopia.

spread of radiometrically dateable samples, enhance predictions of the climatic impact of continental flood basalts, and potentially aid in correlation.

The compositional range of lavas must be related to their facies type. For example, the picritic lavas in the Etendeka (~45 wt% SiO_2) represent low-volume compound pahoehoe sheets. Whereas basaltic andesites form classic ~60 m thick tabular lavas of higher volume, and large-scale silicic flows form similar tabular-classic tabular units (50–150 m thick). Given a certain geochemical composition, can we predict the facies that will form?

The interrelationships of continental flood basalts and their plumbing systems are still poorly understood. Only a few ig-

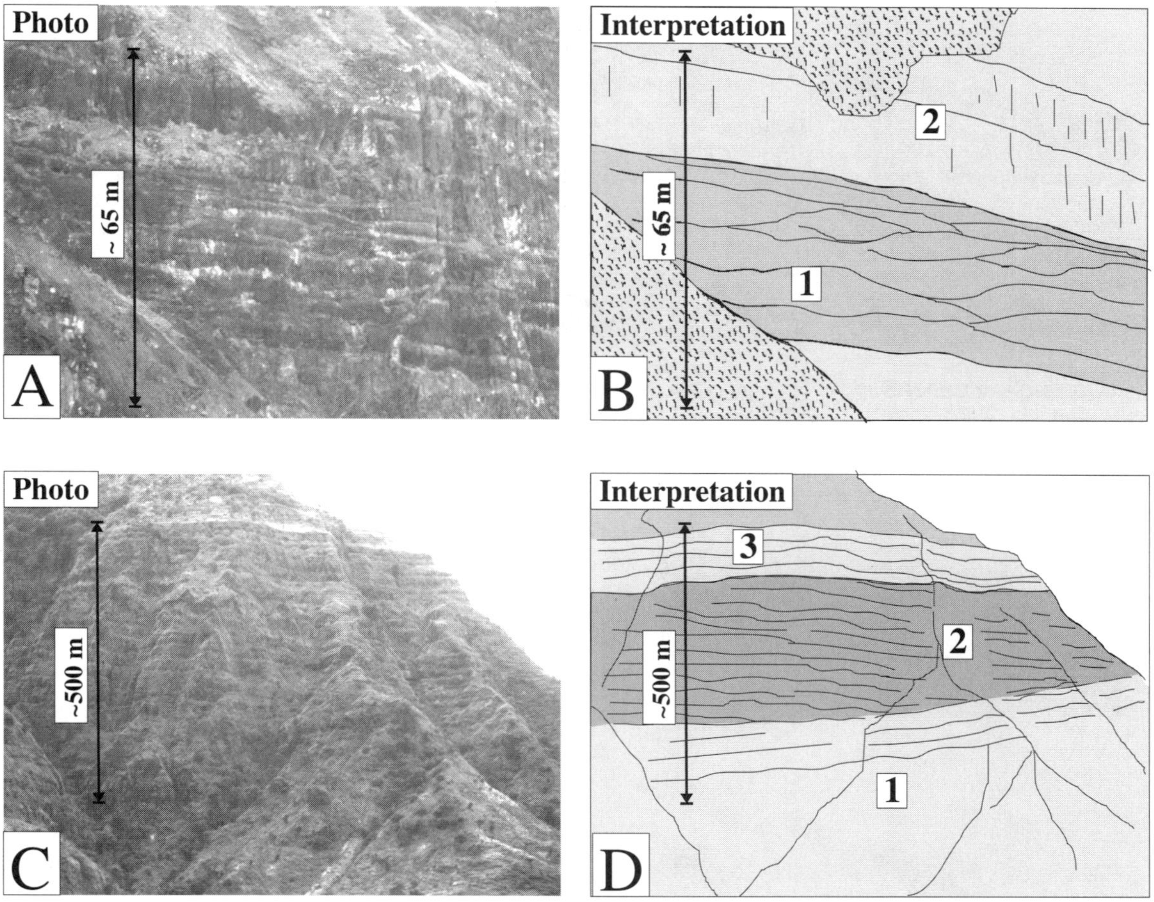

Figure 4. A, B: Example of compound-braided facies overlain by tabular-classic facies, Skye, British Tertiary (North Atlantic igneous province), northwestern Scotland. C, D: Downlap and toplap facies associations, Ethiopian Traps, Africa.

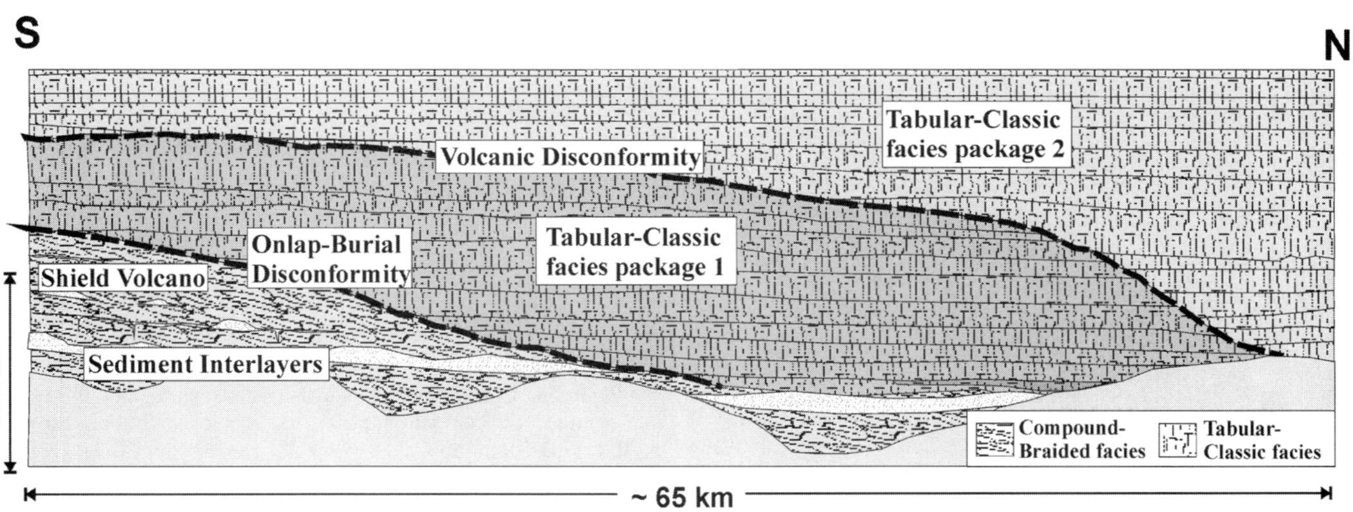

Figure 5. Facies interpretation of detailed section through Etendeka flood basalts, northwestern Namibia (adapted from Fig. 9 of Jerram et al., 1999b).

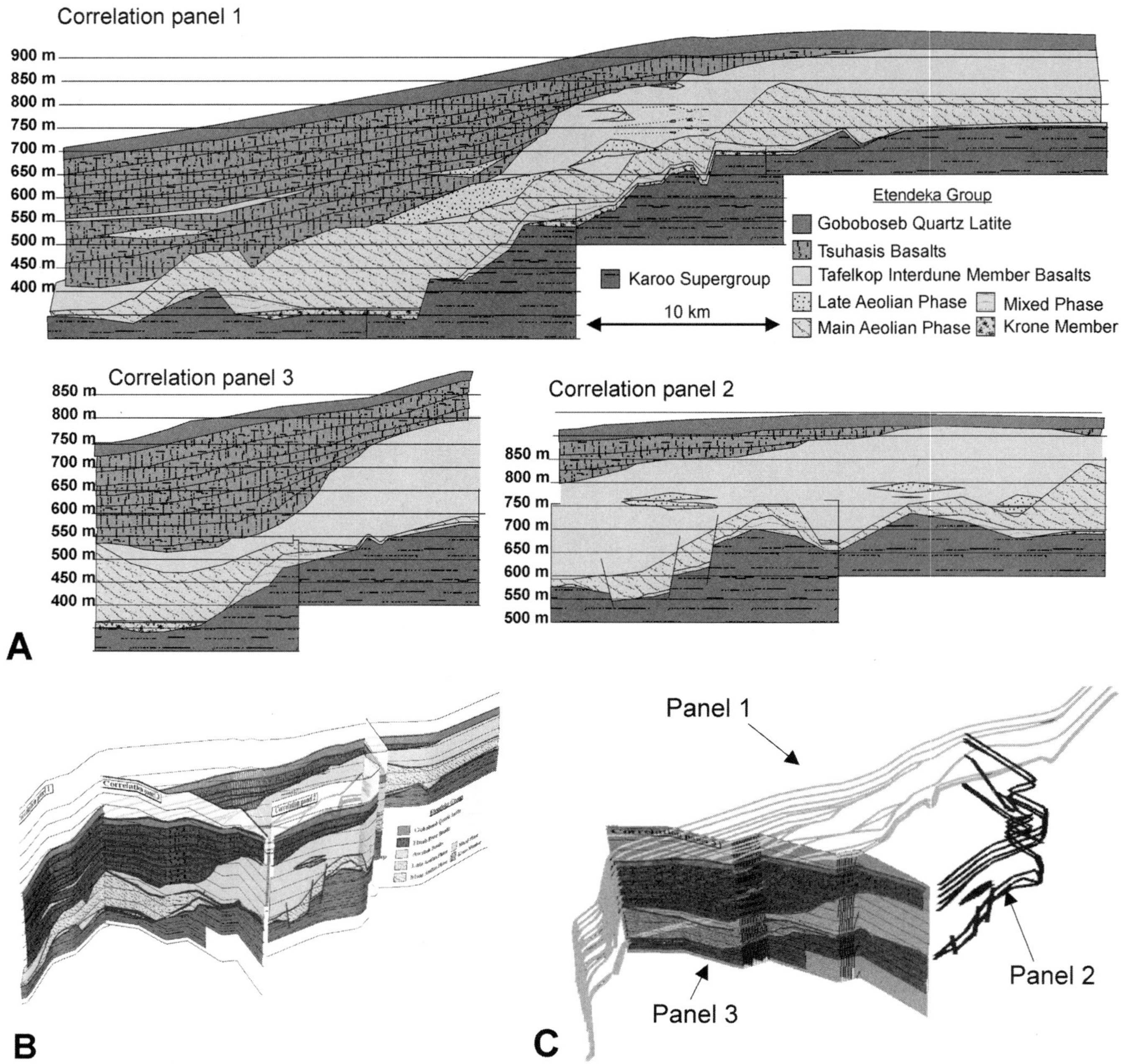

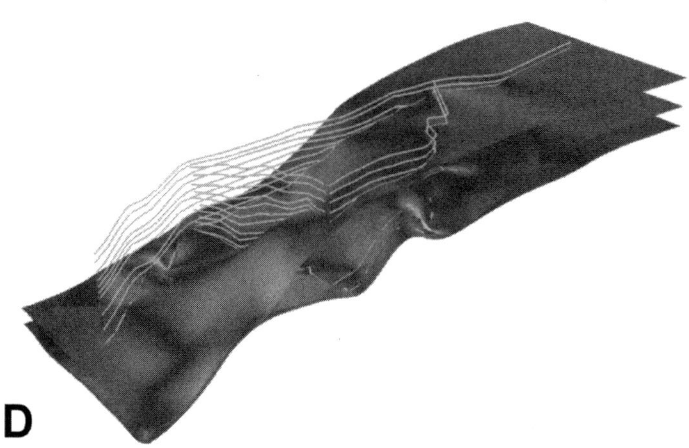

Figure 6. Construction of three-dimensional model from surfaces identified on correlation panels. A: Three correlation panels from Etendeka (Jerram et al., 1999b). B: Panels oriented into true three-dimensional position. C: Key surfaces identified from three correlation panels. D: Three-dimensional surfaces reconstructed as GoCad interpolated surfaces (adapted from Jerram and Robbe, 2001).

neous complexes have been linked directly to the eruptive stratigraphy. The emplacement mechanisms of sill complexes and dike swarms are poorly understood with respect to their eruptive centers.

Better estimates of the gas contents associated with eruptions in continental flood basalt provinces are required to further define estimates of the environmental impact of continental flood basalts. Possible targets for study would include fluid inclusions in trapped phenocrysts.

Detailed three-dimensional models at different scales of observation in flood basalts will prove an important step toward fully integrated data sets of continental flood basalts and as aids in understanding continental flood basalts in offshore settings.

ACKNOWLEDGMENTS

I thank Henry Emeleus, Andy Duncan, Simon Milner, Jon Davidson, Nigel Mountney, Harald Stollhofen, Volker Lorenz, and Frank Holzförster, who have enthused my studies into flood basalts and who have provided valuable insight, help, encouragement; Bjørn Randeberg (Norwegian Petroleum Directorate) and Karl Kravik (Elf Norge) for the seismic image; Henry Emeleus, Lazslo Keszthelyi, Ian Skilling, Paul Wignall, and Mike Widdowson for helpful discussions; and Jon Davidson, Richard Single, Richard England, and Scott Bryan for comments on an earlier version of this manuscript. I also thank J. Gregory McHone and Nicholas Arndt for reviews, Nicholas Arndt for a field visit to Ethiopia funded by Broken Hill Petroleum, Scott Bryan and Nicholas Arndt in particular for very detailed recommendations, and Martin Menzies for editorial assistance.

REFERENCES CI TED

Bell, B.R., and Emeleus, C.H., 1988, A review of silicic pyroclastic rocks of the British Tertiary Volcanic Province, *in* Morton, A.C., and Parson, L.M., eds., Early Tertiary volcanism and the opening of the NE Atlantic: Geological Society [London] Special Publication 39, p. 365–379.

Campbell, I.H., Czamanske, G.K., Fedorenko, V.A., Hill, R.I., and Stepanov, V., 1992, Synchronism of the Siberian traps and the Permian-Triassic boundary: Science, v. 258, no. 5089, p. 1760–1763.

Courtillot, V., Jaupart, C., Manighetti, I., Tapponnier, P., and Besse, J., 1999, On causal links between flood basalts and continental breakup: Earth and Planetary Science Letters, v. 166, no. 3–4, p. 177–195.

Donaldson, C.H., Troll, V.R., and Emeleus, C.H., 2001, Felsites and breccias in the Northern Marginal Zone of the Rum Central Complex: Changing views, c. 1900–2000: Proceedings of the Yorkshire Geological Society, v. 53, pt. 3, p. 167–175.

Dueholm, K.S., and Pedersen, A.K., editors, 1992, Geological analysis and mapping using multi-model photogrammetry: Grønlands Geologiske Undersøgelse, Rapport 156, 72 p.

Ewart, A., Milner, S.C., Armstrong, R.A., and Duncan, A.R., 1998, Etendeka volcanism of the Goboboseb Mountains and Messum Igneous Complex, Namibia. 1. Geochemical evidence of Early Cretaceous Tristan plume melts and the role of crustal contamination in the Paraná-Etendeka CFB: Journal of Petrology, v. 39, no. 2, p. 191–225.

Francis, E.H., 1982, Magma and sediment-1 emplacement mechanism of late Carboniferous tholeiite sills in northern Britain: Journal of the Geological Society of London, v. 139, p. 1–20.

Hamilton, M.A., Pearson, D.G., Thompson, R.N., Kelley, S.P., and Emeleus, C.H., 1998, Rapid eruption of Skye lavas inferred from precise U-Pb and Ar-Ar dating of the Rum and Cuillin plutonic complexes: Nature, v. 394, p. 260–263.

Hawkesworth, C., Mantovani, M., and Peate, D., 1988, Lithosphere remobilisation during Paraná CFB magmatism, *in* Menzies, M.A., and Cox, K.G., eds., Oceanic and continental lithosphere: Similarities and differences: Journal of Petrology Special Issue, p. 205–223.

Hawkesworth, C., Kelly, S., Turner, S., Le Roex, A., and Storey, B., 1999, Mantle processes during Gondwana break-up and dispersal: Journal of African Earth Sciences, v. 28, no. 1, p. 239–261.

Hofman, C., Courtillot, V., Féraud, G., Rochette, P., Yirgu, G., Ketefo, E., and Pik, R., 1997, Timing of the Ethiopian flood basalt event and implications for plume birth and global change: Nature, v. 389, p. 838–841.

Hon, K., Kauahikaua, J., Denlinger, R., and Mackay, K., 1994, Emplacement and inflation of pahoehoe sheet flows: Observations and measurements of active lava flows on Kilauea Volcano, Hawaii: Geological Society of America Bulletin, v. 106, p. 351–370.

Jerram, D.A., and Stollhofen, H., 2002, Lava/sediment interaction in desert settings: Are all peperite-like textures the result of magma-water interaction?: Journal of Volcanology and Geothermal Research, v. 114, p. 231–249.

Jerram, D.A., and Robbe, O., 2001, Building a 3-D geologic model of a flood basalt: An example from the Etendeka, NW Namibia: Electronic Geosciences, v. 6, no. 1.

Jerram, D.A., Mountney, N., and Stollhofen, H., 1999a, Facies architecture of the Etjo Sandstone Formation and its interaction with the Basal Etendeka flood basalts of NW Namibia: Implications for offshore analogues, *in* Cameron, N., Bate, R., and Clure, V., eds., The oil and gas habitats of the South Atlantic: Geological Society [London] Special Publication 153, p. 367–380.

Jerram, D.A., Mountney, N., Holzförster, F., and Stollhofen, H., 1999b, Internal stratigraphic relationships in the Etendeka Group in the Huab Basin, NW Namibia: Understanding the onset of flood volcanism: Journal of Geodynamics, v. 28, p. 393–418.

Jerram, D.A., Mountney, N., Howell, J., Long, D., and Stollhofen, H., 2000, Death of a Sand Sea: An active erg systematically buried by the Etendeka flood basalts of NW Namibia: Journal of the Geological Society of London, v. 157, p. 513–516.

Jervey, M.T., 1988, Quantitative geological modelling of siliciclastic rock sequences and their seismic expression, *in* Wilgus, C.K., Hastings, B.S., Kendall, C.G.St.C., Posamentier, H.W., Ross, C.A., and Van Wagoner, J.C., eds., Sea-level changes: An integrated approach: Society of Economic Palaeontologists and Mineralogists Special Publication 42, p. 47–69.

Kent, W., Saunders, A.D., Kempton, P.D., and Ghose, N.C., 1997, Rajmahal basalts, eastern India: Mantle sources and melt distribution at a volcanic rifted margin, *in* Mahoney, J.J., and Coffin M.F., eds., Large igneous provinces: Continental, oceanic, and planetary flood volcanism: American Geophysical Union Geophysical Monograph 100, p. 145–182.

Keszthelyi, L., and Self, S., 1998, Some physical requirements for the emplacement of long basaltic lava flows: Journal of Geophysical Research, B, Solid Earth and Planets, v. 103, no. B11, p. 27447–27464.

Keszthelyi, L., McEwen, A.S., and Thordarson, T., 2000, Terrestrial analogs and thermal models for Martian flood lavas: Journal of Geophysical Research, E, Planets, v. 105, no. 6, p. 15027–15049.

Lyle, P., 2000, The eruption environment of multi-tiered columnar basalt lava flows: Journal of the Geological Society of London, v. 157, p. 715–722.

Mahoney, J.J., and Coffin, M.F., 1997, Large igneous provinces: Continental, oceanic, and planetary flood volcanism, American Geophysical Union Geophysical Monograph 100.

Marsh, J.S., Ewart, A., Milner, S.C., Duncan, A.R., and Miller, R.McG., 2001, The Etendeka Igneous Province: Magma types and their stratigraphic distribution with implications for the evolution of the Paraná-Etendeka Flood Basalt Province: Bulletin of Volcanology, v. 62, p. 464–486.

Marsh, J.S., Hooper, P.R., Rehacek, J., Duncan, R.A., and Duncan, A.R., 1997, Stratigraphy and age of the Karoos basalts of Lesthotho and implications for correlations within the Karoo Igneous Province, *in* Mahoney, J.J., and Coffin M.F., eds., Large igneous provinces: Continental, oceanic and planetary flood volcanism: American Geophysical Union Geophysical Monograph 100, p. 247–272.

Menzies, M., Baker, J., Chazot, G., and Al'Kadasi, M., 1997, Evolution of the Red Sea volcanic margin, western Yemen, *in* Mahoney, J.J., and Coffin, M.F., eds., Large igneous provinces: American Geophysical Union Geophysical Monograph 100, p. 29–43.

Milner, S.C., Duncan, A.R., and Ewart, A., 1992, Quartz latite rheoignimbrite flows of the Etendeka Formation, north-western Namibia: Bulletin of Volcanology, v. 54, p. 200–219.

Milner, S.C., Duncan, A.R., Whittingham, A.M., and Ewart, A., 1995, Trans-Atlantic correlation of eruptive sequences and individual silicic volcanic units within the Paraná-Etendeka igneous province: Journal of Volcanology and Geothermal Research, v. 69, p. 137–157.

Mitchell, W.I., Cooper, M.R., Hards, V.L., and Meighan, I.G., 1999, An occurrence of silicic volcanic rocks in the early Palaeogene Antrim Lava Group of Northern Ireland: Scottish Journal of Geology, v. 35, p. 179–185.

Peate, D.W., 1997, The Paraná-Etendeka Province, *in* Mahoney, J.J., and Coffin, M.F., eds., Large igneous provinces: Continental, oceanic and planetary flood volcanism: American Geophysical Union Geophysical Monograph 100, p. 217–245.

Pedersen, G.K., Larsen, L.M., Pedersen, A.K., and Hjortkjær, B.F., 1998, The syn-volcanic Naajaat lake, Paleocene of West Greenland: Palaeogeography, Palaeoclimatology, Palaeoecology, v. 140, p. 271–287.

Planke, S., Symonds, P.A., Alvestad, E., and Skogseid, J., 2000, Seismic volcanostratigraphy of large-volume basaltic extrusive complexes on rifted margins: Journal of Geophysical Research–Solid Earth, v. 105, B.8, p. 19335–19351.

Pyle, D.M., 1998, Forecasting sizes and repose times of future extreme volcanic events: Geology, v. 26, p. 367–370.

Reidel, S.P., and Tolan, T.L., 1992, Eruption and emplacement of flood basalt: An example from the large-volume Teepee Butte Member, Columbia River Basalt Group: Geological Society of America Bulletin, v. 104, p. 1650–1671.

Renne, P.R., Glen, J.M., Milner, S.C., and Duncan, A.R., 1996, Age of Etendeka flood volcanism and associated intrusions in southwestern Africa: Geology, v. 24, p. 659–662.

Rochette, P., Tamrat, E., Feraud, G., Pik, R., Courtillot, V., Ketefo, E., Coulon, C., Hoffmann, C., Vandamme, D., and Yirgu, G., 1998, Magnetostratigraphy and timing of the Oligocene Ethiopian traps: Earth and Planetary Science Letters, v. 164, no. 3–4, p. 497–510.

Rose, W.I., and Chesner, C.A., 1990, Worldwide dispersal of ash and gases from Earth's largest known eruption—Toba, Sumatra, 75 ka.: Global and Planetary Change, v. 89, no. 3, p. 269–275.

Self, S., Keszthelyi, L., and Thordarson, Th., 1998, The importance of pahoehoe: Annual Review of Earth and Planetary Sciences, v. 26, p. 81–110.

Self, S., Thordarson, Th., and Keszthelyi, L., 1997, Emplacement of continental flood basalt lava flows, *in* Mahoney, J.J., and Coffin, M.F., eds., Large igneous provinces: Continental, oceanic and planetary flood volcanism: American Geophysical Union Geophysical Monograph 100, p. 381–410.

Self, S., Zhao, J.-X., Holasek, R.E., Torres, R.C., and King, A.J., 1996b, The atmospheric impact of the 1991 Mount Pinatubo eruption, *in* Newhall, C.G., and Punongbayan, R.S., eds., Fire and mud: Eruptions and lahars of Mount Pinatubo, Philippines: Seattle, University of Washington Press, p. 1089–1115.

Self, S., Thordarson, Th., Keszthelyi, L., Walker, G.P.L., Hon, K., Murphy, M.T., Long, P., and Finnemore, S., 1996a, A new model for the emplacement of Columbia River Basalts as large inlated pahoehoe lava flow fields: Geophysical Research Letters, v. 23, no. 19, p. 2689–2692.

Shaw, H.R., and Swanson, D.A., 1970, Eruption and flow rates of flood basalts, *in* Gilmour, E.H., and Stradling, D., eds., Proceedings of the Second Columbia River Basalt Symposium: Cheney, Eastern Washington State College Press, p. 271–299.

Skilling, I.P., 2001, Terrestrial sub ice volcanism and pre-flood basalt hydrovolcanism as models for magma-volatile interaction on Mars: Eos (Transactions, American Geophysical Union), v. 82, no. 20, Spring Meeting Supplement, Abstract, V42A-09.

Storey, M., Mahoney, J.J., Saunders, A.D., Duncan, R.D., Kelley, S.P., and Coffin, M.F., 1995, Timing of hotspot-related volcanism and the break-up of Madagascar and India: Science, v. 267, p. 852–855.

Swanson, D.A., Wrigh, T.L., and Helz, R.T., 1975, Linear vent systems and estimated rates of magma production and eruption for the Yakima Basalt on the Columbia Plateau: American Journal of Science, v. 275, p. 877–905.

Thordarson, Th., and Self, S., 1993, The Laki (Skaftar Fires) and Grimsvotn eruptions in 1783–1785: Bulletin of Volcanology, v. 55, p. 233–263.

Thordarson, Th., and Self, S., 1998, The Roza Member, Columbia River Basalt Group: A gigantic pahoehoe lava flow field formed by endogenous processes?: Journal of Geophysical Research–Solid Earth, v. 103, B11, p. 27411–27445.

Thordarson, Th., Self, S., Oskarsson, N., and Hulsenbosch, T., 1996, Sulphur, chlorine, and fluorine degassing and atmospheric loading by the 1783–1785 AD Laki (Skaftar Fires) eruption in Iceland: Bulletin of Volcanology, v. 58, p. 205–225.

Van Wagoner, J.C., Mitchum, R.M., Campion, K.M., and Rahmanian, V.D., 1990, Siliciclastic sequences in well logs, cores and outcrops: American Association of Petroleum Geologists, Methods in Exploration Series 7, 55 p.

Van Wagoner, J.C., Posamentier, W.H., Mitchum, R.M., Vail, P.R., Sarg, J.F., Loutit, T.S., and Hardenbol, J., 1988, An overview of the fundamentals of sequence stratigraphy and key definitions, *in* Wilgus, C.K., Hastings, B.S., Kendall, C.G.St.C., Posamentier, H.W., Ross, C.A., and Van Wagoner, J.C., eds., Sea-level changes: An integrated approach: Society of Economic Palaeontologists and Mineralogists Special Publication 42, p. 39–45.

White, R.S., and McKenzie, D.P., 1989, Magmatism at rift zones: The generation of volcanic continental rift margins and flood basalts: Journal of Geophysical Research, v. 94, p. 7685–7729.

Widdowson, M., Pringle, M.S., and Fernandez, O.A., 2000, A post K-T boundary (Early Palaeocene) age for Deccan-type feeder dykes, Goa, India: Journal of Petrology, v. 41, no. 7, p. 1177–1194.

Wignall, P.B., 2001, Large igneous provinces and mass extinctions: Earth-Science Reviews, v. 53, p. 1–33.

Williamson, I.T., and Bell, B.R., 1994, The Paleocene lava-field of west-central Skye, Scotland: Stratigraphy, paleogeography and structure: Transactions of The Royal Society of Edinburgh—Earth Sciences, v. 85, p. 39–75.

Manusript Accepted by the Society October 15, 2001

Printed in the U.S.A.

Geological Society of America
Special Paper 362
2002

East Greenland coast-parallel dike swarm and its role in continental breakup

Martin Bromann Klausen
Hans Christian Larsen
Danish Lithosphere Centre, Øster Voldgade 10, DK-1350 Copenhagen K, Denmark

ABSTRACT

The East Greenland coast-parallel dike swarm is exposed along ~250 km of the outermost coast south of the Iceland hotspot track manifest by the Iceland-Greenland Ridge. The dike swarm and associated plutonic centers were emplaced over a period extending from a few million years before final breakup to at least 7 m.y. after breakup ca. 56 Ma. The predominantly mafic dikes are hosted in Precambrian high-grade basement, but locally cut through early Tertiary plutonic complexes. Seaward, continental crust is replaced by 20–30-km-thick Icelandic-type igneous crust, including thick sequences of subaerially erupted lavas that dip oceanward. The orientations of dikes, dike thickness, dike density, field classification, and relative sequence of emplacement of 1410 dikes within this region were studied in the field and by photogrammetry along 6 margin transects.

Statistical treatment of the field data documents a shift in orientation of dikes from predominantly subvertical inland to predominantly landward dipping (as low as 40°) along the outer coast. The earliest dikes, largely intruded prior to breakup, have dip directions at right angles to the bedding of locally preserved lavas and sediments, implying subvertical intrusion of dikes followed by strong seaward rotation of the crust during breakup. Consistent with this, younger generations of dikes emplaced after breakup show steeper to almost vertical dips. The strikes of the earliest dikes suggest that these were intruded in an en echelon pattern and slightly oblique to the margin. These dikes are relatively thick (average 6–8 m) and may have been intruded subvertically from a deep source, possibly at the base of the crust. The other dikes define more curved, coast-parallel trajectories extending out from some of the major plutonic complexes and form partly overlapping dike swarm segments. Together with observed systematic decrease in dike density and increasing underrepresentation of thin dikes away from major igneous complexes, this suggests that with time crustal plutonic complexes became important reservoirs for lateral magma propagation along the margin.

In cross section, the density of dikes increases seaward and reaches ~50% along the outer coast. The average dike thickness decreases from a maximum of 8 m inland to 3–4 m along the coast. A hinge line is located ~15 km inland and is defined by landward rapidly decreasing dike density, maximum dike thickness, and, in general, subvertical orientation of dikes, except for a small population of early dikes that seems to have intruded along preexisting and landward-dipping faults. The parallel increase seaward in dike density and decreasing dike thickness suggests an ophiolitic sheeted dike complex (i.e., continent-ocean boundary) to be present offshore 20–40 km from the inland hinge line. The seaward flexure of the crust is interpreted to reflect initial tectonic extension by rotation of fault blocks followed by monoclinal flexing due to loading by the

Klausen, M.B., and Larsen, H.C., 2002, East Greenland coast-parallel dike swarm and its role in continental breakup, *in* Menzies, M.A., Klemperer, S.L., Ebinger, C.J., and Baker, J., eds., Volcanic Rifted Margins: Boulder, Colorado, Geological Society of America Special Paper 362, p. 133–158.

former lava cover and later margin uplift (coastal region) and subsidence (shelf region). During breakup, the amount of crustal extension through magmatic dilation by dike injection progressively superseded the amount of tectonic extension.

The zone from the inland hinge line to the offshore continent-ocean boundary is interpreted to represent the former (half) width of the rift zone present during continental breakup. Palinspastic reconstruction of the East Greenland part of this rift suggests that it was only 10–13 km wide prior to tectonic extension and magmatic dilation. If a general symmetric development of the conjugate half-rift (Rockall margin) is assumed, a total rift width of 20–26 km is implied. Margin segmentation at ~150 km length scale was introduced during breakup by the major plutonic centers and associated dike swarm segments extending from these. This might have introduced local asymmetry and three-dimensional variation in the margin structure, including small-scale transform offsets of the final line of breakup and increased margin flexure (twisted zones) toward major igneous complexes.

INTRODUCTION

The East Greenland coast-parallel dike swarm (Wager and Deer, 1938; Larsen, 1978; Nielsen, 1978; Myers, 1980; Nielsen and Brooks, 1981) and associated crustal flexure (or monocline) is a striking feature of the East Greenland volcanic rifted margin (e.g., Larsen and Saunders, 1998) and is unique within the Tertiary North Atlantic igneous province. Crustal uplift and deep glacial erosion along the coast provides excellent exposure of an ~350 km length of a coast-parallel dike swarm following the initial line of rifting between East Greenland and Northwest Europe.

Other dike swarms, which in a similar fashion may relate to the breakup of Gondwana (Ernst and Buchan, 1997), are found in the Deccan (Murthy, 1987; Deshmukh and Sehgal, 1988), Karoo (Eales et al., 1984), and Paraná provinces (Sial et al., 1987). The Karoo and Deccan provinces also show similar crustal flexure zones (Gibson, 1966), i.e., the Lebombo monocline (Du Toit, 1929; Watkeys, this volume) and the Panvel flexure (Dessai and Bertrand, 1995; Sheth, 1998). Dike swarms associated with the Neogene opening of the Red Sea are also found in Saudi Arabia and Yemen (Coleman and McGuire, 1988; Voggenreiter et al., 1988; Mohr, 1991). A common feature of all these crustal flexures is that the continental crust is tectonically rotated toward the former rift center and the ocean basin that subsequently formed.

The excellent exposure in East Greenland offers unique opportunities to study both along- and across-margin variations in the dike swarm structure. In order to document this, we have measured simple dike geometry (i.e., strike, dip, and thickness) of 1410 dikes and applied a statistical approach to the presentation and interpretation of this large data set. Sampling focused along selected transects placed at strategic and regular intervals within previously identified dike swarm segments (Myers, 1980), south of the Icelandic-Greenland Ridge (Fig. 1). We have also applied a field classification, including crosscutting relationships, that allows us to address questions related to the temporal evolution of the dike swarm on a statistical basis.

We use the distribution in space and time of dike orientations to determine both the paleostress field during dike emplacement as well as the subsequent crustal deformation recorded by the postintrusive tilting of dikes. Combining the information on dike thickness and density with information on crustal rotation allows us to model the two-dimensional geometry and development of the initial rift in terms of magmatic dilation and tectonic extension, and to compare this model with the geophysically defined continent-ocean boundary (Fig. 1). Variations along the margin in orientation, crustal dilation, and thickness of dikes are used to assess possible lateral propagation and magma flow directions away from major igneous centers, magmatic and tectonic segmentation, and potential three-dimensional geometry of the volcanic rifted margin.

GEOLOGIC SETTING AND PRIOR WORK

The North Atlantic igneous province comprises large areas of subaerially erupted flood basalts emplaced during continental breakup in the Paleocene to earliest Eocene (Saunders et al., 1998). This excessive magmatism is generally related to the arrival of the Proto-Icelandic mantle plume ca. 61 Ma and final breakup between Greenland and northwestern Europe ca. 55.8 Ma over an extensive plume head (e.g., Larsen and Saunders, 1998). The present position of the central plume stem is now below the southeastern part of Iceland (e.g., Wolfe et al., 1997; Breddam et al., 2000). The plume track is represented by the thick crust along the Iceland-Greenland Ridge extending toward the Kangerlussuaq Fjord (Fig. 1).

Early Tertiary flood basalts are now found at elevated positions (e.g., East Greenland, British Tertiary, Faeroe Islands) while other parts have subsided below sea level (e.g., Faeroe Banks; Rockall-Hatton Bank). An extensive sequence of basalts with a special stratigraphic pattern—the seaward-dipping reflector sequences—is found all along the subsided rifted margins of East Greenland and northwestern Europe (Larsen and Jakobsdottír, 1988; White and McKenzie, 1989). These represent the

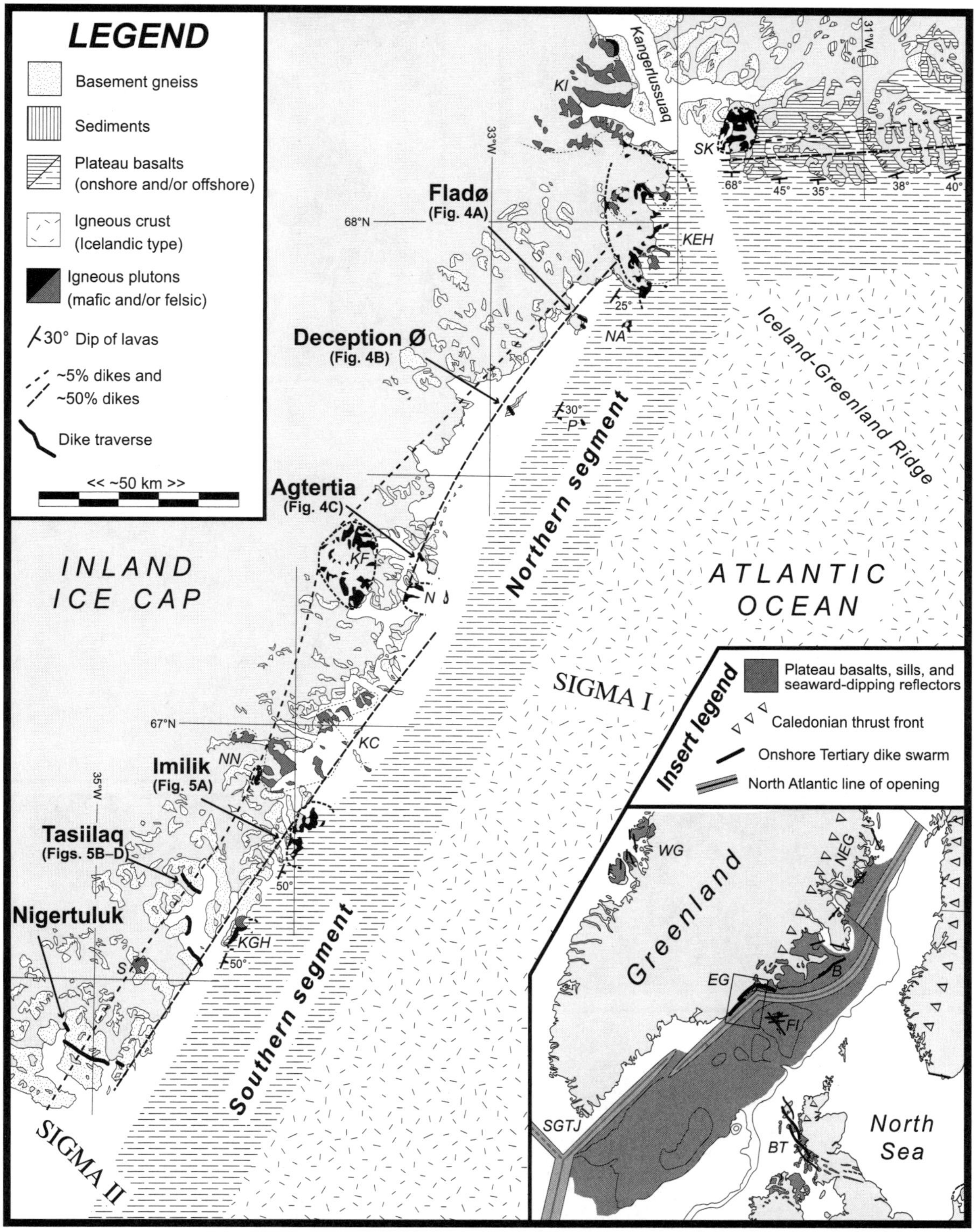

Figure 1. Onshore and offshore geology of East Greenland, including 5% and 50% density contours for coast-parallel dike swarm (after Myers et al., 1988). Six dike transects (profiles) are indicated within studied northern and southern segment of northeast-southwest–trending branch of dike swarm. Sigma I and II are seismic transects shot by Danish Lithosphere Centre (e.g., Korenaga et al., 2000). KEH, Kap Edward Holm; KC, Kialineq complex; KF, Kruuse Fjord intrusion; KGH, Kap Gustav Holm; KI, Kangerlussuaq intrusion; N, Nugalik; NA, Nordre Aputiteq; NN, Noe-Nygaard intrusion; P, Patulajivit; S, Sulugssut; SK, Skaergaard intrusion. Inset: Schematic reconstruction of North Atlantic volcanic province (modified from Larsen et al., 1998) at time of breakup (ca. 56 Ma), with location and general trend of known Tertiary dike swarms. B, Blosseville Kyst; BT, British Tertiary volcanic province; EG, East Greenland; FI, Faeroe Islands; NEG, Northeast Greenland; SGTJ, South Greenland triple junction; WG, West Greenland.

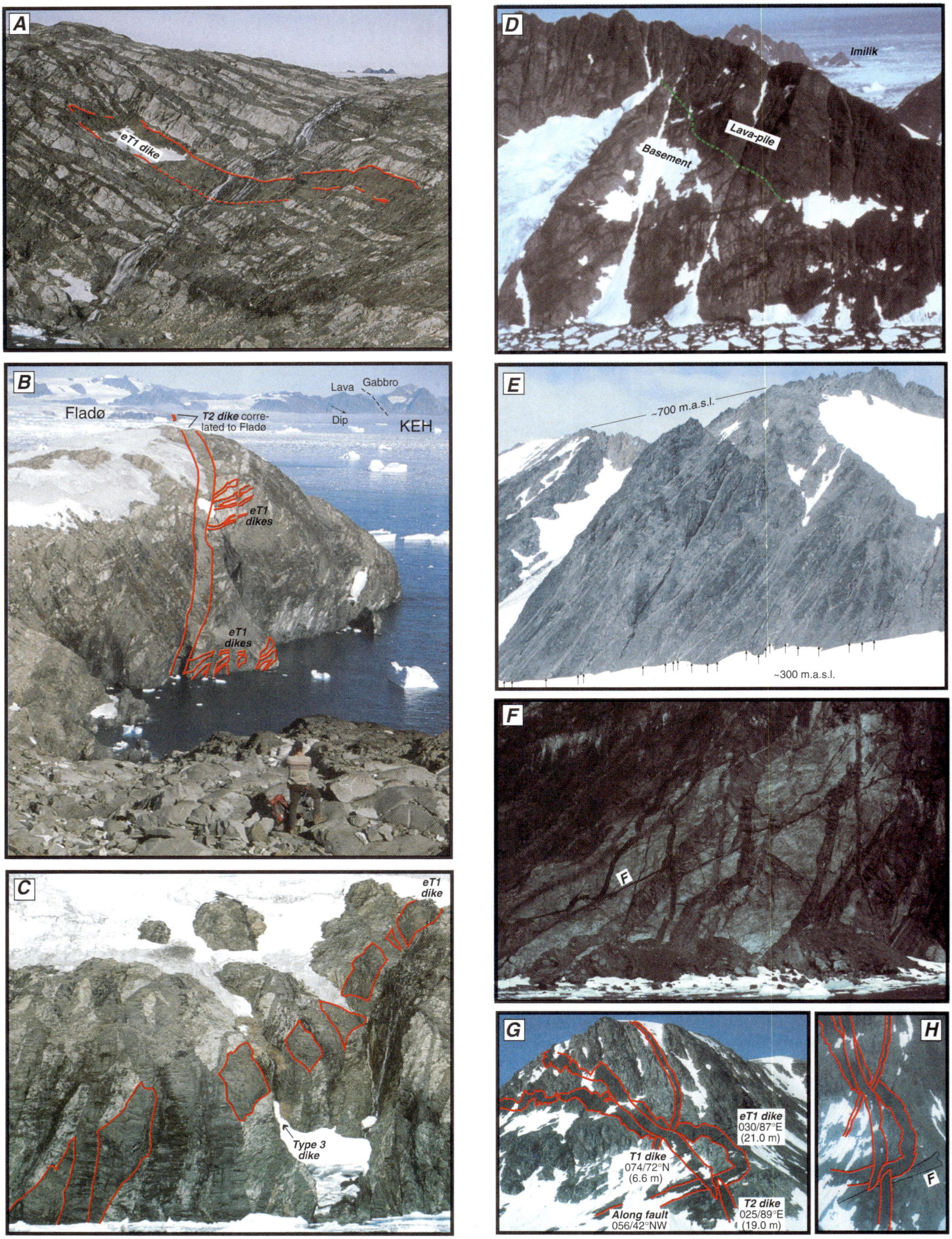
A
eT1 dike
B
Fladø
T2 dike correlated to Fladø
Lava
Gabbro
Dip
KEH
eT1 dikes
eT1 dikes
C
eT1 dike
Type 3 dike
D
Imilik
Lava-pile
Basement
E
~700 m.a.s.l.
~300 m.a.s.l.
F
F
G
eT1 dike
030/87°E
(21.0 m)
T1 dike
074/72°N
(6.6 m)
T2 dike
025/89°E
(19.0 m)
Along fault
056/42°NW
H
F

upper extrusive part of anomalously thick, Icelandic-type, igneous crust that accreted onto the margins during the first few million years of spreading (e.g., Holbrook et al., 2001). Tertiary igneous intrusions are mainly exposed within the uplifted and deeply eroded parts of the North Atlantic igneous province. Classical examples, comprising both central plutonic intrusions and overlapping dike swarms, are known from Scotland (Speight et al., 1982) and from East Greenland (Myers, 1980; Brooks and Nielsen, 1982).

The East Greenland coast-parallel dike swarm is continuously exposed along an ~250-km-long, northeast-southwest–trending branch south of the Kangerlussuaq Fjord and an ~100-km-long, east-west–trending branch to the northeast of the fjord (Fig. 1). Deep glacial dissection provides local three-dimensional outcrops as well as excellent regional exposure of the swarm (Figs. 2, A–H; 3, A–C; and 4, A–D). A similar coast-parallel dike swarm and associated crustal flexure is seen within the flood basalts of the Blosseville Kyst region (Fig. 1, B in inset; Watt, 1975). In the classification of first-order segmentation of the East Greenland margin, Larsen (1990) identified three major segments, the southernmost of which extends from south of Greenland to the Kangerlussuaq area (see also Karson and Brooks, 1999). This segment may be further subdivided into two, slightly en echelon, offset parts (Fig. 1, inset). Our study area is located within the northern end of the northernmost part of this southeast Greenland margin segment.

Figure 2. Coast-parallel dike swarm along northern and southern study segment (Fig. 1). A: Westward view of ~150-m-high, seaward-sloping southeastern side of Fladø, exposing outcrop that is roughly perpendicular to 30°–70° landward-dipping dikes. Note outlined, thick, and rusty-weathered earliest (eT1) dike. B: Northward view along main trajectory of dike swarm across Deception Ø, to Fladø, and seaward-dipping lava and gabbro at Kap Edward Holm (KEH). Outlined ~18-m-thick and ~035/80°NW-oriented late (T2) dike can be correlated to Fladø and is only cut by few dikes. Some thin and exceptionally gentle (~25°NW) landward-dipping earliest dikes are also outlined. C: Detail of transect mapped along north side of the Agtertia Fjord, also shown in Figure 3C. Outlined 16-m-thick earliest (eT1) dike is more east striking (N45°E) than crosscutting dikes, while the deeply eroded slot of an orange-weathered felsic (type 3) dike is more north striking. D: Large section of mapped transect at Poulsen Fjord, also shown in Figure 4A. Note darker ~50°SE seaward-dipping lavas on top of paler Precambrian basement, which were fed by suborthogonally emplaced earliest dikes. E: Northward view of dikes cutting oldest of gabbroic plutons at Imilik. These dikes have remarkably consistent dips (~60°) and are also relatively thin (1–2 m thick), compared to those measured farther inland. Individual dikes are pointed out by arrows along bottom of cliff face, and one of these has been dated as ca. 56 Ma (Tegner et al., 1998). F: Detail of mapped transect along southern side of the Tasiilaq Fjord, also shown in Figure 4D. Note how dikes either cut or follow rotated landward-dipping normal fault (F) within Precambrian basement, locally with pseudotachylite. Photograph is inverted to fit consistent northward perception. G: Northward view of inner part of Nigertuluk, with earliest (eT1) of three crosscutting dikes (outlined in red) following nearly 5-m-thick thrust zone (F, outlined in black), cut by younger (T1 and T2) dikes. H: Same as G, but viewed along strike of youngest (T2) dike.

The mouth of the Kangerlussuaq Fjord is a site of several major plutonic intrusions (Brooks and Nielsen, 1982) and forms a first-order node within the coast-parallel dike swarm (Fig. 1). The two dike swarm branches meet here at an angle of ~120° and with the fjord form a triple junction configuration (Brooks, 1973; Burke and Dewey, 1973). The east-west–striking branch of the dike swarm north of the fjord cuts through a kilometer-thick volcanic cover of early Tertiary age overlying a Cretaceous basin (Brooks and Nielsen, 1982; M. Larsen et al., 1999) and is oblique to the initial direction of opening (Fig. 1, inset). The lava bedding in this area shows an increasing seaward tilt toward the coast and reaches seaward dips as steep as 68° (Fig. 1). Nielsen (1978) demonstrated a magmatic development from tholeiites to transitional basalt compositions within this dike swarm branch. The earliest tholeiites are emplaced at right angles to the lava bedding and are interpreted to predate the tilting of the lavas; younger generations of dikes show progressively less tilting and are interpreted to be synflexural to postflexural (Nielsen, 1978; Gill et al., 1988). Nielsen and Brooks (1981) proposed a model of distributed small offsets along landward-dipping normal faults and seaward rotation of fault blocks (i.e., pervasive domino-block faulting) to explain the crustal flexure, which they interpreted to gradually translate into igneous (oceanic) crust ~100 km offshore.

The focus of this study is the dike swarm branch south of Kangerlussuaq Fjord. As pointed out by Wager and Deer (1938), this part of the swarm is similar to the swarm north of the Kangerlussuaq Fjord, in terms of both associated seaward crustal flexure and increasing dike density toward the coast. However, this part of the dike swarm is exposed within high-grade Precambrian basement (Bridgwater et al., 1978), making the identification of darker dikes particularly easy (Figs. 2, A–H; 3, A–C; and 4, A–D) but the determination of faults and tectonic tilting much more difficult. Only a thin cover of sediment and lava occurs around some of the major plutonic complexes near the coast (Figs. 1, 2D, and 4A; Myers, 1980; Larsen, 1980). In a limited number of locations, Myers (1980) on the basis of dike density, subdivided the southern branch of the coast-parallel dike swarm into two sinistrally offset en echelon swarms; these roughly correspond to what we refer to as the northern and southern study segment in Figure 1. Gabbroic and more evolved Tertiary plutons are found along the coast-parallel dike swarm south of Kangerlussuaq (Figs. 1 and 2E; Brooks and Nielsen, 1982; Bernstein et al., 1998). These both cut dikes and are cut by dikes, and offer geochronological data that are difficult to obtain from the dikes (Tegner et al., 1998). The large gabbroic Kap Edward Holm complex (Bernstein et al., 1996) at the mouth of the Kangerlussuaq Fjord, marks the northern end of the northeast-southwest–striking branch of the dike swarm (Fig. 1). The most southerly exposed gabbroic intrusion at Kap Gustav Holm shows indisputable evidence of progressive tilting during its intrusion (Myers et al., 1993); consistent with highly (50°) seaward-dipping sediments and lavas preserved around the intrusion. The early dikes were intruded at right angle to these locally

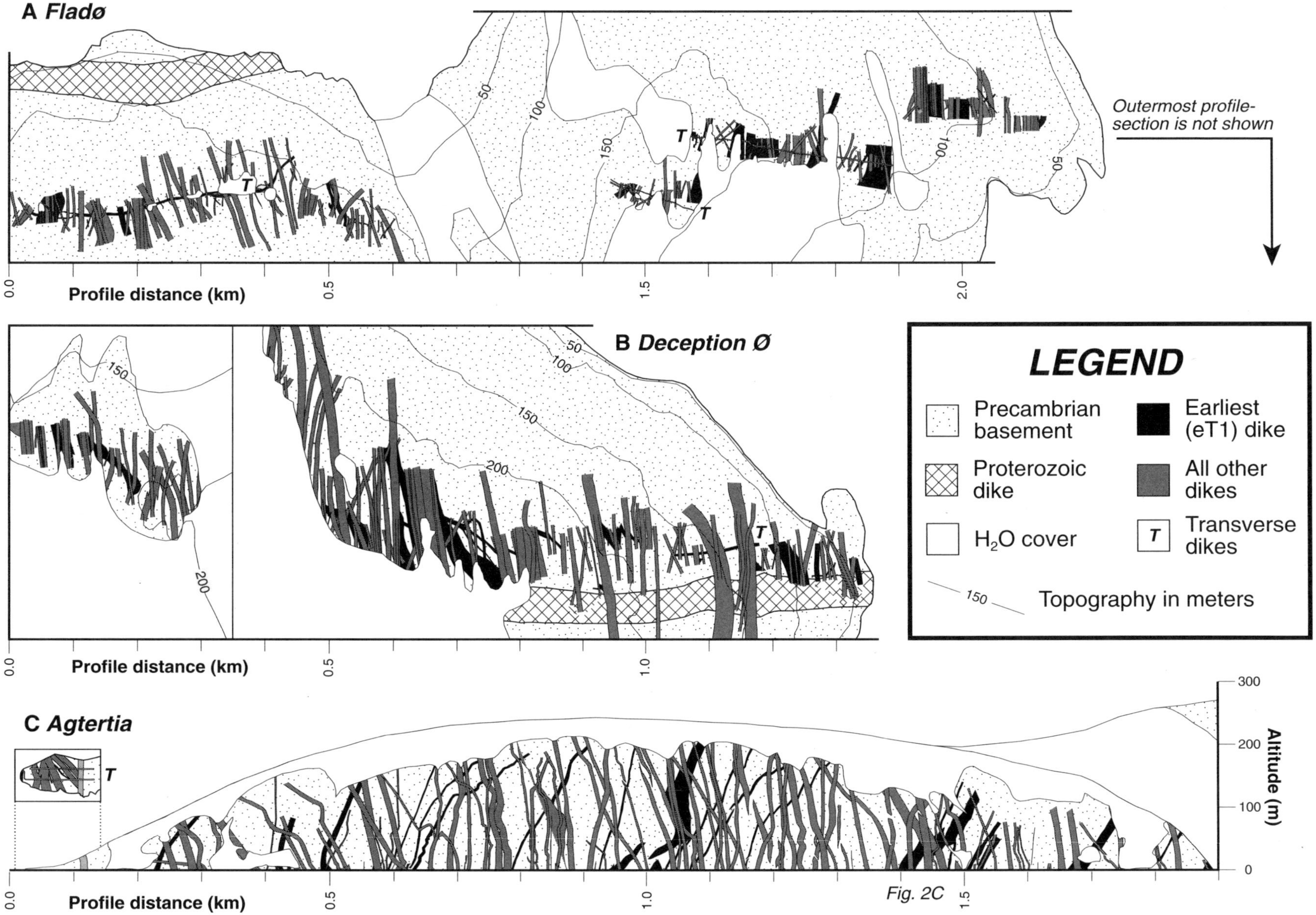

Figure 3. Maps and profiles of dike-transect across (A) Fladø, (B) Deception Ø, and (C) Agtertia. All profile distances are from landward (northwest) to seaward (southeast) positions. Dikes in A and B are only shown as far as these were mapped in field and from aerial photographs (~1:35 000). Thicker Proterozoic dikes are thus severely sliced by Tertiary coast-parallel dike swarm and are not as coherent as shown here. Agtertia profile in C is mapped in field and by multimodel photogrammetry (Dueholm, 1992). Note that this orthogonal projection causes more east-striking earliest (eT1) dikes to appear more gently landward dipping, while other more northerly striking and nearly vertical dikes appear to be seaward dipping. Amphibolite-gneiss structure within Precambrian basement is omitted for clarity.

preserved cover rocks, but are now gently dipping and cut by successively younger and more steeply dipping dikes. The gabbroic Imilik complex north of Kap Gustav Holm (Fig. 1) shows similar internal structures, tracking a history of dike intrusion and seaward crustal rotation that commenced prior to 56 Ma (i.e., prebreakup) and continued well beyond final breakup, to at least 47 Ma (Tegner et al., 1998). The 47–49 Ma gabbroic intrusions exposed on relatively small islands between Imilik and Kap Edward Holm (Fig. 1; Table 1) are only cut by few dikes, and provide a likely minimum age on the main part of the onshore dike swarm.

Although the seaward rotation of crust and its approximate timing are evident from field relationships and geochronology, identification of possible fault blocks and their bounding faults is very difficult. This is because there is no continuous cover rock preserved from which fault traces and fault offsets can be easily identified. Locally, however, fault zones including pseudotachylites radiometrically dated to ca. 60 Ma are found and testify to early Tertiary fault activity (Fig. 2F; Karson et al., 1998; Curewitz and Karson, 1999). In addition, small-scale offsets of Tertiary dikes have been reported (Karson and Brooks, 1999), and some gently dipping dikes most likely follow fault planes of uncertain ages (e.g., Figs. 2, G–H, and 4B). However, the overall lack of evidence for large-scale faulting is remarkable, despite the inherent difficulties in tracing faults within the high-grade basement, although consistent with Nielsen's (1978) findings northeast of Kangerlussuaq. It is possible that major fault zones actually are located in coast-parallel fjords or ice-covered valleys, as demonstrated within the Blosseville Kyst region of flood basalts (Fig 1, inset; Watt, 1975; Larsen et al., 1989; Pedersen et al., 1997).

In addition to the roughly coast-parallel early Tertiary dikes (locally up to 25 m thick), the basement also includes a limited number of thicker subvertical dikes (up to 60 m thick). These strike N50°W, along the main tectonic grain of the Precambrian basement (Bridgwater et al., 1978), and at a high angle to the margin. They are also, despite their fresh appearance in outcrop, often partially amphibolized and consistently cut by all coast-parallel dikes. These thick and margin-perpendicular dikes are probably of Early Proterozoic age. However, there are also a few thin (1–2 m) and subvertical Tertiary dikes, which also strike at a high angle to the coast; we refer to these as transverse dikes (Figs. 3, A–B, and 4, A–D).

The present-day coastline and innermost shelf seem to mark the approximate transition between uplift and erosion of the land area and subsidence of the shelf (e.g., Larsen, 1990; Hansen, 1996; Larsen and Saunders, 1998), where an extensive wedge of seaward-dipping basaltic lavas is present (Fig. 1). Locally, seaward-dipping sediments overlying continental basement are present close to the coast, cropping out below the early Tertiary seaward-dipping lavas (Larsen, M. et al., 1999). Rafts of a former lava cover are also present within the Noe-Nygaard intrusion located ~30 km inland (Fig. 1; Bernstein and Bird, 2000) and, together with locally preserved lavas around the coastal intrusions, suggest that the flows initially extended at least that far inland before uplift and erosion commenced. Seismic images also suggest that 20–30-km-thick igneous (Icelandic type) crust is present offshore from the continental shelf (Fig. 1; Larsen and Saunders, 1998; Korenaga et al., 2000). The uplift of the inland region and the thermal subsidence of the new igneous crust below the shelf region has added some postbreakup flexure to this zone of breakup deformation.

SYSTEMATIC STRUCTURAL MAPPING ALONG DIKE PROFILES

To define the role of dike emplacement in the process of continental breakup and the more detailed nature of the continent-ocean transition, 497 and 913 dikes were mapped (Klausen, 1999) along selected transects within the northern (Fig. 2, A–C) and southern (Fig. 2, D–H) study segments of the dike swarm, respectively (Fig. 1). This field work was combined with vectorized multimodel photogrammetry (Dueholm, 1992) in order to achieve the most complete mapping of dikes along these transects (Figs. 3, A–C, and 4, A–D). The geometry of each dike is described by its measured orientation (i.e., strike and dip) and thickness. In addition, a crude field classification of the dikes is made from crosscutting relationships and differences in appearance.

This large database allows us, on a statistical basis, to (1) identify subpopulations of dikes with statistically different orientations; (2) examine the distribution of dike thickness; (3) calculate dike densities (equivalent to dilation caused by dike injection); and (4) identify spatial or temporal variations in dike orientation, dike thickness, and dilation by dikes. See Karson and Brooks (1999) for specific studies of fault and deformation zones associated with the dike swarm and the coastal flexure.

Field classification and chronology

The coast-parallel dike swarm consists of a large variety of compositionally different dikes. On the basis of geochemistry, Nielsen (1978) and Gill et al. (1988) distinguished early tholeiitic suites followed by transitional and, eventually, minor local alkaline series, all within the area close to Kangerlussuaq. Hanghøj (1998), on the basis of detailed geochemical work, confirmed that roughly the same pattern exists within the branch of the dike swarm extending south of Kangerlussuaq. These chemical trends most likely reflect an overall progression toward lower degrees of mantle melting with time. Klausen (1999), on the basis of field observations, recognized the following three main types of dikes.

Type 1 dikes. The oldest group of relatively dark gray mafic dikes is systematically cut by other dikes and commonly represents about one-third of the total number of dikes. These early dikes often weather to a rusty-brown color (Fig. 2A) and vary from being aphyric to having variable amounts of phenocrysts (plagioclase, ±mafic minerals).

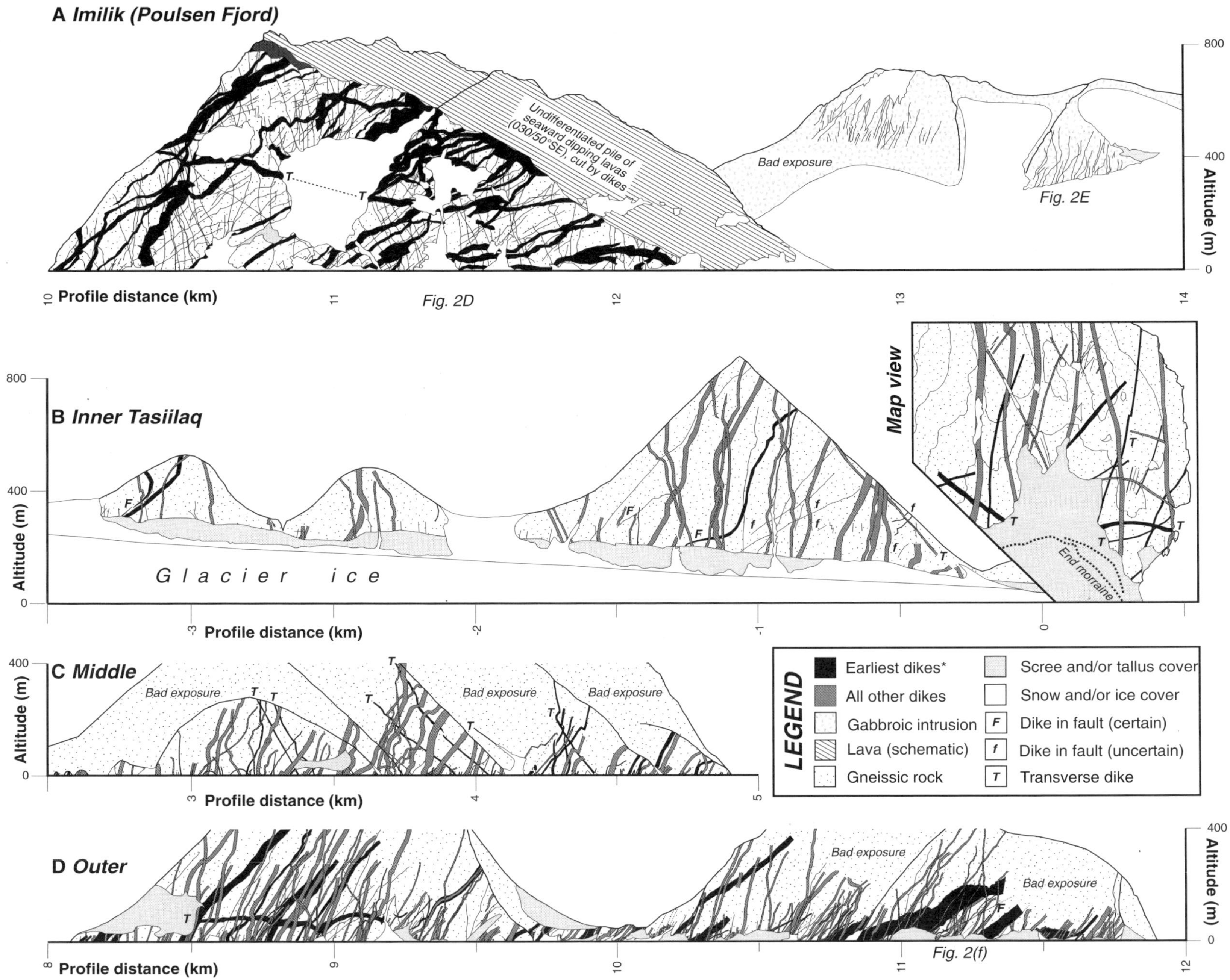

Figure 4. Profiles of dike swarm transect along (A) Poulsen Fjord, and (B–D) inner to outer sections of Tasiilaq. All profile distances are from landward (northwest) to seaward (southeast) positions. A was only mapped by multimodel photogrammetry (Dueholm, 1992); B–D include field observations. Note that N56°E view in A causes both earliest dikes and lava-gneiss contact to appear more gently dipping than these structures really are, and that later subvertical dikes appear to be seaward dipping. Poor contrast in texture and color prevented us from tracing dikes within the lava pile. Middle and outer sections of Tasiilaq are inverted to fit consistent northward view. Internal structures within amphibolite gneiss, lava, and gabbro are omitted for clarity.

TABLE 1. FIELD CLASSIFICATION OF DIKES AND TENTATIVE CORRELATION TO OTHER IGNEOUS STRUCTURES

Structural groups*	Outcrop character†	Composition§	Extrusives**	Central intrusions††	Age span (Ma)†††
± Latest dikes MA	Type 3	Lamprophyre#	No observations	Kialineq complex	38–35§§§
	Chocolate-brown-weathered type 2			Wehrlitic intrusions§§ within gabbroic intrusions	49–47
Main swarm of sub-parallel orientated dikes	Main group of volatile-saturated type 2	Trans-1	No observations	Late KEH, Nordre Aputiteq, Kruuse Fjord, Noe-Nygaard##, late Imilik	54–49?
	Precursor type 2				
	Late type 1	Thol-2	Middle-Upper Lavas	Skaergaarden, early KEH, Nugalik***, early Imilik	56–54
Earliest dikes	Early type 1	Thol-1	Lower Lavas		61–58

Note: KEH = Kap Edward Holm gabbroic complex. Superscripts correspond to the following sources: * = This paper; † = Klausen (1999); § = Nielsen (1978) and Hanghøj (1998); # = Rucklidge et al. (1980); ** = L. Larsen et al. (1999); †† = Bernstein et al. (1998); §§ = Tegner et al. (1993); ## = Bernstein and Bird (2000); *** = Rex et al. (1979); ††† = Tegner et al. (1998) and Tegner and Duncan (1999); §§§ = Noble et al. (1988) and Tegner (1999, personal communication).

Nearly half of these type 1 dikes constitute a statistically valid subgroup (~10%–20% of the coast-parallel dike swarm) with a significantly different mean orientation on a 5% level of confidence. Field relationships consistently show these dikes to be the very earliest; they dip more gently landward (Figs. 2B and 4, A–D), strike more eastward (Figs. 2C and 4B), and are thicker than any other group of dikes (e.g., Fig. 5, A–F; Klausen, 1999). In the following we refer to these as a subgroup named "earliest dikes." The dip directions of these earliest dikes form poles to the locally preserved, seaward-dipping peneplain and overlying sediments and lavas of prebreakup age (Figs. 2D and 4A). This shows that in general, these earliest dikes were emplaced subvertically prior to breakup ca. 56 Ma and that they underwent the entire seaward crustal rotation during breakup. The younger part of the type 1 dikes structurally (i.e., orientation and thickness) overlaps more with type 2 dikes than with the earliest dikes (e.g., Fig. 5, A–F), with which they are grouped on the basis of their field appearance.

Type 2 dikes. The remaining dikes along the northern study segment are distinctly different in appearance from the type 1 dikes. They are generally paler, less fractured, and often exhibit volatile-induced textures (e.g., zones of amygdales or pegmatites).

Within this large group, there appears to be an evolution from (1) an initial phase of somewhat thinner dikes with marked flow textures, through (2) a dominant group of more massive, uniformly thick and pale dikes, to (3) a late phase of characteristically chocolate-brown weathered dikes. Except for differing grain size, the latter dikes resemble the late wehrlitic intrusions within the gabbroic intrusions (Tegner et al., 1993, 1998; C. Tegner, 1999, personal commun.).

Type 3 dikes. Subswarms of relatively young and more exotic (alkaline or felsic) looking dikes may occur within the dense part of the coast-parallel dike swarm, and locally may compose as much as 10% of the dikes. Type 3 dikes are often easily recognized by distinctly pale weathering and are often deeply eroded (e.g., Fig. 2C). Rucklidge et al. (1980) noted that such a subswarm close to Nigertuluk included lamprophyric dikes.

The presented field classification can tentatively be correlated to the main tholeiitic and transitional series, as well as local subswarms of alkaline or felsic dikes, respectively (Table 1). Tholeiitic dikes are slightly more numerous than transitional basaltic dikes, according to the geochemical classification of statistically sampled populations by Hanghøj (1998). The earliest dikes of our study may also correlate with the preflexural Thol-1 dikes of Nielsen (1978) and Gill et al. (1988). Consistent with this, the Thol-1 dikes have been related to the prebreakup lavas in the Kangerlussuaq area (Lower Lavas; Nielsen et al., 1981). The remaining part of the main group of tholeiites (late type 1 and some of type 2) is most likely related to the breakup phase (i.e., ca. 56–54 Ma); one dike from this group has been dated to ca. 56 Ma (Tegner et al., 1998). The $^{40}Ar/^{39}Ar$ age determination of the tholeiitic dikes has otherwise largely failed because of their low potassium content (<0.5%) and high degree of metamorphism (Rose and Bird, 1994). One transitional dike (late type 2) is dated as ca. 52 Ma (R. Duncan, 2000, personal commun.) and places this subgroup after breakup. The wehrlitic intrusions within young gabbroic intrusions are as young as 47–49 Ma (Tegner et al., 1998), and thereby mark the minimum age of the coast-parallel dike swarm. However, some type 3 dikes may be related to the even younger (ca. 35 Ma) alkaline or felsic plutons within the Kialineq area (Fig. 1; Noble et al., 1988; C. Tegner, 2001, personal commun.). Table 1 summarizes the most likely chronology.

Orientation of dikes

Dike orientations (strike and dip) were measured in the field by hand compass, within an accuracy of ±1°–10°, depending on

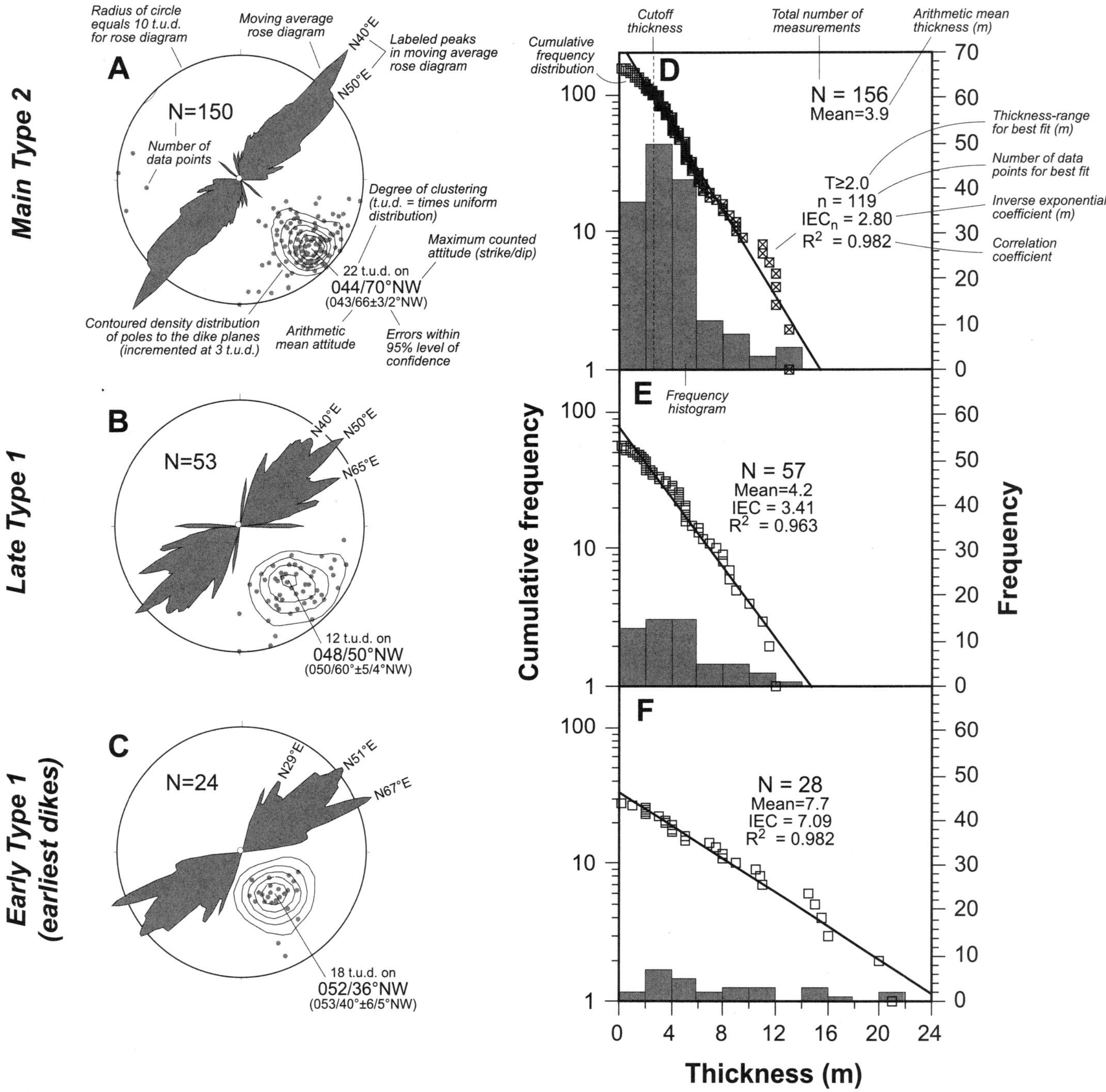

Figure 5. Methods of presentation and statistical analysis of dike orientations (A–C) and dike thicknesses (D–F), for three groups of field-classified dikes mapped on Fladø: main group of type 2 dikes (A, D); late type 1 dikes (B, E); and early type 1 (earliest) dikes. Rose diagrams show moving average density distribution in strike, within 10° sections that are shifted at 1° intervals. Distinct peaks are labeled. Contoured density distributions of poles to dike planes are calculated within cones about grid points (weighed by factor of two) with opening angles such that, for random distribution, average point should be found within each cone. Cumulative frequency-thickness diagrams with logarithmic y-axis and linear x-axis show negative exponential distributions as straight segments (Jolly and Sanderson, 1995), which are quantified by regression. In D, cutoff thickness is evident from relatively low frequency of thin dikes. Notice systematic (1) clockwise rotation in maximum counted mode attitude, (2) less negative exponential regression, and (3) less marked cutoff thickness for progressively older generations of dikes. Arithmetic mean values (±errors on 5% level of confidence) are added for comparison. Individual poles (gray dots in A–C) and ordinary frequency-thickness histograms (D–F) are only shown here as reference, and omitted from other plots in this chapter.

accessibility and quality of the exposure. However, in addition to the uncertainty of the measurements, the geological complexity (e.g., irregular or highly dissected dikes) introduces an additional error in the determination of typical orientations. Local complexities were avoided in the field as much as possible. Furthermore, by collecting a large set of data and plotting them in rose diagrams and density-contoured plots of poles to dike planes (Fig. 5, A–C), we have minimized the effects of local variations and imprecise observations (e.g., Leyshon and Lisle, 1996). This statistical approach also allows us to resolve structurally significant subgroups, as shown by the three subdivided groups in Figure 5 (A–C). Where possible, we use the maximum counted density distribution of the poles to the planes, because it defines a more representative orientation mode that is less sensitive than the arithmetic mean to a few highly deviating dike orientations.

The three subgroups in Figure 5 (A–C) resolve a successive counterclockwise change in orientation that characterizes individual outcrops along the margin. However, only the earliest dikes consistently exhibit significantly different average orientations compared to the other groups. To map spatial variations along margin-crossing transects, we use a moving average of strike or dip with a window length of 30 samples and a step length of one sample. This approach provides an effective filtering of the local data variation, while still allowing us to resolve spatial variations across the margin.

Dike thickness

We analyzed the frequency-thickness distribution (Fig. 5, D–F): source errors in the field are simple errors (within 10%) in the measurement (e.g., difficult conditions or poor exposure) or multiple and/or sheeted dikes mistakenly recorded as a single intrusion.

Ideally, a single population of dikes would show a spread in thickness with number of dikes increasing with decreasing thickness. Using the method presented by Jolly and Sanderson (1995), Klausen (1999) found that for most populations a remarkably good negative exponential correlation exists between the frequency and thickness of dikes. However, the thinnest dikes occasionally fall below the exponential distribution defined by the regression line, as shown by Figure 5D. Dikes thinner than what we take as the cutoff thickness are thus underrepresented, although not absent.

The two groups of earlier dikes in Figure 5 (E–F) show no clearly developed cutoff thicknesses, suggesting that these represent single populations. However, in all three cases (Fig. 5, D–F), the slope of the regression line provides a measure of the average dike thickness within the swarm, if this were to follow the same negative exponential distribution throughout. We calculate and thus refer to this characteristic average as the inverse exponential coefficient (IEC) thickness. The three subgroups furthermore show a temporal change toward a successively more marked cutoff thickness and steeper regressional fit (i.e., smaller IEC thickness), or tendency to cluster around a mode thickness.

Spatial variations in dike thickness along margin transects were also analyzed by the simple moving average thickness described herein.

Dike density and tectonic extension

The mapping of dike thickness along coherent transects across the margin allows us to calculate the amount of crustal dilation by magma emplacement, and the amount of crustal rotation (from dike orientations) allows us to estimate the amount of tectonic extension by assuming simple domino-faulted block rotation. The principles behind these calculations are illustrated in Figure 6 (A–D).

Figure 6A shows a crustal block cut by two generations of dike swarms and tectonically rotated along normal faults on either side of the block. The cumulative thickness of the dikes is summed from field observations, and the dilation by dikes is calculated by dividing this sum by the length of the block. However, the appropriate block length is the one measured at right angle to the dikes, i.e., the length prior to tilting. In order to correct for this we define a mean tilt angle. We have also chosen to use a weighted mean tilt of the dikes, i.e., each dike is weighted by its corresponding thickness, simply because thicker dikes make a bigger contribution to dilation than thinner dikes. In order to calculate the dilation by the earliest generation of dikes, we first reduce the length of the cross section by subtracting the dilation caused by later dikes (Fig. 6B) and then repeat the steps outlined herein.

The total amount of tectonic extension is calculated from the maximum degree of rotation, recorded either by the dip of the lowermost lavas or corresponding tilt of the earliest dikes. We assume that this extension is accommodated through simple domino-block rotation along initially 60°-dipping normal faults, as schematically illustrated in Figure 6C. If crustal rotation by domino-block faulting were to exceed 30°, we initiate a new set of normal faults instead of maintaining slip along the first set of faults at lower angles. However, the amount of fault-block rotation rarely exceeds 30° once correction for differential movements in the vertical plane, such as postbreakup continental uplift and shelf subsidence, is applied (Fig. 6D). We convert tectonic extension to tectonic dilation according to the formula given in Figure 6C in order for this to be directly comparable with finite range of dike dilation (between 0% and 100%).

NORTHERN DIKE SWARM SEGMENT

The northern segment extends south from the gabbroic Kap Edward Holm igneous complex at the mouth of the Kangerlussuaq Fjord (Fig. 1). Several other plutonic complexes are present around this fjord, including the famous Skaergaard Intrusion and the large alkaline Kangerlussuaq Intrusion. The exposure of the dike swarm and associated flexure within this segment is largely

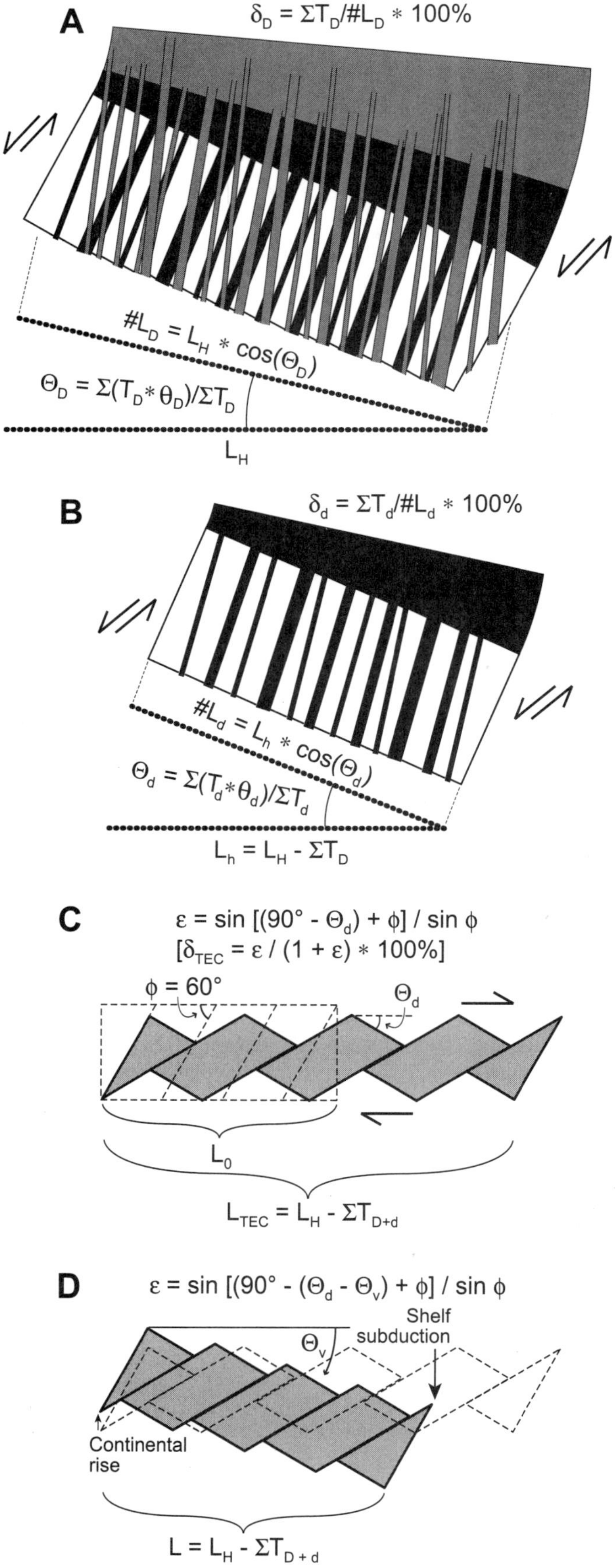

restricted to some islands (Figs. 1 and 7A) with relatively high dike densities (40%–70%) within basement rock. Substantially lesser densities of dikes are also observed along the heavily ice-covered mainland, but we did not investigate these further.

All Tertiary dikes were mapped along three main profiles at Fladø, Deception Ø, and Agtertia, respectively (Figs. 2, A–C; 3, A–C). The three profiles are correlated on the basis of dike trajectories defined by the strike mode of all dikes at each profile location, shown in Figure 7A. The distribution of outcrops along this northern segment, and hence the three profiles, places these close to the same dike swarm trajectory. It is likely that these islands owe their presence to a rapid seaward increase in dike density, creating a zone of particular resistance to erosion.

The correlation along the main dike trajectory does not account for the slightly more east-striking earliest dikes (e.g., Fig. 5, A–C; Gill et al., 1988). Effectively, the implication of our approach is that the dike trajectory that extends through the three profiles is not likely to truly represent the earliest dikes (~10% of the total), but to be highly dominated by the main suite of dikes; and these seem to show a very consistent pattern. We therefore analyze the orientations, frequency-thickness distribution, and dilation by dikes within each study location for the total population of dikes as well as for the population of the earliest dikes, before we compare these locations along the defined dike trajectory.

Dike orientations

The dike orientations (all dikes in Fig. 7, B–D) show well-defined mode values (16–28 times uniform distribution). These show a systematic counterclockwise rotation in strike of the dikes (N44°E–N31°E) and a parallel increase in the dip of dikes (69°NW–87°NW), going south from Fladø to Agtertia. The skewed density distributions for Fladø and Agtertia (Fig. 7, B and D) reflect a counterclockwise shift in orientation from the earliest dikes (Fig. 7, E and G) to later dikes, quantified by the mode orientation of all dikes. That these furthermore reflect a continuous rotation with time is clearly supported by the three subpopulations of dikes at Fladø (Fig. 5, A–C).

Figure 6. Methods to calculate average magmatic dilation of (A) all and (B) earliest dikes, as well as modeled tectonic extension through domino-faulted block rotation of earliest dikes, assuming either (C) horizontal or (D) combined horizontal and vertical simple shear. Note that uppercase/lowercase subscripts in the following abbreviations refer to results by either all dikes or to subgroup of earliest dikes, respectively. $T_{D/d}$: thickness of dike. $\theta_{D/d}$: Tilt angle of dike. $L_{H/h}$: Horizontal length across mapped outcrop. $\#L_{H/h}$: Length perpendicular to weighted average orientation of dikes. $\Theta_{D/d}$: Angular difference between $L_{H/h}$ and $\#L_{H/h}$ (resolved into components of strike and dip). $\delta_{D/d}$: Dilation by dikes. ϕ: Initial angle of faults (assumed to be 60°). L_0: Initial length of undeformed crust. L_{TEC}: Horizontal length of tectonically extended crust. Θ_V: Degree of differential vertical movement. ε: Horizontal extension, where tectonic dilation $\delta_{TEC} = \varepsilon/(1 + \varepsilon)$.

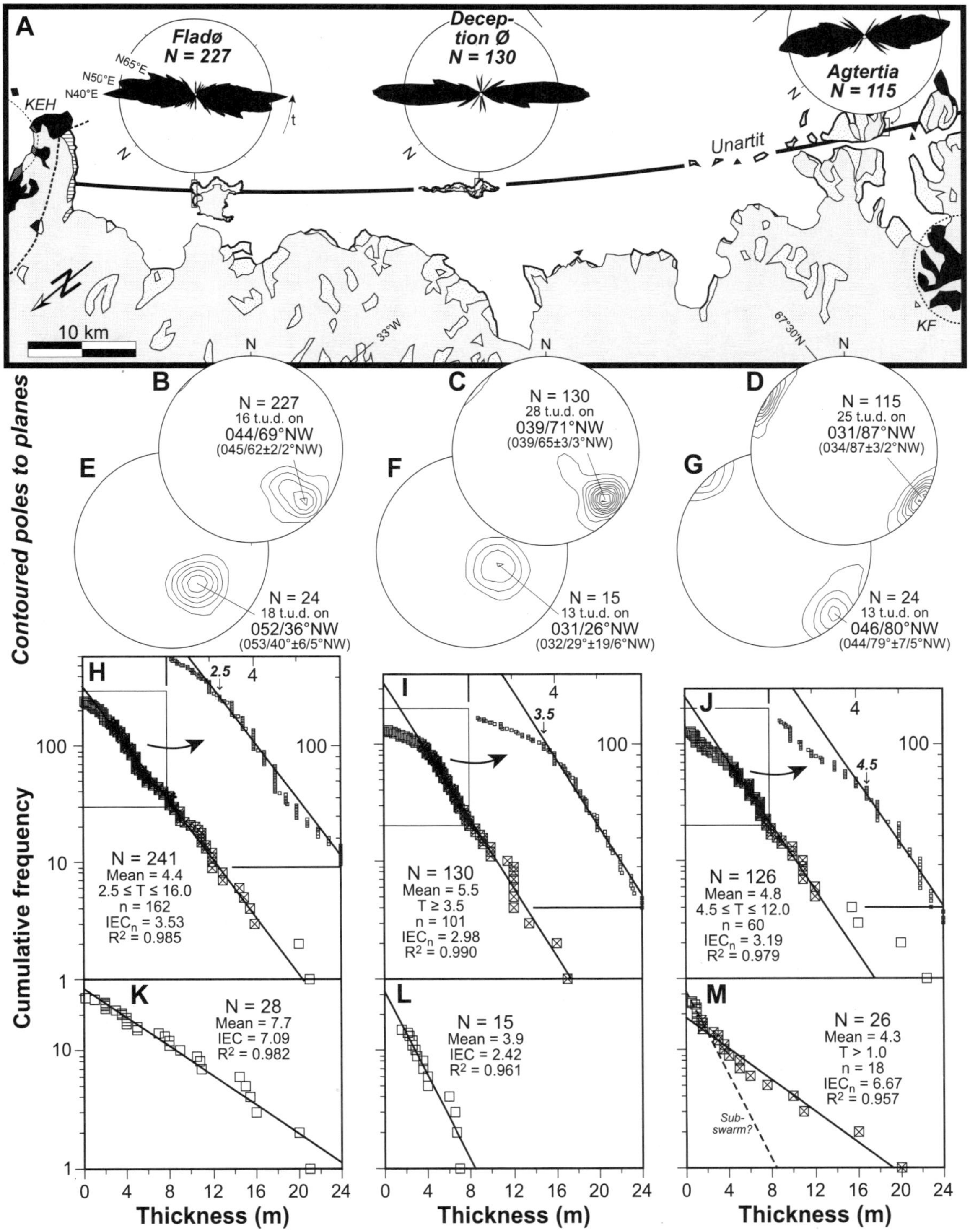

Figure 7. Statistical compilation of structure of all dikes and subgroup of earliest dikes along northern dike swarm segment. A: Map with rose diagrams showing dominant dike trajectories at each transect. Abbreviations are as in Figure 1. Arrow labeled with "t" indicates systematic temporal rotation in strike on Fladø (Fig. 5, A–C). Contoured density distribution of poles to planes of all dikes (B–D), as well as earliest dikes (E–G), around mode orientation. Cumulative frequency-thickness distribution of all dikes (H–J) and earliest dikes (K–M), showing exponential regression and corresponding inverse exponential coefficient (IEC) thickness. Inset enlargements of thickness range between 0 and 8 m enhance distinct cutoff thicknesses (labeled), observed for all dikes. All diagrams are constructed according to Figure 5 (A–F). Note that B, E, H, and K refer to Fladø, C, F, I, and L refer to Deception Ø, and D, G, J, and M refer to Agtertia.

Only the data from Deception Ø deviate from this pattern by showing a tendency toward a more bimodal distribution (Fig. 7C). It is introduced by the population of earliest dikes (Fig. 7F), which show unusually shallow dip mode (26°NW) and northward strike mode (N31°E).

Frequency-thickness distributions

Marked cutoff thickness (all dikes) is found at all three localities (Fig. 7, H–J). Furthermore, these increase systematically from ~2.5 m at Fladø, to ~3.5 m at Deception Ø, to ~4.5 m at Agtertia, i.e., away from the Kangerlussuaq area and the Kap Edward Holm igneous complex. The earliest dikes (Fig. 7, K–M) show no distinct cutoff thickness. On the contrary, Agtertia shows a slight overrepresentation of dikes thinner than ~1 m. This most likely reflects mixing of the regional swarm with a more local swarm of thinner dikes. The cutoff thicknesses defined by the entire set of dikes therefore clearly developed with time, as indicated by the three subpopulations in Figure 5 (D–F).

All three localities show (Fig. 7, H–J) good negative exponential regressional fits ($R^2 > 0.957$), with remarkably similar IEC thicknesses (3.0–3.5 m). The arithmetic mean thicknesses of all dikes are more variable and 1.0–2.5 m greater than its corresponding IEC thickness, because the latter values ignore dikes thinner than the cutoff thickness. The earliest dikes (Fig. 7, K–M) at Fladø and Agtertia have nearly similar IEC thicknesses (6.7 m and 7.1 m) that are twice as thick as the IEC thickness of all dikes. However, the IEC thickness of the earliest dikes at Deception Ø is only 2.4 m, and somewhat similar to the subgroup at Agtertia. This is exceptionally thin for such early dikes (e.g., Fig. 5, D–F), also compared to the data from the southern swarm segment (6–8 m; Klausen, 1999).

The southern segment shows a significant and systematic variation in the average dike thickness (all dikes) from thicker inland dikes to thinner dikes seaward. Thus, the approximately similar IEC thicknesses (all dikes) at Fladø, Deception Ø, and Agtertia are consistent with their suggested position along the same dike trajectory.

Along-margin variation in dike orientation, thickness, and density

In Figure 8 (A–D) we show the along-margin variations in strike mode, dip mode, IEC thickness, and average dilation by the entire population of dikes as well as the earliest dikes, shown as a separate group.

The populations of all dikes show a counterclockwise rotation in the strike mode from north to south (Fig. 8A), defining the curved trajectory in Figure 7A. The dip mode of all dikes increases from just below 70°NW at Fladø to nearly vertical at Agtertia (Fig. 8B). We find as earlier noted a systematic southward increase in cutoff thickness and a slight southward decrease in the IEC thickness. We note a linear decrease in dilation, from >70% in the north to ~40% in the south (Fig. 8D). When linearly extrapolated, this latter trend reaches 100% at 0 km (i.e., at the mouth of the Kangerlussuaq Fjord) and 0% at ~150 km offset. In addition, the earliest dikes show such a decreasing dilation away from the Kap Edward Holm complex (Table 2; Fig. 8D).

It is evident from Figure 8 (A–D) that data from the subgroup of earliest dikes at Fladø and Agtertia are consistent with these trends. However, as previously noted, both the thickness and orientation of the earliest dikes at Deception Ø are anomalous and deviate from the systematic along-margin variation defined by all dikes. This discrepancy may reflect a local swarm or complexity, like that indicated by the overrepresentation of thinner (earliest) dikes in Figure 7M. However, an alternative and more attractive explanation is that these earliest dikes at Deception Ø simply were injected along preexisting landward-dipping normal faults and underwent some flattening by vertical subsidence of the hanging wall. In Figure 8 (E, F) we illustrate the average effect on the final orientation of these dikes if they were injected along preexisting faults dipping 62°NW (i.e., consistent with Andersonian normal faults) and striking parallel to the margin (N31°E). We furthermore assume that subsidence of the hanging-wall block on average led to flattening of the dike thickness from 6.9 m to the observed 2.4 m (Fig. 8G). The dike dilation is also calculated on the basis of this model in Figure 8H.

It is striking how well this single mechanism (intrusion along fault planes) can explain the deviations we see in all parameters, including dilation by the earliest dikes (Fig. 8D). However, there is no field evidence from Deception Ø (cf. Fig. 2B) to support this hypothesis. In any case, the relatively small population (12%) of deviating earliest dikes on Deception Ø does not alter the broader picture of all dikes showing systematic variations away from the Kap Edward Holm gabbroic complex.

SOUTHERN DIKE SWARM SEGMENT

Exposure across the margin is more coherent within the southern dike swarm segment than within the northern segment. However, in no place is there any continuous exposure from the most inland areas of minimal dike injection and tectonic disturbance to the most seaward zone with high dike density and strong tectonic deformation. We therefore piece together a margin-wide master profile from the three main study areas Nigertuluk (farthest inland), Tasiilaq (central), and Imilik (farthest seaward).

Figure 4 (A–D) shows our mapping of Tertiary igneous rocks along the profiles at Tasiilaq and Imilik (Poulsen Fjord), where a seaward increase in the density of dikes and a parallel decrease in their landward dips is most evident. The Imilik exposure clearly demonstrates that the more gently landward dipping dikes were injected normal to the erosional peneplain and the overlying pile of seaward-dipping lavas (Figs. 2D and 4A). It also shows that these more gently dipping dikes belong to the group of earliest dikes cut by steeper and younger dikes. These observations are consistent with the general assumption that a majority of dikes were emplaced subvertically and subsequently have been rotated to their present position; i.e., younger dikes

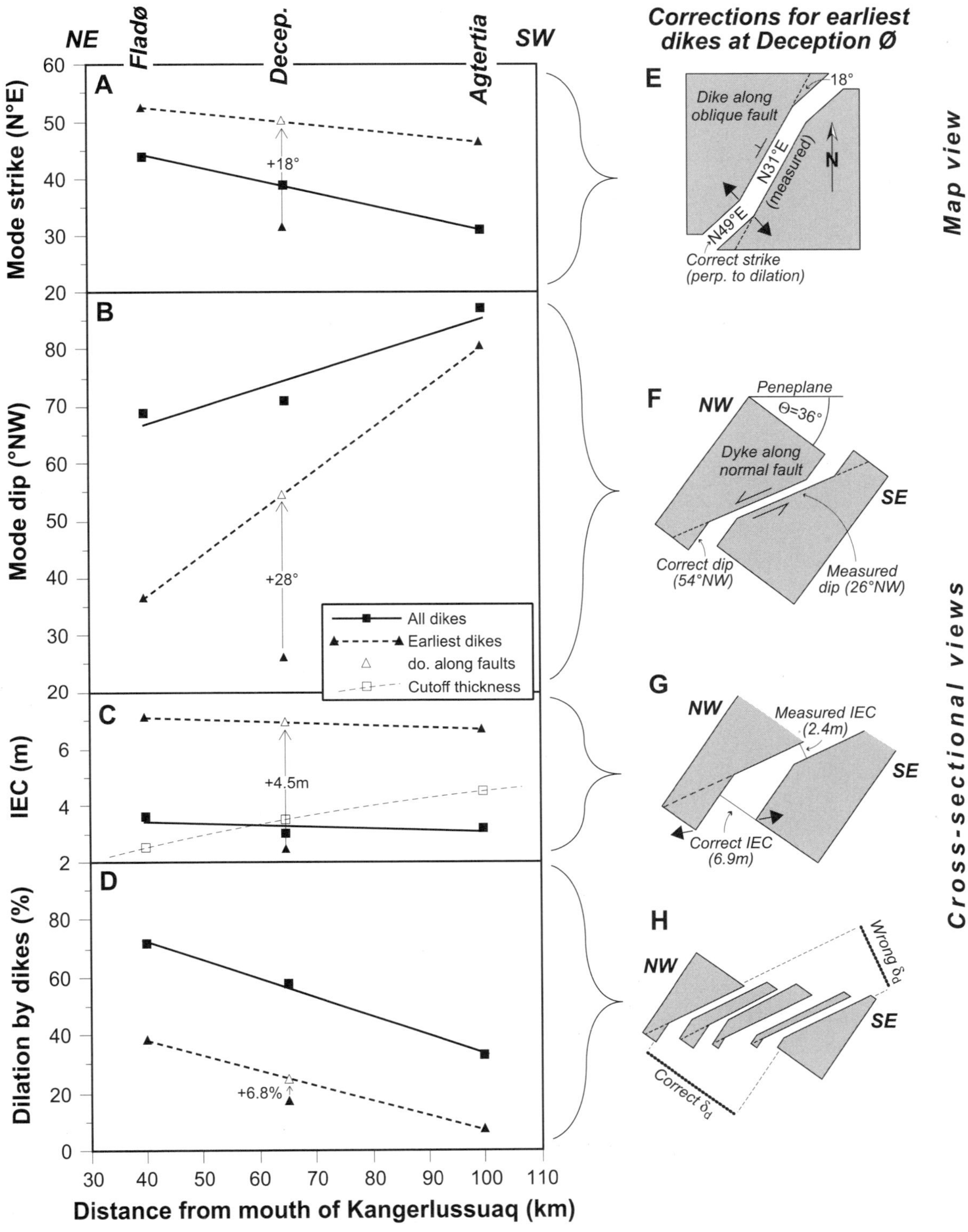

Figure 8. Structural variations along correlated main trajectory of coast-parallel dike swarm for total population of dikes (squares) as well as earliest dikes (triangles) through Fladø, Deception Ø, and Agtertia. A: Mode strike. B: Mode dip. C: Inverse exponential coefficient (IEC) and cutoff thickness. D: Dilation by dikes. A–D are shown as function of distance from mouth of the Kangerlussuaq Fjord (Fig. 1). All values are from Figure 7 (B-M) and Table 2. Solid lines represent linear regressions through three statistical values on all dikes, and dashed lines interpolate statistical values of earliest dikes at Fladø and Agtertia. Consistently deviating earliest dikes at Deception Ø are corrected according to geometrical model shown in E–H, i.e., assuming that these dikes intruded along landward-dipping normal faults (see text). Labeled arrows to open triangles (i.e., corrected values) in A–D quantify amount of correction involved. δ_d: Dilation by earliest dikes.

TABLE 2. FIRST-ORDER DIKE STRUCTURES ALONG THE NORTHERN DIKE SWARM SEGMENT

	N (dikes)	Mode strike† (N°E)	Mode dip† (°NW)	IEC-thickness§ (m)	Dike dilation# (%)
All dikes					
Fladø	241	44	69	3.5	71.8
Deception Ø	130	39	71	3.0	58.0
Agtertia	126	31	87	3.2	33.0
Earliest dikes					
Fladø	24	52	36	7.1	38.9
Deception Ø (along faults*)	15	31	26	2.4	18.0
Deception Ø (corrected**)	15	49	54	6.9	24.4
Agtertia	26	46	80	6.7	8.1

Note: Statistically constrained structural values from † Fig. 7 (B–F) and § Fig. 7 (G–M), and calculated according to # Fig. 6 (A, B).

* The earliest dikes at Deception Ø were probably injected inclined, along pre-existing faults with an average orientation of 031/62°NW. ** Corrections explained in text and Figs. 8(A–H).

have undergone less tectonic rotation. However, 10 dikes show direct evidence of having intruded along preexisting landward-dipping normal faults (e.g., Fig. 2, F–H; F in Fig. 4, B–D); other very gently dipping dikes are subparallel to these and are also assumed to have intruded inclined (e.g., F in Fig. 4, B–D).

Dike orientations and composite master profile

Rose diagrams summarizing the strike of dikes at different profile sections are shown in Figure 9A. In sections 3 and 4 only one common strike mode is defined; in other areas (sections 6, 8, and 9) 2–3 populations with slightly different strike modes are present. In the Nigertuluk profile these three slightly different strike populations represent a temporal change from the more east-striking earliest dikes to younger dike groups trending more northward. Similar systematic temporal variation was observed along the northern dike swarm segment (e.g., Fladø; Fig. 5, A–C). In order to define dike trajectories representing the main swarm, we therefore ignore the more east-striking subpopulations, which we interpret to represent the earliest dikes. At section 9 (outer part of Nigertuluk in Fig. 9A), where three subpopulations are present, we furthermore use the central peak in the rose diagram for definition of dike trajectories, because the most northward mode strike is defined by lamprophyric (type 3) dikes (Rucklidge et al., 1980).

The peaks (mode values) follow an overall curved pattern similar to that of the northern segment and extending south from the Imilik igneous complex. A population extending northward from Imilik (section 1) also indicates a marked change in dike trajectory pattern across the Imilik complex. This apparent convergence in the strike pattern, on approaching the Imilik gabbroic complex, suggests that this area of plutonic intrusion has controlled the pattern of dike intrusion within the main swarm in the way modeled by Vann (1978).

In order to construct one master profile across the margin, we have projected observations from the different sections along the main trend of these dike trajectories onto a composite profile centered on the Tasiilaq profile. Our choice of projections implies an ~8 km overlap between the landward part of Tasiilaq profile and the seaward part of Nigertuluk and ~2 km of interpolation between the most seaward extent of the central Tasiilaq profile to the seaward Imilik profile (Fig. 9A; Klausen, 1999). In the following we describe variations within strike, dip, thickness, and dilation by dikes from local profiles and within the composite master profile (Fig. 10, A–D).

Moving average of dike orientation

Figure 10A shows a 10° difference between the two moving average strike curves for Tasiilaq and Nigertuluk, as well as a 5° difference between Tasiilaq and Imilik; both of these features are consistent with dikes striking more northward in the south (i.e., curved trajectory; Fig. 9A). In addition, the moving average strikes along the Nigertuluk and Tasiilaq profiles exhibit a 1°/km counterclockwise rotation toward the coast and are fully consistent with the overall convergence onto the Imilik complex. However, part of this systematic change in the moving average strike could also partly be caused by a relatively larger fraction of older and more eastward-striking dikes farther inland (e.g., Fig. 9B).

There is a marked and systematic change across the margin in the moving average dip (Fig. 10B): we interpret this to directly reflect the change from an apparently undeformed inland area to progressively larger degrees of deformation and crustal rotation seaward. In detail, the moving average dip of all dikes across Nigertuluk shows an inland zone of nearly constantly landward-dipping (~88°) dikes. The slight deviation (~2°) from vertical is caused by the inclusion of some very gently dipping dikes (open triangles in Fig. 10B). Proportional to the population of all dikes this is a minor group that has little effect on the moving average of all dikes, which generally are subvertically oriented in this inland area. We interpret the inland and gently

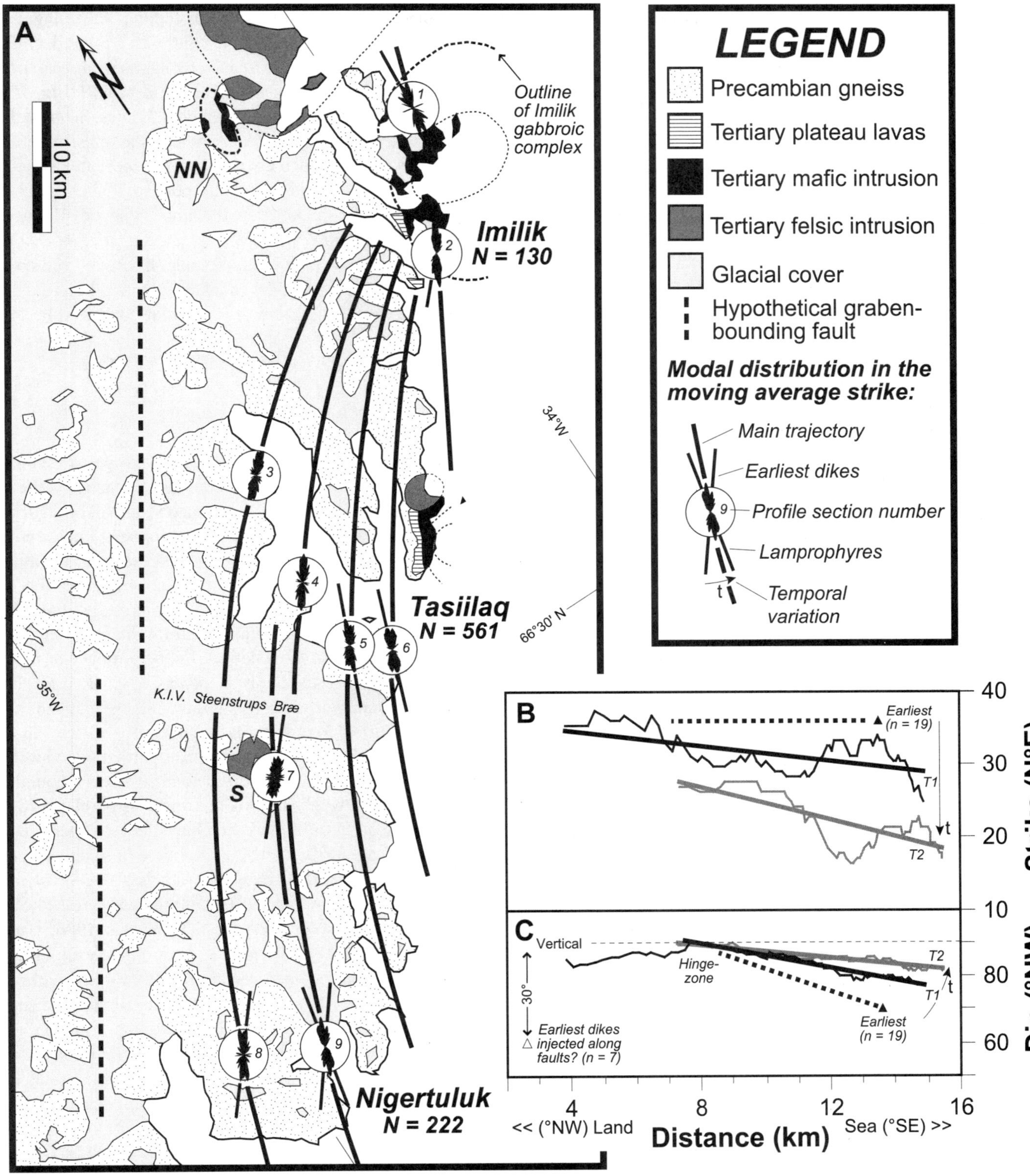

Figure 9. A: Dike trajectory pattern along mapped southern study segment, deduced from rose diagrams (constructed as in Fig. 5, A–C) of all dikes within profile sections. Thick solid lines through major strike populations emphasize dominantly curved and converging pattern of coast-parallel dike swarm. Secondary peaks, emphasized by short bars, are interpreted as subswarms of either more east-striking earliest dikes or more north-striking lamprophyric (type 3) dikes. Data from Sulugssut (section 7) and northern part of Imilik (section 1) were collected by Stefan Bernstein and Christian Tegner. Abbreviations are as in Figure 1. Inset: Moving average (B) strike and (C) dip for tentatively classified younger type 2 and older type 1 dikes across Nigertuluk, as well as single mean value for 19 earliest dikes (triangle). These results clearly reflect systematic decrease in average strike, both seaward and with time, as well as parallel increase in average dip, both landward and with time (arrows labeled with "t").

dipping dikes to have intruded nonvertically, most likely along preexisting fault planes (e.g., Figs. 2, G–H, and 4B; Karson and Brooks, 1999).

A hinge zone defined by a fairly abrupt seaward decrease in dip is present at both the Nigertuluk and Tasiilaq profiles. Within the hinge zone, the average dip quickly drops more than 5° and continues to decrease seaward by close to 1°/km to a dip of nearly 60° landward (i.e., nearly 30° seaward rotation) close to the coast (e.g., Imilik dikes in Fig. 10B). However, the locally preserved lavas at Imilik and Kap Gustav Holm show considerably stronger seaward rotation (~50°, shown as poles to the bedding in Fig. 10B). The simple explanation for this misfit is that only the earliest dikes (cut by the Imilik complex; e.g., Fig. 2D) have undergone the maximum seaward rotation, younger dikes (cutting the old gabbroic host; e.g., Fig. 2E) being intruded after some of the early tectonic deformation took place.

The individual dips of the 103 earliest dikes at Nigertuluk and Tasiilaq are shown in Figure 10B. With very few exceptions, they show distinctly lower dips than the moving average of all dikes, and the lower envelope of the field of earliest dike data can be interpreted to define the total amount of crustal rotation (Fig. 10B). This curve shows excellent correlation with the poles to the lava bedding at Imilik and Kap Gustav Holm (Fig. 10B) and seems to level off offshore at ~30°NW, representing 60° of crustal rotation. Of the earliest dikes, 22 define a subpopulation (open triangles in Fig. 10B) largely restricted to the inland zone. These more gently inclined dikes are interpreted not to have been intruded close to vertical and are ignored in this context.

A group of latest dikes (open diamonds in Fig. 10B; largely restricted near to the coast) plots with few exceptions above the moving average dip of all dikes but still shows some landward dip. Taken together, these two features suggest that crustal rotation continued to take place after breakup and emplacement of the main dike swarm. Nearly half of the crustal rotation indicated by the dip of dikes might have taken place following breakup (Fig. 10B). This would be consistent with Tegner et al. (1998) reporting continued seaward rotation of gabbroic intrusions to at least 47–49 Ma, and we will discuss possible models for such crustal deformation.

Moving average thickness and dilation by dikes

The average thickness of dikes and the crustal dilation associated with dike injection (i.e., density) provide additional details on the nature of the volcanic rift (Fig. 10, C and D). A marked maximum in the average dike thickness (8–10 m) coincides with the hinge zone (Fig. 10C) as defined above. Seaward of the hinge zone, the moving average thickness of all dikes clearly decreases along the composite profile toward a level of 1–2 m farther offshore. The average is ~1.5 m at the Imilik locality. This fairly low average thickness is close to that found within ophiolites (e.g., Pallister, 1981; Rosencrantz, 1983; Baragar et al., 1987).

A slight decrease in dike thickness inland of the hinge zone is also indicated (Fig. 10C), coincident with a rapid decline in dilation by dikes that is more marked along the inland part of the Tasiilaq profile than along the Nigertuluk profile (Fig. 10D). However, both profiles show exponential increase in crustal dilation by dikes toward 20% dilation within the hinge zone. Seaward of this zone, the increase in dilation is more moderate and reaches ~50% around the coast. Extrapolated offshore this trend would reach 100% ~30 km from the hinge zone, or ~15 km offshore in this location.

Thus, the three parameters, dip, dike thickness, and crustal dilation by dikes show parallel and consistent trends, all suggesting the presence of full igneous crust not far offshore (Fig. 1).

DISCUSSION AND MODEL IMPLICATIONS

Nature of the continent-ocean boundary

The inland hinge zone is defined by the general concordance of: (1) maximum thickness of dikes; (2) change in the rate of seaward increase in the dilation by dikes; and (3) onset of progressively more seaward rotation of the crust (Fig. 10, B–D). This zone can be considered to mark the inland limit of significant crustal deformation and magmatism during breakup. Some dikes are emplaced landward of this zone, but crustal dilation by these rapidly levels off landward. The outer limit of this continent-ocean transition is defined by the presence of full igneous crust. The extrapolation of increasing dilation by dikes places this boundary in the range of only ~30 km seaward of the hinge zone (Fig. 10D), not far from the coast.

Figure 10B shows the amount of crustal rotation to level off at ~60° on approaching the inferred continent-ocean boundary. However, approximately half of this rotation is related to the postbreakup margin uplift and shelf subsidence, which created a late, monoclinal flexure with an amplitude of several kilometers (e.g., Larsen, 1990; Larsen and Saunders, 1998). In addition, seaward flexing of the crust is an inherent part of the formation of Icelandic-type crust (e.g., Palmason, 1986; Larsen and Jakobsdottír, 1988); this type of flexure may well have extended as far inland as the coastal zone. However, neither of these flexural processes reflects tectonic extension per se and in particular not crustal thinning prior to breakup. Thus only the early (0°–30°) of crustal rotation, from the hinge zone to the seaward part of Figure 10B, reflects tectonic extension related to breakup and initial formation of the continent-ocean transition. This is consistent with domino-block rotation along only one initial set of landward-dipping normal faults (i.e., Fig. 6, C and D), after which magmatic dilation took over as the dominant rift process.

Figure 11A shows a schematic interpretation of the cross section we mapped, and Figure 11B places this within the larger context of the continent-ocean transition. According to the extrapolation of dike dilation, full igneous crust is located ~30 km

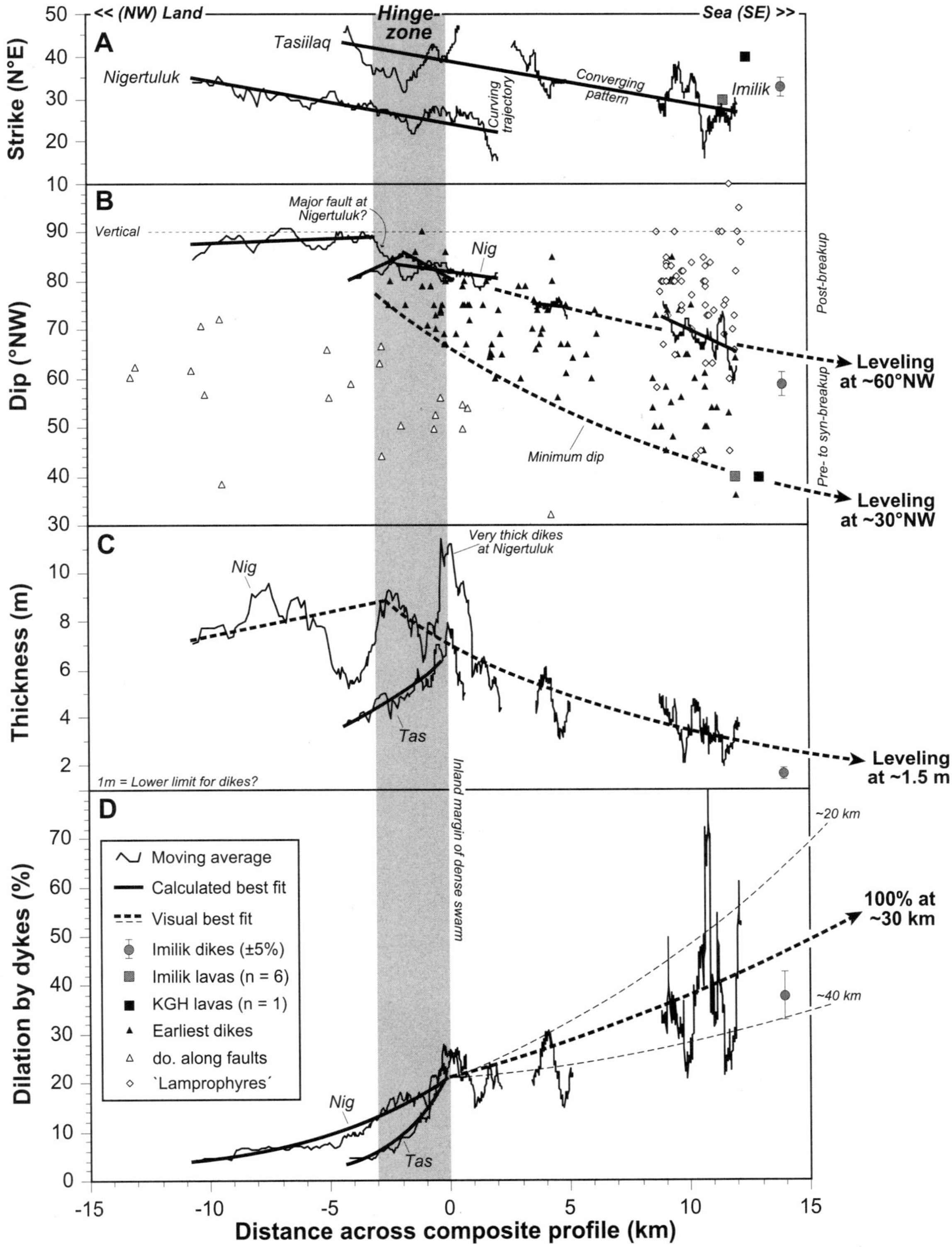

Figure 10. Moving average (A) strike, (B) dip, (C) thickness, and (D) crustal dilation by dikes, across Nigertuluk (Nig) and Tasiilaq (Tas), as well as single mode values from Imilik (circle). Poles to lava remnants at Kap Gustav Holm (KGH) and Imilik (n = 6) are also shown (black and gray squares, respectively). Calculated (solid) and visual (dashed) best fits indicate first-order structural variations across composite profile, pointing (extrapolated) toward offshore continent-ocean boundary. Hinge zone (gray shading) represents inland margin of both tectonically rotated dikes (i.e., dips < 90°) and dense part of coast-parallel dike swarm (i.e., dilation > 20%), which also coincides with maximum average dike thickness and location of inland plutons. Individual dips of earliest dikes are assumed to have been injected either subvertically (filled triangles) or along landward-dipping normal faults (open triangles). Extracted individual dips of lamprophyric (type 3) dikes (diamonds) along outer parts of Tasiilaq are shown as indication of amount of postbreakup flexure of margin, compared to amount of prerift to synbreakup rotation recorded by earliest dikes.

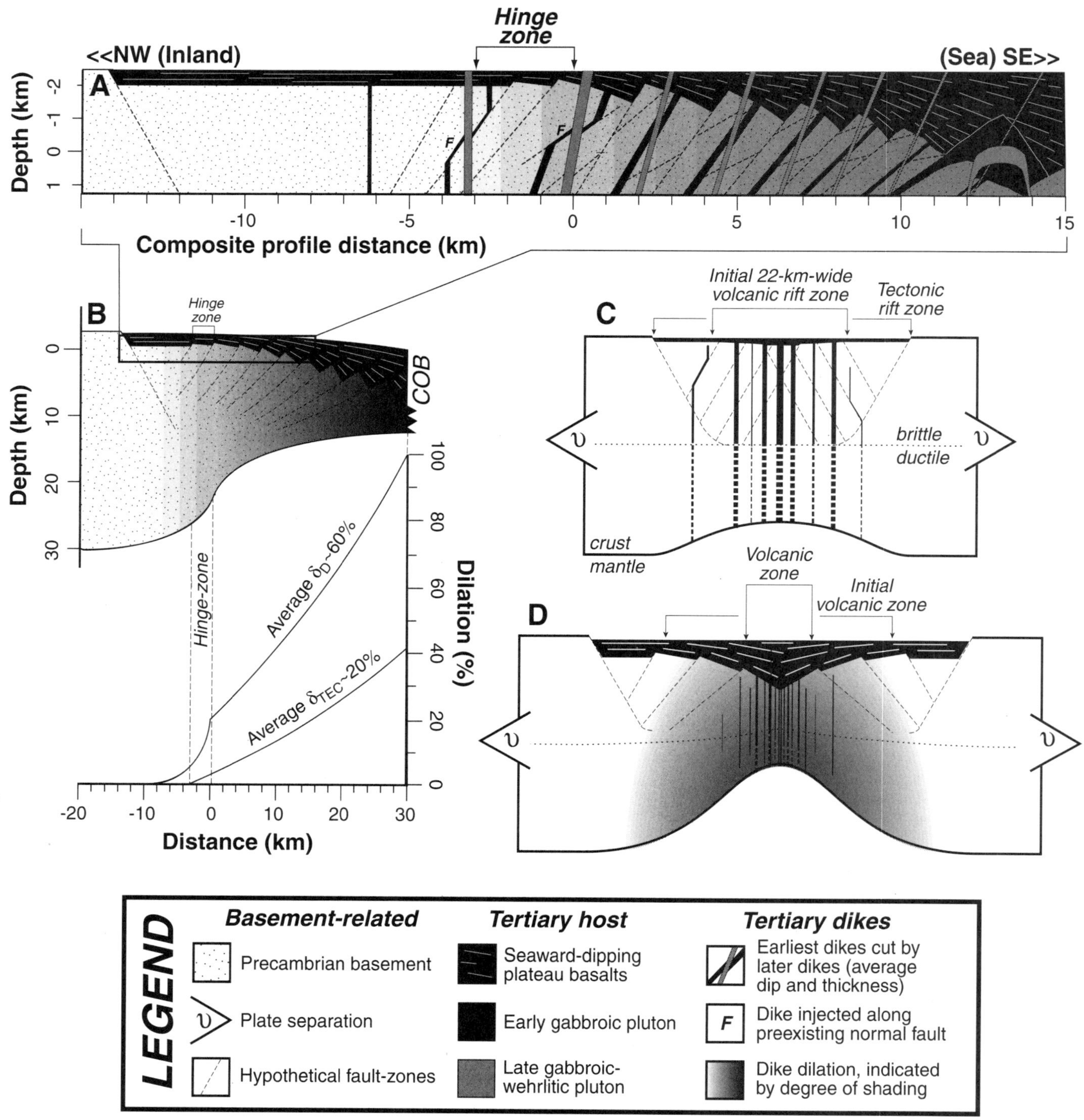

Figure 11. Cross-sectional margin geometry and simple palinspastic reconstruction for southern study segment. A: Mapped section, constructed from first-order variations in Figure 10 (B–D), showing: minimum dip of earliest dikes (black dikes and orthogonal dip of gneiss-lava interface), moving average dip of all dikes (gray dikes), relative average thickness (exaggerated ~100 times), and dilation by dikes (degree of shading). We assume overall model of domino-block faulting, with 1–3-km-wide crustal blocks, separated by major fault zones along coast-parallel fjords and glaciers. B: Crustal-scale section, where extrapolation of dilation (δ_D) to 100% dikes (degree of shading and upper curve in diagram, from Fig. 10D) positions continent-ocean boundary (COB) ~30 km seaward of hinge zone. Lower curve shows additional amount of tectonic dilation (δ_{TEC}) modeled from simple domino-faulted block rotation (Fig. 6, C and D) of 0°–30° across flexured part of margin. C, D: Assuming simple symmetrical rift model, we model cross-sectional geometry of (C) initial volcanic rift zone during emplacement of earliest dikes and (D) during breakup phase. Note that tectonic rift zone between hypothetical graben-bounding faults is wider than volcanic rift zone between hinge zones. Mid-crustal magma reservoirs are not included in this model; mainly for clarity but also because dikes propagated laterally from more focused igneous centers along margin.

from the inland hinge zone. This is likely to be a minimum distance, and offshore geophysical data indicate the southeast Greenland continent-ocean boundary to be located farther seaward (~40 km) from the hinge zone (Fig. 1; Larsen et al., 1998). Figure 11B shows how the crustal dilation by dikes varies across the margin, with a steep gradient up to ~20% dilation within the hinge zone followed by gradual seaward increase toward full igneous crust. A model calculation of the tectonic extension (also expressed as dilation) is shown to vary from 0% to 40% across the continent-ocean transition using the model assumptions discussed here and illustrated in Figure 6 (C and D).

Initial rift zone

The tectonic extension associated with breakup seems to have taken place fairly early during this process, because the average degree of rotation is much less than the maximum rotation of the earliest dikes (Fig. 10B). On average, the tectonic extension only provided what corresponds to ~20% dilation across the former rift zone, and significantly less than the average ~60% dilation by dikes (Fig. 11B). Overall, it therefore seems that tectonic extension during breakup was relatively limited and mainly took place early in the breakup process (ca. 61–55 Ma), and that crustal dilation by dikes took over with time and toward what became the center of the rift.

In order to reconstruct the width of the former rift zone prior to significant crustal extension we first remove the average dilation by dikes (~60%), reducing a 30-km-wide continent-ocean transition (from hinge zone to full igneous crust) to a 12 km width. If further reduced by the average ~20% of tectonic dilation, the original width of the part of the rift that is preserved along the East Greenland margin was only ~9.6 km. Assuming symmetry, this implies that the total width of the initial volcanic rift zone was no more than ~20 km. If a slightly wider East Greenland continent-ocean transition of 40 km (i.e., Larsen and Saunders, 1998) is assumed, then the total width of the initial volcanic rift zone would be ~26 km.

Figure 11 (C and D) shows how a development of the initial rift could have progressed. We have indicated the possible presence of an inland boundary fault similar to those present along many continental rifts (e.g., Rosendahl, 1987) and analogue experiments (e.g., McClay, 1996). However, whether such boundary faults form in highly magmatic rifts, such as East Greenland, is unknown. There is a very noticeable rift-parallel topographic lineament ~10 km inland of the hinge zone along the southern swarm segment (Fig. 9A), which might mark a former fault zone less resistant to glacial erosion. Thus, the initial width between such a conjugate set of boundary faults (i.e., again assuming symmetry) would be 40–46 km wide. However, even if the tectonic rift was this wide, our interpretation suggests that the most dominant extension by far took place within the narrower (20–26 km wide) volcanic rift zone, dominated by magmatic dilation by dikes.

Margin symmetry

Rift models involving large-scale low-angle detachment (e.g., Wernicke and Burchfield, 1982; Lister et al., 1986) inherently imply an asymmetry within the initial tectonic rift zone and subsequent margin formation (e.g., Driscoll and Karner, 1998). Karson and Brooks (1999) suggested that the East Greenland dike swarm and flexure is an example of such a crustal-scale, asymmetric flexure. However, the conjugate margin to East Greenland is submerged and covered by flood basalts (Faeroe Platform; Fig. 1), so no direct comparison of margin structure can be made. The East Greenland and the Faeroe-Hatton Bank margins are symmetric regarding the upper crust, in the sense that seaward-dipping reflectors are found along both margins (White and McKenzie, 1989). These probably formed from eruption within a narrow (<10 km wide), symmetrical volcanic rift zone similar to that of Iceland (Larsen and Jakobsdottìr, 1988; Larsen and Saunders, 1998).

The greater loading by the volcanic edifice building along the center of the rift led to the flexure of the upper crust in a fairly symmetric way along the flanks of the rift (Palmason, 1986). Thus, there can be little doubt that the narrow initial rift zone quickly developed symmetry, i.e., that two opposing and largely symmetrical flexures formed (Fig. 11D). Any asymmetric rift development would therefore have had to take place early in the rift process and be converted into an overall symmetric process of magmatic dilation and loading by extrusives within a maximum of ~5 m.y. from the initiation of magmatism (ca. 60–61 Ma).

Our observations only delimit the nature of the continent-ocean transition within the upper crust. Mid-crustal magma chambers may have formed and extension within the physically different lower crust could thus have been detached from the upper crust. This is largely undefined by our data, but seismic studies from the southeast Greenland margin at ~63°N suggest that delamination of the lower crust may have occurred (Larsen et al., 1998; Larsen and Saunders, 1998). Continental contamination of dikes and indications of relatively low pressure fractionation, in support of mid-crustal magma chambers that may have acted as decollement zones for upper crustal extension, were reported by Hanghøj (1998).

Igneous centers and emplacement of dikes

The group of earliest (prebreakup) dikes defines a regional pattern of intrusion shown in Figure 12A. These en echelon offset groups of swarms are similar to the structure envisaged by Myers (1980), by striking more eastward (~N40°E) than the later margin-parallel swarm and slightly oblique to the eventual line of breakup. If this trend was controlled by the regional stress field, the sinistral en echelon pattern would indicate that a dextral transtensional stress field was active during this initial stage of breakup. However, these initial dikes by and large are restricted to the area seaward of the hinge zone. This suggests that the lateral distribution of these earliest dikes was controlled by the subsurface distribution of source magma, i.e., centrally below the incipient rift zone. This would be consistent with the largely subvertical propagation of these dikes, suggested by the lack of clearly developed cutoff thicknesses in Fig. 7 (K–M).

The strikes within the later main group of dikes (Figs. 7A and 9A) define two sets of curved dike trajectories, which extend from the Kap Edward Holm and Imilik gabbroic complex, respectively (Fig. 12B). A third hypothetical center may be located offshore and farther south along the margin (marked with a question mark in Fig. 12B). The systematic along-margin variation we mapped along the northern swarm segment (Fig. 8, A–D) strongly indicates that, with time, these upper crustal magma chambers played an important role in shaping the pattern of these main swarms. Thus, lateral propagation of dikes (Fig. 12D) is suggested by the slight decrease in IEC thickness and parallel increase in cutoff thickness, observed along the segment and away from the Kap Edward Holm igneous center (Fig. 8C). This is in accordance with the lesser ability of thin dikes to propagate far through a relatively cool crust (e.g., Lister and Kerr, 1991; Walker et al., 1995; Fialko and Rubin, 1998).

Lateral propagation of dikes is not an uncommon mechanism of intrusion (e.g., Halls and Fahrig, 1987; Ernst et al., 1995). Furthermore, recent studies of the Troodos ophiolite (Staudigel et al., 1999; Abelson et al., 2001) concluded a temporal evolution in the magma flow direction of dikes similar to that we propose for the East Greenland coast-parallel dike swarm (Fig. 12, C and D), i.e., a temporal change from steep to upward-flowing magma in earlier dikes to later dikes with predominantly lateral magma flow directions. In East Greenland this is interpreted to reflect a growth and importance of major igneous centers, within overlapping groups of dike swarm segments (Fig. 12B), in much the same way as proposed for the en echelon array of dike swarms in Scotland (Vann, 1978; Fig. 12B, inset map of Scotland). This pattern also mimics the modeled geometry of propagating hydrofractures (Olson and Pollard, 1991). In order for these igneous centers to exert such a prominent control on the stress field, they must have been fairly deep rooted and probably shaped as semicylindrical diapirs (e.g., Chadwick and Dieterich, 1995).

Magmatic and tectonic segmentation

We can only observe the East Greenland leg of the two overlapping dike swarm segments we envisage to have formed during breakup. The former offshore position of the other overlapping northwestern European leg is indicated in Figure 12B, but this is likely to have been rifted away and transposed to the conjugate margin off the Faeroe Islands (FI in Fig. 12E). If such overlapping dike swarm segments were present during breakup, the total magmatic dilation that took place along the initial rift zone between these two igneous centers would be the sum of the dilation from both legs, effectively acting as two overlapping rifts. Consistent with this, the dilation by dikes along the East Greenland leg decreases with distance from its respective center (i.e., Fig. 8D). A structural accommodation zone (e.g., Faulds and Varga, 1998) must therefore have developed within the area between the two rift legs, in order to relay extension from one to the other. This could ultimately lead to formation of a small transform offset of the continent-ocean boundary and the initial spreading axis (Fig. 12E).

In summary, formation of dike swarms around igneous centers provides a mechanism for magmatically driven margin segmentation. In the area of overlap between magmatic segments, accommodation zones are likely to develop and contribute to tectonic margin segmentation. The total length of the overlapping magmatic segments in East Greenland is ~300 km (Fig. 12B). However, because of the large overlap, the tectonic segmentation these may generate with time is only half of that (i.e., ~150 km). Both the spacing and possible location of the accommodation zone we propose are roughly consistent with the second order of margin segmentation proposed by Karson and Brooks (1999). The relative displacement within such an accommodation zone (Fig. 12E) is consistent with the main deformation zone near Agtertia reported by Karson and Brooks (1999).

In addition to the complexities introduced by overlapping magmatic rifts, flexural response to differential loading of the crust may lead to significant variations along the margin, such as increasing seaward downwarping, or twisting, of the margin flexure toward the Kap Edward Holm igneous center (Fig. 12E). A similar structure may also have formed along the other, overlapping swarm segment extending north from Imilik, but is now transposed to the conjugate margin. Such twisting of the crustal flexure may be accommodated by relatively small, transverse offset zones (Fig. 12E). These could contribute to an even lower order of tectonic segmentation similar to the third-order segmentation by Karson and Brooks (1999).

The first-order segmentation proposed by Karson and Brooks (1999) is related to the spacing of triple junctions along the margin. These authors, following many others (e.g., Burke and Dewey, 1973; Brooks, 1973; Ernst and Buchan, 1997), consider the Kangerlussuaq Fjord to represent a failed rift arm and the mouth of the fjord to be the location of a triple junction. However, while the length of the fjord region and associated fjord-parallel dikes is close to the typical half-length of a magmatic segment, the total amount of dilation by dikes along the fjord (<200 m; Barfod, 1998) is almost negligible compared to the dike dilation along the coast and hardly supports the notion of a true rift-rift-rift triple junction.

CONCLUSIONS

The spatial distribution and orientation of dikes within the coast-parallel dike swarm and associated crustal flexure south of Kangerlussuaq show the presence of a narrow (~30–40 km wide) continent-ocean transition, ranging from undeformed high-grade Precambrian crust inland to an almost sheeted dike complex toward the sea. Tectonic extension seems limited to ≤30° of seaward crustal rotation (domino-faulted blocks?) and

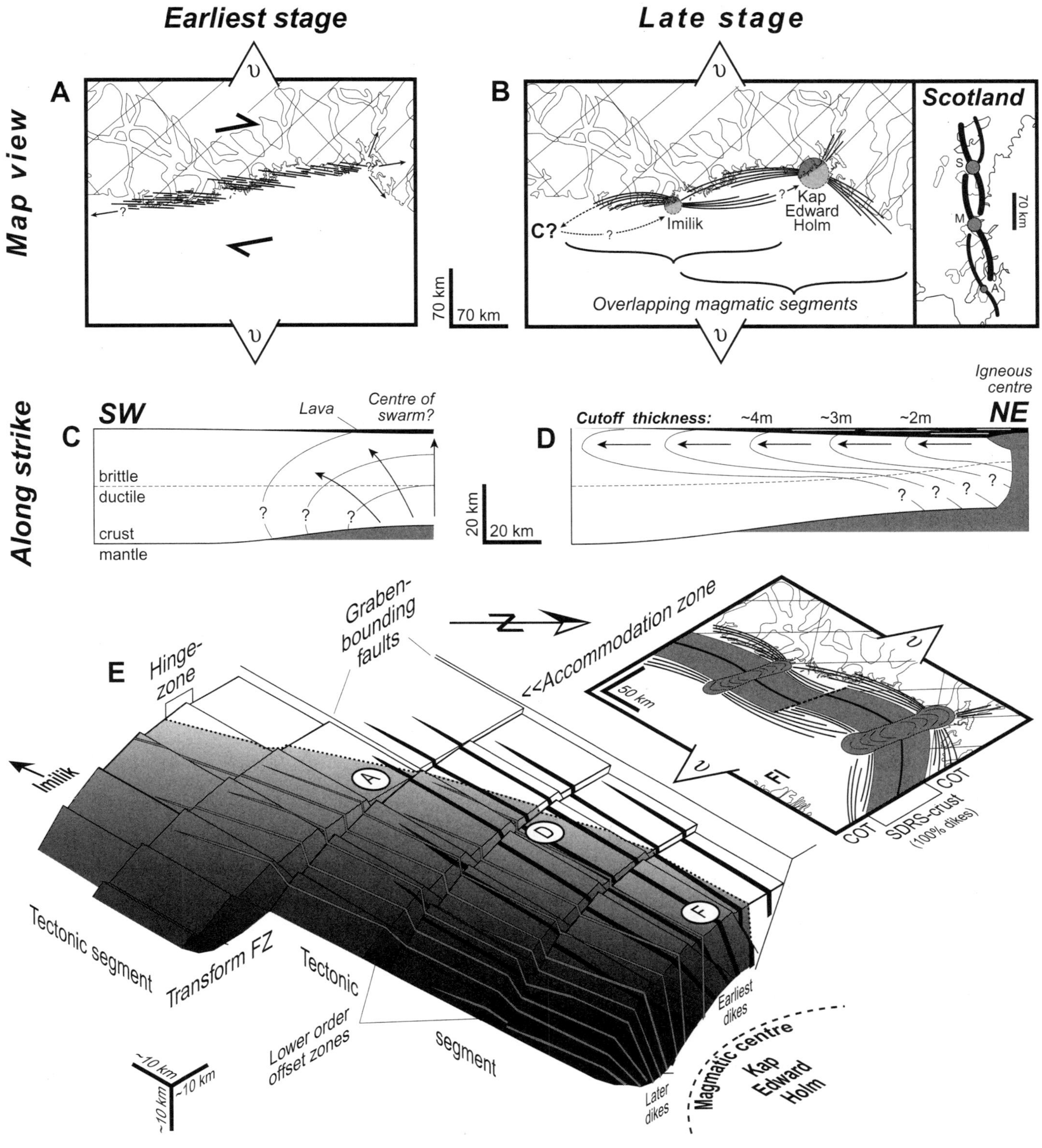

Figure 12. Schematic models of (A, B) sinistral en echelon pattern and (C, D) in-plane ellipsoidal shape of propagating earliest dikes and later dikes, as well as (E) composite three-dimensional view of one leg of East Greenland volcanic rifted margin, between 66°30′N and 68°30′N. Note that these stylistic dikes represent many hundreds of ≤25-m-thick dikes, within much denser swarm indicated by degree of shading in E. Earliest dikes only represent 10%–15% of mapped transects at Fladø (F), Deception Ø (D), and Agtertia (A), and proportion of later dikes furthermore increases toward continent-ocean boundary. Note that curvature and trend of dikes in B suggest that another intrusive center may be located farther south and offshore (marked by C?). Smaller-scaled model of Tertiary dike swarm in Scotland (S is Skye, M is Mull, A is Arran; adapted from Speight et al., 1982) is attached to B for comparison to our East Greenland model. In E, twisted geometry of margin is emphasized by initially horizontal peneplain that is stripped of its Tertiary sediment and lava cover. We assume that crustal rotation was accommodated through domino-faulted block rotation across margin, as well as small-scale transform fault zones along margin. COT, continent-ocean transition; FI, Faeroe Islands; FZ, fault zone; SDRS, seaward-dipping reflector sequence; υ, plate separation.

to mainly have taken place early in the process of breakup. An equal amount of nonextensional flexure of the crust took place during and after final breakup because of differential loading by extrusives, uplift of the continental crust, and subsidence of new igneous crust below the shelf.

Magmatic dilation as a mechanism of crustal extension increased in importance during breakup and rapidly became the dominant mode of extension leading to breakup. Removing first the effect of magmatic dilation and then the estimated early tectonic extension shows that the part of the initial rift zone preserved on the East Greenland margin was very narrow (~10–15 km). The predominantly magmatic mode of rifting is likely to initially have generated a full ~20–30-km-wide symmetrical volcanic rift, which later became focused into a spreading ridge of the type seen in present-day Iceland. This mechanism of rifting and volcanic margin development is closer to the models of focused and symmetrical rifting of Bohannon and Eittreim (1991), Nicolas et al. (1994), and Larsen and Saunders (1998) than the models of broad asymmetric margins controlled by crustal scale, low-angle detachment faults (e.g., Wernicke 1985; Lister et al., 1986; Voggenreiter et al., 1988; Driscoll and Karner, 1998).

The earliest dikes show the same amount of crustal rotation as the earliest volcanics and the semiconformably underlying prerift sediments. This documents that significant rifting, tectonic extension, and crustal rotation took place contemporaneous with initial magmatism. Within the rift zone exposed along the East Greenland coast this development from early dike emplacement, through virtually nonrifted and high-grade Precambrian crust, to final breakup, took place in only 5–6 m.y., starting ca. 61 Ma. An earlier start of rifting within the now submerged part of the margin cannot be ruled out, but is not indicated by the available data.

As the degree of magmatism increased, large igneous centers formed within the crust along the margin, and some of these seem to have acted as foci of magmatism and a source for lateral dike propagation. A consequence of this development is a magmatic segmentation of the margin, with highly overlapping dike swarms as long as 300 km. Overlapping rifts or embryonic spreading centers most likely were linked by accommodation zones, which transferred extension between overlapping rift legs. This potentially led to tectonic segmentation, with (1) small-scale transform offsets of the initial spreading centers, (2) internally twisted along-margin variation in crustal flexure, and (3) local asymmetry between conjugate margin second- and third-order segments, not unlike the geometry within the hanging-wall block of half-grabens. Eventually, full igneous crust characterized by seaward-dipping reflector crust (Icelandic-type crust) was formed in an overall symmetrical fashion.

ACKNOWLEDGMENTS

This work is part of Klausen's Ph.D. thesis, and he is particularly grateful for the field and academic supervision of Jeffrey A. Karson, C. Kent Brooks, Hans Christian Larsen, and Agust Gudmundsson. Klausen also thanks Asger Ken Pedersen for help with all aspects of the multi-model photogrammetry; Stefan Bernstein and Christian Tegner for supplementing structural field data, and many others for their team effort during the Danish Lithosphere Centre field expeditions to East Greenland (1994, 1995); We appreciate the constructive reviews by Richard Ernst and Paul Mohr, as well as editorial comments by Martin Menzies and Adam Kent. The work was carried out at the Danish Lithosphere Centre, funded by the Danish National Research Foundation.

REFERENCES CITED

Abelson, M., Baer, G., and Agnon, A., 2001, Evidence from gabbro of the Troodos ophiolite for lateral magma transport along a slow-spreading mid-ocean ridge: Nature, v. 409, p. 72–75.

Arnason, J.G., Bird, D.K., Bernstein, S., Rose, N.M., and Manning, C.E., 1997, Petrology and geochemistry of the Kruuse Fjord Gabbro Complex, East Greenland: Geological Magazine, v. 134, p. 67–89.

Baragar, W.R., Lambert, M.B., Baglow, N., and Gibson, I., 1987, Sheeted dykes of the Troodos Ophiolite, Cyprus, *in* Halls, H.C., and Fahrig, W.F., eds., Mafic dyke swarms: Geological Association of Canada Special Paper 34, p. 257–272.

Barfod, G.H., 1998, Magma generation and evolution in dyke geochemistry in the inner Kangerlussuaq Fjord area, East Greenland [M.S. thesis]: Copenhagen, University of Copenhagen, 121 p.

Bernstein, S., and Bird, D.K., 2000, Formation of wehrlites through dehydration of metabasalt xenoliths in layered gabbros of the Noe-Nygaard Intrusion, Southeast Greenland: Geological Magazine, v. 137, p. 109–128.

Bernstein, S., Kelemen, P.B., and Brooks, C.K., 1996, Evolution of the Kap Edward Holm Complex: A mafic intrusion at a rifted continental margin: Journal of Petrology, v. 37, no. 3, p. 497–519.

Bernstein, S., Kelemen, P.B., Tegner, T., Kurz, M.D., Blusztajn, J., and Brooks, C.K., 1998, Plume-related, post-rift basaltic magmatism along the East Greenland margin: Earth and Planetary Science Letters, v. 160, p. 845–862.

Bohannon, R.G., and Eittreim, S.L., 1991, Tectonic development of passive continental margins of the southern and central Red Sea with a comparison to Wilkes Land, Antarctica: Tectonophysics, v. 198, p. 129–154.

Breddam, K., Kurz, M.D., and Storey, M., 2000, Mapping out the conduit of the Iceland mantle plume with helium isotopes: Earth and Planetary Science Letters, v. 176, p. 45–55.

Bridgwater, D., Davies, F.B., Gill, R.C.O., Gorman, B.E., Myers, J.S., Pedersen, S., and Taylor, P., 1978, Precambrian and Tertiary geology between Kangerdlugssuaq and Angmagssalik, East Greenland: Grønlands Geologiske Undersøgelser Rapport, v. 83, p. 1–17.

Brooks, C.K., 1973. Rifting and dooming in southern East Greenland: Nature of Physical Sciences, v. 244, p. 23–25.

Brooks, C.K., and Nielsen, T.F.D., 1982, The Phanerozoic development of the Kangerdlugssuaq area, East Greenland: Meddelelser om Grønland: Geosciences, v. 9, p. 1–30.

Burke, K., and Dewey, J.F., 1973. Plume-generated triple junctions: Key indicators in applying plate tectonics to old rocks: Journal of Geology, v. 81, p. 406–433.

Chadwick, W.W., and Dieterich, J.H., 1995, Mechanical modelling of circumferential and radial dike intrusion on Galapagos volcanoes: Journal of Volcanology and Geothermal Research, v. 66, p. 37–52.

Coleman, R.G., and McGuire, A.V., 1988, Magma systems related to the Red Sea opening, *in* Bonatti, E., ed., Zabargad Island and the Red Sea rift: Tectonophysics, v. 150, p. 77–100.

Curewitz, D., and Karson, J.A., 1999. Ultracataclasis, sintering, and frictional melting in pseudotachylites from East Greenland: Journal of Structural Geology, v. 21, p. 1693–1713.

Deschmuk, S.S., and Sehgal, M.N., 1988, Mafic Dyke Swarms in Deccan Volcanic Province of Madhya Pradesh and Maharashtra, *in* Subbarao, K.V., ed., Deccan flood basalts: Memoir—Geological Society of India, v. 10, p. 323–340.

Dessai, A.G., and Bertrand, H., 1995, The "Panvel Flexure" along the Western Indian continental margin: An extensional fault structure related to Deccan magmatism: Tectonophysics, v. 241, p. 165–178.

Driscoll, N.W., and Karner, G.D., 1998, Lower crustal extension across the northern Carnarvon Basin, Australia: Evidence for an eastward dipping detachment: Journal Geophysical Research, v. 103, p. 4975–4991.

Dueholm, K.S., 1992, Geologic photogrammetry using standard small-frame cameras: Grønlands Geologiske Undersøgelser Rapport, v. 156, p. 7–17.

Du Toit, A.L., 1929, The volcanic belt of the Lebombo: A region of tension: Transactions of the Royal Society of South Africa, v. 18, p. 189–217.

Eales, H.V., Marsh, J.S., and Cox, K.G., 1984, The Karroo Igneous Province: An introduction: Geological Society of South Africa Special Publication 13, p. 1–26.

Ernst, R.E., and Buchan, K.L., 1997, Giant radiating dyke swarms: Their use in identifying pre-Mesozoic large igneous provinces and mantle plumes, *in* Mahoney, J.J., and Coffin, M.F., eds., Large igneous provinces: Continental, oceanic, and planetary flood volcanism: American Geophysical Union Geophysical Monograph 100, p. 297–333.

Ernst, R.E., Head, J.W., Parfitt, E., Grosfils, E., and Wilson, L., 1995, Giant radiating dyke swarms on Earth and Venus: Earth-Science Reviews, v. 39, p. 1–58.

Faulds, J.E., and Varga, R.J., 1998, The role of accommodation zones and transfer zones on the regional segmentation of extended terrains, *in* Faulds, J.E., and Stewart, J.H., eds., Accommodation zones and transfer zones: The regional segmentation of the Basin and Range province: Geological Society of America Special Paper 323, p. 1–45.

Fialko, Y.A., and Rubin, A.M., 1998, Thermodynamics of lateral dike propagation: Implications for crustal accretion at slow spreading mid-ocean ridges: Journal of Geophysical Research, v. 103, p. 2501–2514.

Gibson, I.L., 1966, Crustal flexures and flood basalts: Tectonophysics, v. 3, p. 447–456.

Gill, R.C.O., Nielsen, T.D.F., Brooks, C.K., and Ingram, G.A., 1988, Tertiary volcanism in the Kangerlugssuaq region, E Greenland: Trace-element geochemistry of the Lower Basalts and tholeiitic dyke swarms, *in* Morton, A.C. and Parson, L.M., eds., Early Tertiary volcanism and the opening of the NE Atlantic: Geological Society [London] Special Publication 39, p. 161–179.

Halls, H.C., and Fahrig, W.F., 1987, Dyke swarms and continental rifting: Some concluding remarks, *in* Halls, H.C. and Fahrig, W.F., eds., Mafic dyke swarms: Geological Association of Canada Special Publication 34, p. 483–492.

Hanghøj, K., 1998, Magmatic evolution during continental rifting in the North Atlantic: Constraints from geochemistry of the East Greenland coastal dyke swarm. [Ph.D. thesis]: Copenhagen, University of Copenhagen, 137 p.

Hansen, K., 1996. Thermotectonic evolution of a rifted continental margin: Fission track evidence from the Kangerlussuaq area, SE Greenland: Terra Nova, v. 8, p. 458–469.

Holbrook, W.S., Larsen, H.C., Korenaga, J., Dahl-Jensen, T., Reid, I.D., Kelemen, P.B., Hopper, J.R., Kent, G.M., Lizarralde, D., Bernstein, S., and Detrick, R.S., 2001, Mantle thermal structure and active upwelling during continental breakup in the North Atlantic: Earth and Planetary Science Letters, v. 190, p. 251–266.

Jolly, R.J.H., and Sanderson, D.J., 1995, Variations in the form and distribution of dykes in the Mull swarm, Scotland: Journal of Structural Geology, v. 17, p. 1543–1557.

Nicolas, A., Achauer, U., and Daignieres, M., 1994, Rift initiation by lithospheric rupture: Earth and Planetary Science Letters, v. 123, p. 281–298.

Karson, J.A., and Brooks, C.K., 1999, Structural and magmatic segmentation of the Tertiary East Greenland volcanic rifted margin, *in* Ryan, P., and MacNiovaill, C., eds., J.F. Dewey volume on continental tectonics: Geological Society [London] Special Publication 164, p. 313–338.

Karson, J.A., Brooks, C.K., Storey, M., and Pringle, M.S., 1998, Tertiary faulting and pseudotachylytes in the East Greenland volcanic rifted margin: Seismogenic faulting during magmatic construction: Geology, v. 26, p. 39–42.

Klausen, M.B., 1999, Structure of rift-related igneous systems and associated crustal flexures [Ph.D. thesis]: Copenhagen, University of Copenhagen, 283 p.

Korenaga, J., Holbrook, W.S., Kent, G.M., Kelemen, P.B., Detrick, R.S., Larsen, H.C., Hopper, J.R., and Dahl-Jensen, T., 2000, Crustal structure of the southeast Greenland margin from joint refraction and reflection seismic tomography: Journal of Geophysical Research, v. 105, p. 21591–21614.

Larsen, H.C., 1978, Offshore continuation of East Greenland dyke swarm and North Atlantic Ocean formation: Nature, v. 274, p. 220–223.

Larsen, H.C., 1980, Geological perspectives of the East Greenland continental margin: Geological Society of Denmark Bulletin, v. 29, p. 77–101.

Larsen, H.C., 1990, The East Greenland shelf, *in* Grantz, A., Johnson, L., and Sweeney, J.F., eds., The Arctic Ocean region: Boulder, Colorado, Geological Society of America, Geology of North America, v. L, p. 185–210.

Larsen, H.C., and Jakobsdottír, S., 1988, Distribution, crustal properties and significance of seaward-dipping sub-basement reflectors off E Greenland, *in* Morton, A.C. and Parson, L.M., eds., Early Tertiary volcanism and the opening of the NE Atlantic: Geological Society [London] Special Publication 39, p. 95–114.

Larsen, H.C., and Saunders, A.D., 1998, Tectonism and volcanism at the southeast Greenland rifted margin: A record of plume impact and later continental rupture, *in* Saunders, A.D., Larsen, H.C., and Wise, S.W.J., eds., Scientific Results, Ocean Drilling Program, Leg 152: College Station, Texas, Ocean Drilling Program, p. 503–533.

Larsen, H.C., Dahl-Jensen, T., and Hopper, J.R., 1998, Crustal structure along the leg 152 drilling transect, *in* Saunders, A.D., Larsen, H.C., and Wise, S.W.J., eds., Scientific Results, Ocean Drilling Program, Leg 152: College Station, Texas, Ocean Drilling Program, p. 463–475.

Larsen, L.M., Watt, W.S., and Watt, M., 1989, Geology and petrology of the Lower Tertiary basalts of the Scoresby Sund region, East Greenland: Grønlands Geologiske Undersøgelser Rapport, v. 157, p. 1–164.

Larsen, L.M., Waagstein, R., Pedersen, A.K., and Storey, M., 1999, Trans-Atlantic correlation of the Palaeogene volcanic successions in the Faeroe Islands and East Greenland: Journal of the Geological Society of London, v. 156, p. 1081–1095.

Larsen, M., Hamberg, L., Olaussen, S., Nørgaard-Pedersen, N., and Stemmerik, L., 1999, Basin evolution in southern East Greenland and outcrop analog for Cretaceous-Paleogene basins on the North Atlantic margins: American Association of Petroleum Geologists Bulletin, v. 83, p. 1236–1261.

Leyshon, P.R., and Lisle, R.J., 1996, Stereographic projection techniques in structural geology: Oxford, Butterworth-Heinemann, p. 104 p.

Lister, J.R., and Kerr, R.C., 1991, Fluid-mechanical models of crack propagation and their application to magma transport in dykes: Journal of Geophysical Research, v. 96, p. 10049–10077.

Lister, G.S., Etheridge, M.A., and Symonds, P.A., 1986, Detachment faulting and the evolution of passive continental margins: Geology, v. 14, p. 246–250.

McClay, K.R., 1996, Recent advances in analogue modelling: Uses in section interpretation and validation, *in* Buchanan, P.G. and Nieuwland, D.A., eds., Modern developments in structural geology: Geological Society [London] Special Publication 99, p. 201–225.

Mohr, P., 1991, Structure of Yemeni dike swarms, and emplacement of coeval granite plutons: Tectonophysics, v. 198, p. 203–221.

Murthy, N.G.K., 1987, Mafic dyke swarms in the Indian Shield, *in* Halls, H.C., and Fahrig, W.F., eds., Mafic dyke swarms: Geological Association of Canada Special Paper 34, p. 393–400.

Myers, J.S., 1980, Structure of the coastal dyke swarm and associated central intrusionic intrusions of East Greenland: Earth and Planetary Science Letters, v. 46, p. 407–418.

Myers, J.S., Dawes, P.R., and Nielsen, T.F.D., 1988, Geological map of Greenland Sheet 13, scale 1:500 000: Kangerdlugssuaq, East Greenland, Geological Survey of Greenland and Geodetic Institute of Denmark, Copenhagen, 1 sheet.

Myers, J.S., Gill, R.C.O., Rex, D.C., and Charnley, N.R., 1993, The Kap Gustav Holm Tertiary Plutonic Centre, East Greenland: Journal of the Geological Society of London, v. 150, p. 259–276.

Nicolas, A., Achauer, U., and Daignieres, M., 1994, Rift initiation by lithospheric rupture: Earth and Planetary Science Letters, v. 123, p. 281–298.

Nielsen, T.F.D., 1978, The Tertiary dike swarm of the Kangerdlugssuaq Area, East Greenland: Contributions to Mineralogy and Petrology, v. 67, p. 63–78.

Nielsen, T.F.D., and Brooks, C.K., 1981, The E Greenland rifted continental margin: An examination of the coastal flexure: Journal of the Geological Society of London, v. 138, p. 559–568.

Nielsen, T.F.D., Soper, N.J., Brooks, C.K., Faller, A.M., Higgins, A.C., and Matthews, D.W., 1981, The pre-basaltic sediments and the Lower Basalts at Kangerdlugssuaq, East Greenland: Their stratigraphy, lithology, palaeomagnetism and petrology: Meddelelser om Grønland: Geosciences, v. 6, p. 3–25.

Noble, R.H., Macintyre, R.M., and Brown, P.E., 1988, Age constraints on Atlantic evolution: Timing of magmatic activity along East Greenland continental margin, *in* Morton, A.C., and Parson, L.M., eds., Early Tertiary volcanism and the opening of the Northeast Atlantic: Geological Society [London] Special Publication 39, p. 201–214.

Olson, J.E., and Pollard, D.D., 1991, The initiation and growth of en échelon veins: Journal of Structural Geology, v. 13, p. 595–608.

Pallister, J.S., 1981, Structure of the sheeted dike complex of the Samail Ophiolite near Ibra, Oman: Journal of Geophysical Research, v. 86, p. 2661–2672.

Palmason, G., 1986, Model of crustal formation in Iceland, and application to submarine mid-ocean ridges, *in* Vogt, P.R., and Tucholke, B.E., eds., The western North Atlantic region: Boulder, Colorado, Geological Society of America, Geology of North America, v. M, p. 87–97.

Pedersen, A.K., Watt, M., Watt, W.S., and Larsen, L.M., 1997, Structure and stratigraphy of the Early Tertiary basalts of the Blosseville Kyst, East Greenland: Journal of the Geological Society of London, v. 154, p. 565–570.

Rex, D.C., Gledhill, A.R., Bridgwater, D., and Myers, J.S., 1979, A Rb-Sr whole rock age of 55±7 m.y. from the Nualik plutonic centre, East Greenland, *in* Report of activities: Grønlands Geologiske Undersøgelser Rapport, v. 100, p. 102–105.

Rose, N.M., and Bird, D.K., 1994, Hydrothermally altered dolerite dykes in East Greenland: Implications for Ca-metasomatism of basic protoliths: Contributions to Mineralogy and Petrology, v. 116, p. 420–432.

Rosencrantz, E., 1983, The structure of sheeted dikes and associated rocks in North Arm massif, Bay of Islands ophiolite complex, and the intrusive process at oceanic spreading centres: Canadian Journal of Earth Science, v. 20, p. 787–801.

Rosendahl, B.R., 1987, Architecture of continental rifts with special reference to East Africa: Annual Review of Earth and Planetary Sciences, v. 15, p. 445–503.

Rucklidge, J.C., Brooks, C.K., and Nielsen, T.F.D., 1980, Petrology of the coastal dykes at Tugtilik, southern East Greenland: Meddelelser om Grønland: Geosciences, v. 3, p. 1–17.

Saunders, A.D., Fitton, J.G., Kerr, A.C., Norry, M.J., and Kent, R.W., 1998, The North Atlantic igneous province, *in* Coffin, M.F., and Mahoney, J.J., eds., Large igneous provinces: Continental, oceanic, and planetary flood volcanism: American Geophysical Union Geophysical Monograph 100, p. 45–93.

Sheth, H.C., 1998, A reappraisal of the coastal Panvel flexure, Deccan Traps, as a listric-fault-controlled reverse drag structure: Tectonophysics, v. 294, p. 143–149.

Sial, A.N., Oliverira, E.P., and Chudhuri, A., 1987, Mafic dyke swarms in Brazil, *in* Halls, H.C., and Fahrig, W.F., eds., Mafic dyke swarms: Geological Association of Canada Special Paper 34, p. 257–272.

Speight, J.M., Skelhorn, R.R., Sloan, T., and Knaap, R.J., 1982, The dyke swarms of Scotland, *in* Sutherland, D.S., ed., Igneous rocks of the British Isles: London, John Wiley & Sons, p. 449–459.

Staudigel, H., Tauxe, L., Gee, J.S., Bogaard, P., Haspels, J., Kale, G., Leenders, A., Meijer, P., Swaak, B., Tuin, M., Van Soest, M.C., Verdurmen, E.A.Th., and Zevenhuizen, A., 1999, Geochemistry and intrusive directions in sheeted dikes in Troodos ophiolite: Implications for mid-ocean ridge spreading centres: Geochemistry, Geophysics, Geosystems. v. 1 [online: 1999GC000001].

Tegner, C., and Duncan, R.A., 1999, ^{40}Ar-^{39}Ar chronology for the volcanic history of the Southeast Greenland rifted margin, *in* Saunders, A.D., Larsen, H.C., et al., eds., Scientific Results, Ocean Drilling Program, Leg 163: College Station, Texas, Ocean Drilling Program, p. 53–62.

Tegner, C., Wilson, J.R., and Brooks, C.K., 1993, Intraplutonic quench zones in the Kap Edward Holm layered gabbro complex, East Greenland: Journal of Petrology, v. 34, p. 681–712.

Tegner, C., Duncan, R.A., Bernstein, S., Brooks, C.K., Bird, D.K., and Storey, M., 1998, ^{40}Ar-^{39}Ar geochronology of Tertiary mafic intrusions along the East Greenland rifted margin: Relation to flood basalts and the Iceland hotspot track: Earth and Planetary Science Letters, v. 156, p. 75–88.

Vann, I.R., 1978, The siting of Tertiary volcanicity, *in* Bowes, D.R., and Leake, B.E., eds., Crustal evolution in northwestern Britain and adjacent regions: Journal of the Geological of London, Special Issue, v. 10, p. 393–414.

Voggenreiter, W., Hötzl, H., and Mechie, J., 1988, Low-angle detachment origin for the Red Sea Rift System?: Tectonophysics, v. 150, p. 51–75.

Wager, L.R., and Deer, W.A., 1938, A dyke swarm and crustal flexure in East Greenland: Geological Magazine, v. 75, p. 39–46.

Walker, G.P.L., Eyre, P.R., Spengler, S.R., Knight, M.D., and Kennedy, K., 1995, Congruent dyke-widths in large basaltic volcanoes, *in* Baer, G., and Heimann A., eds., Physics and chemistry of dykes: Rotterdam, A.A. Balkema, p. 35–40.

Watt, M., 1975, Photo-reconnaissance of the Blosseville Kyst between Steward Ø and Søkongens Bugt, central East Greenland: Grønlands Geologiske Undersøgelser Rapport, v. 75, p. 91–95.

Wernicke, B., 1985, Uniform-sense normal simple shear of the continental lithosphere: Canadian Journal of Earth Science, v. 22, p. 108–125.

Wernicke, B., and Burchfield, B.C., 1982, Modes of extensional tectonics: Journal of Structural Geology, v. 4, p. 101–115.

White, R.S., and McKenzie, D.P., 1989, Magmatism at rift zones: The generation of volcanic continental margins and flood basalts: Journal of Geophysical Research, v. 94, p. 7685–7729.

Wolfe, C.J., Bjarnason, I.T., VanDecar, J.C., and Solomon, S.C., 1997, Seismic structure of the Iceland mantle plume: Nature, v. 385, p. 245–247.

MANUSCRIPT ACCEPTED BY THE SOCIETY OCTOBER 15, 2001

Printed in the U.S.A.

Geological Society of America
Special Paper 362
2002

Crustal architecture of South Atlantic volcanic margins

W.U. Mohriak
Petroleo Brasileiro S.A., Exploração e Produção, Gerência de Exploração, Gerência de Sistemas Petroliferos, 20035-900 Rio de Janeiro, Brazil, and Universidade do Estado do Rio de Janeiro, Rio de Janeiro, Brazil
B.R. Rosendahl
University of Miami, Rosenstiel School of Marine and Atmospheric Science, Miami, Florida 33149, USA
J.P. Turner
University of Birmingham, School of Earth Sciences, Birmingham B15 2TT, UK
S.C. Valente
Universidade Federal Rural do Rio de Janeiro, BR 465 Km7 Seropédica, 23890-000 Rio de Janeiro, Brazil

ABSTRACT

Integration of seismic, potential field, and borehole data from the conjugate margins of eastern Brazil and West Africa indicates that the rift architecture varied along strike, and that volcanic episodes (or lack thereof) may substantially affect petroleum exploration of the deep-water provinces. In this chapter we discuss various pairs of conjugate sedimentary basins, from Pelotas and Namibia in the south to the Sergipe-Alagoas and Rio Muni counterparts in the north. The following aspects are emphasized: (1) rift depocenters are controlled by border faults subparallel to the margin and by transverse faults that may continue as transform fractures in the oceanic crust; (2) in many basins along the southernmost segment of the South Atlantic, Early Cretaceous volcanics underlie continental lacustrine synrift sediments of Neocomian age; (3) in the northern segment of the salt basin, prerift sediments with no volcanic material underlie the synrift sediments; (4) in some segments of the margin, the transition from outer rift blocks to oceanic crust is characterized by wedges of seaward-dipping reflectors with a possible origin associated with the initial phases of oceanic crust emplacement; (5) locally, the outermost rift blocks seem to be highly eroded by postrift uplift caused by shearing or by magmatic underplating; (6) subsequent to the rift sequence, a quiescent period marked by a sag basin above a regional unconformity predated the deposition of Aptian evaporites; (7) the South Atlantic salt basin along both margins was controlled by tectonic and volcanic elements, and locally, salt was deposited directly on the volcanic substratum; (8) volcanic ridges formed before and after salt deposition, and separated portions of the rift and salt basins before the final breakup installed a divergent regime with pure oceanic crust; (9) basement-involved extensional faults, volcanic activity, and enhanced continental margin uplift and denudation are indicative of reactivation of rift-phase faults after salt deposition in some segments of the margin; and (10) tectono-magmatic episodes climaxed in the later Cretaceous–early Tertiary, forming large volcanic complexes along the conjugate margins.

Three main episodes of magmatic activity are observed in the South Atlantic salt basins: the Late Jurassic–Early Cretaceous event is related to the Paraná-Etendeka flood basalts and the volcanic rocks that occur in the offshore basins; it is followed by thick wedges of volcanic rocks interpreted as seaward-dipping reflectors; and the Late Cretaceous–early Tertiary event is related to hotspot and leaking fracture zones.

Mohriak, W.U., Rosendahl, B.R., Turner, J.P., and Valente, S.C., 2002, Crustal architecture of South Atlantic volcanic margins, *in* Menzies, M.A., Klemperer, S.L., Ebinger, C.J., and Baker, J., eds., Volcanic Rifted Margins: Boulder, Colorado, Geological Society of America Special Paper 362, p. 159–202.

Fitting together of deep-imaging, multichannel seismic profiles at approximate conjugate margin positions south of the Maceió-Ascension-Kribi Fracture Zone system shows a high degree of asymmetry for the crustal architecture and the rift sequences, suggestive of a simple shear mechanism of lithospheric stretching. Although there is no evidence for synrift volcanism at this segment of the margin, the development of seaward-dipping reflector packages observed in the Brazilian side suggests that magmatic activity in this region may also be asymmetric.

INTRODUCTION

Understanding the crustal architecture of passive continental margins, particularly in the ultradeep-water exploration provinces (bathymetries exceeding 2000 m), is a major challenge for regional basin analysis and petroleum exploration in frontier regions, especially in the petroleum-rich South Atlantic (Wannesson et al., 1991; Mohriak et al., 1995a; Rosendahl, 1997). The roots of these passive margins were initially located at intraplate continental positions, which subsequently rifted apart and ultimately moved the margins to opposite (conjugate) edges of newly created oceanic basins. The general progression is almost incontrovertible, but many key elements remain controversial, in particular the timing, types, and provenance of magmatic activity loosely associated with rifting.

Some of the problems faced by explorationists working on divergent margins include the position and nature of the continent-ocean boundary; the delimitation of the outermost rift blocks in the deep-water provinces; the identification of presalt lacustrine source rocks; and the differentiation between clastic rift fill, flood basalts, seaward-dipping volcanic wedges, mid-ocean ridge basalt (MORB) volcanics, and postbreakup intrusive and extrusive rocks.

This chapter analyzes certain aspects of these and related problems along the rifted margins in the South Atlantic, with a concentration on the volcanic-rich basins and the role of igneous rocks in continental breakup. We review the geochronology of subaerial flood basalts that occur on continental crust of the Paraná Basin and in the Etendeka province, and correlate these rocks to the volcanic sequences below the synrift siliciclastic rocks of the continental margin. On the basis of seismic interpretation and tectonic models, we differentiate these basalt layers from the seaward-dipping wedges that occur near the transition from the continental to oceanic crust, and also from the MORB-type basalts of typical oceanic crust. We also discuss the petrogenesis of the igneous rocks along the offshore segment of the Paraná Basin, including a review of the chemistry of selected samples.

We analyze salt tectonics in these ultradeep-water regions and the role that breakup volcanics have played in delimiting the South Atlantic salt basin. We emphasize that presalt source rocks and postsalt reservoirs are usually linked by migration pathways related to salt tectonics, which may be triggered by basement-involved faults. These basement-involved faults are usually associated with tectonic reactivation phases and volcanic episodes that affect the rift-, transitional-, and drift-phase postsalt sequences.

Any interpretation of the rift architecture in the South Atlantic is hampered by the lack of well-defined cross-sectional structural models of the conjugate margins. The problem arises from a paucity of reliable databases, compounded by uncertainties in the conjugate fits. The problem is exacerbated by the fragmentary nature of the work efforts, both geographically and topically. Different groups working on opposing continental margins tend to develop independent models based upon independent databases. Although tectonic analysis of the South Atlantic basins is significantly improved by integrating potential field data (gravity and magnetics) with regional seismic interpretations, our discussion of the South Atlantic is necessarily broad and generalized. The tectonic models developed, although nonunique, are based on a very large potential field and seismic datasets that were cooperatively analyzed to assess alternative interpretations.

We also compare regional transects of conjugate margins using a unique set of deep-imaging multichannel seismic (DIMCS) profiles and ancillary industry seismic data from subequatorial offshore West Africa and northeastern Brazil. Here the deep cross-sectional structural controls are much better than in the meridional South Atlantic; however, the role of volcanism in rifting is more controversial. The equatorial conjugate comparisons emphasize the extreme variability of structural styles along the Atlantic margins, in both latitudinal and paired margin senses.

Highlights of the geodynamic evolution of the South Atlantic

Several recent works have analyzed the geodynamic evolution of the South Atlantic continental margins, based on the integration of geophysical and geological data (e.g., Popoff, 1988; Chang et al., 1992; Karner et al., 1997; Mohriak et al., 1998a, 1998b, 2000a; Cainelli and Mohriak, 1998, 1999; Davison, 1999; Coward et al., 1999; Matos, 2000; Jackson et al., 2000).

The South Atlantic continental margins (Fig. 1) are marked by sedimentary basins with grossly similar geodynamic mechanisms of formation and repetitive patterns of stratigraphic evolution within a regional framework (e.g., Kusznir et al., 1977; Guardado et al., 1989; Teisserenc and Villermin, 1989; Mohriak et al., 1990; Chang et al., 1992; Meyers et al., 1996a, 1996b;

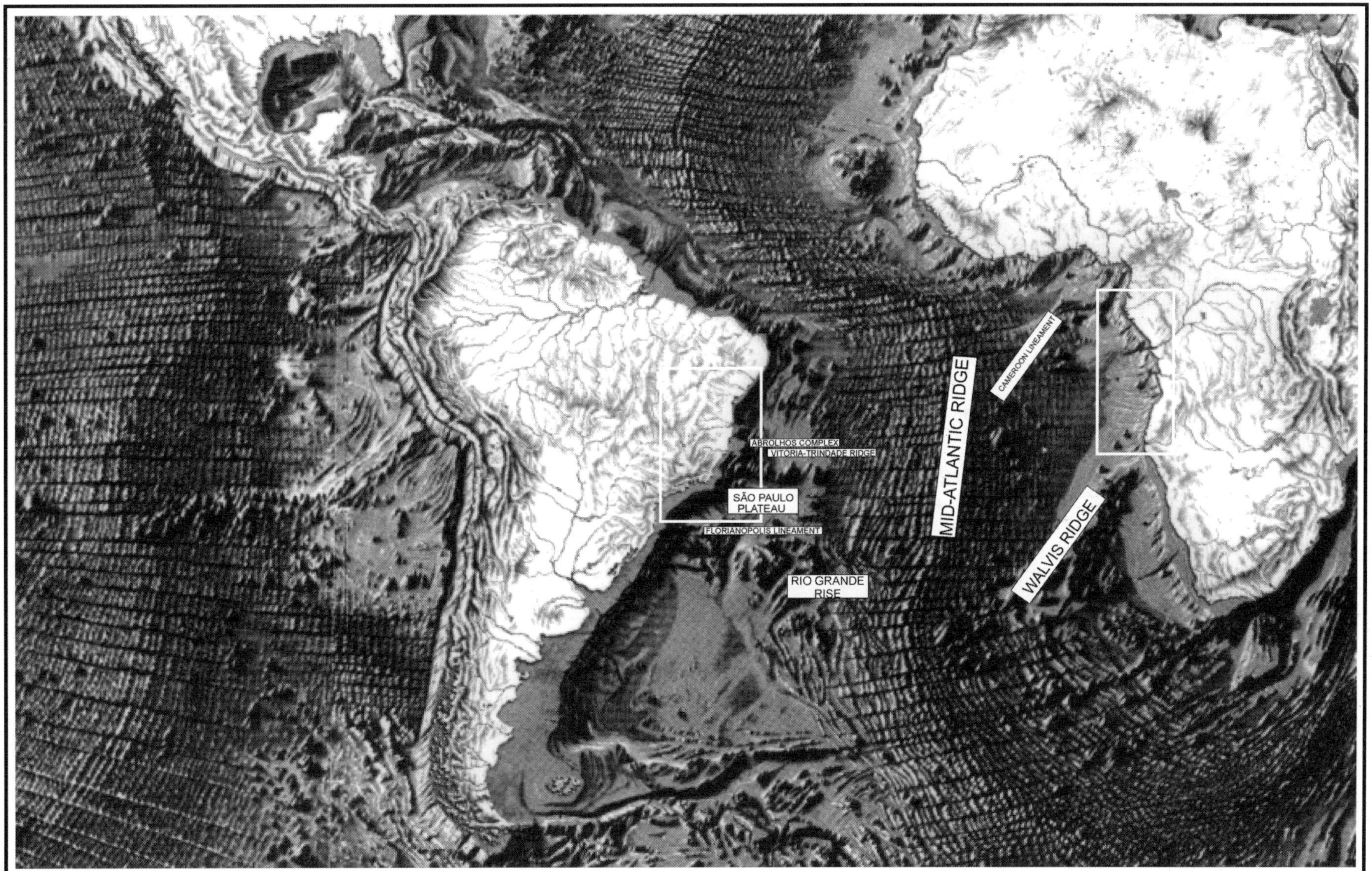

Figure 1. Morphological map of South Atlantic continental margins and sedimentary basins, showing mid-Atlantic spreading ridge, Rio Grande Rise, Walvis Ridge, São Paulo Plateau, Abrolhos volcanic complex, and Cameroon volcanic line. The Aptian salt basin along continental margins is mainly encompassed within area marked by rectangles.

Cainelli and Mohriak, 1999). Nonetheless, the conjugate margins are characterized by significant peculiarities in their geological evolution (Bradley and Fernandez, 1991; Henry et al., 1996).

Gravity anomalies derived from the Geosat database (Sandwell and Smith, 1997; Smith and Sandwell, 1997) show that the South Atlantic basins are characterized by a negative Bouguer anomaly near depocenters in the platform areas, and by increasing positive anomalies from the shelf toward the deep-water region (Karner et al., 1997; Karner and Driscoll, 1999; Cainelli and Mohriak, 1998). The continent-ocean boundary also has been interpreted by modeling the gravity (free air or Bouguer) anomalies and comparing with the regional seismic profiles (e.g., Wannesson et al., 1991; Meyers, 1995; Groschel-Becker, 1996; Meyers and Rosendahl, 1996a; Mohriak et al., 1998a, 2000a).

Major gravity and magnetic anomalies also indicate that different basement domains are crossed along the margin, and some fracture zones have been interpreted to extend from the oceanic crust domain toward the platform, controlling different compartments of the rift basins (Wannesson et al., 1991; Meyers et al., 1996a, 1996b; Rosendahl and Groschel-Becker, 2000).

The South Atlantic is segmented into different tectonic provinces, and various geologic and tectonic criteria have been used to define their boundaries. One of the simplest approaches is to divide the South Atlantic into three main segments on the basis of fracture zones and magmatic hotspot traces (Fig. 2). The subequatorial South Atlantic, north of the St. Helen hotspot, includes the Cameroon volcanic line, and the region south of the Fernando de Noronha Fracture Zone on the Brazilian side. The South Atlantic salt basin along the eastern Brazil and western Africa margins is limited to the south by the Tristan hotspot (which is related to the Walvis Ridge–Rio Grande Rise and the Florianópolis Fracture Zone (Cainelli and Mohriak, 1998). The southernmost South Atlantic, south of the Rio Grande Rise–Walvis Ridge, includes the Brazilian Pelotas Basin and the offshore Argentina basins, corresponding to the South Africa–Namibe segment along the African margin.

The breakup of Gondwana and the genesis of the South Atlantic are generally associated with three main phases of opening. There is evidence that the equatorial margin of South America and the southernmost North Atlantic were formed by rifting from Jurassic to Cretaceous time, and the oldest rocks associated with magmatic activity also range from Jurassic to Cretaceous (Coward et al., 1999; Jackson et al., 2000). Jurassic to Cretaceous rifting and magmatism are further defined by boreholes drilled on the Argentinian basins, indicating precursory events along the southern segments of the South Atlantic, oceanic crust propagators advancing from south to north like an opening zipper (Hinz et al., 1999; Jackson et al., 2000). This resulted in a later onset of rift sedimentation and seafloor spreading along the eastern Brazilian margin segment (Nürnberg and Müller, 1991; Cainelli and Mohriak, 1998). In the Santos and Campos basins, the volcanics underlie the synrift layers and there are rare examples of synrift sediments with volcaniclastic intercalations; however, in the Espírito Santo basin a number of exploratory boreholes indicate that the synrift sequence is interbedded with volcanic extrusive rocks postdating the Serra Geral volcanics and predating the Aptian salt sequence.

The onset of rifting in West Africa was 220–200 Ma in the Cape Basin (Light et al., 1992); 160–144 Ma in the Orange Basin (Erlank et al., 1984; Guiraud and Maurin, 1992); 144–126 Ma in the Benguela and Kwanza Basins (Brice et al., 1982; Teisserenc and Villemin, 1989; Nürnberg and Müller; 1991; Guiraud and Maurin, 1992); and 125–119 Ma in North Gabon and Rio Muni Basins (Teisserenc and Villemin, 1989; Nürnberg and Müller, 1991). Along the conjugate Brazilian margin, dates for the earliest continental rift sequences (145–130 Ma) have been obtained in the onshore Recôncavo, Espírito Santo, and Potiguar Basins, and the earliest synrift fill in the offshore basins has been dated as 130–120 Ma (Cainelli and Mohriak, 1998), thus suggesting an early phase of synrift sedimentation in the onshore branches of these basins.

The end of rifting can be estimated by identifying the youngest sedimentary clearly affected by basement-involved faulting, by dating the breakup unconformity (or unconformities, considering them as diachronous events along the margin), by late syn-rift or sag or sag basins below the Aptian salt (Henry et al., 1996; Karner et al., 1998), and by the identification of magnetic anomalies in the adjoining oceanic crust (Marton et al., 2000). Estimates for the end of rifting and the onset of seafloor spreading range from 137 to 130 Ma in the Orange River–Lüderitz–Walvis basin (Austin and Uchupi, 1982; Nürnberg and Müller, 1991; Gladczenko et al., 1997; Peate, 1997). The end of rifting is diachronous and slightly younger (127–117 Ma) in the region between the Tristan and the St. Helen hotspot, which corresponds to the Benguela-Kwanza–Gabon–Rio Muni salt basins in West Africa (Brice et al., 1982; Teisserenc and Villemin, 1989; Nürnberg and Müller, 1991; Guiraud and Maurin, 1992; Gladczenko et al., 1997; Karner and Driscoll, 1998; Coward et al., 1999). Synrift faults seem to be active until the Albian in the northeastern Brazilian margin (Cainelli and Mohriak, 1998).

Recent interpretation of seismic and potential field data suggests that fossil volcanic ridges (probably associated with early oceanic spreading centers) indicate a mechanism of formation of volcanic rifted margins by progressive oceanic spreading center indentation and propagation along rifted continental crust. This mechanism, which shows some similarities to the present structures observed along the Afar region, is associated with propagators advancing from the southern (Pelotas) toward the northern (Campos) provinces, with right-lateral steps. The marked indentation of the meridional part of the Santos basin is associated with the Abimael propagator (Mohriak, 2001).

The Aptian salt basin was developed on both conjugate margins, from slightly south of the St. Helen hotspot (Sergipe-Alagoas and the conjugate Rio Muni Basins) to the Tristan hotspot (Santos and the conjugate Namibe Basins). Carbonate and siliciclastic sediments predominated from Albian to Cenomanian time, and subsequently, when the continental margins

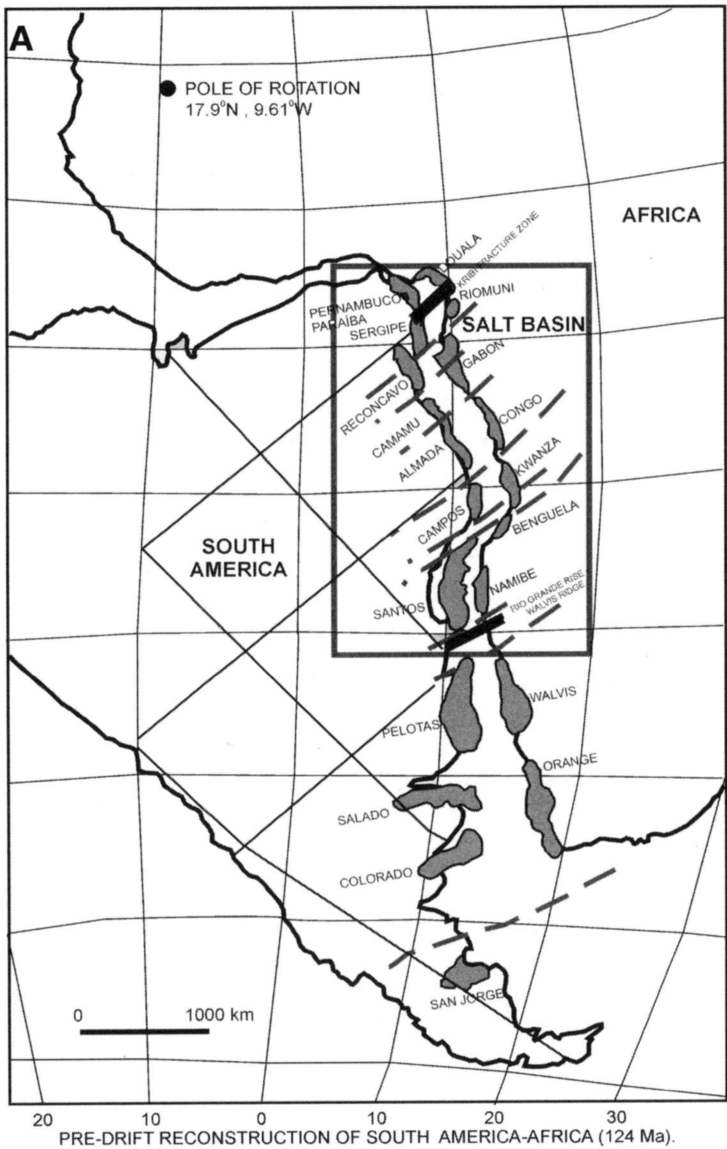

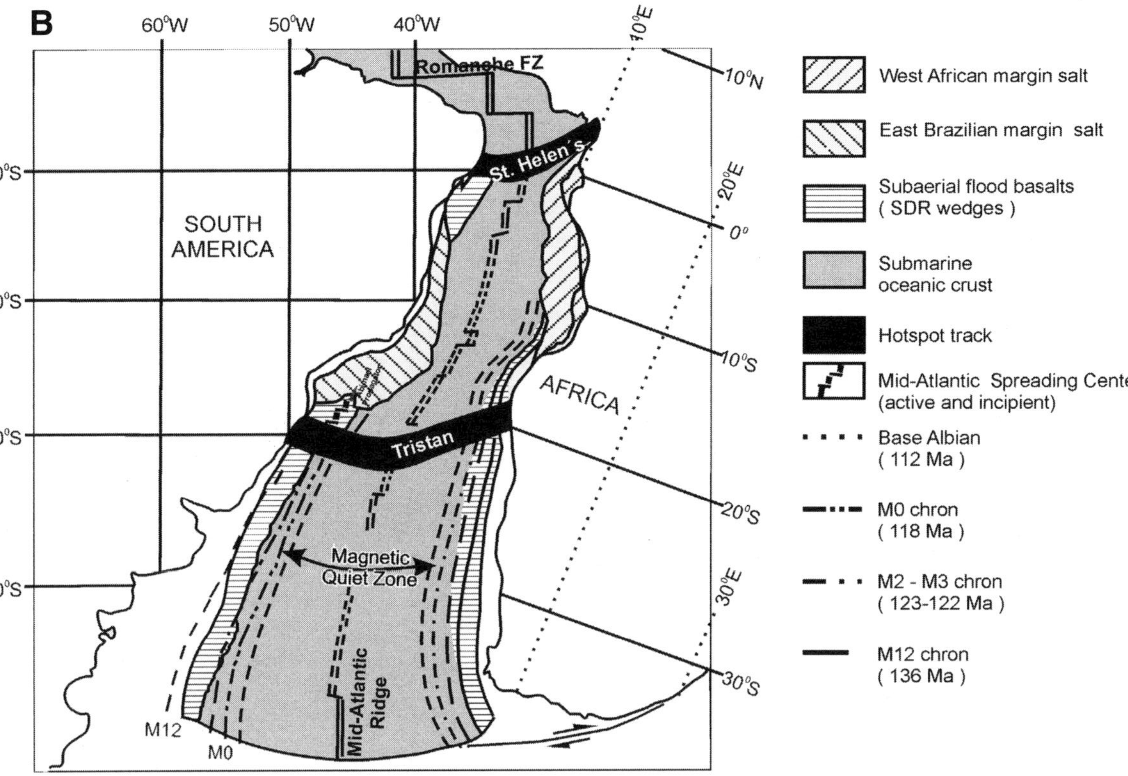

Figure 2. A: South Atlantic prebreakup reconstruction at Neocomian time. The inset shows the location of the Aptian salt basin. B: South Atlantic reconstruction at Aptian time, showing asymmetric distribution of salt basins along eastern Brazil and West Africa margins. Seaward-dipping reflectors (SDR) have been identified mainly in the region south of the Tristan hotspot. The Abimael failed spreading center corresponds to a propagator (Mohriak, 2001) that advanced from the Pelotas Basin in the south toward the meridional part of the Santos Basin. It was probably aborted by Aptian-Albian time, and the active spreading axis shifted to the Mid-Atlantic Ridge. Modified from Jackson et al. (2000).

were subjected to regional thermal subsidence, deepening of the environment of deposition resulted in predominance of siliciclastic deposition. The thermal subsidence pattern was modified by a few tectono-magmatic events that caused continental uplift and erosion, increased sedimentation rates, and enhanced salt mobilization. Radiometric dating of igneous rocks (e.g., Misuzaki et al., 1998) and apatite fission-track dating (Gallagher and Brown, 1999) indicate that the mid-Cretaceous event (ca. 100–80 Ma) and the early Tertiary event (ca. 50 Ma) are related to periods of magmatic activity and enhanced denudation of the rift flanks.

The sequential evolution of the South Atlantic is highly controversial and probably variable both spatially and temporally. A possible generalized model is shown in Figure 3, which is a pure shear variant with five main phases marking different patterns of tectonics and sedimentation (Cainelli and Mohriak, 1998).

The first phase (I) is marked by the beginning of extensional processes that led to the separation of the South American and African continents. The onset of the next phase (II), which is characterized by increasing lithospheric stretching and extrusion of continental flood basalts in the southern basins, coincides with large faults affecting the continental crust, and formation of half-grabens filled with continental lacustrine sediments in the northern basins. By the end of the rifting episode, there is an increase in the lithospheric extension in the offshore region, marked by large faults that rotate the rift blocks and the sedimentary layers previously deposited. This phase is associated with continental and oceanic volcanism, including the formation of the seaward-dipping reflectors, reactivation of large faults, and erosion of rift blocks by a regional unconformity that levels the topography. Above the breakup unconformity and below the evaporite transitional sequence, some sedimentary basins register substantial thicknesses of Aptian siliciclastic and carbonate rocks (III). Subsequent to salt deposition in the Aptian, sedimentation becomes predominantly carbonate (IV). An increase in the bathymetry resulted in the deepening of the environment of deposition by the end of the Albian, with demise of the shallow-water carbonates (V).

Although certain tectonic-depositional aspects of this simplified schematic model may be applicable throughout the South Atlantic, it is unclear where and to what extent a pure-shear–type rifting mechanism (e.g., McKenzie, 1978) should be applied. We show that the subequatorial margins of Brazil and West Africa almost certainly did not rift apart in this fashion, and an evolution model assuming simple-shear–type rifting mechanisms (Wernicke, 1985) may be more applicable.

Magmatic activity in the South Atlantic

The South Atlantic is characterized by some of the largest igneous provinces worldwide (Fig. 4). Following Coffin and Eldholm (1992, 1994) and Jackson et al. (2000), large igneous provinces are characterized by massive and rapid production of volcanic rocks. These onshore flood basalts may occur along the continental margins, flooring, and infilling synrift troughs, and may extend along the adjacent ocean basins. Ocean-basin flood basalts may constitute proto-oceanic volcanic rocks imaged as thick wedges of seaward-dipping reflectors and also form thicker than normal (exceeding 20 km) oceanic crust, oceanic plateaus, and aseismic ridges, and isolated or linear chains of seamounts derived from hotspots or from leaking fracture zones. Large igneous provinces also include the intrusive equivalent (mainly gabbros) of the extrusive sequences, and the middle to lower crustal, underplated igneous bodies having high density and high P-wave velocities (Furlong and Fountain, 1986; Coffin and Eldholm, 1994; Eldholm et al., 1995).

Although large igneous province basalts may share several compositional characteristics with MORB-type basalts, they may be distinguished from these mafic rocks by major and trace elements geochemistry. Flood basalts and seaward-dipping volcanics may be characterized in seismic data by some peculiarities that allow distinction from the basalts formed by normal seafloor spreading.

The large igneous provinces probably formed when the lithosphere was extended in the presence of mantle plumes or thermal anomalies, resulting in widespread magmatism along continental margins, particularly at rift triple junctions (Burke and Dewey, 1973; Morgan, 1983; Keen, 1985; Coffin and Eldholm, 1994; White and McKenzie, 1995). The areal distribution of basalts capping Paleozoic rocks in the Paraná-Etendeka onshore province is asymmetric; the Paraná sector encompasses the largest portion, extending from the Brazilian continental landmass toward the offshore region. Near the city of Torres, the basalts crop out near the continental margin, and are downfaulted basinward, flooring the rifts of the Pelotas and Santos Basins. The Etendeka volcanic rocks occur onland in scattered patches from Walvis Bay to Luanda, although they may extend offshore for more than 1500 km (Peate, 1997). The asymmetry may be explained by the off-center Tristan plume (O'Connor and Duncan, 1990) or by asymmetric lithospheric simple shear processes during rifting (Peate, 1990). Alternative hypotheses include (1) thinner crust and higher magmatic activity below the Paraná Basin due to previous lithospheric thinning (Thompson and Gibson, 1991); (2) a horizontal pressure gradient in the asthenosphere between the Paraná Basin and the axial rift zone along the continental margin (Garland et al., 1996); and (3) preexisting weakness zones in the crust reactivated during extension and resulting in oceanward shifting of the magmatic sources (Harry and Sawyer, 1992).

Geochemistry of onshore and offshore Jurassic to Cretaceous magmatism

Petrogenetic studies indicate that melt is generated rapidly by adiabatic decompression of a rising mantle plume during extensional processes (Furlong and Fountain, 1986; White and McKen-

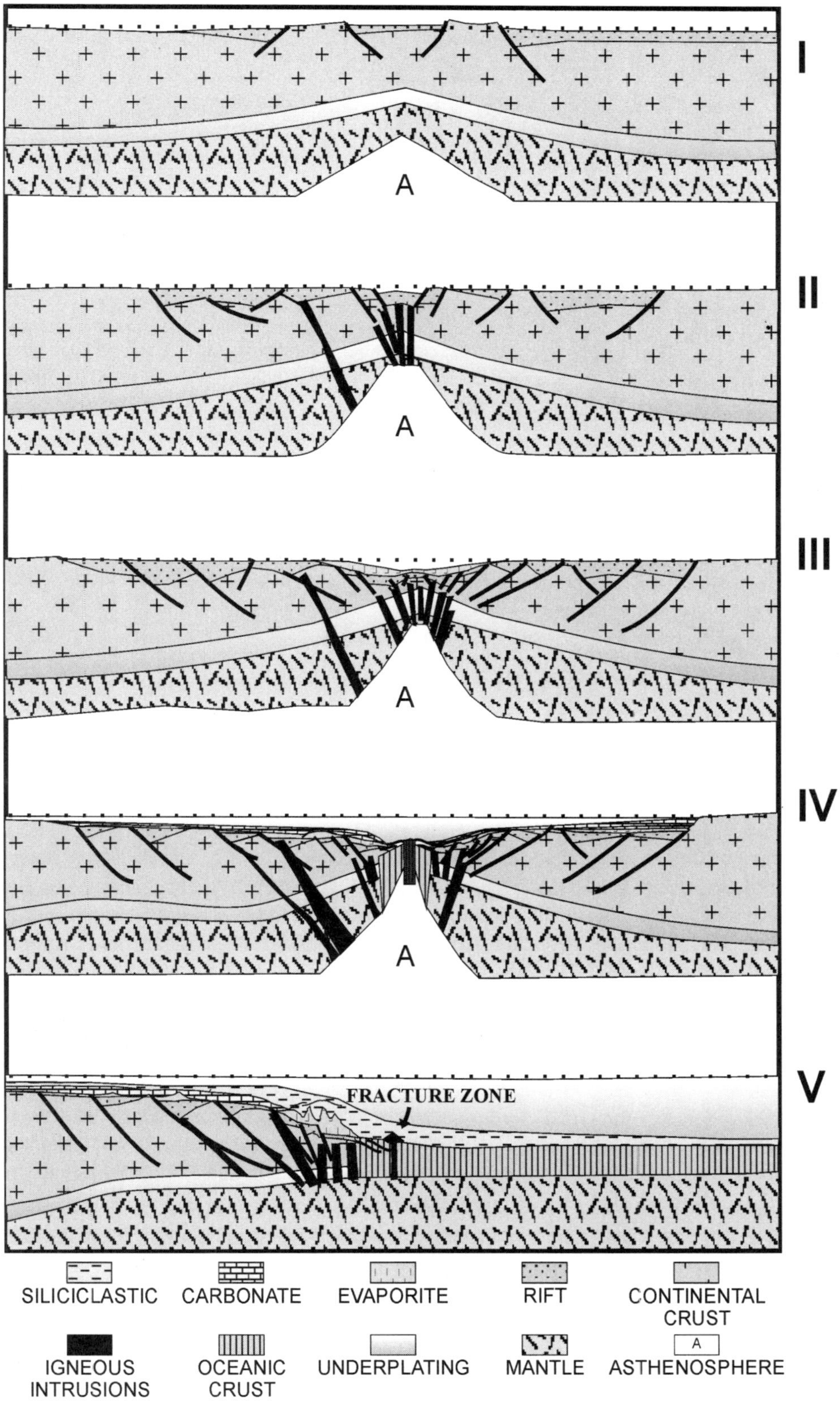

Figure 3. Schematic model showing geodynamic evolution of South Atlantic continental margin rift basins.

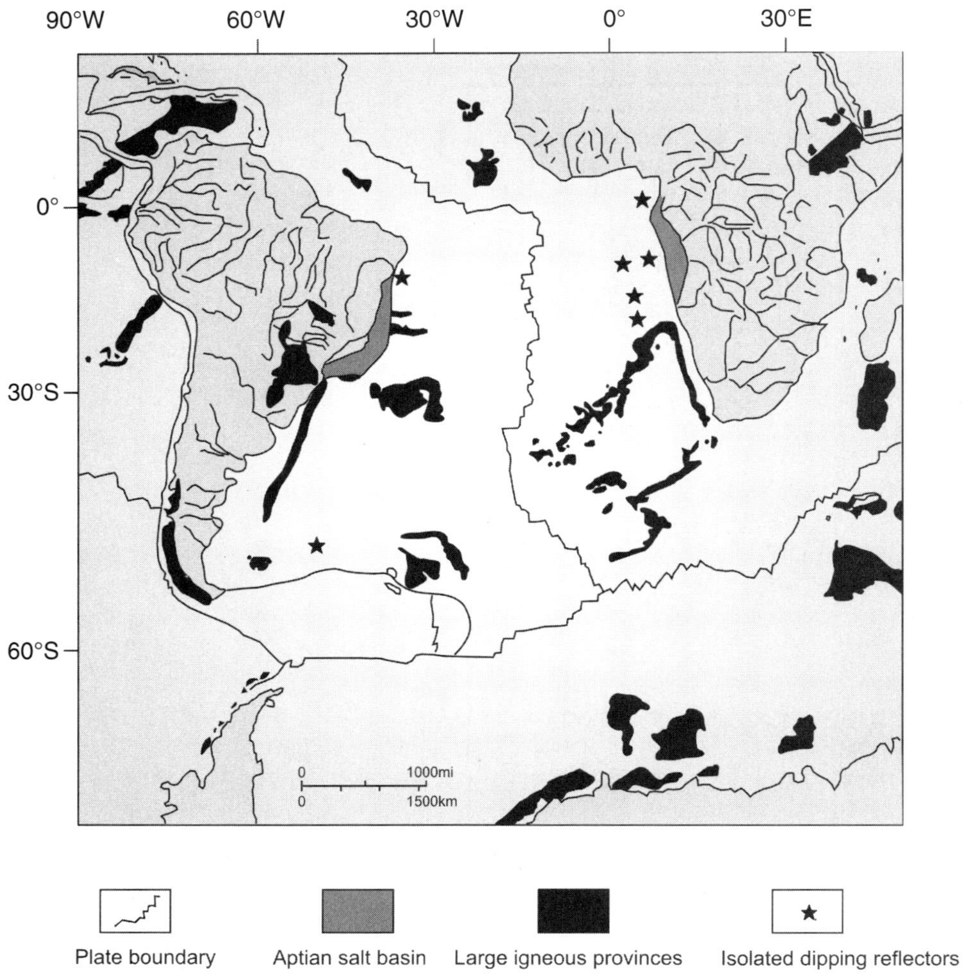

Figure 4. Simplified tectonic map showing Aptian salt basins and large igneous provinces in South Atlantic. Modified from Jackson et al. (2000).

zie, 1989). Thermal modeling indicates that the quantity of melt depends on the thickness of lithosphere before rifting, the amount of lithospheric extension, the duration of rifting, and subtle temperature increases (50–100 °C) above the normal background in the asthenosphere (White and McKenzie, 1989; Keen et al., 1994; Bown and White, 1994; Eldholm et al., 1995). The thermal anomaly in southeastern Brazil and the conjugate margin is probably associated with the flow lines of the Tristan hotspot (which currently corresponds to the volcanically active Tristan da Cunha Island) and the Gough hotspot (now active around Gough Island).

The main period of magmatic activity within the South Atlantic continental margin basins peaked in Late Jurassic–Early Cretaceous time, and is generally related with the rifting of the Gondwana (Fodor et al., 1985a, 1985b; Piccirillo and Melfi, 1988; Piccirillo et al., 1990; Hawkesworth et al., 1992, 2000; Mizusaki et al., 1992; Garland et al., 1996; Peate and Hawkesworth, 1996). It is mostly represented by aphyric tholeiitic basalts (e.g., Fodor et al., 1983; Fodor and Vetter, 1984; Fodor, 1987; Mizusaki et al., 1992; Lobo, 2000).

The geochronological data presented in Table 1 indicate that the volcanism in the offshore rift basins can be generally considered coeval with that in Paraná-Etendeka continental flood basalt province (Renné et al., 1992, 1996; Mizusaki et al., 1992; Turner et al., 1994, 1999; Stewart et al., 1996; Peate, 1997) as well as with the Ponta Grossa and Serra do Mar dike swarms (ca. 137–127 Ma). The oldest dated basalt sample is from Santos and is similar in age to the oldest samples in the Paraná-Etendeka continental flood basalt province (whole-rock $^{40}Ar/^{39}Ar$ 138.4 ± 1.3; Stewart et al., 1996).

Major and trace element and Sr-Nd isotope analyses have been published for the Campos basalts (Mizusaki et al., 1992), but (with the exception of geochronology) Santos and Espírito Santo lack isotope data (Fodor and Vetter, 1984). Major and trace element geochemical data of the Pelotas basalts have been analyzed (Lobo, 2000).

TABLE 1. RELEVANT WHOLE-ROCK K/Ar GEOCHRONOLOGICAL DATA FOR SANTOS, CAMPOS, AND ESPÍRITO SANTO BASALTS

Sample	Age	Basin	Source
SPS-4A	138.1 ± 3.5	Santos	1
BD-1A	134.0 ± 4.0	Campos	2
IP1ES	129.8 ± 7.0	Espírito Santo	1

Sources: (1) Fodor et al., 1983; (2) Mizusaki et al., 1992.

More than 50 samples (30 from Campos, 11 from Pelotas, 9 from Espírito Santo, and only 2 from Santos) have been analyzed from the available database (Fodor and Vetter, 1984; Mizusaki et al., 1992; Lobo, 2000). The samples are mostly represented by basalt and basaltic trachyandesite and minor trachybasalt, basaltic andesite, and andesite (Fig. 5).

The Pelotas, Campos, and Espírito Santo basalts comprise transitional series; samples straddle the alkaline-subalkaline boundary in the total alkali versus silica (TAS) diagram (Fig. 5). The Espírito Santo trend in the alkaline, iron-rich, and magnesium-rich members (AFM) diagram is clearly tholeiitic, but iron enrichment is suppressed in Pelotas and Campos (Fig. 5, inset). Nevertheless, the hypersthene normative composition of all samples denotes the tholeiitic affinity of the transitional series.

Bivariant plots (taking MgO as a differentiation index) show coherent variations for Campos and Espírito Santo that may represent true liquid lines of descent, although a compositional gap ($4.50 < MgO < 7.00$ wt%) is observed in Espírito Santo. In both basins, second-order polynomial correlations give better results than linear ones, with values between 95% and >99.9% significance levels for most elements (Table 2). These features denote that samples in the Campos and Espírito Santo Basins comprise two single series that did not evolve either by magma mixing or simple contamination. Lobo (2000) pointed out that fractional crystallization (either with or without concomitant assimilation, i.e., AFC) is the most likely process. However, both linear and second-order polynomial coefficient correlations give poor results for Pelotas, with significance levels below 95% for the number of samples (n = 11). This indicates that Pelotas may not be represented by a single basaltic suite. Thus, the Pelotas basalts seem to comprise two suites unrelated by any evolutionary process, i.e., a low-TiO_2 suite ($TiO_2 = 1.19 \pm 0.02$ wt% and $Ti/Y = 284 \pm 4$) and a high-TiO_2 suite ($TiO_2 = 2.10 \pm 0.19$ wt% and $Ti/Y = 400 \pm 22$), the latter being relatively less restricted ($2.67 < MgO < 6.58$ wt%) than the former ($6.19 < MgO < 7.08$ wt%).

The restricted MgO range of the low-TiO_2 suite in Pelotas prevents a detailed assessment of its evolutionary processes; thus only the high-TiO_2 suite may be analyzed for differentiation processes. In general, second-order polynomial correlation coefficients in the Pelotas high-TiO_2 suite are better than linear ones and above 95% significance level, although some elements show dispersion (Table 2). These data and trace element ratios variations close to unity (e.g., $La/Yb_N = 1.33$, Ta/Hf

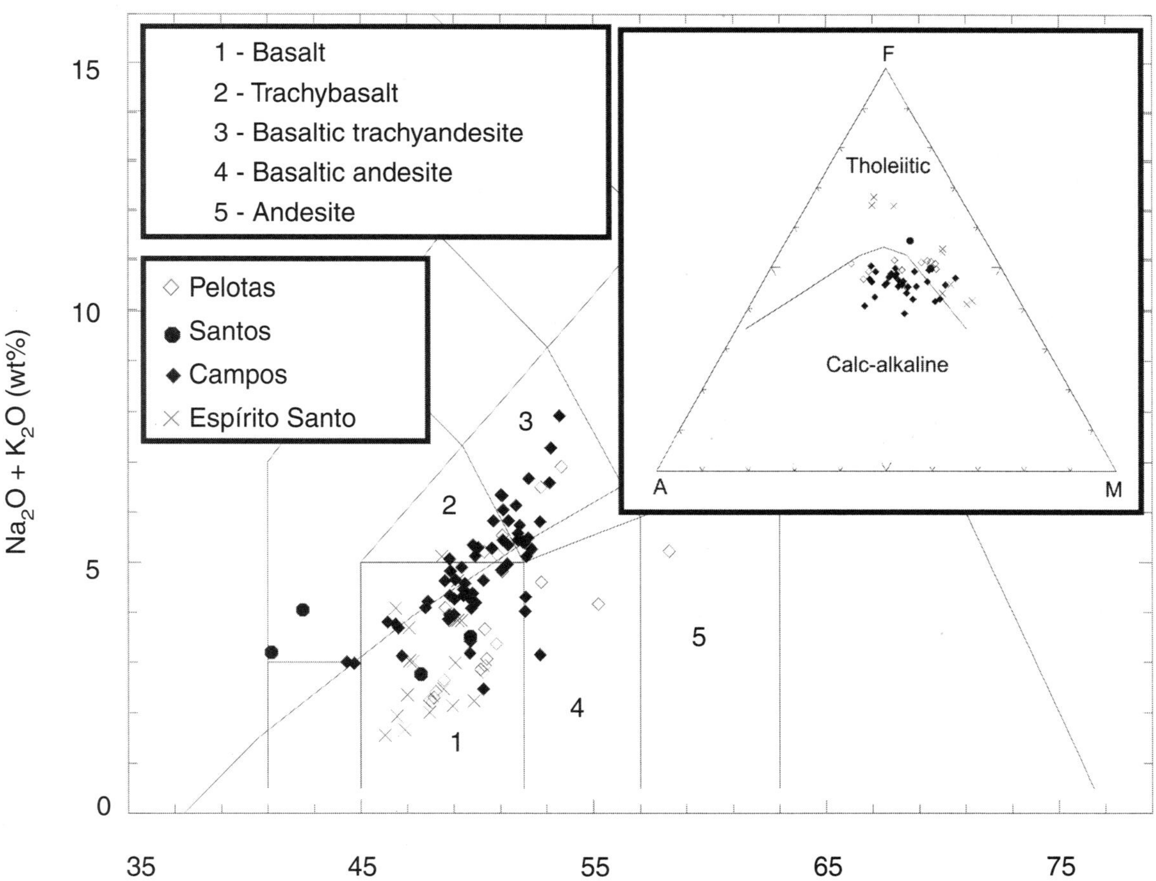

Figure 5. Series discrimination and rock classification of rift basins in southeastern Brazil. Inset: AFM diagram: $A = Na_2O + K_2O$, $F = FeO + 0.08998\ Fe_2O_3$, $M = MgO$ (from LeMaitre, 1989; Irvine and Baragar, 1971).

TABLE 2. SQUARED VALUES OF SECOND-ORDER POLYNOMIAL CORRELATION COEFFICIENTS (R^2) AND LEVELS OF SIGNIFICANCE (%) FOR THE NUMBER OF SAMPLES (N) FOR SELECTED ELEMENTS IN THE HIGH-TIO_2 PELOTAS, CAMPOS, AND ESPÍRITO SANTO BASALTIC SUITES

High-TiO_2 Pelotas (n = 5)	R^2	% (n = 30)	Campos	R^2	%	Espírito Santo (n = 9)	R^2	%
SiO_2	0.97	99.0–99.9	SiO_2	0.37	>99.9	SiO_2	0.44	95.0
TiO_2	0.91	>99.0	TiO_2	0.13	95.0	TiO_2	0.97	>99.9
Al_2O_3	0.34	<80.0	Al_2O_3	0.59	>99.9	Al_2O_3	0.62	95.0–99.0
Fe_2O_3	0.33	<80.0	FeO^t	0.50	>99.9	FeO^t	0.89	>99.9
CaO	0.86	90.0–95.0	CaO	0.14	95.0	CaO	0.82	>99.9
Na_2O	0.98	99.0–99.9	Na_2O	0.13	95.0	Na_2O	0.52	95.0–99.0
K_2O	0.87	90.0–95.0	K_2O	0.22	99.0	K_2O	0.90	>99.9
P_2O_5	0.82	90.0–95.0	P_2O_5	0.03	<80.0	P_2O_5	0.99	>99.9
Ba	0.65	80.0–90.0	Ba	0.08	80.0	Ba	0.91	>99.9
Rb	0.71	80.0–90.0	Rb	0.14	95.0	Rb	0.26	80.0–90.0
Sr	0.61	80.0	Sr	0.39	>99.9	Sr	0.66	99.0–99.9
Y	0.78	80.0–90.0	Y	0.28	99.0	Y	0.86	>99.9
Zr	0.80	90.0	Zr	0.15	95.0	Zr	0.92	>99.9
Nb	0.99	>99.9	Nb	0.27	99.0	Nb	0.95	>99.9
Ni	0.84	90.0–95.0	Ni	0.13	95.0	Ni	0.48	95.0–99.0
Cr	0.96	90.0–95.0	Cr	0.11	90.0	Cr	0.88	>99.9

= 1.15, Nb/Y = 1.07) indicate that fractional crystallization is the most likely evolutionary process to have operated in the high-TiO_2 suite.

Geochronology and geochemistry

Younger phases of magmatic activity having more rhyolitic compositions are dated as 128–120 Ma in the southern South Atlantic onshore volcanic provinces (Peate, 1997). The composition of the lavas shows distinct types according to different percentages of Ti and rare earths (Turner et al., 1994; Stewart et al., 1996). These units dip toward the north, indicating northward-migrating rifting and magma sources, a pattern that is also repeated for the seaward-dipping wedges (Peate et al., 1992; Turner et al., 1994; Talwani and Abreu, 2000).

The role of plagioclase as a main fractionating phase in the Pelotas high-TiO_2 basalts can be depicted by its dominant abundance in their phenocryst assemblage (50% plagioclase, 45% augite, 5% olivine) as well as by the pronounced negative Sr anomaly (e.g., Nd/Sr_N = 5.37; La/Sr_N = 9.14) of the most evolved high-TiO_2 sample.

Campos basalts comprise a single low-TiO_2 suite (TiO_2 = 1.20 ± 0.12 wt% and Ti/Y = 249 ± 64). In general, second-order polynomial correlation coefficients in Campos are above 95%, although some diagrams show dispersion (Table 3). These data and trace element ratio variations well above unity (e.g., Zr/Nb = 5.70, Y/Nb = 4.19, Sr/Nb = 2.42) point to AFC as the most likely evolutionary process to have operated in Campos. Geochemical modeling (De Paolo, 1981) indicates that Campos basalts evolved by ~30% AFC with a fractionating assemblage composed of 5% olivine, 30% augite, and 65% plagioclase, closely approximating the phenocryst assemblage of the samples. The modeling also indicates that the AFC process took place in a lower crust magma chamber represented by a local granulite contaminant (Lobo et al., 1999a, 1999b). The AFC may have led to a relatively early enrichment in alkalies during the crystallization process, resulting in the apparent iron-enrichment suppression depicted in the AFM trend of the suite (Fig. 5, inset).

Refined petrogenetic interpretations for the Espírito Santo basaltic suite are difficult due to various interpretations that could arise from its compositional gap. Such a compositional gap may simply result from the restricted sampling in the basin (Fodor and Vetter, 1984). However, the gap highlights the possibility of two distinctive suites, low-TiO_2 and high-TiO_2, in Espírito Santo, similar to Pelotas and analogous to well-known regional geochemical provincialities such as that in the Paraná-Etendeka continental flood basalt province (e.g., Gibson et al., 1995; Marzoli et al., 1999). Nevertheless, we prefer the hypothesis of a single basaltic suite because the samples were collected from boreholes located only 6 km apart (IP-11ES: 19.5°S, 39.95°W and N-12ES: 19.4°S, 40.0°W; Fodor and Vetter, 1984) and thus too close to allow the characterization of possible geochemical provinces in the Espírito Santo basin. In addition, the robust second-order polynomial correlation coefficients obtained for its major and selected trace element bivariant diagrams (Table 3) support the hypothesis of a single suite in Espírito Santo. Such correlations and trace element ratio variations well above unity (e.g., Ba/Rb = 6.37; Th/Ta = 10.36; Zr/Y = 1.98; La/Yb = 4.37) indicate AFC as the most likely evolutionary process for the Espírito Santo basalts, similar to Campos.

In conclusion, the Espírito Santo and Campos basalts comprise two single suites that may have evolved by AFC, while two distinctive suites, i.e., low-TiO_2 and high-TiO_2, occur in Pelotas;

TABLE 3. TRACE ELEMENT RATIOS FOR THE PELOTAS, CAMPOS, AND ESPIRITO SANTO BASINS

Basin	La/Yb_N	La/Nb_N	Sample	Source
Espírito Santo	1.22	0.54	IP1ES 22 4/4	1
Campos	5.78	0.97	RJS-156	2
High-TiO_2 Pelotas	5.05	2.41	2BPS6A-T7	3
Low-TiO_2 Pelotas	2.50	1.46	RSS3-T1 1/9	3

Sources: (1) Fodor and Vetter, 1984; (2) Mizusaki et al., 1992; (3) Lobo, 2000. Normalization factors: Thompson et al., 1984.

the high-TiO_2 most probably evolved by simple fractional crystallization under anomalous oxidizing conditions. The restricted number of samples (n = 2) in Santos prevents the assessment of possible evolutionary processes and likely mantle sources.

Mantle sources

The distinctive La/Yb_N and La/Nb_N values of the least-evolved samples in the low-TiO_2 and high-TiO_2 suites in Pelotas as well as in the Campos and Espírito Santo basaltic suites (Table 3) indicate possible contributions from different mantle sources.

The La/Yb_N ratios indicate that Espírito Santo, Campos, and Pelotas basalts are unlikely to be derived exclusively from a normal (N) MORB source (e.g., $La/Yb_N = 0.57$; Bevins et al., 1984). Coupled $La/Yb_N > 1$ and $La/Nb_N > 1$ in Pelotas indicate at least contributions from an enriched mantle source likely to be found in the subcontinental lithospheric mantle. However, values of $La/Yb_N > 1$ and $La/Nb_N < 1$ in Espírito Santo and Campos indicate contributions from an asthenospheric, undepleted source, possibly a plume-like one.

The low-TiO_2 and high-TiO_2 suites in Pelotas cannot be related to any evolutionary process, and thus the geochemical provinciality has to be explained either by generation from different mantle sources or generation from different amounts of partial melting from a single mantle source. These two hypotheses are difficult to test because the possible evolutionary processes for the low-TiO_2 suite could not be assessed due to its narrow MgO range. The petrogenesis of the low-TiO_2 and high-TiO_2 suites in Pelotas may be attributed to generation from different mantle sources. The best results of the modeling (using the same Kd^{Yb}_{garnet} values; 6.1670; Arth, 1976) showed that the petrogenesis of the low-TiO_2 basalts in Pelotas is related to higher amounts of partial melting and less garnet-bearing residual mantle sources than the high-TiO_2 basalts (Fig. 6).

Binary mixing calculations (Faure, 1986) were performed to test for possible contributions of different mantle sources in the petrogenesis of the Pelotas, Campos, and Espírito Santo basalts: (1) subcontinental lithospheric mantle (SCLM; average lamproite; Rock, 1991), (2) N-MORB (Sun and McDonough, 1989), and (3) plume like (Tristan da Cunha lavas; LeRoex et al., 1990). Multielement variation diagrams patterns show that the petrogenesis of the Pelotas, Campos, and Espírito Santo basalts cannot be related to mixing between the SCLM and Tristan da Cunha components (Fig. 7).

Calculations showed that minimum contributions (~2%) from a SCLM source in the mixing result in $La/Nb_N > 1$. Therefore, such a component cannot be involved in the petrogenesis of Campos and Espírito Santo basalts that have $La/Nb_N < 1$ (Table 3), N-MORB and Tristan da Cunha being the likely sources. Samples among the least evolved basalts in both basins with similar MgO contents were used in the modeling (Fig. 8). The results showed that a N-MORB mantle source largely contributed (~90%) to the petrogenesis of the Espírito Santo basalts;

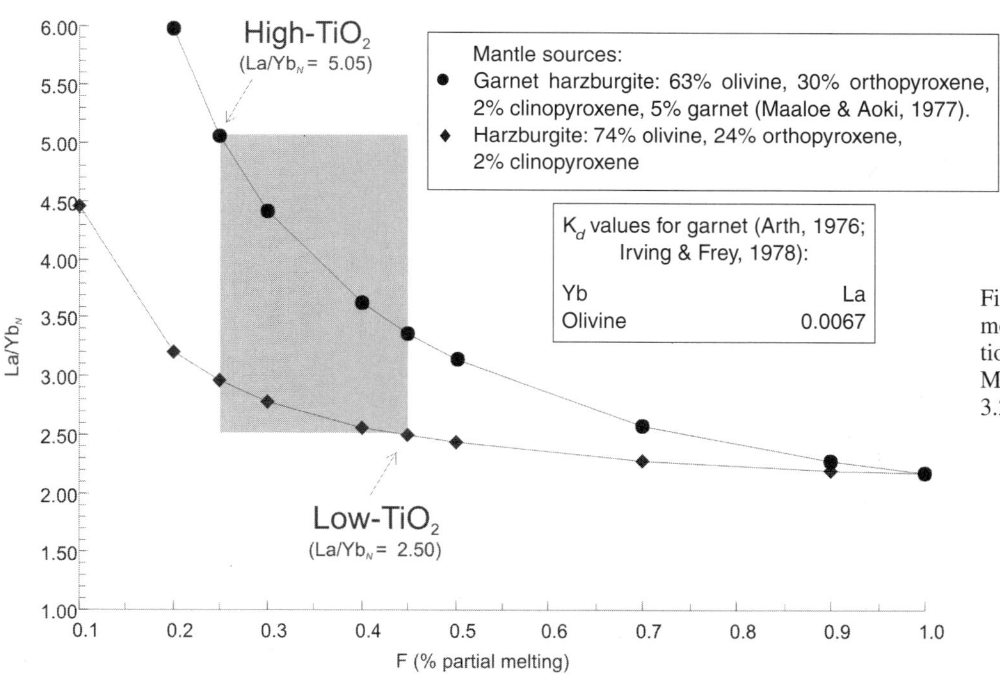

Figure 6. Modal batch partial melting modeling for Pelotas basalts. Normalization factors are from Nakamura (1974). Mantle source: harzburgite with La/Yb = 3.26.

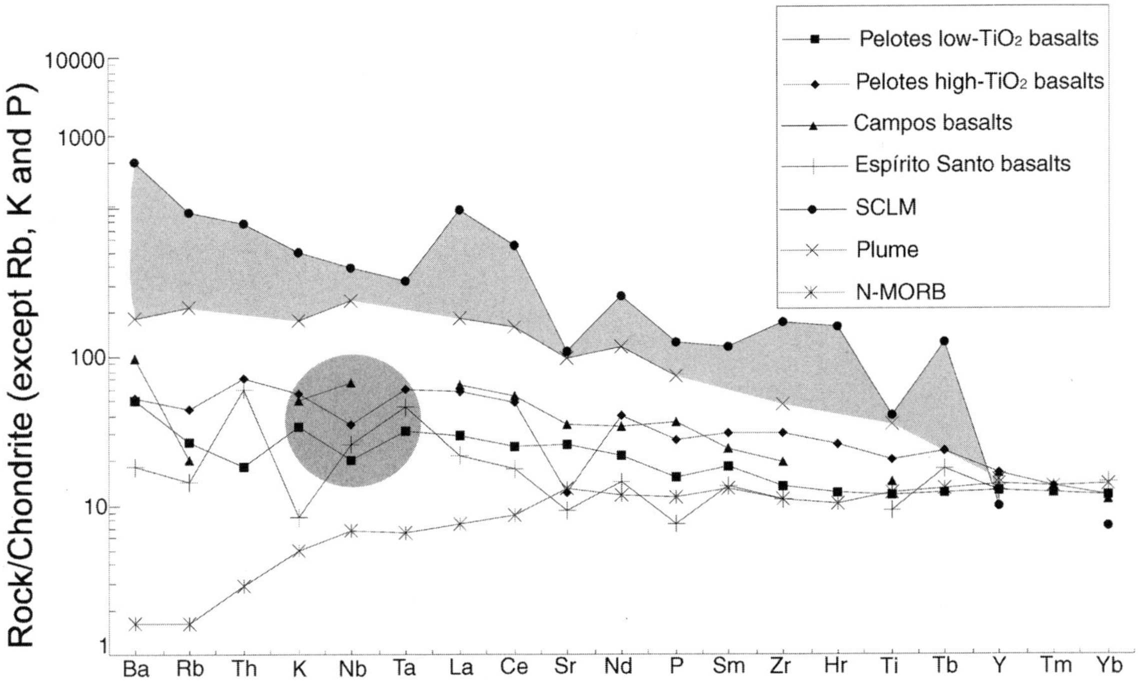

Figure 7. Multielement variation diagram patterns for rift basin basalts (Fodor and Vetter, 1984; Mizusaki et al., 1992) and plume, subcontinental lithospheric mantle (SCLM), and normal mid-ocean ridge basalt (N-MORB) source components. SCLM: average lamproite (Rock, 1991); plume: Tristan da Cunha basanite TDC61 (LeRoex et al., 1990); N-MORB: Sun and McDonough (1989). Normalization factors: Thompson et al. (1984), with Ba = 3.85 (Hawkesworth et al., 1984). Rb, K, and P are from primitive mantle of Sun (1980). Circle includes Nb anomalies for rift basin basalts. Shaded area is SCLM-plume mixing field.

there were fewer contributions (~10%) from a plume-like source (i.e., Tristan da Cunha). However, the petrogenesis of the Campos basalts cannot be attributed to mixing between components with compositions similar to those involved in the petrogenesis of the Espírito Santo basalts. The geochemical data and the modeling also showed that enriched (E) MORB and the SCLM cannot be considered possible mantle sources for the petrogenesis of the Campos basalts. The Tristan da Cunha component used in the mixing calculations shown in Figure 8 is represented by basanite TDC61 (LeRoex et al., 1990). Tests with other Tristan da Cunha basanites and ankaramitic basanites gave very similar results for Espírito Santo, but poor results were obtained for Campos. One possible explanation is that the Tristan da Cunha compositions used in the modeling do not represent the composition of the plume component involved in the petrogenesis of the Campos basalts.

In conclusion, the petrogenesis of the low-TiO_2 and high-TiO_2 basalts in Pelotas, Campos, and Espírito Santo basalts are related to different contributions of distinctive mantle sources. Modeling showed that N-MORB is a major source component in the three basins, and there are subordinate contributions from Tristan da Cunha (Espírito Santo and Campos) and from the SCLM (high-TiO_2 Pelotas). If the La/Nb_N ratio of the least evolved sample in Pelotas is also a source characteristic, geo-

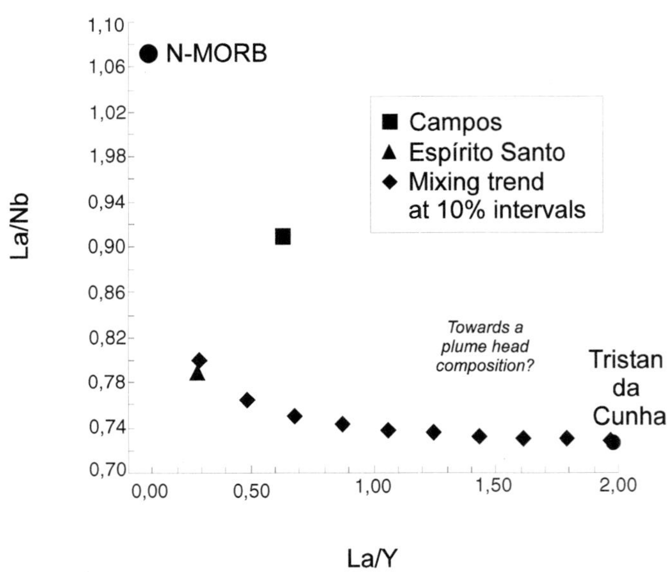

Figure 8. Bivariant diagram showing results of binary mixing calculations between normal mid-ocean ridge basalt (N-MORB) (Sun and McDonough, 1989) and Tristan da Cunha basanite TDC61 (LeRoex et al., 1990). Potential parental compositions in Campos (Mizusaki et al., 1992) and Espírito Santo (Fodor and Vetter, 1984) are shown for comparisons.

chemical provinciality in Pelotas is probably a result of mantle heterogeneities.

Geodynamic implications

The South Atlantic volcanism within the rift basins can be generally associated with the extensional processes that ultimately led to the opening of the South Atlantic Ocean. However, attempts to draw possible geodynamic scenarios will have to be considered only preliminary due to the lack of substantial geochemical data.

Failure in obtaining good results in N-MORB–plume binary mixing models for Campos may be due to the involvement of the head of the plume in the partial melting process, the composition of which may not be represented by the Tristan da Cunha lavas (Fig. 8), as suggested for the Paraná-Etendeka continental flood basalt province elsewhere (Gibson et al., 1995). However, it is relevant that the petrogenesis of the Espírito Santo basalts is most probably related to a N-MORB–plume mixing, in which many of the basanitic and ankaramitic lavas in Tristan da Cunha represent likely plume compositions (Fig. 8). It could be that the Espírito Santo basalt petrogenesis is related to a Gondwana rifting stage in which most of the plume head had already melted beneath Campos. This would leave only the tail, deeper parts of the plume to be melted in Espírito Santo, resulting in the alkalic-like liquid compositions represented by the Tristan da Cunha lavas. Such a rifting stage had to be related to a time of advanced lithospheric attenuation so as to allow for the large contributions from the N-MORB component. Previous melting episodes involving the plume head in Espírito Santo may have been prevented due to the presence of an abnormally thick lithosphere or due to inefficient lithospheric thinning. Lithospheric differential thinning has already been proposed as a likely process during the Gondwana breakup in order to explain the distinctive rare earth element patterns of the Ponta Grossa and the Serra do Mar dike swarms (Valente et al., 1999).

Another consideration to be made regards the geochemical provinciality observed in Pelotas. Continental flood basalt–related geochemical provincialities have usually been attributed to chemical heterogeneities within the SCLM. Therefore, it is relevant that the involvement of such a source component in the petrogenesis of the rift basin–related basalts in southeastern Brazil was found in Pelotas, the only basin that showed evidence for the presence of a low-TiO_2 and a high-TiO_2 suite. The geochemical provinciality in Pelotas can be considered analogous to that described for major continental flood basalt provinces. For example, the low-TiO_2 and high-TiO_2 basalt suites in Pelotas were sampled in the southern and northern sectors of the rift basin and thus reproduce the geochemical provinciality described in the adjoining Paraná-Etendeka continental flood basalt province (e.g., Gibson et al., 1995; Marzoli et al., 1999). However, in Pelotas such a provinciality is described for a relatively small area (~500 times smaller than Paraná-Etendeka) in comparison to other major continental flood basalt provinces. The general implication here is that geochemical heterogeneities may have been imprinted on both regional and local scales. Alternatively, local mantle heterogeneities could be simply reflecting vertical geochemical variations within the SCLM.

Cretaceous SDR wedges

Figure 9 (top) shows a cross section of the Paraná basin from northwest to southeast, illustrating the main aspects of the Paleozoic and Mesozoic sedimentary and volcanic sequences onshore, and the proposed geometry of the rift architecture of the Santos Basin along the continental margin (Macedo, 1990). Figure 9 (bottom) shows the architecture of the Santos Basin rift as alternatively interpreted by Gladczenko et al. (1997), which permits the presence of a large extrusive complex related to the breakup of Gondwana, corresponding to a seaward-dipping reflector wedge.

The volcanic sequences that occur in the continental margin are less well known than the onshore equivalents and comprise mostly tholeiitic basalts and volcaniclastic rocks. Although some exploratory boreholes have penetrated the flood basalts, particularly in the Campos Basin, where they are associated with minor oil production (Guardado et al., 1989), their true thickness, age, and petrogenesis, particularly in the deep-water region, are still conjectural. Seismic interpretation suggests that the offshore basalts may reach thicknesses of several kilometers, thus exceeding the average lava thickness (<1 km) in the onshore Paraná province (Leinz et al., 1968; Peate et al., 1992). In addition, the volcanic sequences are characterized by a peculiar geometry corresponding to sigmoidal reflectors that dip and thicken basinward, with no apparent fault control, forming the seaward-dipping reflectors often imaged along volcanic rifted margins (Hinz, 1981).

Seaward-dipping reflectors (SDRs) observed on seismic profiles are interpreted as volcanic rocks (flood basalts) rapidly extruded during the latest phases of rifting, or during the onset of subaerial seafloor spreading (Hinz and Weber, 1976; Hinz, 1981; Mutter et al., 1982; Roberts et al., 1984). They have been drilled in some margins (e.g., offshore Norway and Greenland) and their geochemical analyses indicate that their composition is marked by an oceanic affinity; thus, the term "proto-oceanic crust" (Meyers et al., 1996a; Talwani and Abreu, 2000) has often been applied. They may occur in both shallow- and deep-water settings, as exemplified by Iceland in the North Atlantic and continental margin analogs (Hinz et al., 1987; Skogseid and Eldholm, 1995). In the South Atlantic, SDRs have been interpretationally imaged in regions from the Argentine shelf to the northeast Brazilian margin (Souza et al., 1993; Hinz et al., 1999, Mohriak et al., 1998a) and in the conjugate basins in West Africa south of the Walvis Ridge (Hinz, 1981; Gladczenko et al., 1997).

Based on seismic interpretations and analogies with ophiolites (e.g., Christensen and Salisbury, 1982; Eldhom et al., 1995),

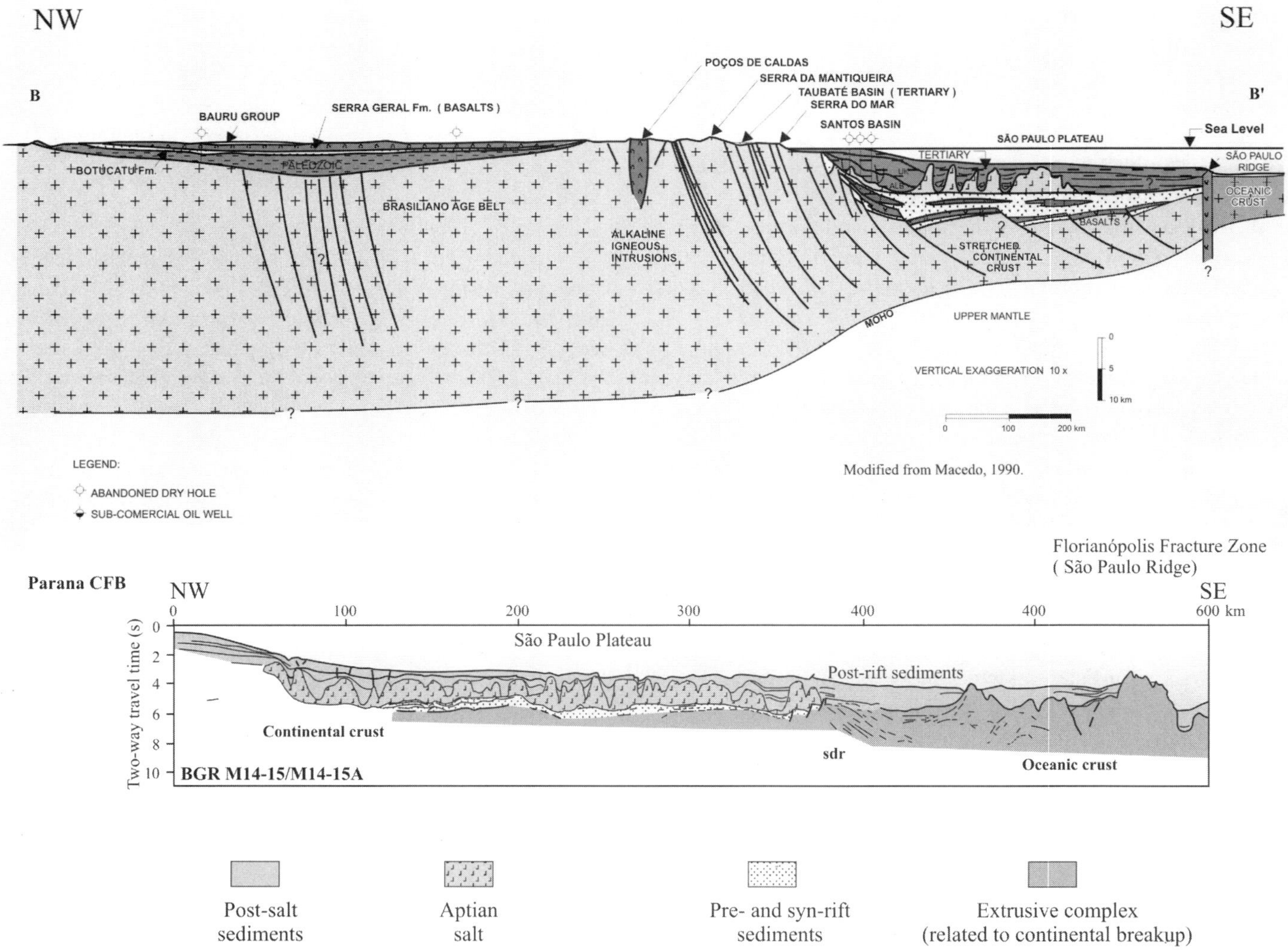

Figure 9. Regional schematic geologic sections showing alternative interpretations in Paraná and Santos Basins. Top: Schematic crustal section from Paraná Basin toward Santos Basin showing extension of rift beyond salt diapir province and extending to Florianópolis Fracture Zone (São Paulo Ridge). Bottom: Line-drawing interpretation of regional profile along Santos Basin showing volcanic complex related to proto-oceanic crust in São Paulo Plateau. CFB = continental flood basalt. Modified from Gladczenko et al. (1997).

the SDR wedges are interpreted to comprise three main layers of crust. The upper layers, which correspond to the dipping reflectors observed on seismic profiles, are associated with flood basalts and overlain by some interbedded siliclastic sediments (Fig. 10). P-wave velocity increases rapidly downward from ~3–4 km/s in the sediments and pillow basalts to more than 6 km/s in the deepest portions of the wedges. The intermediate layer occurs in the middle to lower crust, and may consist of vertical dikes overlying gabbro; seismic velocities range from 6 km/s at the top to more than 7 km/s at the bottom. The lower layer of the SDR wedges may be associated with lenses of high-density and high-velocity ultramafic material (>7 km/s) in the lower crust. These mafic bodies may form lenses that some have identified as underplated material (e.g., Lase Study Group, 1986).

There is still controversy as to whether SDR wedges occur exclusively on the continental or oceanic crust, or represent proto-oceanic material (Mutter et al., 1982; Talwani and Abreu, 2000). Seismic interpretation indicates that SDRs may partly overlie stretched continental crust (Skogseid and Eldholm, 1995; Fontana, 1996), but rotated rift blocks under these volcanics are rarely imaged. Jackson et al. (2000) suggested that these rift blocks have been thermally or tectonically uplifted and eroded, or because the high temperature and melts formed during lithospheric extension, may have weakened and remobilized the crust. In addition, the SDRs may form multiple proto-oceanic wedges. Landward, these belts may wedge out and onlap continental crust, indicating that their source was seaward, although the present-day geometry suggests an apparent basinward progradation. Toward the platform, volcaniclastic deltas have been imaged in some margins (Planke et al., 2000). Basinward, the SDRs may correspond to flat-lying basalt layers forming volcanic plateaus often intruded by igneous plugs (Barton

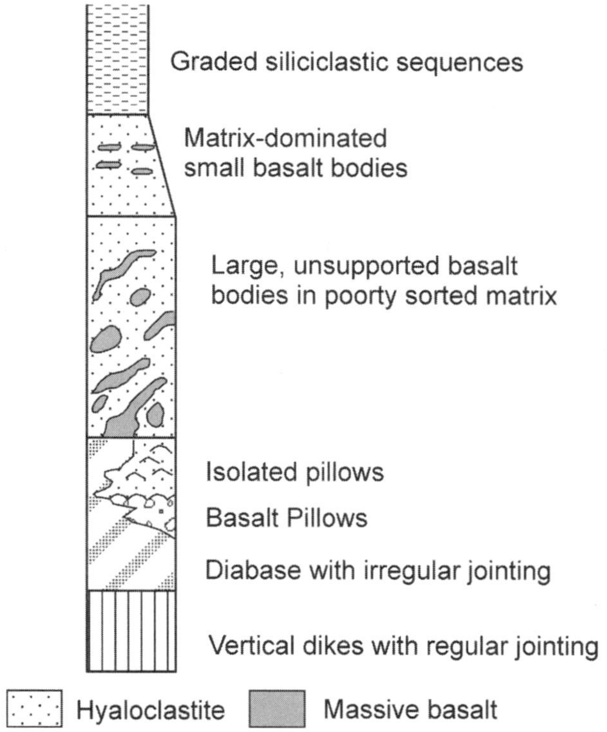

Figure 10. Schematic seaward-dipping reflector (SDR) stratigraphic column (modified from Planke et al., 2000).

and White, 1997; Talwani and Abreu, 2000; Planke et al., 2000). The oceanward SDR belts extend from the stretched continental crust and may amalgamate with typical oceanic crust. The oceanic crust may contain short or chaotic reflectors that dip both seaward and landward (Hinz et al., 1999), and it may also include relict structures from magmatic chambers (Groschel-Becker, 1996), wedges of isolated SDRs (Bassetto et al., 2000), and steep-sided volcanic mounds and igneous intrusions along leaking fracture zones (Rosendahl et al., 1992).

On the South American continent, the SDR province is ~3000 km long and 60–120 km wide, extending from the Argentine shelf to the Santos and Campos Basins (Souza et al., 1993; Lohmann et al., 1995; Gladczenko et al., 1997; Hinz et al., 1999). On the African continent, the SDR wedges extend for equivalent lengths south of the Walvis Ridge and may reach widths of 200 km (Hinz, 1981; Gerrard and Smith, 1982; Abreu, 1998; Talwani and Abreu, 2000).

Although some recent works have proposed a ubiquitous presence of SDRs along the South Atlantic continental margins, and suggested that they correspond to continental flood basalts drilled onland and in the platform (e.g., Jackson et al., 2000), we propose a differentiation in both timing and environment of formation of these volcanics during continental breakup and inception of oceanic spreading centers. The depositional geometry of the continental flood basalts in the volcanic sequences along the Brazilian and West African margins, as well as in other volcanic margins, has a seismic signature characterized by flat-lying to chaotic reflectors, whereas the SDR reflectors exhibit a peculiar style of sigmoidal geometries (Hinz, 1981; Planke et al., 2000). Oceanic crust is typically ~1–2 s two-way traveltime (TWTT) with average thickness ~7 km (White and McKenzie, 1995) and generally has a hummocky surface obscured by diffractions on seismic data.

Based on seismic interpretation of several regions worldwide, and incorporating insights from modern analogs (such as Iceland; Pálmason, 1980), and scientific drilling (e.g., Leg 152 Shipboard Party, 1994), an evolutionary model for SDR formation is presented in the following. The model involves lithospheric extension and a thermal anomaly that dissipated with time, magmatism being related to early, intermediate, and late stages (Fig. 11).

During the early stage (I in Fig. 11), lithospheric extension in regions affected by mantle plumes or abnormal mantle temperatures leads to fracturing of the continental crust and the formation of extensive volcanism. These flood basalts, fed from subaerial volcanoes and fissure swarms parallel to the rift axis, were formed at very high (up to 5 km/m.y.) extrusion rates (Gudmundsson, 1995; Barton and White, 1997), but lasted only for a few (<5) million years. As the volcanic margin drifted away from the spreading center and migrated off the radius of influence of the thermal anomaly, the lengths of the lava flows were reduced. Cooling of the thermal anomaly resulted in subsidence, which eventually submerged the evolving margin and spreading center (II in Fig. 11). The adjacent borderland blocks may remain isostatically elevated because of underplating (White and McKenzie, 1989), resulting in shallow bathymetry for the initial period, thus allowing continued erosion of the outermost rift blocks if they were not covered by transitional phase siliciclastic rocks and evaporites. Subsequently, the slow-decaying thermal anomaly and drifting away from the oceanic spreading center eventually led to submersion of the axis, resulting in deep submarine volcanism and very rapid freezing of the lavas. At this stage (III in Fig. 11), a more typical oceanic crust was formed at normal extrusion rates, leading to the onset of typical magnetic stripes caused by magnetic reversals (Jackson et al., 2000).

Onshore and offshore Cretaceous-Tertiary magmatism

Two other phases of postrift magmatic activity are registered in the eastern Brazilian margin (Cainelli and Mohriak, 1998): the mid-Cretaceous (100–80 Ma), and early Tertiary (60–50 Ma) phases. Both are characterized by onshore and offshore igneous activity with a predominantly alkaline composition (Ulbrich and Gomes, 1981). Some of these plugs are apparently aligned with a hotspot trend, e.g., the east-west intrusions from Poços de Caldas toward the Morro de São João alkaline plugs, which extend from the northern part of the Paraná Basin toward the Cabo Frio region of the Campos and Santos Basins (Sadowski and Dias-Neto, 1981; Mohriak et al., 1995c; Szatmari et al., 2000). Some fracture zones in the east-

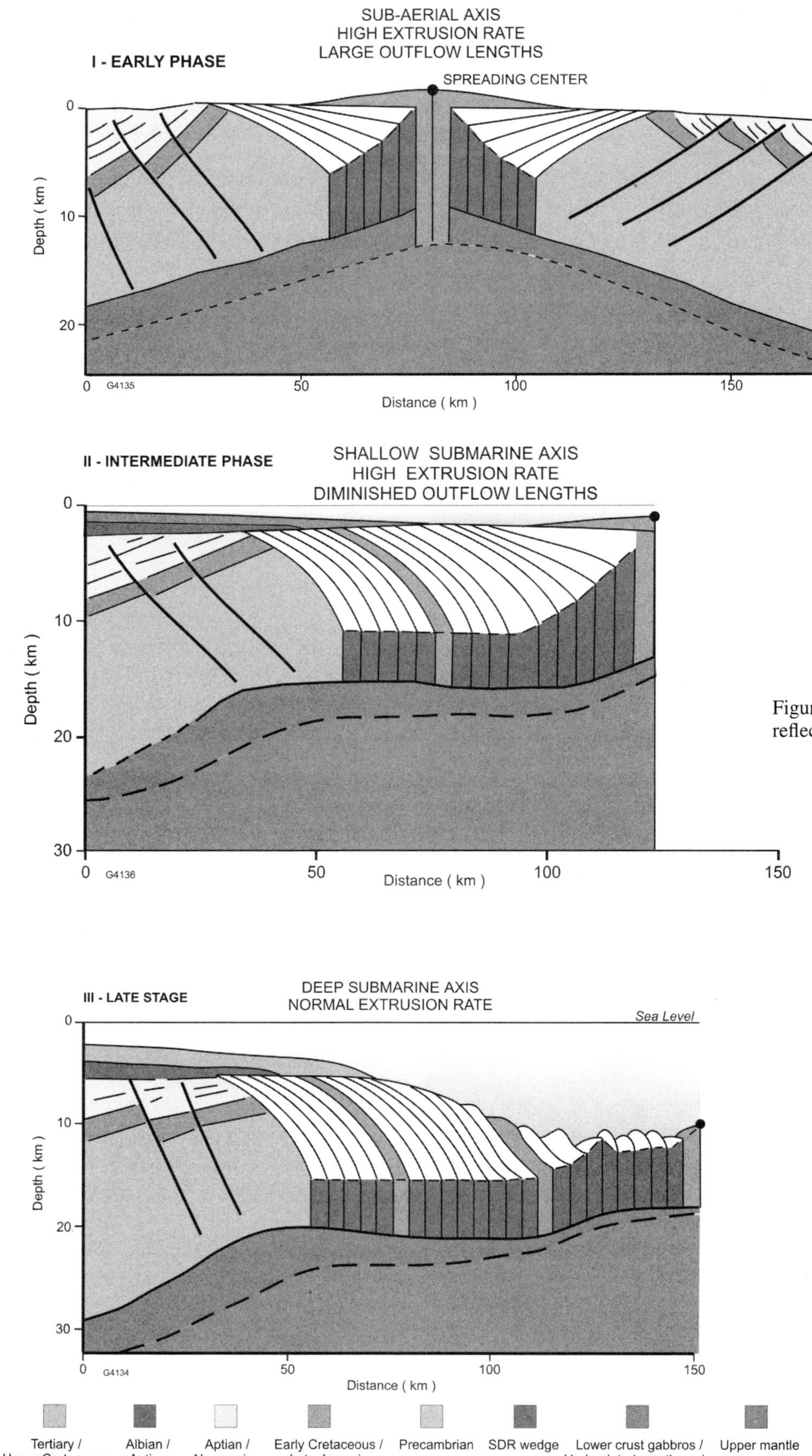

Figure 11. Schematic model of seaward-dipping reflector formation.

ern Brazilian continental margin also seem to be associated with volcanic features of Late Cretaceous to Tertiary age, particularly the Maceió and Sergipe fracture zones in the northeastern region, the Vitória-Trindade chain, and the Florianópolis fracture zone (Cainelli and Mohriak, 1998). Other fracture zones (e.g., in Gabon and Rio Muni) seem to be largely devoid of these volcanics. Still younger phases of magmatism (Miocene and locally reaching the present day) are registered along the Cameroon lineament (Meyers and Rosendahl, 1991).

REGIONAL TRANSECTS IN THE SOUTH ATLANTIC

Basins south of the Rio Grande Rise–Walvis Ridge

The continental margin of South Africa and Namibia (Fig. 12) is characterized by large volcanic wedges (Austin and Uchupi, 1982; Gerrard and Smith, 1982; Hinz et al., 1999) formed by the latest stages of rifting in the Early Cretaceous (134–129 Ma). Volcanic activity lasted until postbreakup time in the Late Cretaceous (110–80 Ma).

Gladczenko et al. (1997) interpreted the crustal architecture of offshore Namibia to comprise four domains: (1) oceanic crust of normal thickness; (2) thickened oceanic crust (to 15 km thick) including most of the SDR sequence overlying the breakup unconformity; (3) a wide (>100 km) Late Jurassic–Early Cretaceous rift zone partly covered by the feather edge of the SDRs; and (4) thicker, unrifted continental crust, partly affected by Paleozoic extension in a landward direction. Figure 13 shows a regional cross section illustrating the crustal architecture of the Namibia margin, based on seismic interpretation. Volcanic plateau areas occur onshore (Figs. 12 and 13) as well as offshore in the Walvis Ridge (south of the Rio Grande–Walvis Fracture Zone; see Fig. 2). The slope and deep basin are also characterized by several igneous plugs and sedimentary fold belts, particularly in the deep-water region of the South African continental margin (Bray and Lawrence, 1999).

The Argentinian counterpart margin is characterized by Jurassic to Early Cretaceous rift basins and by volcanic features interpreted as SDR sequences in the boundary between continental and oceanic crust (Hinz, 1990).

The Pelotas Basin offshore southern Brazil (Fig. 14) is characterized by proximal rift blocks controlled by landward-dipping normal faults and by SDR sequences in the platform and deep water (Souza et al., 1993; Fontana, 1996; Cainelli and Mohriak, 1998; Talwani and Abreu, 2000). Figure 15 illustrates the seismic expression of features interpreted as SDR wedges marking the transition to a pure oceanic crust (Talwani and Abreu, 2000). However, there are alternative interpretations that assume that these features represent a Neocomian rift sequence that thickens basinward and is controlled by major antithetic faults (Dias et al., 1994).

Figure 16 shows a schematic transect illustrating the distribution of the volcanic sequences typical of the segment south of Walvis Ridge–Rio Grande Rise, with volcaniclastic flows in the platform and seaward-dipping wedges in the transition from continental to oceanic crust.

Salt basins south of the Abrolhos volcanic complex

The Santos and Campos salt basins (Fig. 14) extend north of the Florianópolis Fracture Zone and constitute the most prolific hydrocarbon province offshore Brazil (Cainelli and Mohriak, 1998).

Figure 17 (modified from Cainelli and Mohriak, 1998) shows the interpretation of a regional deep seismic profile corresponding to a transect across the southern Campos and northern Santos Basins (Fig. 14), which are characterized by very large salt diapir provinces that are impressive even at crustal-scale transects.

Four tectonic compartments associated with salt tectonics can be identified along the profile (Mohriak et al., 1999). The first or proximal domain (shelf to upper slope) is characterized by variable thicknesses of Aptian salt overlying Aptian siliciclastic rocks, or locally continental volcanic rocks. Small normal faults, which predominantly dip basinward, form prerafts of moderately extended Albian strata. The second or intermediate domain (middle to lower slope) is characterized by major extension of the overburden, forming turtle structures and rafts separated by low, reactive salt walls that eventually subsided because of extension. A major step in the basement (Atlantic hinge zone) appears to separate two subbasins with presalt sediments. The presalt sag basin, overlying continental to transitional crust, indicates oceanward shifting of the extensional axis (Mohriak et al., 2000a). The third or basinal domain corresponds to a major salt diapir province (lower slope to deep basin) characterized by tall extensional and contractional structures cored by Aptian diapirs that may affect the seafloor (Cainelli and Mohriak, 1998). The salt diapirs are laterally squeezed or welded, forming diapiric walls and antiformal duplex stacks, which are separated by broader minibasins. Basinward-migrating contraction resulted in buckling, inversion of synkinematic wedges, and rejuvenation of tall diapirs. The crustal limit province marks the boundary with the fourth domain (deep basin to oceanic crust), which is characterized by thick Aptian salt that may directly overlie volcanic crust (Jackson et al., 1998; Mohriak et al., 1999; Marton et al., 2000). This region is also locally characterized by seaward-dipping reflector wedges and igneous structures typical of oceanic crust (Gladczenko et al., 1997; Cainelli and Mohriak, 1998).

Equivalent structural features are also present in the Kwanza Basin offshore West Africa, but there is some dispute on the presence of SDRs underlying the rift-phase sediments. One alternative interpretation assumes that the volcanic rocks drilled by exploratory boreholes underlying rift sediments correspond to Early Cretaceous volcanics (equivalent to Paraná-Etendeka tholeiitic basalts), which constitute the floor of the sedimentary deposits and precede the main extensional phase associated with the rift blocks (Karner et al., 1997; Cainelli and

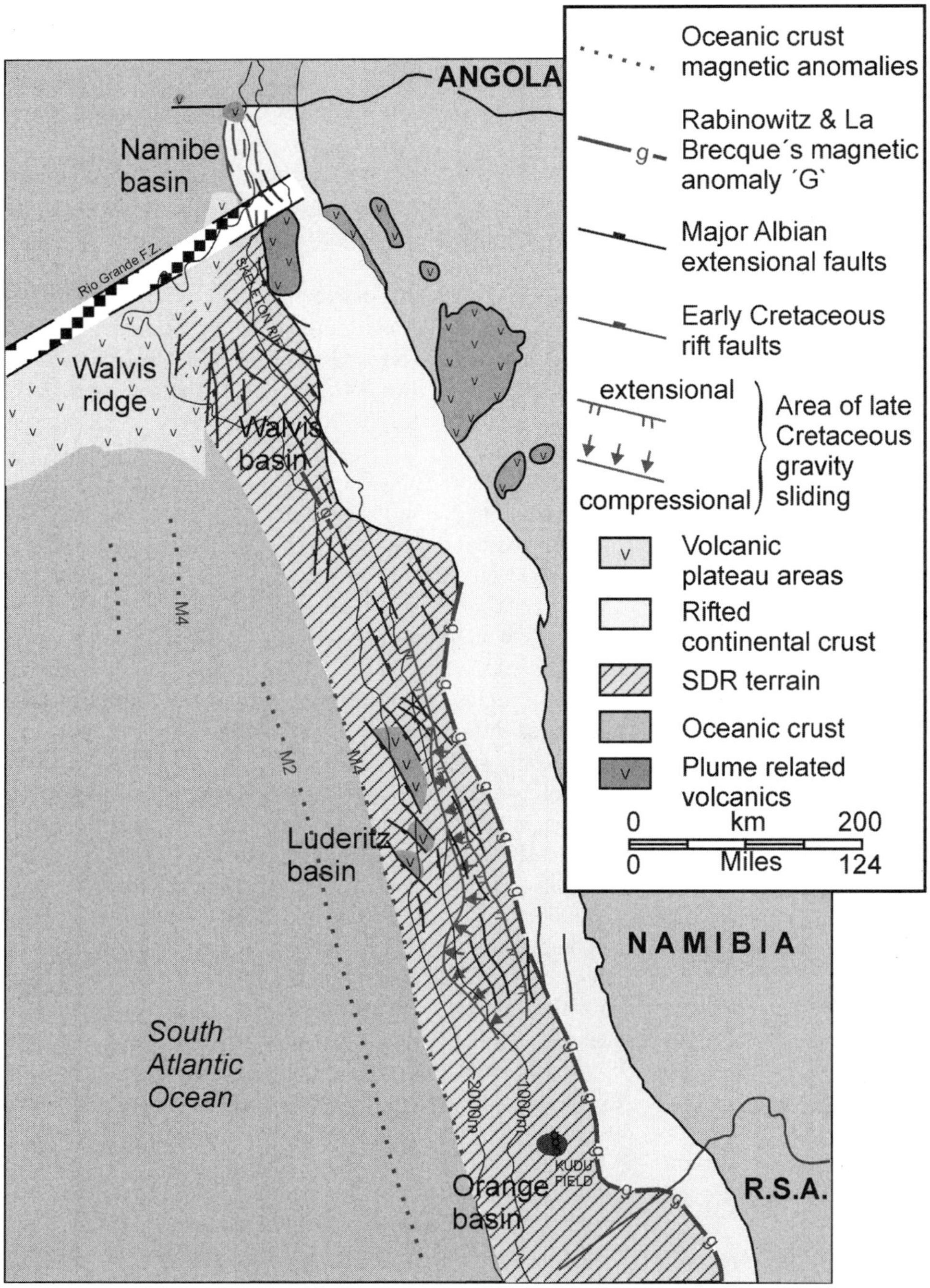

Figure 12. Simplified tectonic map of southwestern Africa–Namibe Basin. SDR = seaward-dipping reflector. Modified from Bray and Lawrence (1999).

Mohriak, 1998). The other hypothesis (Fig. 18) assumes that these volcanics also correspond to SDR wedges, which are not imaged in the deep-water region because salt structures have distorted the continuity of deeper reflectors (Jackson et al., 2000). However, we suggest that these volcanics in the Kwanza Basin are older than the SDRs, and may actually have preceded the inception of spreading centers by a few million years.

There is some seismic evidence pointing to the widespread occurrence of SDRs below the distal portion of the salt diapir province (Fonck et al., 1998; Marton et al., 2000), the salt basin limit probably being controlled by volcanic ridges associated with incipient oceanic crust. Palinspastic reconstructions of the salt basins in the South Atlantic at the beginning of ocean spreading indicate a bulge ~2000 km wide over the Tristan da Cunha plume (White and McKenzie, 1989; Hinz et al., 1999; Jackson et al., 2000). These large dimensions suggest that mantle plumes could have influenced both the West African and southeastern Brazilian margins, and thus SDRs could underlie

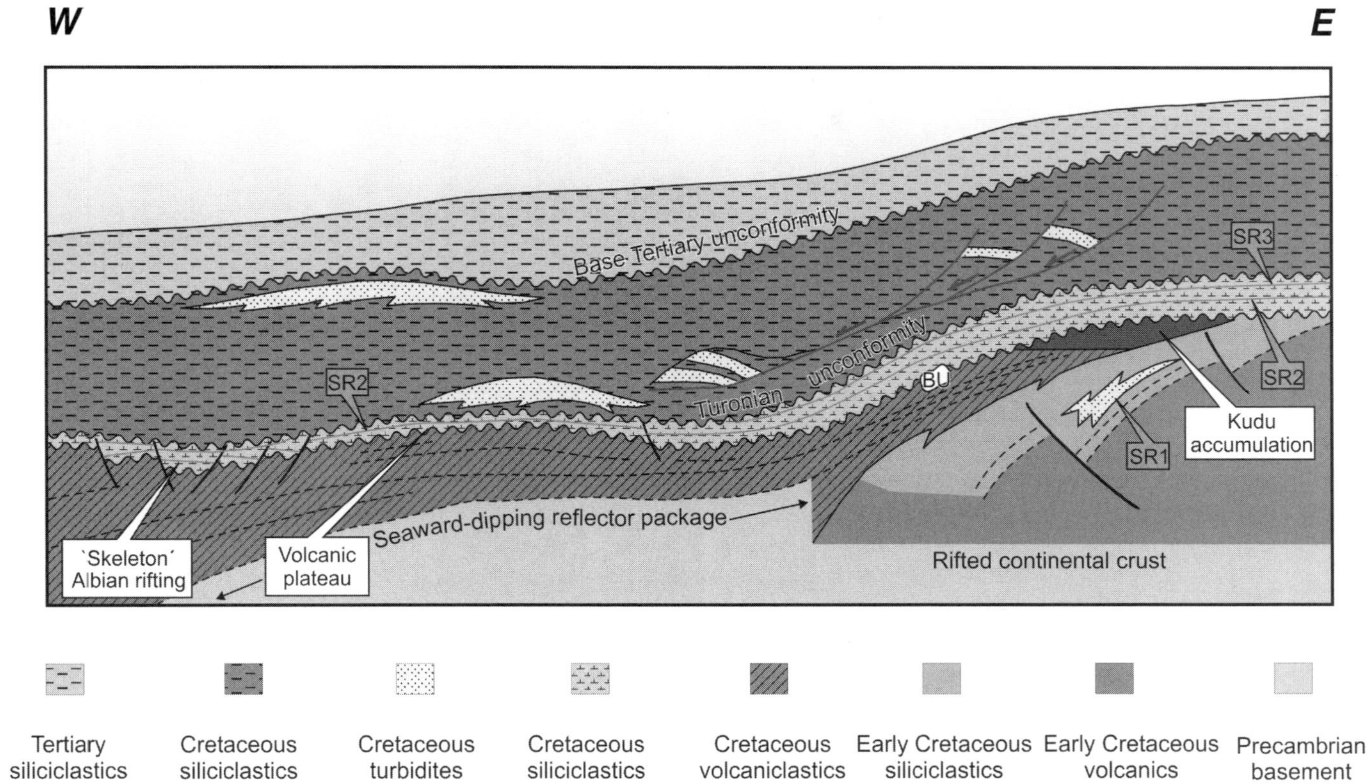

Figure 13. Line-drawing interpretation of regional profile, offshore Namibia. Modified from Bray and Lawrence (1999).

much of the sedimentary basins in the deep-water and ultradeep-water regions of the South Atlantic (Jackson et al., 1998, 2000; Hinz et al., 1999; Marton et al., 2000).

Subequatorial South Atlantic

Some extensional models developed in the past few decades assumed that continental margin rift troughs should thicken basinward toward the boundary between continental and oceanic crust, where the lithospheric stretching factor should reach a maximum (e.g., McKenzie, 1978). Such a model of a basinward thickening of the rift system is illustrated on a regional transect across the West African margin (Fig. 19), which is based on integration of industry seismic and exploratory boreholes in the platform and shelf regions of the Gabon Basin (Palagi, 1998). This interpretation of the Gabon margin as a convergent trough (basinward-thickening) is highly conjectural; it was mainly based on seismic data of a quality not adequate for imaging the structures underlying the salt diapir province, which require seismic profiles with deep resolution (Mohriak et al., 1998a).

Some recent models for continental margin development assume that the rifts may be controlled by depth-dependent stretching factors (e.g., Karner et al., 1997), and depending on these factors, the presalt sequences may pinch out toward the continental–oceanic crust boundary. This implies a landward thickening of the synrift sedimentary layers, and the basin would correspond to a divergent trough (Fig. 20), in a marked contrast with the crustal architecture presented in Figure 19.

With the aim of providing a common framework to analyze the rift architecture of the basins north of the Abrolhos Complex and south of the St. Helen hotspot, we have selected transects from the Petrobras (Fig. 21), Probe, and Spog datasets (Fig. 22) in the subequatorial South Atlantic, and compared the alternative interpretations of the Jacuípe-Sergipe-Alagoas Basins with the conjugate margin in West Africa (Rosendahl et al., 2000; Mohriak et al., 2000b).

The Probe and Petrobras deep-imaging seismic data have been extensively discussed in the context of the widely separated margins (e.g., Rosendahl et al., 1991, 1992; Rosendahl, 1997; Rosendahl and Groschel-Becker, 2000; Meyers, 1995; Meyers et al., 1996a, 1996b; Groschel-Becker, 1996; Mohriak, 1995; Mohriak and Latge, 1991; Mohriak et al., 1993, 1995a, 1995b, 1998a, 2000a).

Parts of these data sets have been examined in a conjugate margin sense (Rosendahl et al., 2000; Mohriak et al., 2000b) to provide cross-sectional structural proxies at conjugate margin positions.

The correlation of geological structures depends on a reasonably accurate conjugate margin fit, and although more than a dozen closures have been presented for this part of the Atlantic, none have been adequately definitive to select matching DIMCS profiles with a high degree of confidence. We have attempted to

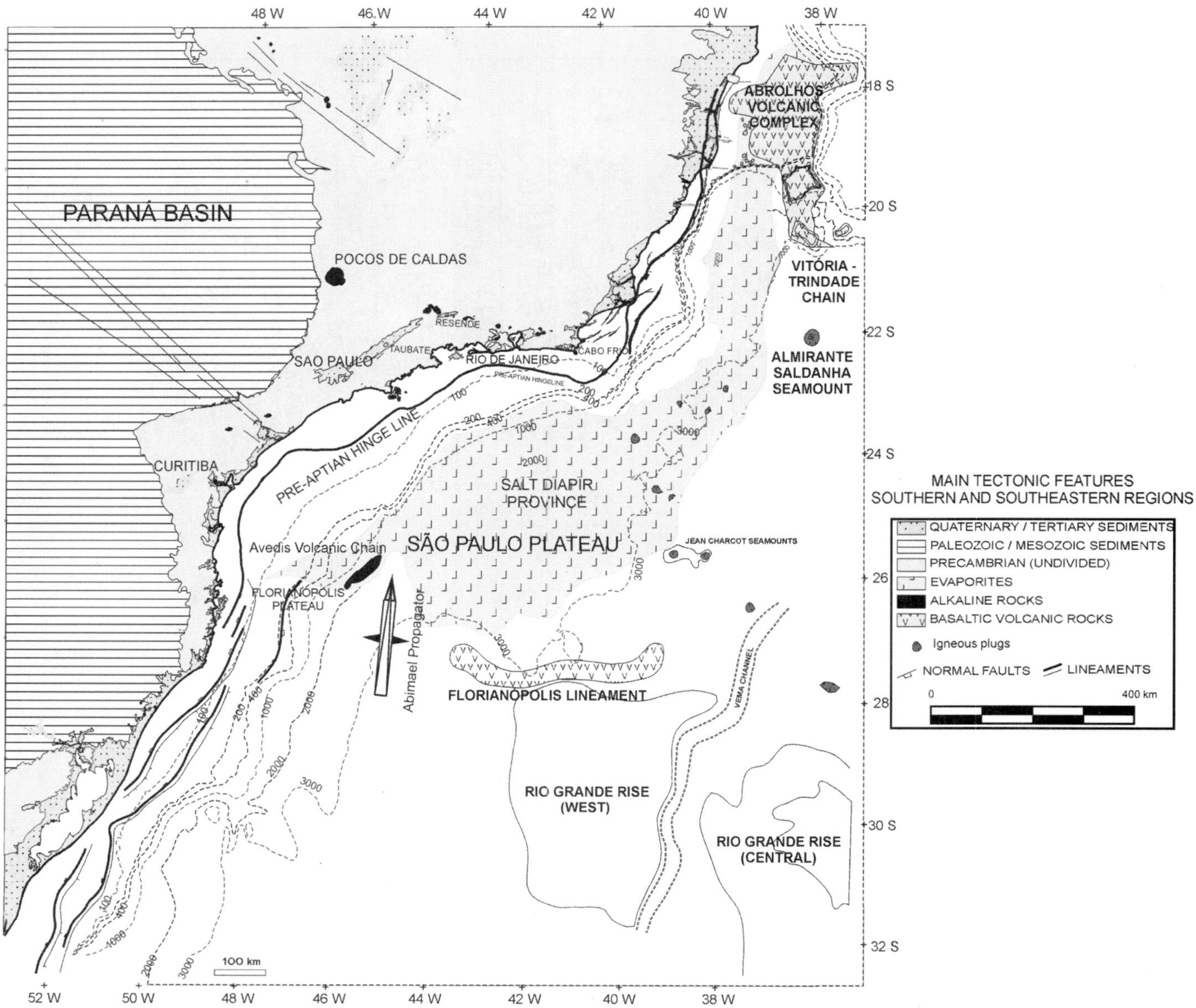

Figure 14. Simplified tectonic map of southeastern Brazil, encompassing Pelotas Basin (south of Florianópolis platform), and Santos, Campos, and Espírito Santo Basins (south of Abrolhos volcanic complex). Modified from Schobbenhaus et al. (1984).

produce a computer-enhanced rendition of the Smith and Sandwell (1997) global ocean bathymetry map based upon satellite altimetry and shipboard bathymetry, supplemented with the DEOS Altimetry Atlas (e.g., Wisse et al., 1994), which was then closed to magnetic anomaly 34, or ca. 80 Ma, using the magnetic lineation maps of Cande et al. (1988) and Nürnberg and Müller (1991).

A major difficulty is that magnetic anomaly identifications in this part of the Atlantic become progressively worse toward the equator and the placement of anomaly 34 should be considered approximate. The major relations in the reconstructed palinspastic map of the subequatorial South Atlantic (Fig. 23) are clear, particularly the importance of the Precambrian fabrics in controlling the structural grain of the rifts and the location of fracture zones (Fig. 24). The boundary between continental and oceanic crust is still a puzzle; some researchers advocate the extension of the rift sequence toward the ultradeep-water region (e.g., Daily, 2000), whereas others suggest that the transition is marked by wedges of SDRs (e.g., Fig. 25; Bray and Lawrence, 1999).

The fracture zones may divide the continental margin into several compartments with independent depocenters, affect the rift architecture, and also separate crustal domains along oceanic transforms. Although possible links between onshore and offshore structures in West Africa have been suggested, e.g., by Francheteau and Le Pichon (1972), Tesserence and Villemin

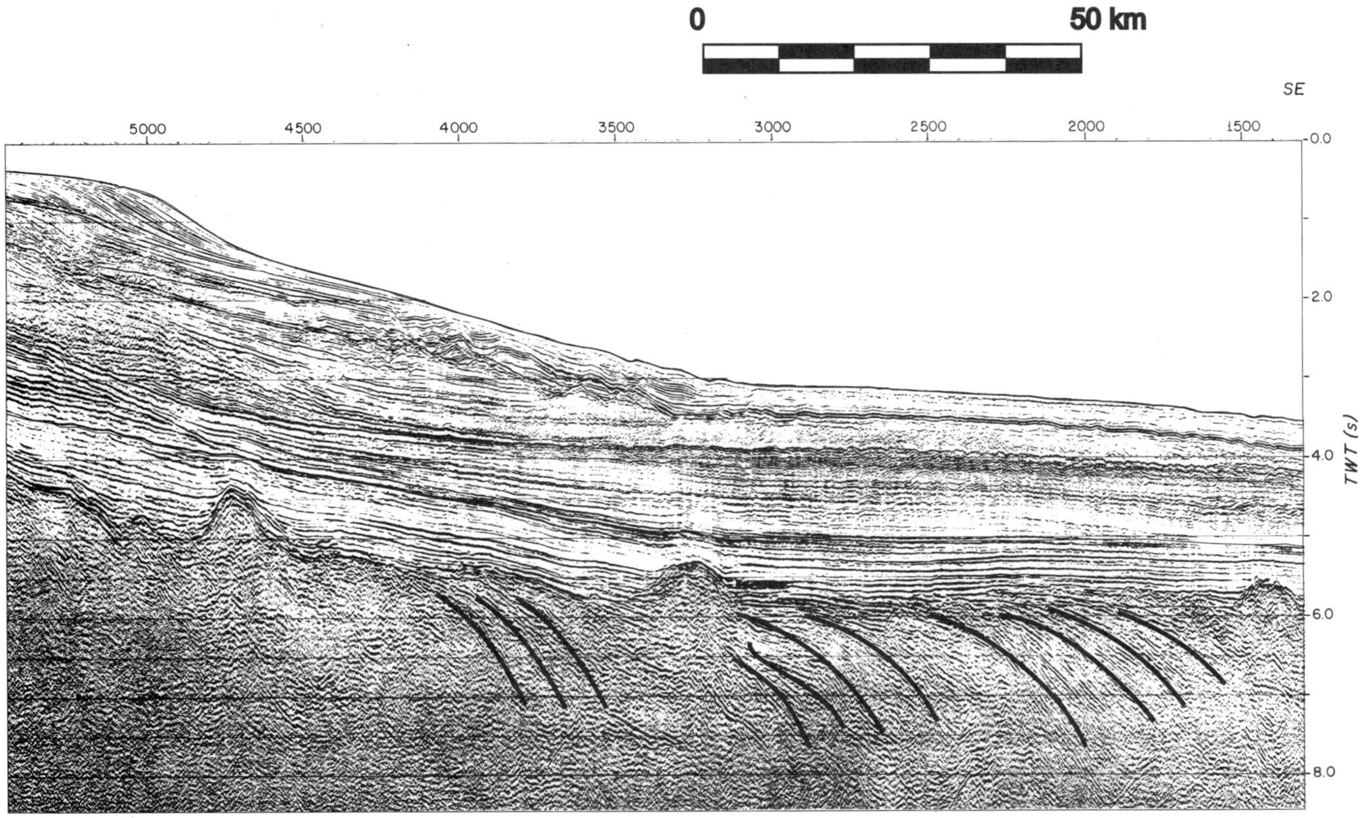

Figure 15. Regional seismic profile in Pelotas Basin with interpretation. TWT = two-way traveltime. Modified from Abreu (1998); Cainelli and Mohriak (1998); Talwani and Abreu (2000).

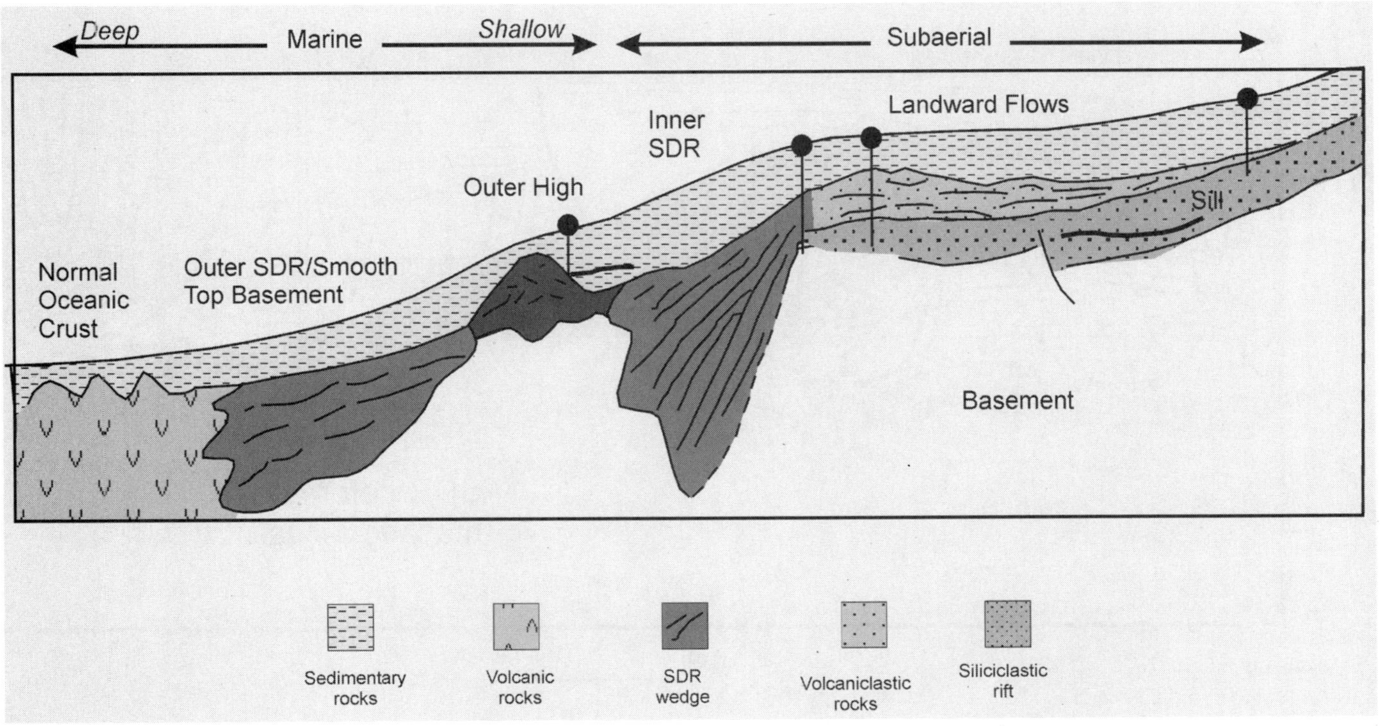

Figure 16. Schematic model of volcanic margin with seaward-dipping wedges. SDR = seaward-dipping reflector. Modified from Planke et al. (2000).

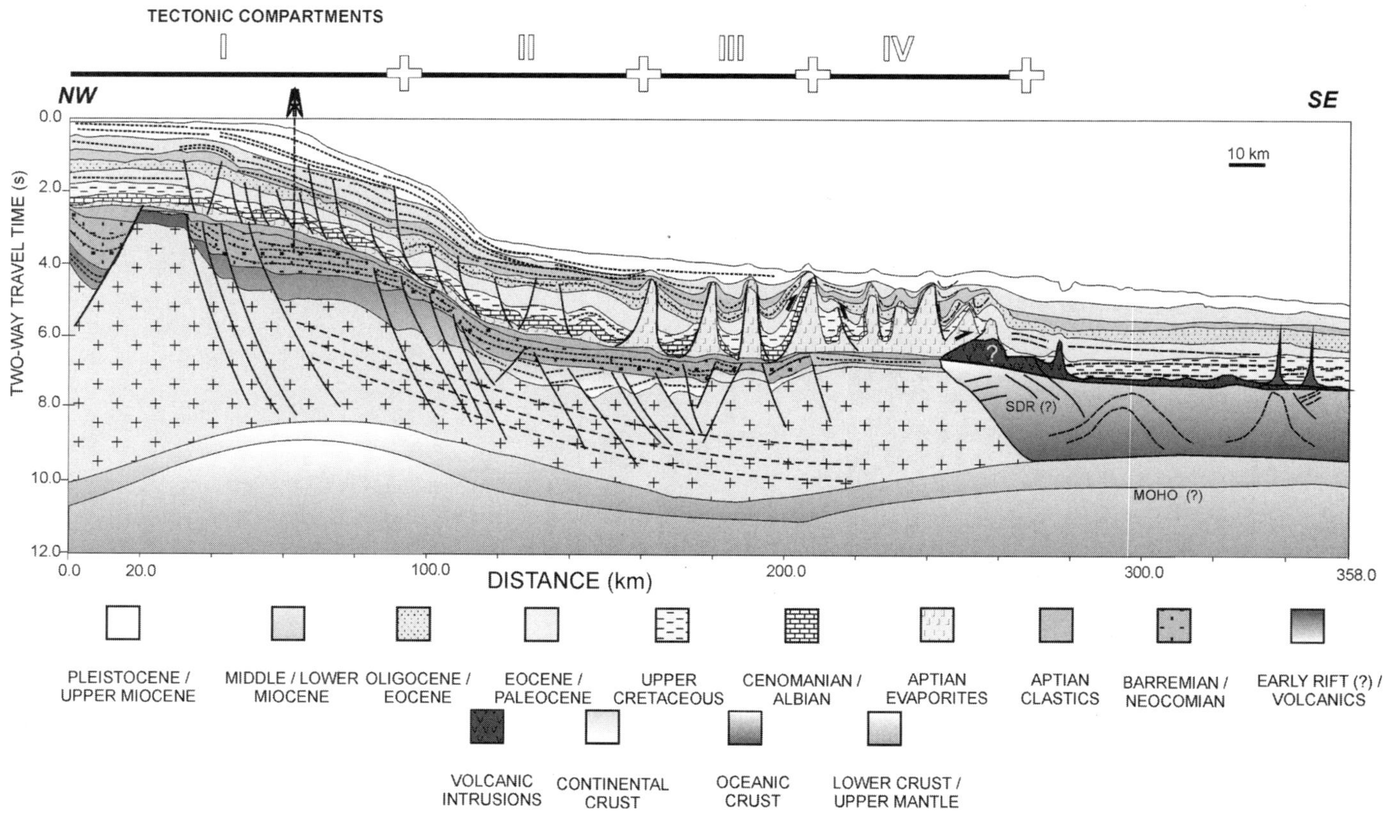

Figure 17. Schematic line-drawing interpretation of regional transect across Campos and Santos Basins, showing crustal architecture and tectonic compartments related to salt tectonic provinces (I–IV). Modified from Mohriak et al. (1998b).

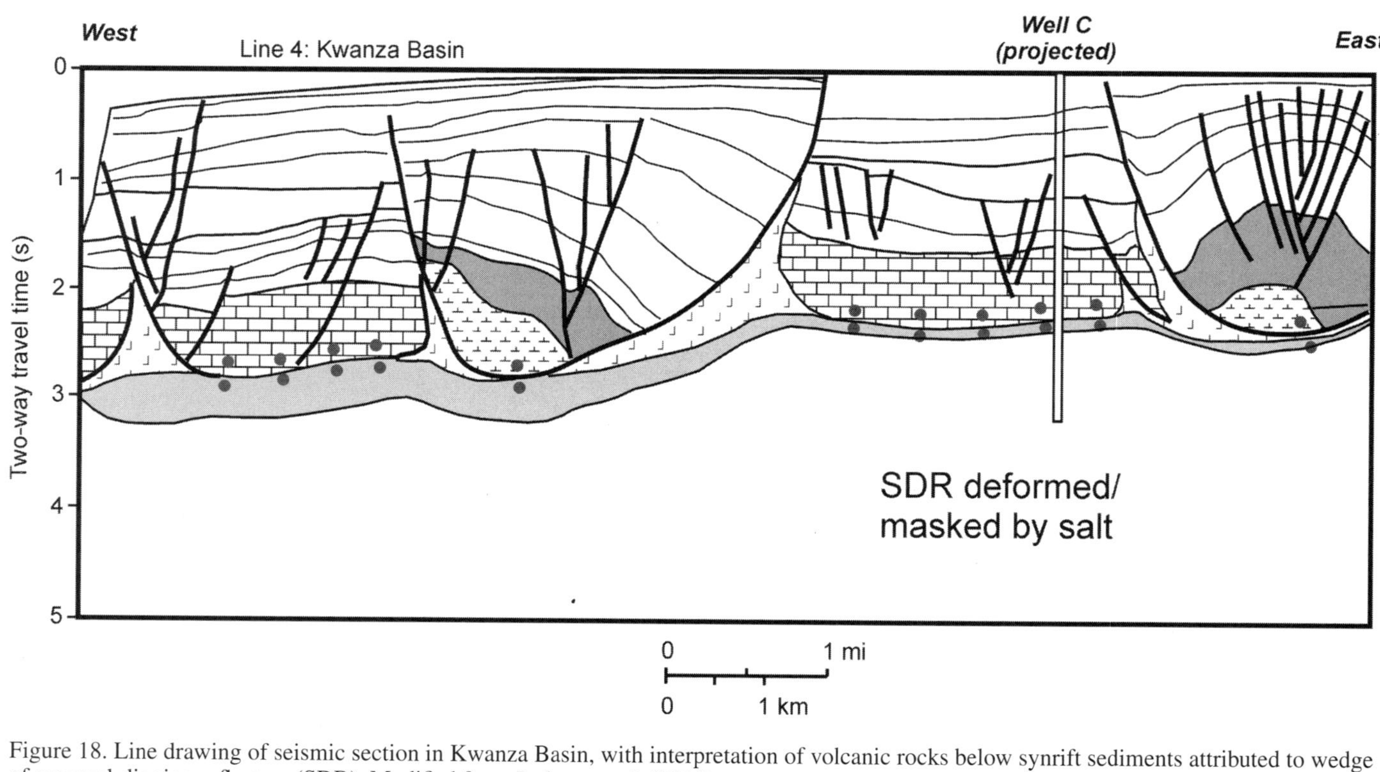

Figure 18. Line drawing of seismic section in Kwanza Basin, with interpretation of volcanic rocks below synrift sediments attributed to wedge of seaward-dipping reflectors (SDR). Modified from Jackson et al. (2000).

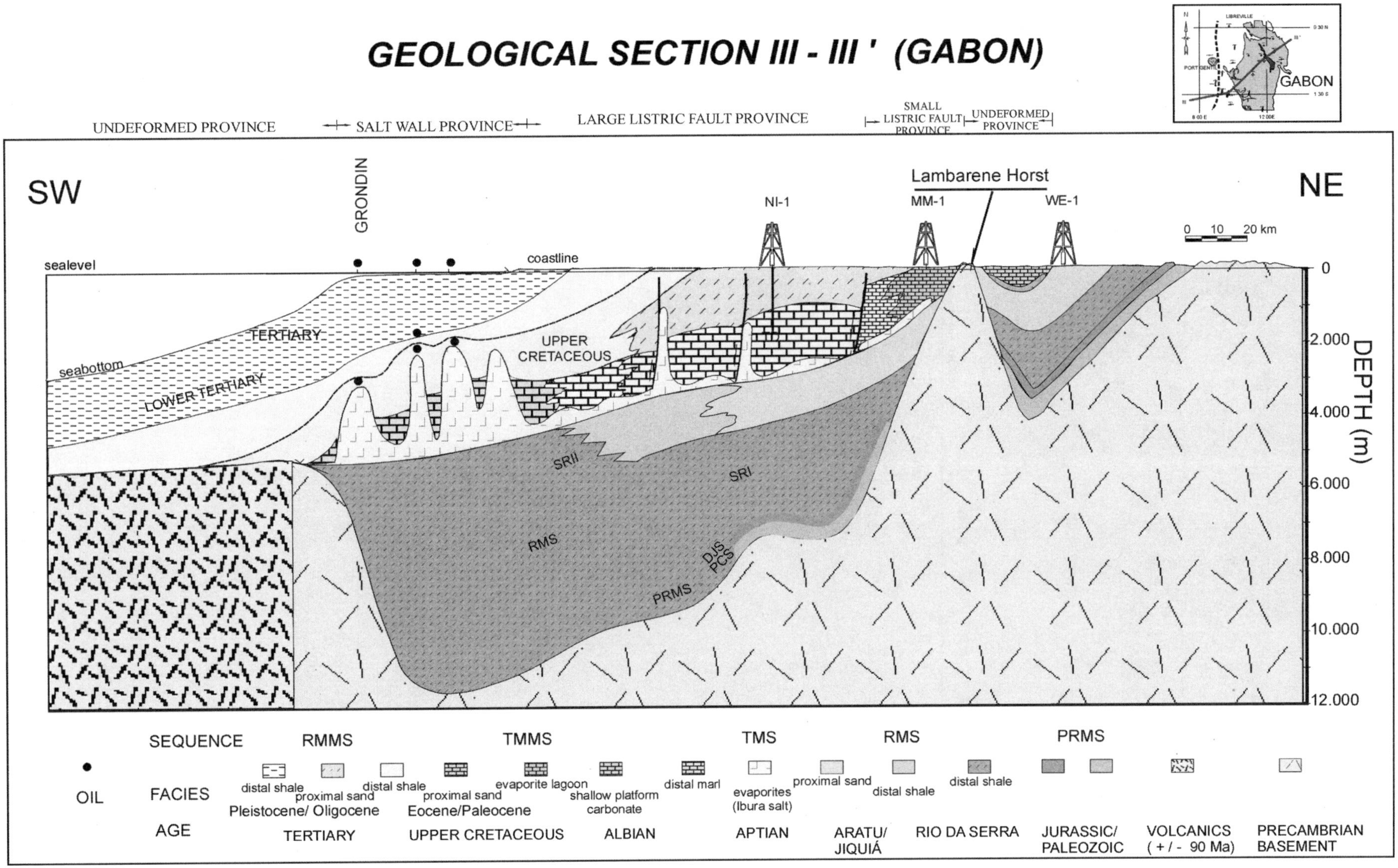

Figure 19. Schematic geological section in Gabon Basin showing interpretation of rift architecture as basinward-thickening trough (according to Palagi, 1998).

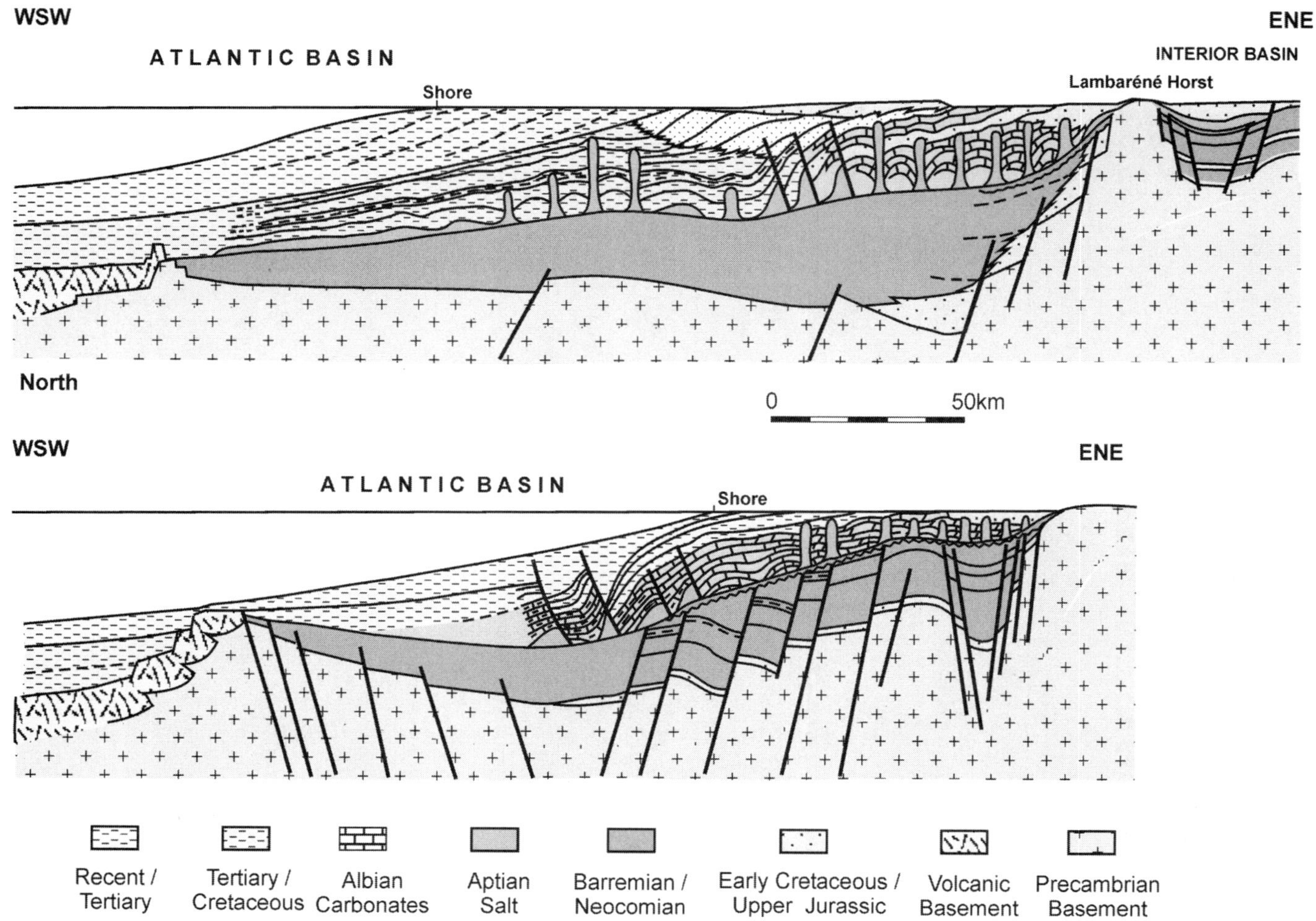

Figure 20. Schematic geological sections in Gabon Basin showing alternative interpretation of rift architecture as landward-thickening trough. Modified from Schlumberger (1983).

(1989), Wannesson et al. (1991), Groschel-Becker (1996), Meyers (1995), and Meyers et al. (1996b), the imprint of the prerift fabric on synrift tectonics may have been understated.

The Ascension Fracture Zone is one of the most important fracture zones in West Africa (Fig. 24). It is a double or multiple fracture zone in plan view that indisputably continues into the Kribi Fracture Zone (KFZ) in Rio Muni (Burke, 1969). There is a close association of the Kribi Fracture Zone with the onshore boundary of the Pan-African tectonic complex and the Archean Congo craton (Figs. 23 and 24).

The Kribi Fracture Zone continuation is well displayed in the Probe DIMCS data and its eastern end is crossed by several grids of conventional MCS data (Turner, 1999). In the Probe tectonic models (e.g., Rosendahl and Groschel-Becker, 2000) the Kribi Fracture Zone constitutes a transform fault margin dividing oceanic crust to the north from stretched continental crust to the south. It originally divided distinct styles of nearly contemporaneous extensional deformation. North of the Ascension Fracture Zone, the lithosphere ripped apart cleanly and abruptly, resulting in a very narrow margin that subsided almost catastrophically as seafloor spreading commenced. This is particularly evident off southern Cameroon and Equatorial Guinea (Rosendahl and Groschel-Becker, 2000). Extension south of the Ascension Fracture Zone produced a broad swath of attenuated continental lithosphere. An important consequence of these models is that continental stretching factors varied dramatically between adjacent transform-bounded segments of this part of the South Atlantic margin.

The influence of the transform faults on the margin architecture is clearly illustrated in Figure 26, which shows uplifted rift blocks along the N'Komi Fracture Zone in Gabon. Figure 27 shows a detailed view of this Fracture Zone, which is responsible for uplift and erosion of the rift blocks in the deep-water region. The N'Komi Fracture Zone (Fig. 26) is a strand of the complex Ascension transform system, which is crossed at high angles in whole or part by at least seven different Probe lines (lines 11 and 13).

Two major transforms are observed between the Salvador and the Pernambuco-Paraíba fracture zones (Fig. 21) along the Brazilian conjugate margin: the Sergipe Fracture Zone in the

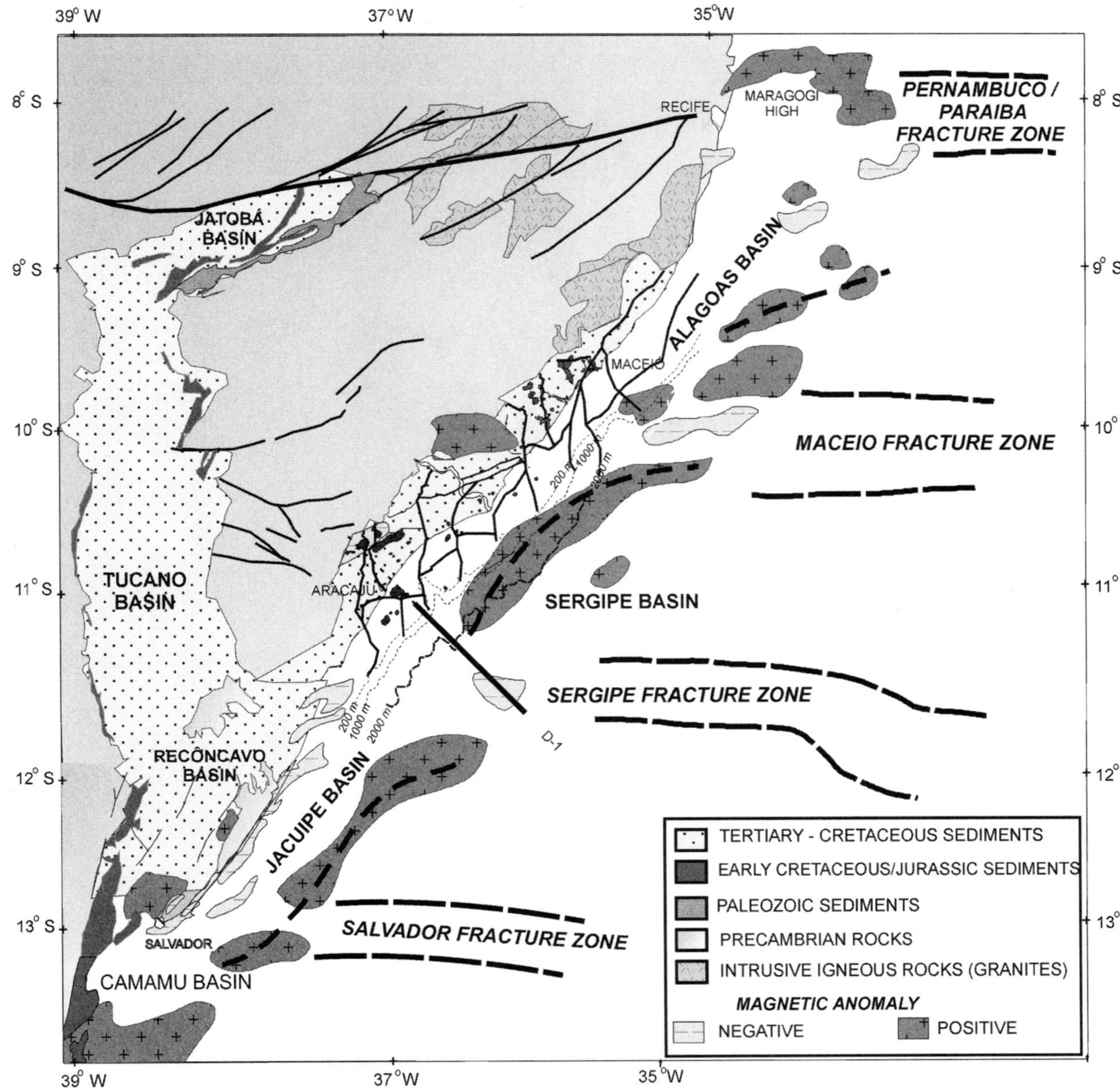

Figure 21. Simplified tectonic map of northeastern Brazil with main oceanic fracture zones, magnetic anomalies, and location of regional deep seismic profiles.

southern segment of the Sergipe-Alagoas basin, and the Maceió Fracture Zone in the northern segment. The Sergipe Fracture Zone deflects to the northwest and continues on land as a major fault controlling the rift depocenter in the Mosqueiro low and extends toward the onshore Tucano rift, where it may coincide with an accommodation zone (Mohriak et al., 2000a).

The Maceió Fracture Zone separates the Maceió subbasin from the Pernambuco-Paraíba basin, and is associated with a cluster of igneous plugs aligned in an east-west direction. The Maceió Fracture Zone is the landward continuation of the Ascension Fracture Zone, and thus loosely correlates to the Kribi Fracture Zone in West Africa.

The Maceió and Sergipe fracture zones have been imaged by only a few Petrobras DIMCS profiles, but they are crossed in whole or part by numerous conventional MCS lines. The fracture zone is also well covered by magnetic and gravimetric data sets. Figure 28 illustrates the seismic signature of the Sergipe Fracture Zone, which apparently offsets the seismic reflection Moho and is marked by an igneous plug near the boundary between the continental and oceanic crust. The profile segment in Figure 28 is comparable, in an isochronal sense, to Probe lines 11 and 13.

A regional deep seismic profile in the Gabon Basin (Probe line 23) is shown as Figure 29 (uninterpreted seismic) and Fig-

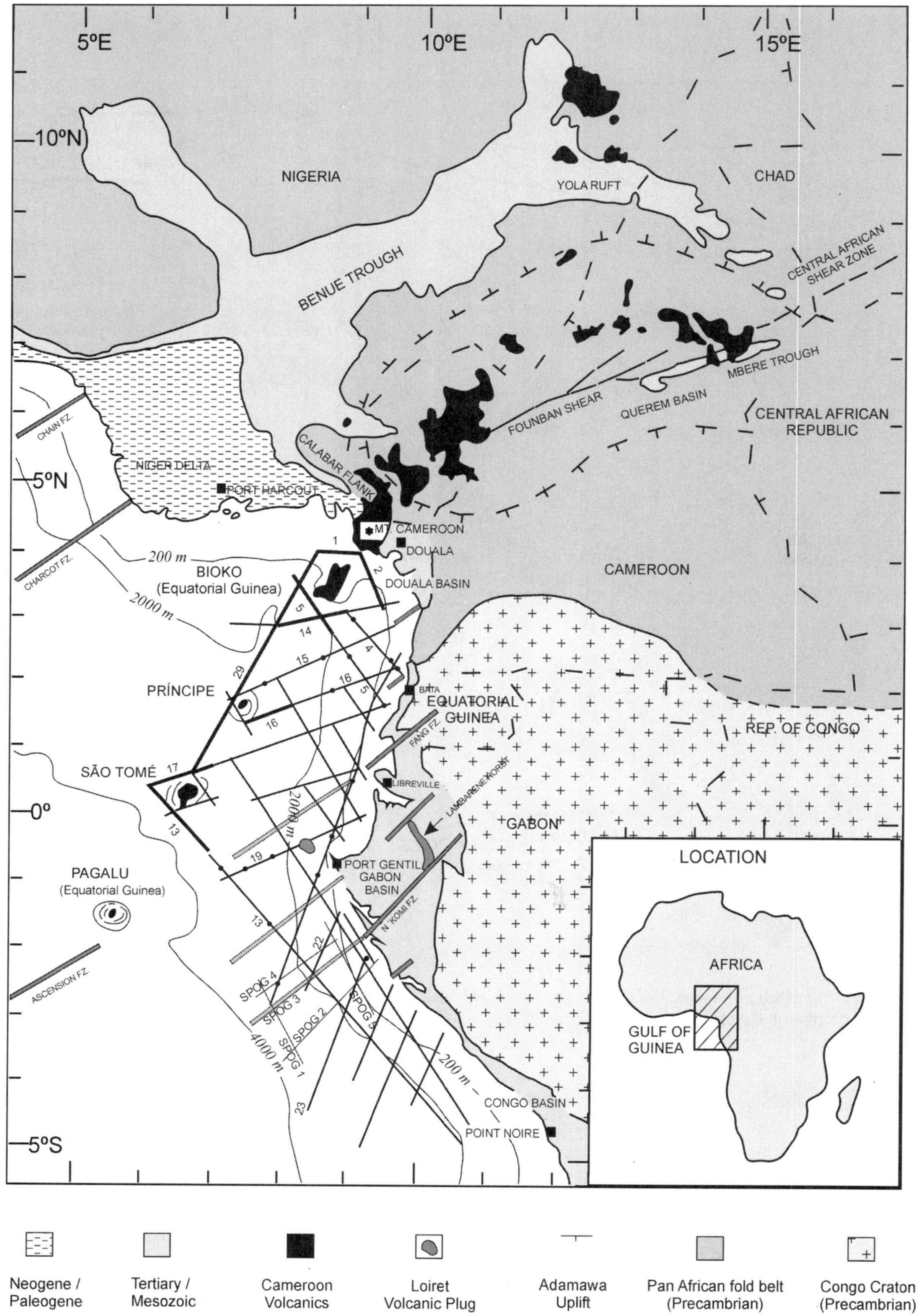

Figure 22. Simplified tectonic map of northwestern Africa with main oceanic fracture zones, volcanic centers, and locations of regional deep seismic profiles (Probe and SPOG data sets).

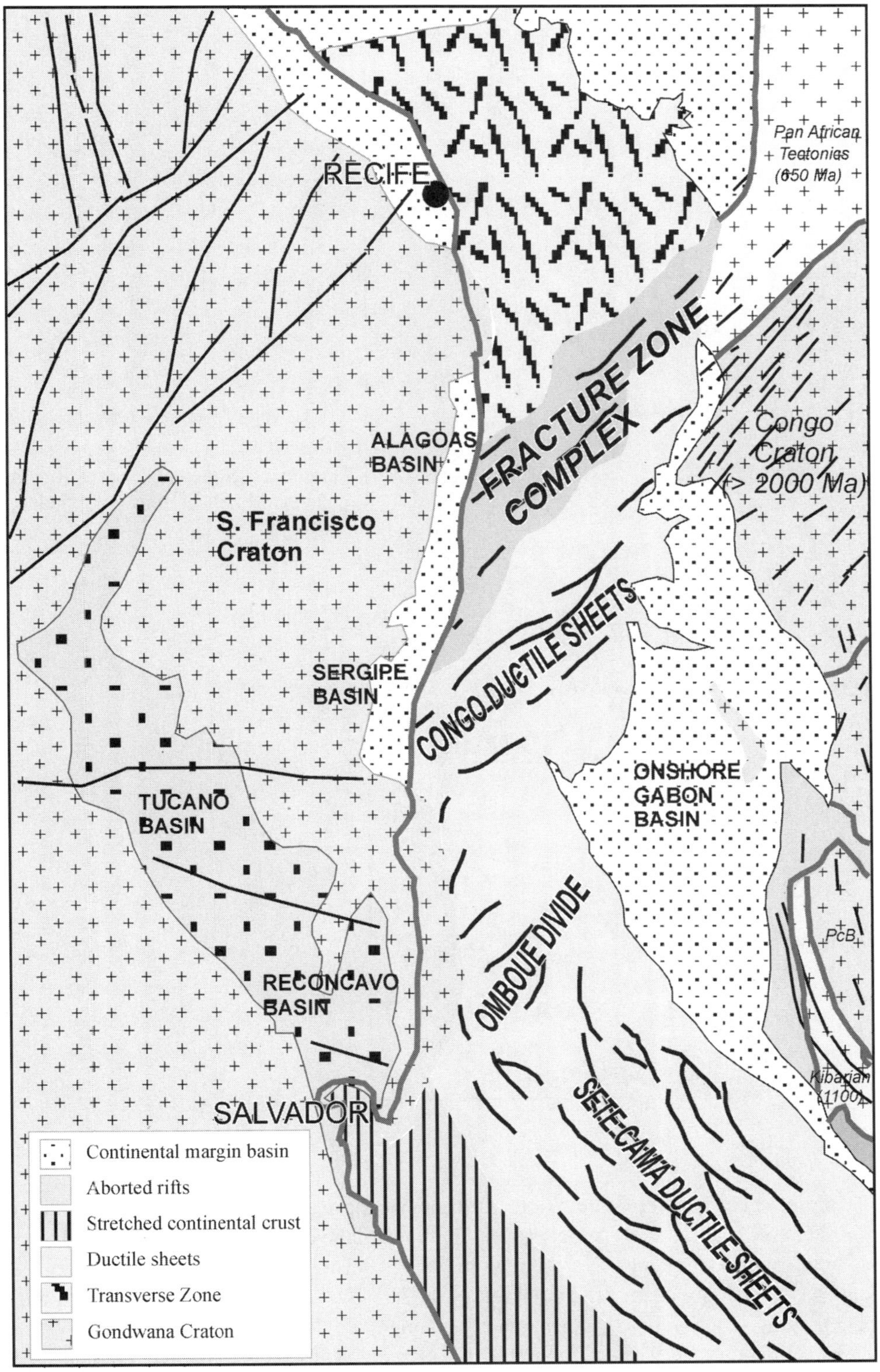

Figure 23. Palinspastic reconstruction of subequatorial South Atlantic.

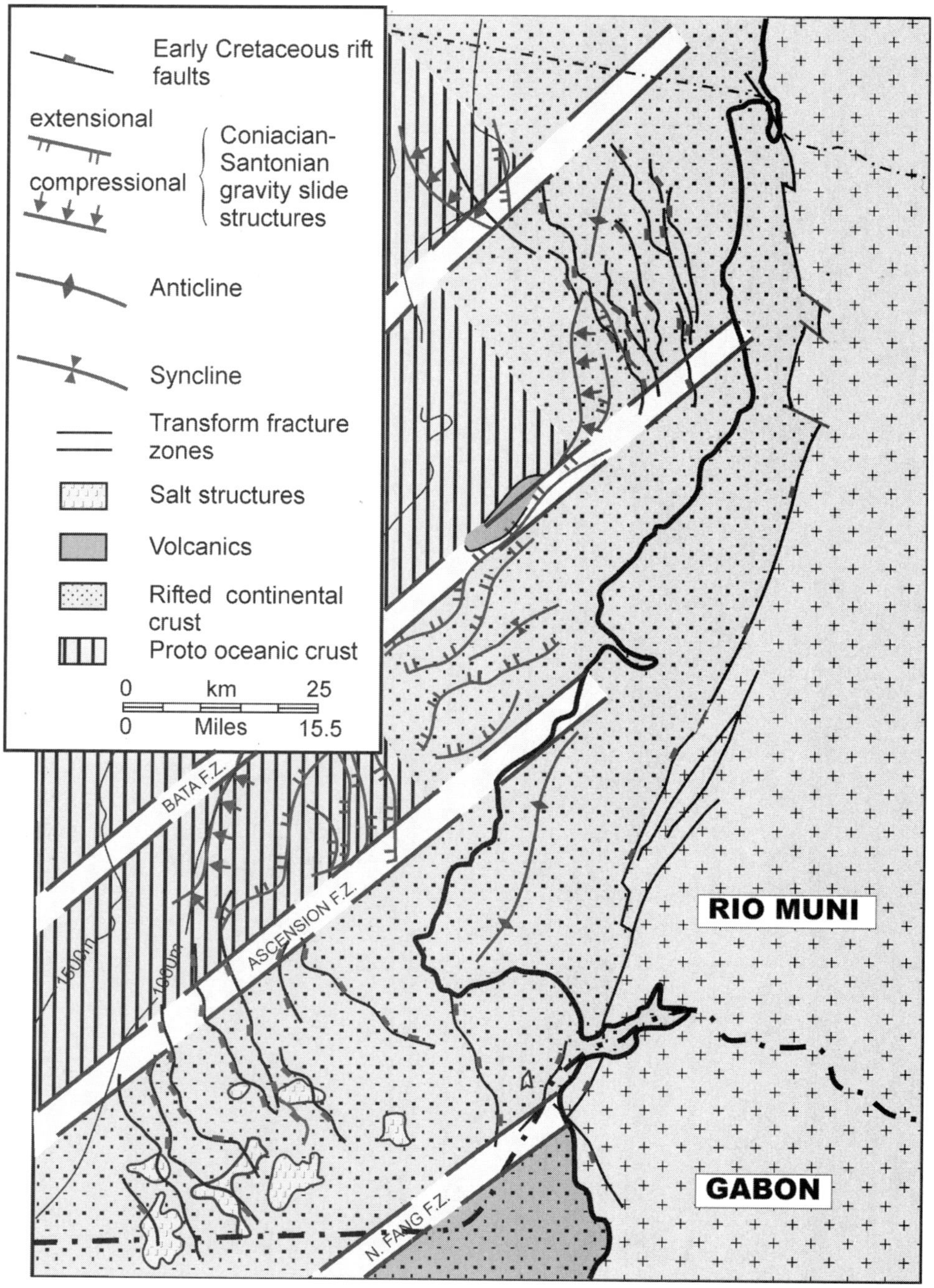

Figure 24. Structural map of Rio Muni region. Modified from Bray and Lawrence (1999).

ure 30 (schematic line interpretation). This profile is characterized by an array of strong reflectors beginning at ~7 s along the eastern panel, which evolve (or coalesce) seaward into the event at 8–10s. The top of this reflector package has previously been interpreted to mark a lower crustal to upper mantle transition between brittle and ductilely deformed rocks (Rosendahl, 1997; Rosendahl and Groschel-Becker, 2000). The interpretation describes this event as a detachment zone that evolves seaward from relatively unstretched lenses of Congo fold belt structures and toward the oceanic crust, apparently amalgamating with the seismic Moho. Crust above the detachment appears to irregularly but progressively thin toward the oceanic-continental boundary.

A regional deep seismic profile in the Sergipe Basin (Mohriak et al., 1998a, 2000a) on the Brazilian margin that is approximately conjugate to the Gabon line 23 profile is shown in Figure 31 (location map with palinspastic reconstruction) and Figure 32 (seismic data).

The Sergipe seismic line extends from the Mosqueiro low on the shelfal platform to the deep-water province. A detailed analysis of all geological and geophysical data indicates several

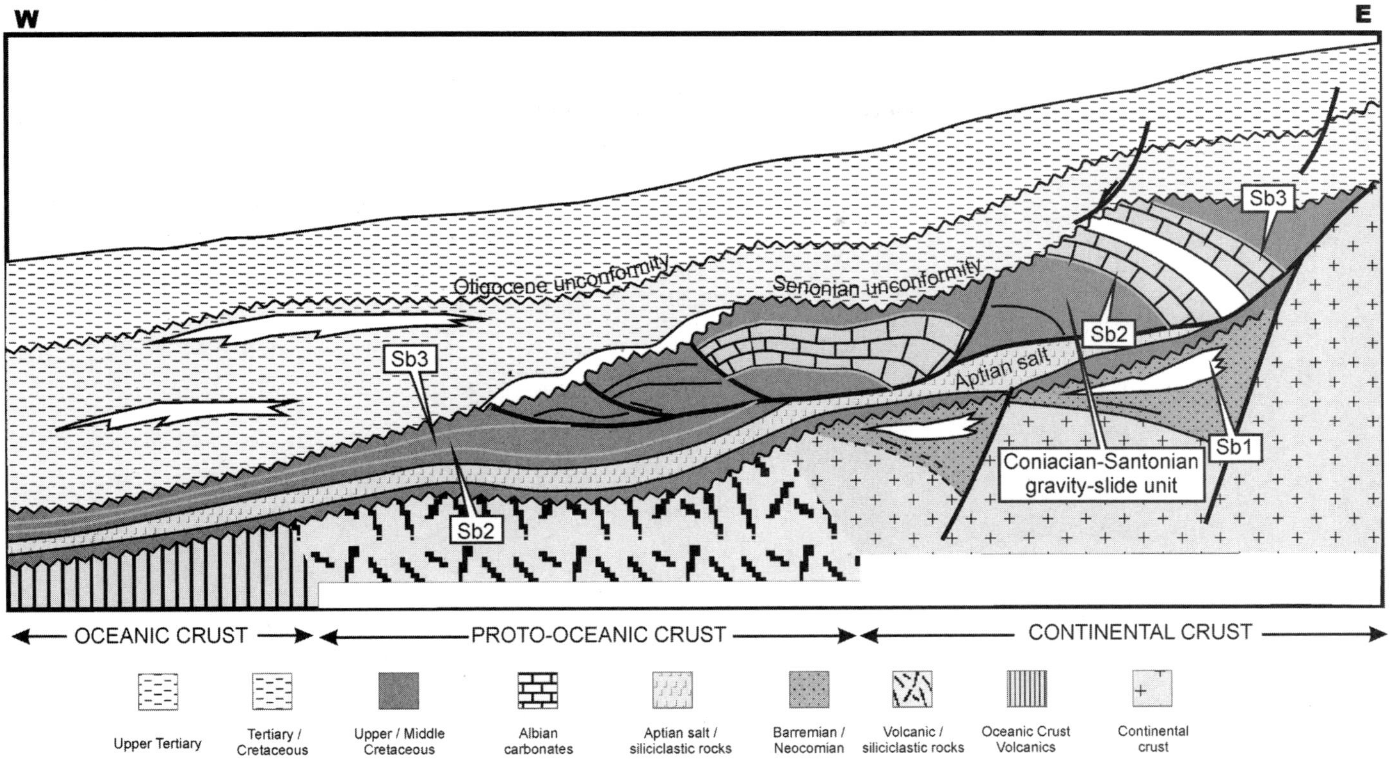

Figure 25. Rio Muni section. Modified from Bray and Lawrence (1999).

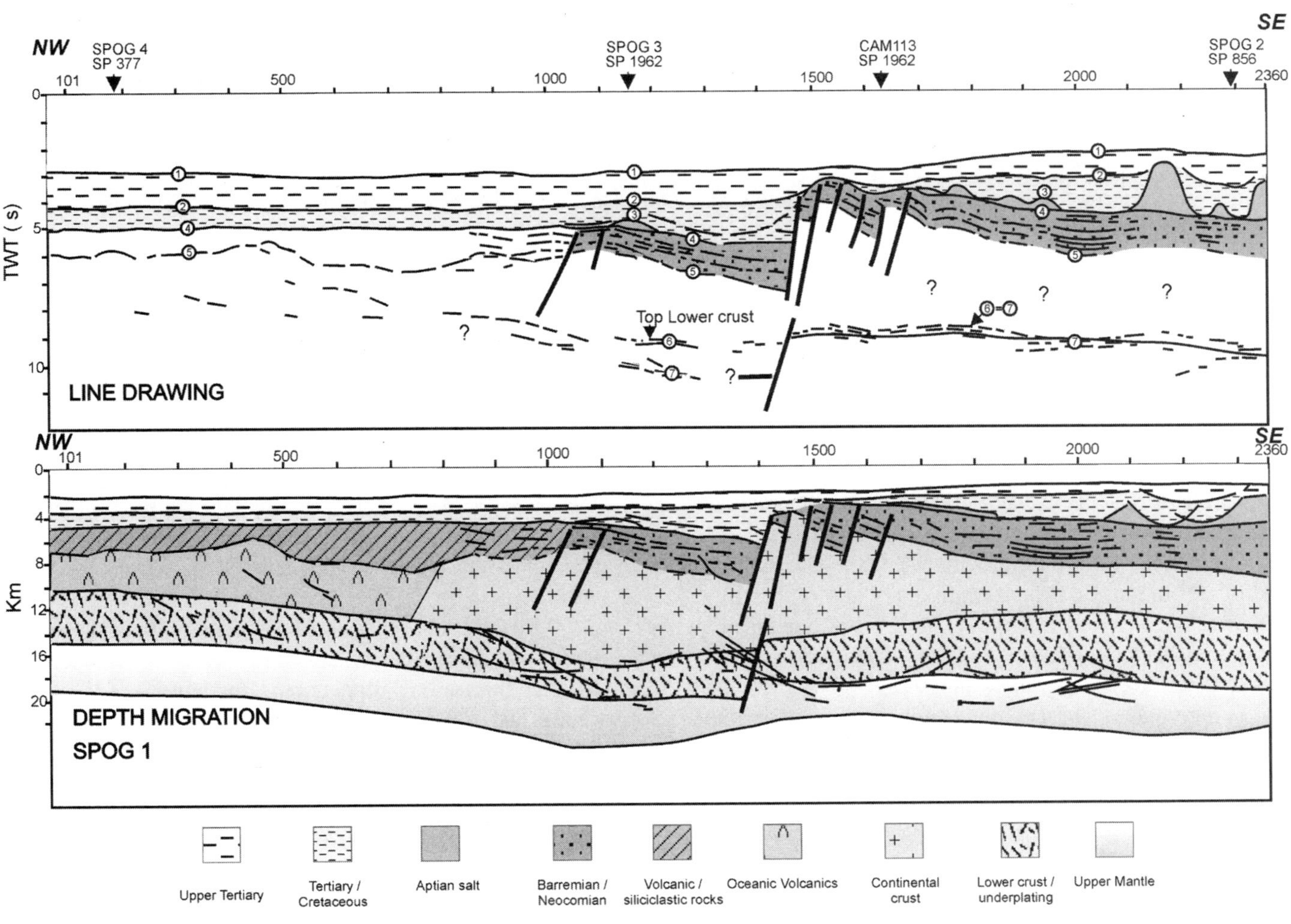

Figure 26. Schematic line drawing interpretation of regional deep seismic profile (SPOG profile 1) in Gabon Basin (top), with depth-converted interpretation (bottom). Modified from Wannesson et al. (1991).

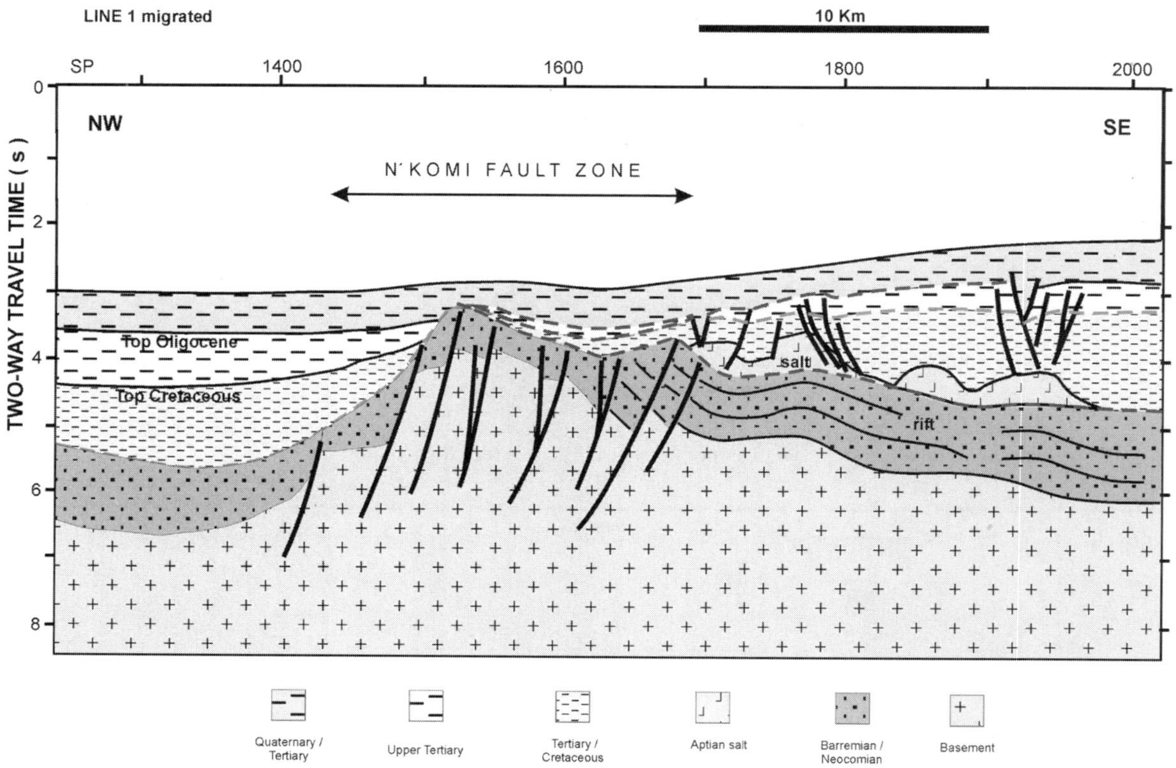

Figure 27. Schematic line drawing interpretation of regional deep seismic profile (SPOG profile 1) showing detail of N'Komi fracture zone. Modified from Wannesson et al. (1991).

alternative interpretations for the margin (Pontes et al., 1991; Mohriak et al., 1995b).

One (Fig. 33, top) suggests a very thick syn-rift trough basinward of the shelf break, equivalent to the rift sediments that have been drilled by several exploratory boreholes onshore and on the platform; the other (Fig. 33, bottom) interprets a basinward pinchout of the rift sequence adjoining a seismic reflector package interpreted as a seaward-dipping wedge.

These two end-member models are each bolstered by several favorable and unfavorable arguments, such as the apparent corroboration of the seaward-dipping wedge by gravimetric and magnetic modeling (Mohriak et al., 1998a).

Obviously, these different interpretations have profoundly different implications concerning the width of the rift system, thermal maturation of possible source rocks, and the proper approaches to be taken to petroleum exploration in these deep-water provinces (Mohriak et al., 1995b).

Figure 34 shows a possible conjugate fit of Probe line 23 to the Petrobras profile, juxtaposing the preferred interpretation of the Sergipe Basin transect in Figure 32, presented in Mohriak et al. (1998a), with the preferred interpretation of the Probe line by Rosendahl et al. (2000).

The differences in interpretation here mainly pertain to the preferred pairs of lines to conjugate, the preferred positioning of the oceanic-continental boundary on the Petrobras lines, and the nature of the material interpreted as SDR, LDR (landward-dipping reflectors), and Moho (Mohriak et al., 1998a).

Rather than attempt to resolve the differences here, we prefer to let them stand as an illustration of the types and degrees of uncertainty that currently exist in fitting together these conjugate margins with deep-imaging seismic data. Regardless of the differences, it seems abundantly clear that the Sergipe and middle to southern Gabonese margins rifted apart in a highly asymmetric fashion, suggestive of a simple shear mechanism replete with one or more low-angle detachments. In the context of this special volume, it is interesting to note that the symmetric (pure shear) mechanism for SDR development shown in Figure 6, and possibly applicable farther south, may not be appropriate in this region.

Both the eastern Brazilian and the West African models recognize that the crust is probably intruded during or after rifting, although the seaward-dipping reflectors interpreted as volcanic wedges and other structures inferred in the Sergipe Basin (LDR, plugs) are not interpretationally justified in the Gabon data set. The seismic data do not suggest or require voluminous emplacement of synrift igneous rocks within the continental crust in either model. To the contrary, the models imply an apparent dearth of prerift and synrift magmatism at these margins. The interpretation of underplating in both models, although purely subjective, is consistent with the observation of strong arrays of

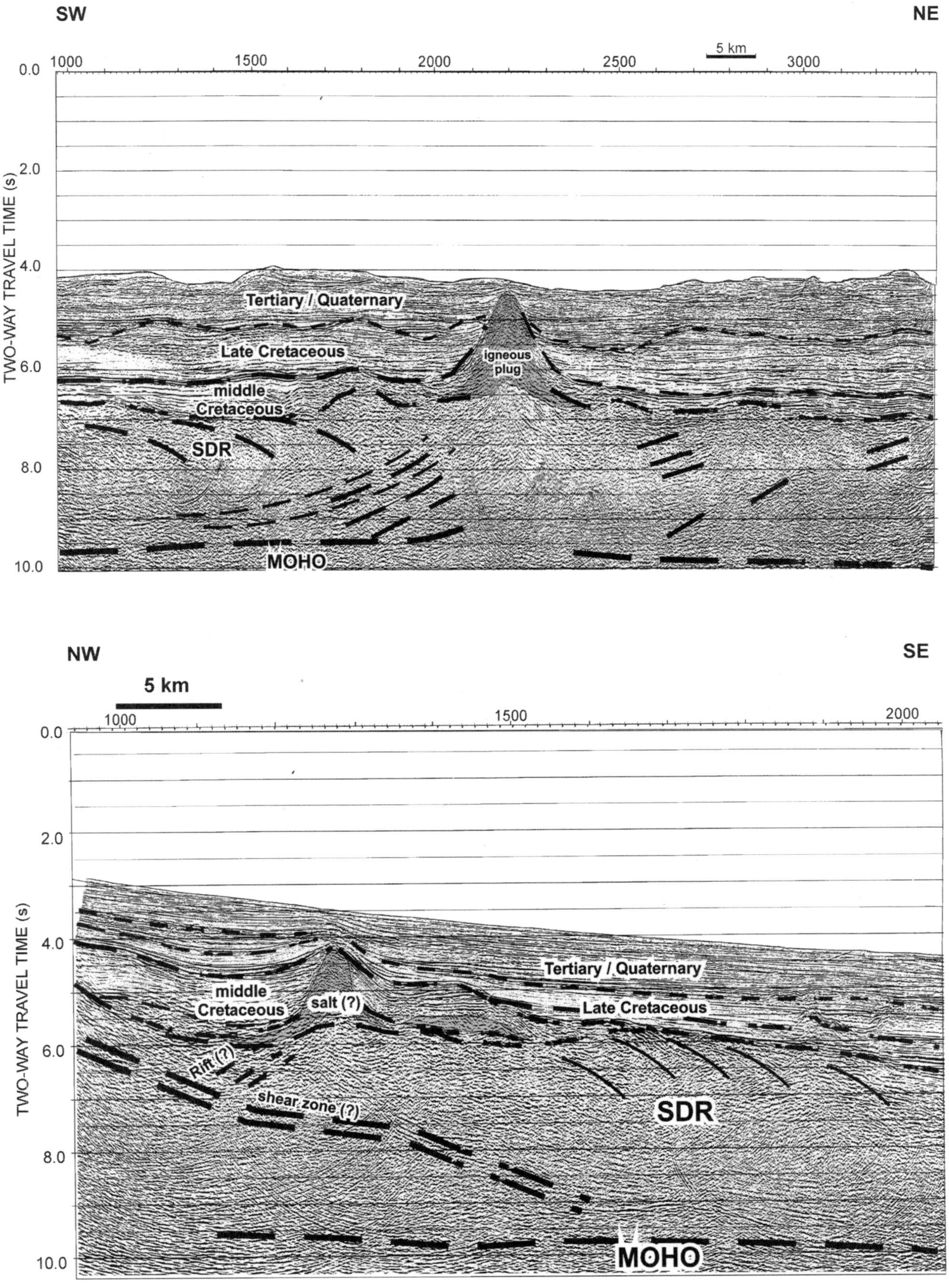

Figure 28. Part of regional deep seismic profile in Sergipe Basin with interpretation showing volcanic plug associated with Sergipe fracture zone (top) and possible salt diapir near the boundary between continental and oceanic crust (bottom).

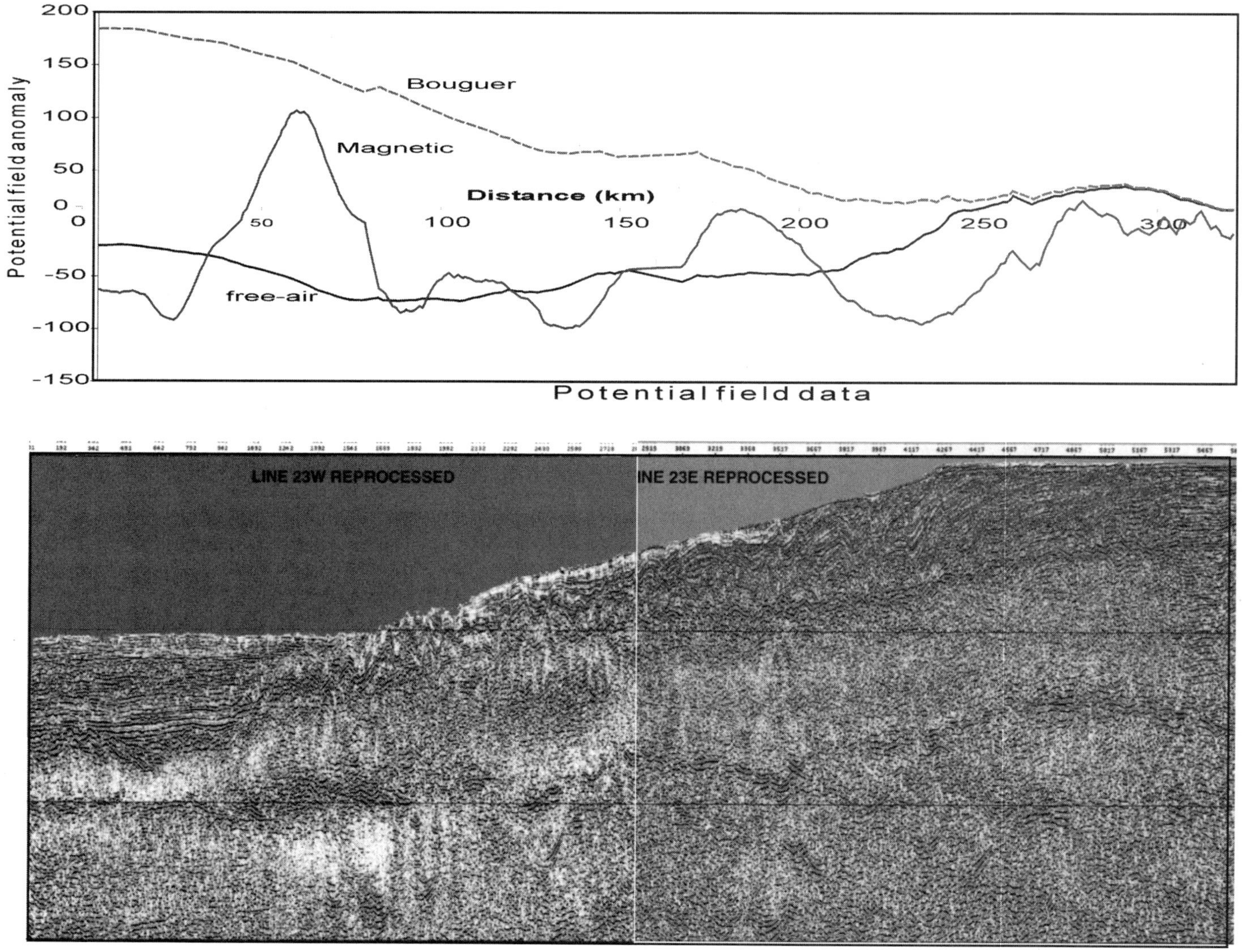

Figure 29. Regional deep seismic profile (Probe Line 23) in the Gabon Basin (top: earlier processing; bottom: reprocessed).

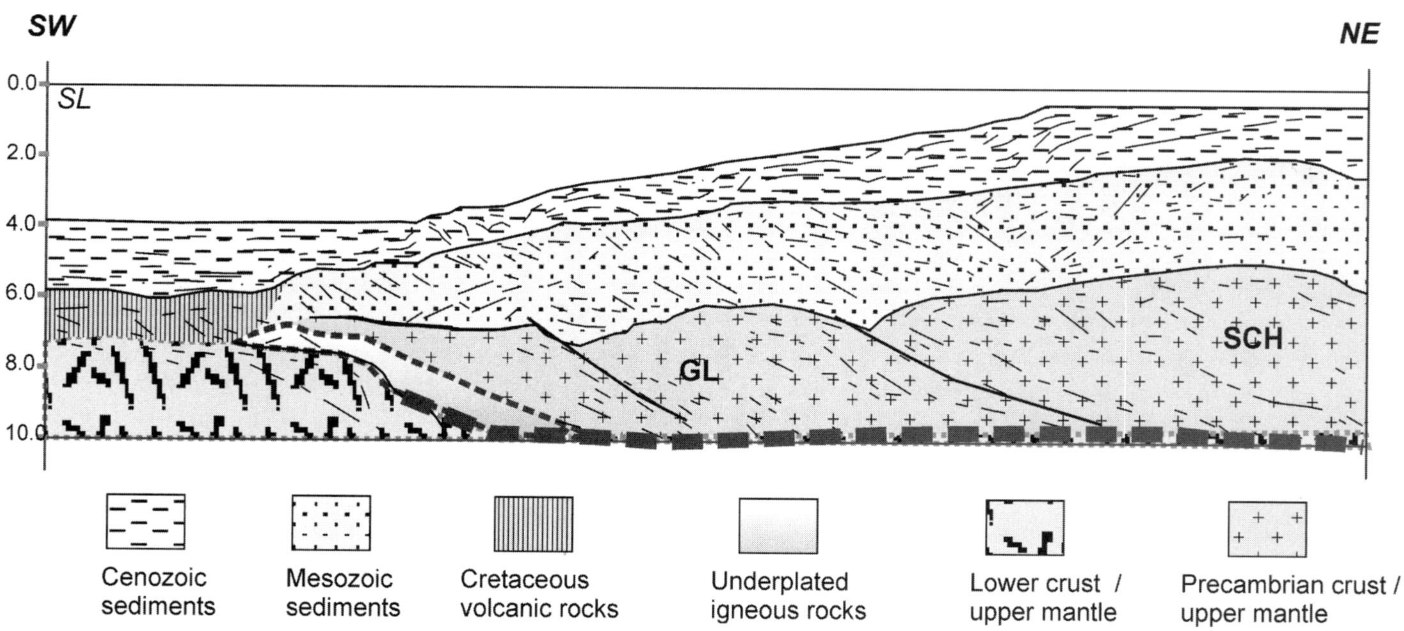

Figure 30. Schematic line drawing interpretation of regional deep seismic profile (Probe Line 23) in the Gabon Basin. The upper portion of the profile is characterized by postrift sediments, the intermediate portion corresponds to the presalt sediments and continental crust, and the lower portion corresponds to lenses of lower crust and upper mantle material forming ductile sheets bounded by landward-dipping shear zones. The oceanic Moho is interpreted at ~10 s two-way traveltime in the western end of the section. Modified from Rosendahl et al. (2000).

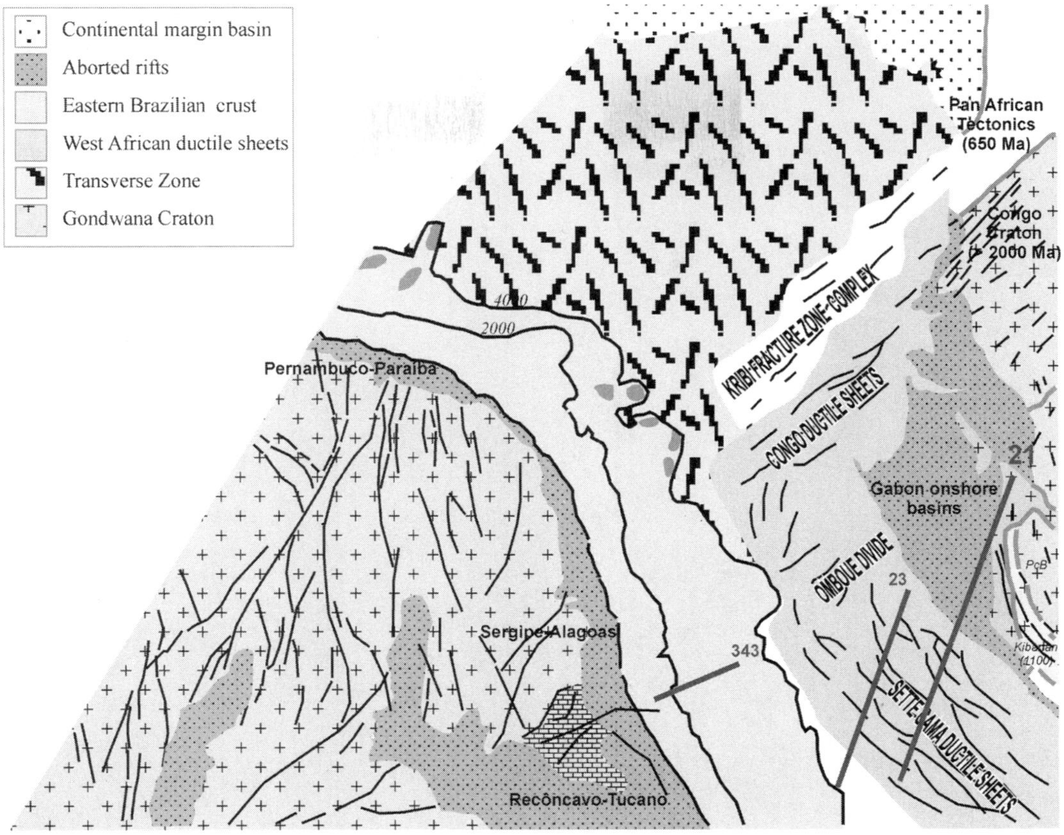

Figure 31. Palinspastic map of subequatorial South Atlantic rift system with location of conjugate transects (lines 343 and 341 offshore Brazil and Probe lines 21 and 23) (from Mohriak et al., 1998).

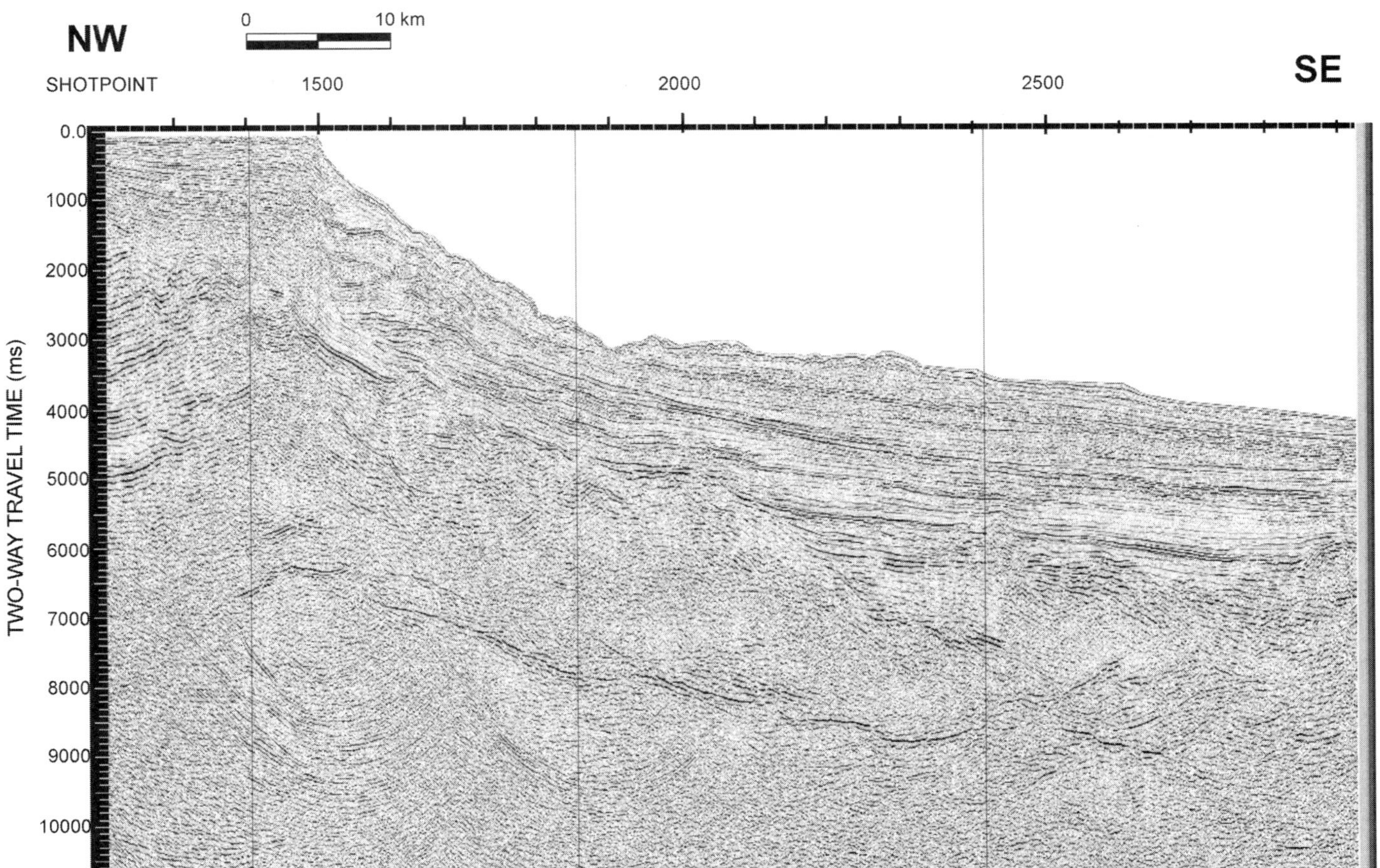

Figure 32. Regional deep seismic profile in Sergipe Basin (line 343).

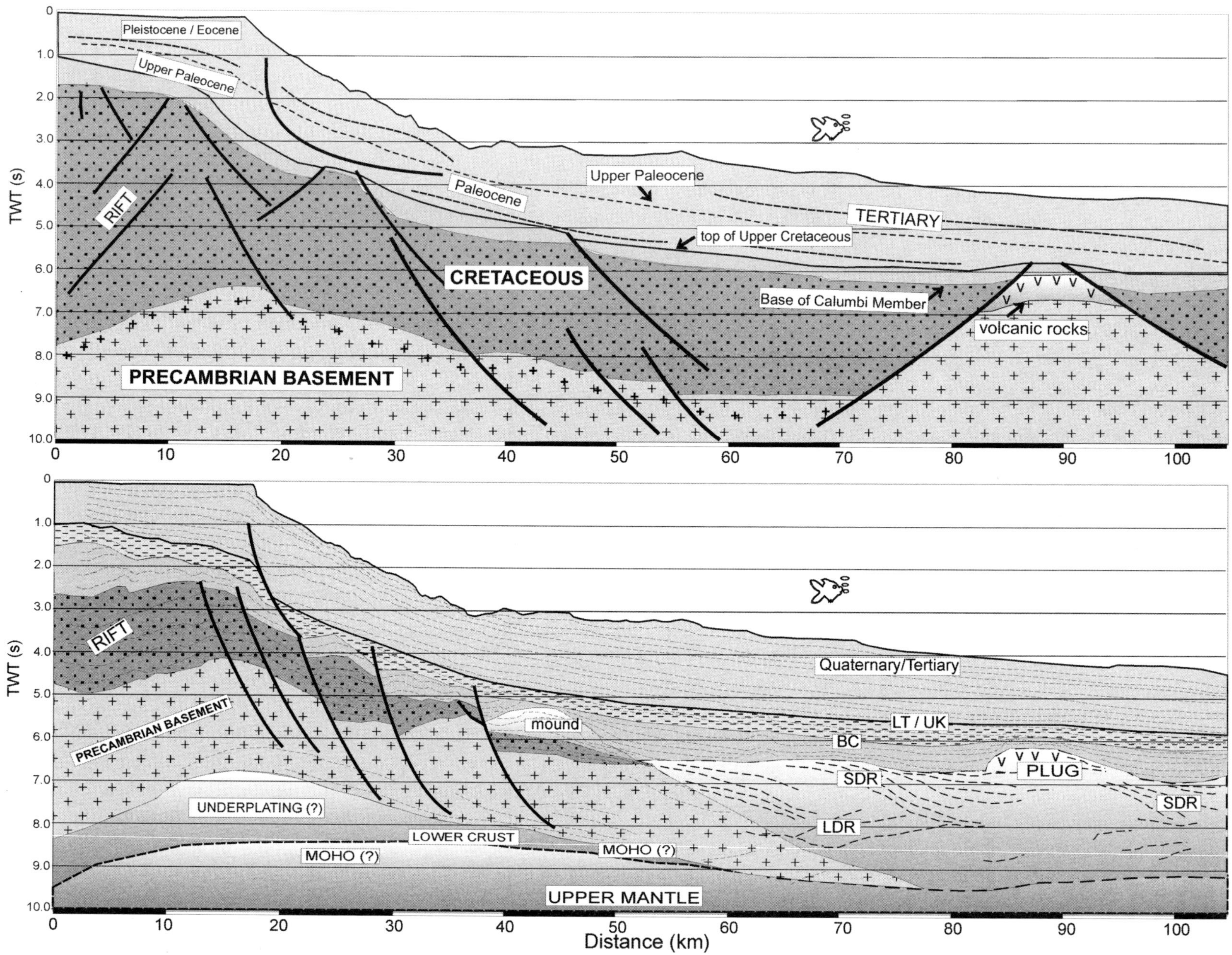

Figure 33. Schematic line drawing interpretations of regional deep seismic profile in Sergipe Basin. Top figure assumes thick rift sequence in deep-water region; bottom figure assumes basinward pinchout of rift sequence and presence of wedges of seaward-dipping reflectors and other igneous structures. SDR = seaward-dipping reflectors; LDR = landward-dipping reflectors; LT/UK = lower Tertiary–Upper Cretaceous; BC = base of Calumbi Formation (Santonian-Turonian unconformity); TWT = two-way traveltime. Modified from Pontes et al. (1991).

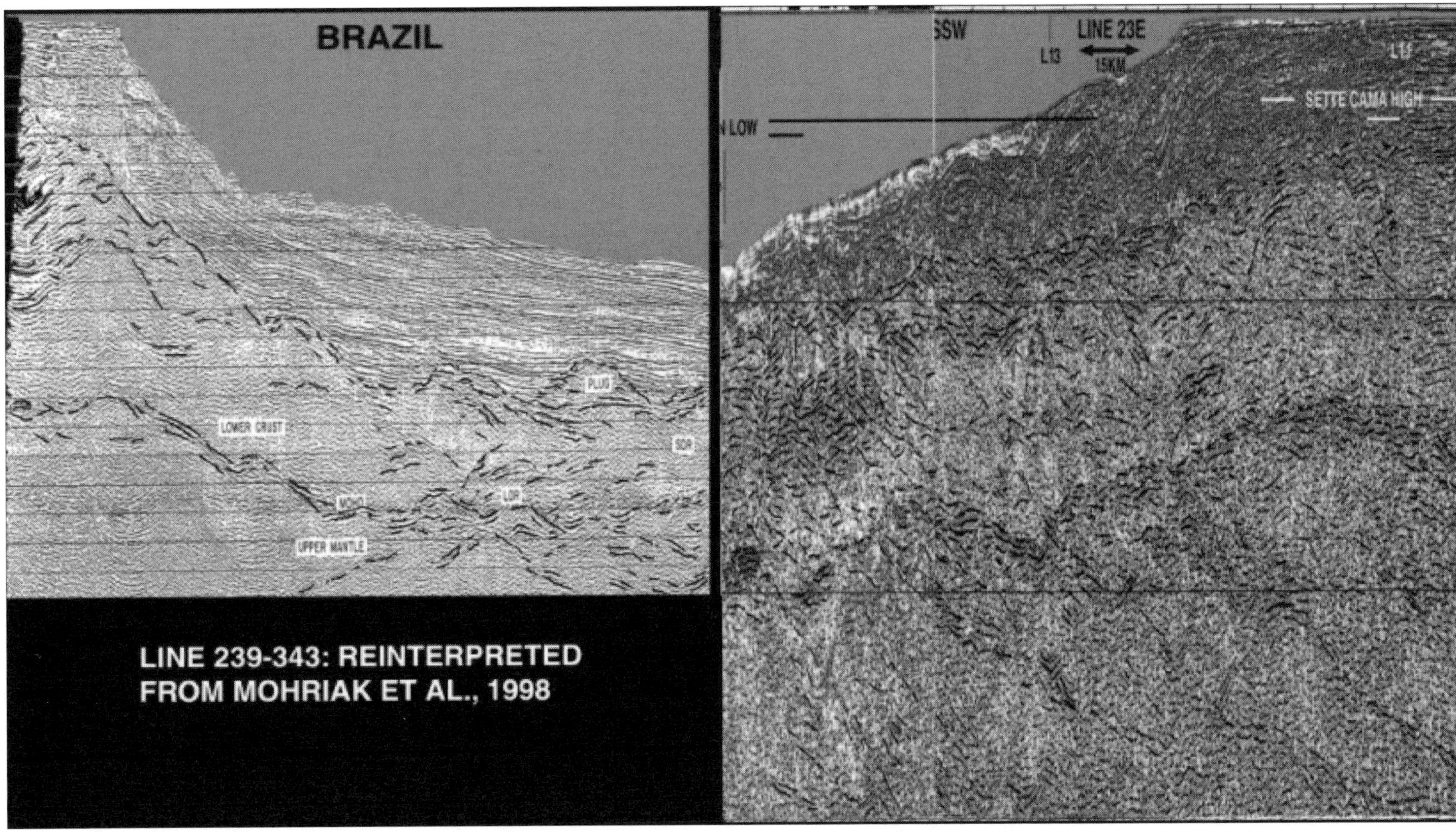

Figure 34. Comparison between Sergipe line 344 (Mohriak et al., 1988a) and Probe line 23 (Rosendahl and Groschel-Becker, 2000).

reflectors in the lower crust and the gravity models (Mohriak et al., 1998a).

A more important difference is the placement of the continental crust–oceanic crust boundary. This is largely definitional: by the Gabon model definition, the oceanic crust–continental crust boundary in the Sergipe model must occur seaward of the end of the regional profile, and by the Sergipe definition, the oceanic crust–continental crust boundary would be placed well within the brittle crustal unit of the Gabon model. The crust landward of normal oceanic crust may be designed as marginal extensional crust, and is often associated with outer highs of uncertain provenance (e.g., see Schuepbach and Vail, 1980), which apparently correspond to highly attenuated, terminal continental crust. The SPOG profiles in the Gabon Basin (Fig. 35) also

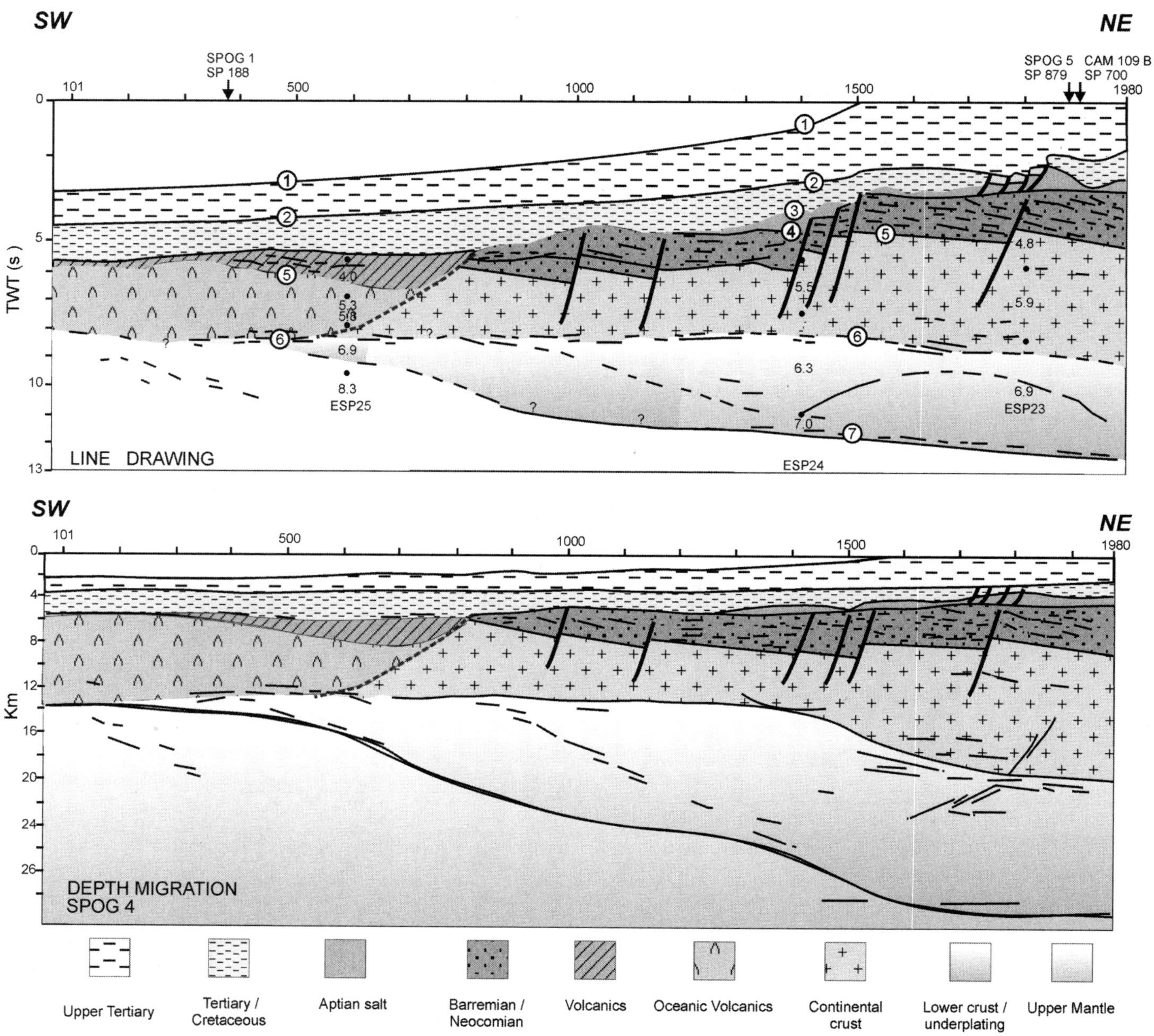

Figure 35. SPOG line 4. Top: Schematic line drawing of two-way time seismic reflection profile. Modified from Wannesson et al. (1991). Bottom: Line interpretation of depth migration profile. Shotpoint numbers are indicated along the profile, and seismic refraction velocities are indicated on top figure. Main seismic reflectors identified correspond to (1) top of Quaternary, (2) top of Cretaceous, (3) top of Aptian salt, (4) base of Aptian salt, (5) top of basement, (6) lower crust/top of underplating, and (7) Moho discontinuity/upper mantle transition.

illustrate the geometry of the rift structures and the transition from continental to oceanic crust incorporating refraction data (Wannesson et al., 1991).

In summary, the continental crust–oceanic crust boundary may be characterized by the first persistent appearance of igneous crust with a thickness of ~1.75 s (TWTT) and seismically by an acoustic character that is typical of Atlantic oceanic crust. This definition essentially separates 100% bona fide oceanic crust from the amalgamation of crustal materials farther landward, which is designated transitional crust. Along the Gabon margin the consequent boundary placement along Lines 23 and 25 is essentially coincident with where the Smith and Sandwell (1997) and DEOS Altimetry Atlas (Wisse et al., 1994) gravity data become negative. This also coincides with where the sediment Bouguer anomaly (density = 2.2 g/cm^3) becomes negative (Meyers et al., 1996). The gravity profile along the Sergipe and Jacuípe Basins (Mohriak et al., 1998a) also indicates a marked flattening of the gravity anomaly basinward of the outermost rift blocks, indicating that the region beyond the SDRs may be interpreted as a very thinned proto-oceanic crust, similar to the interpretation of regional transects along the Greenland margin (Larsen and Jakobsdóttir, 1988).

In the Brazilian continental margin, the seaward-dipping reflectors are clearly imaged and below these igneous features there is a marked strong reflection that apparently coincides with a reflector that in the oceanic crust region is most probably the seismic Moho, as interpreted in the Leplac and other profiles (Mohriak et al., 1998a; Gomes et al., 2000).

Taking both the West African and the eastern Brazilian interpretations, a general model depicting the crustal architecture along a transect of a South Atlantic volcanic margins is presented in Figure 36.

Another relevant element concerns the reality of the deep, landward-dipping reflections interpreted as ductile shear zones in the Gabon model. These reflections were first indicated in Rosendahl et al. (1991), largely ignored in Groschel-Becker (1996), Meyers (1995), and Meyers et al. (1996b), and then resurrected in Rosendahl and Groschel-Becker (1999, 2000) based in part upon poststack reprocessing. These deep reflections are weak on any given Probe line, but they occur along every line from 21 south, and on most of the lines south of the Kribi Fracture Zone. They also occur on tie lines. A comparable fabric is not interpretationally justified in the data from the Brazilian margin, although there are whimsical hints of a possible deep

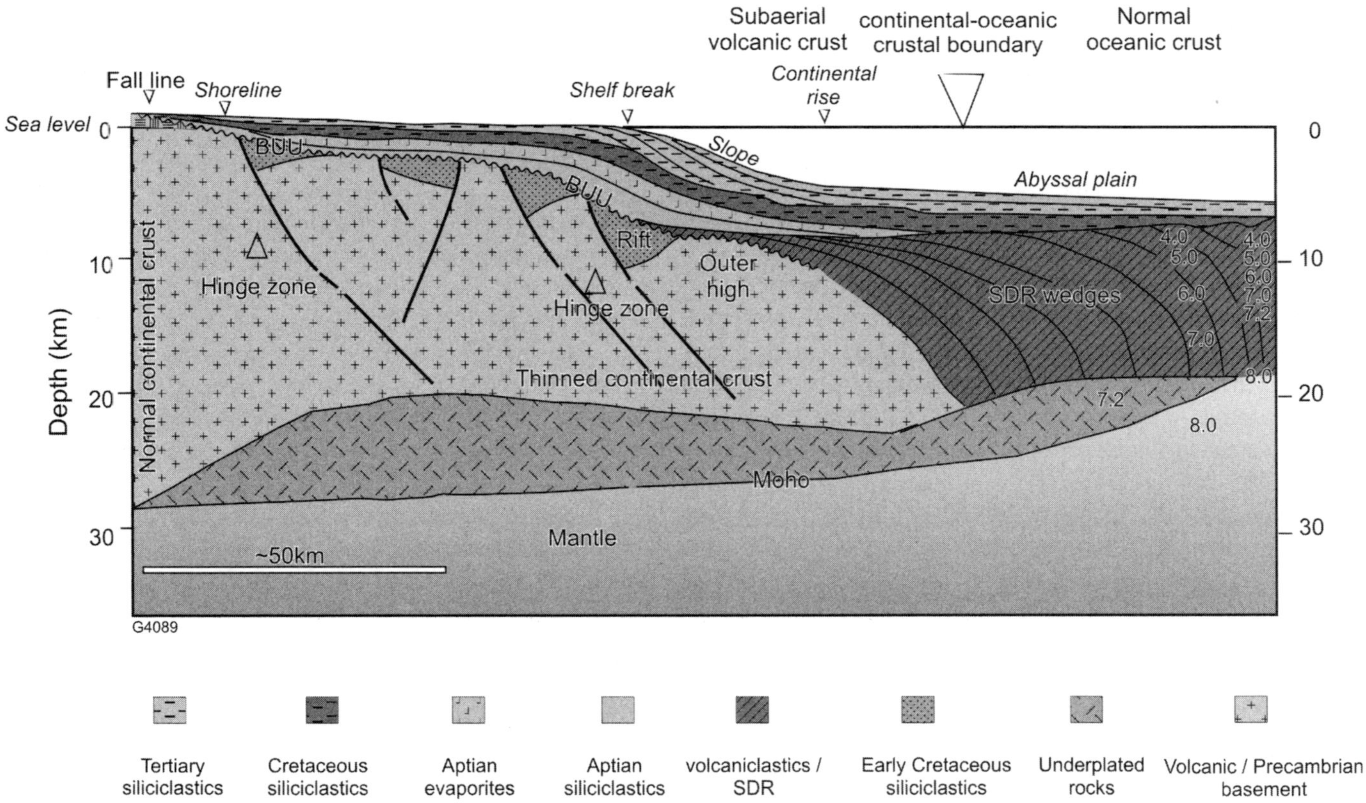

Figure 36. Schematic section showing crustal architecture of South Atlantic basins. SDR = seaward-dipping reflector.

fabric in the Sergipe profile in either or both dip directions, and some of them are interpreted as landward-dipping reflectors with multiple origins (Mohriak et al., 1998a).

These features have been imaged in oceanic basins, and they seem to be associated with seafloor spreading ridges, particularly in the Cape Verde Islands (McBride et al., 1994), or interpreted as relict magma chambers in the North Atlantic (McCarthy et al., 1988).

CONCLUSIONS

Hotspot traces and fracture zones define major segments of the South Atlantic rift margin along the Brazilian and West African continental edges. These segments, north and south of the Abrolhos volcanic complex in eastern Brazil, are characterized by different distribution of prerift and synrift volcanism, synrift sedimentary troughs, thick wedges of seaward-dipping reflectors, and also evaporite deposition.

The volcanic activity in the southern segment is related to the breakup of Gondwana, and is expressed both onshore (Serra Geral Formation of the Paraná Basin in southern Brazil, and the Etendeka province of West Africa) and offshore (tholeiitic basalts flooring several of the South Atlantic rift basins, from Argentina to Brazil and from South Africa to Angola). North of the Abrolhos volcanic complex, the Neocomian synrift sequences are largely devoid of volcanic material (Chang et al., 1992; Dias, 1993; Szatmari, 1998).

The main episode of flood basaltic volcanism lasted from 134 to 129 Ma, coinciding with the end of the rift phase in the Argentine offshore basins (Hinz et al., 1999), and slightly preceding the synrift phase in the Pelotas, Santos, and Campos Basins offshore Brazil (Mizusaki et al., 1992; Cainelli and Mohriak, 1998). Thus, these Late Jurassic–Early Cretaceous basalts constitute the floor of the Neocomian continental rift sequence. They are affected by synrift faults and may actually be interpreted as a volcanic basement, or a prerift sequence.

Most models presented for regional continental flood basalt provinces, such as in Paraná-Etendeka, highlight a subcontinental lithospheric mantle (SCLM) and/or a plume-like component (often referred to as Tristan da Cunha) and the general lack or subordinate contribution from a N-MORB source (e.g., Gibson et al., 1995; Marques et al., 1999). However, in the case of the petrogenesis of the rift basin–related basalts in southeastern Brazil, geochemical data and mixing calculations emphasize the large contribution from a N-MORB component. One possible explanation is that the petrogenesis of the rift basins basalts spreads over a wider range of time than is indicated from the available, nonrefined geochronological data. A lithospheric to asthenospheric transition was described for the petrogenesis of the low-TiO_2 basalts in southern Paraná-Etendeka (Peate and Hawkesworth, 1996), where basaltic volcanism may have been active over ~10 m.y. (e.g., Turner et al., 1994). It is possible that the rift basin basalts represent the last stages in basalt generation, possibly just prior to the opening of the South Atlantic Ocean, when large contributions from a N-MORB component are to be expected.

The rift architecture along the strike of conjugate basins varies significantly along the margin. South of the Walvis Ridge–Rio Grande Rise, controversial interpretations have been suggested for the deep-water extension of the rift troughs. In this region, the transition from the outermost rift blocks to oceanic crust is characterized by peculiar wedges of seaward-dipping reflectors, interpreted as volcanic features formed during initial stages of oceanic crust emplacement, and igneous features, particularly intrusive plugs and volcanoes. However, such SDR wedges are not ubiquitous, insofar as the West African margin from Cameroon to southernmost Gabon seems to be largely devoid of such features (Rosendahl et al., 2000), as is the Pernambuco basinal area (Gomes et al., 2000).

In the Sergipe-Alagoas area of the Brazilian margin, the prerift or synrift volcanics are absent, and these basins were historically considered classical examples of nonvolcanic margins (Chang et al., 1992). However, seismic interpretation of the deep-water province indicates that these basins also may be at least locally characterized by thick wedges of seaward-dipping reflectors, here believed to be postrift in age and formed during the earliest stages of oceanic crustal genesis (Mohriak et al., 1998a), thus constituting a second large magmatic event in the South Atlantic.

There is a third episode of magmatic activity in the South Atlantic, which on the Brazilian margin has been dated as Late Cretaceous (Santonian to Maastrichtian in the Santos Basin), early Tertiary in the Campos and Espírito Santo Basins (Cainelli and Mohriak, 1998). This episode is characterized by volcanic plugs both in the oceanic crust, forming clusters of seamounts and linear volcanic chains along oceanic fracture zones, and by intrusive rocks (mainly with alkaline affinity) in the continental crust, with ages ranging from Late Cretaceous to Early Tertiary, similar to their offshore equivalents. It is arguable whether these plugs actually constitute the manifestation of hotspots, as proposed by several researchers, because there is no clear geochronological decrease in radiometric ages from the continental to oceanic regions. However, some of these lineaments follow a consistent decrease in radiometric age from the continent toward the margin (e.g., Cabo Frio lineament, Szatmari et al., 2000).

Following the rift sequence, a quiescent period marked by a sag basin above a regional unconformity predated the deposition of the Aptian evaporites. The South Atlantic salt basin extended along both margins, controlled by major structural elements, including tectonic and volcanic barriers.

Some similarities in salt tectonic styles are observed in conjugate margins across the South Atlantic, if seismic data in the eastern Brazilian margin are compared with equivalent regional profiles along the West African margin (Jackson et al., 1998; Marton et al., 2000; Mohriak et al., 1998a).

Several geodynamic models for evaporite deposition in the South Atlantic suggested that synrift sediments and salt layers were contemporaneously deposited across the conjugate margins, before continental breakup, and the salt basins subsequently split apart as a consequence of oceanic crust inception and continental drift (e.g., Asmus, 1984). More recently, regional profiles in the ultradeep-water region of conjugate basins indicate that the deposition of Aptian salt postdates the rift phase along the Brazilian and West African margins (Fonck et al., 1998; Marton et al., 2000; Jackson et al., 2000). The salt distribution is highly asymmetric, some segments being characterized by very wide salt provinces (e.g., Campos and Santos Basins), as a function of preferential positioning of oceanic ridges along one of the margins (Mohriak et al., 1998a).

The subequatorial South Atlantic encompasses the region north of the Abrolhos volcanic complex, including the Aptian salt basin that is limited by the Ascension Fracture Zone, and the part of the northeastern Brazilian region that is more properly characterized as a transform margin (Matos, 2000).

The basins immediately south of the Maceió-Ascension-Kribi Fracture Zones are characterized by Early Cretaceous synrift siliciclastic sediments deposited in continental lacustrine environments, along both the Brazilian and the West African sides. These sequences are underlain either by prerift sedimentary sequences (Paleozoic to Mesozoic) or by Precambrian rocks. In this region, synrift volcanic process of any kind prior to the onset of seafloor spreading are at least volumetrically insignificant and perhaps nonexistent. Concomitant with the inception of oceanic crust (Aptian-Albian time), there is seismic evidence of volcanic activity associated with wedges of seaward-dipping reflectors in the Jacuípe and Sergipe-Alagoas Basins along the Brazilian side (Mohriak et al., 1995a).

These features are not present in the Rio Muni basin along the West African side to any extensive degree, although there are some alternative interpretations of SDR wedges in the deep-water region (Bray and Lawrence, 1999; Daily, 2000).

The northern segment of the South Atlantic clearly broke apart asymmetrically, the broader upper plate margin being located on the African side. Lithospheric stretching and rupture of the Sergipe-Gabon margins is a highly asymmetric process involving progressive stretching and thinning of Congo fold belt rocks. It is certain that this segment of the conjugate margin pair was strongly influenced by transform fault development that ultimately led to the Ascension Fracture Zone, which extends toward the Maceió Fracture Zone in Brazil and the Kribi Fracture Zone in West Africa.

The lithosphere ruptured cleanly and abruptly along these fracture zones, creating a steep, narrow margin pair dividing relatively unstretched and elevated continental crust to the south from nearly fully subsided oceanic crust to the north. South of the Ascension Fracture Zone, in the African side, the Congo fold belt was progressively and ductilely stretched for more than 200 km (Rosendahl and Groschel-Becker, 2000), forming the South Atlantic salt basin with the conjugate rifts along the Brazilian side. North of the Ascension Fracture Zone, the breakup leaves a 50-km-wide, subsided, rotated, cantilevered half-graben that now resides dominantly on the African side, and a residual rift on the Pernambuco Plateau (the original elevated rift flank) on the Brazil side. The controlling detachment for these basins shoals seaward and is represented on deep seismic reflector profiles as a seaward-dipping horizon that merges basinward with the Moho, and probably represents one of the best imaged simple-shear conjugate margin pairs worldwide.

ACKNOWLEDGMENTS

We thank Petrobras management for allowing the publication of data collected from the Brazilian margin. We express our appreciation to Dr. A. Spadini for supporting the participation in the Penrose Conference. We also thank the governments of Cameroon, Equatorial Guinea, Gabon, and São Tomé–Príncipe for permitting the Probe fieldwork. This manuscript has benefited from contributions from many geoscientists (particularly A.B. Tankard and Peter Szatmari), and we also thank several explorationists at Petrobras Exploração e Produção, Universidade do Estado do Rio de Janeiro, and Universidade Federal Rural do Rio de Janeiro for enlightening discussions. We thank Agip, Amoco, Ark Geophysical, Petrobras, Geco, and Jebco for technical support during this work.

REFERENCES CITED

Abreu, V.S., 1998, Geologic evolution of conjugate volcanic passive margins: Pelotas Basin (Brazil) and Offshore Namibia (Africa): Implications for global sea-level changes [Ph.D. thesis]: Houston, Texas, Rice University, 355 p.

Arth, J.G., 1976, Behaviour of trace elements during magmatic processes: A summary of theoretical models and their applications: Journal of Research of the United States Geological Survey, v. 4, p. 41–47.

Asmus, H.E., 1984, Geologia da margem continental brasileira, *in* Schobbenhaus, C., Campos, D.A., Derze, G.R., and Asmus, H.E., eds., Geologia do Brasil: Ministério das Minas e Energia/DNPM, Brasília, p. 443–472.

Austin, J.A., and Uchupi, E., 1982, Continental-oceanic crustal transition off southwest Africa: American Association of Petroleum Geologists Bulletin, v. 66, p. 1328–1347.

Barton, A.J., and White, R.S., 1997, Volcanism on the Rockall continental margin: Journal of the Geological Society of London, v. 154, p. 531–536.

Bassetto, M., Alkmin, F.F., Szatmari, P., and Mohriak, W.U., 2000, The oceanic segment of the southern Brazilian margin: Morpho-structural domains and their tectonic significance, *in* Mohriak, W.U., and Talwani, M., eds., Atlantic rifts and continental margins, American Geophysical Union Geophysical Monograph 115, p. 235–259.

Beslier, M.O., Ask, M., and Boillot, G., 1993, Ocean-continent boundary in the Iberian Abyssal Plain from multichannel seismic data: Tectonophysics, v. 218, p. 383–393.

Bown, J.W., and White, R.S., 1994, Variation with spreading rate of oceanic crustal thickness and geochemistry: Earth and Planetary Science Letters, v. 121, p. 435–449.

Bradley, C.H., and Fernandez, M.N., 1991, Early Cretaceous paleogeography of Gabon/Northeastern Brazil: A tectono-stratigraphic model based on propagating rifts, *in* Curnelle, R., ed., Géologie Africaine: Mémoire 13, Elf Aquitaine, p. 17–30.

Bray, R., and Lawrence, S., 1999, Nearby finds brighten outlook: The Leading Edge, May 1993, p. 608–614.

Brice, S.E., Cochran M.D., Pardo, G., Edwards, A.D., 1982, Tectonics and sedimentation of the south Atlantic rift sequence: Cabinda, Angola, *in* Watkins, J. S., and Drake, C.L., eds., Studies in continental margin geology: American Association of Petroleum Geologists Memoir, v. 34, p. 5–18.

Burke, K., 1969, Seismic areas of the Guinea coast where Atlantic fracture zone reach Africa: Nature, v. 222, p. 655–657.

Burke, K., and Dewey, J.F., 1973, Plume-generated triple junctions: Key indicators in applying plate tectonics to old rocks: Journal of Geology, v. 81, p. 406–433.

Cainelli, C., and Mohriak, W.U., 1998, Geology of Atlantic Eastern Brazilian basins: American Association of Petroleum Geologists International Conference and Exhibition Short Course, Brazilian Geology Part 2: Rio de Janeiro, Brazil, p. 9.

Cainelli, C., and Mohriak, W.U., 1999, Some remarks on the evolution of sedimentary basins along the Eastern Brazilian continental margin: Episodes, v. 22, no. 3, p. 206–216.

Cande, S.C., LaBrecque, J.L., and Haxby, W,F., 1988, Plate kinematics of the South Atlantic: Chron C34 to present: Journal of Geophysical Research, v. 93, p. 13479–13492.

Castro Jr., A.C.M., 1987, The Northeastern Brazil and Gabon basins: A double rifting system associated with multiple crustal detachment surfaces: Tectonics, v. 6, no. 6, p. 727–738.

Chang, H. K., Kowsmann, R.O., Figueiredo, A.M.F., and Bender, A., 1992, Tectonics and stratigraphy of the East Brazil Rift system: An overview: Tectonophysics, v. 213, p. 97–138.

Christensen, N.I., and Salisbury, M.H., 1982, Lateral heterogeneity in the seismic structure of the oceanic crust inferred from velocity studies in the Bay of Islands Ophiolite, Newfoundland: Geophysical Journal of the Royal Astronomical Society, v. 68, p. 675–688.

Coffin, M.F., and Eldholm, O., 1992, Volcanism and continental breakup: A global compilation of large igneous provinces, *in* Storey, B.C., Alabaster, T., and Pankhurst, R.J., eds., Magmatism and the causes of continental break-up: Geological Society [London] Special Publication 68, p. 21–34.

Coffin, M.F., and Eldholm, O., 1994, Large igneous provinces: Crustal structure, dimensions, and external consequences: Reviews of Geophysics, v. 32, p. 1–36.

Cordani, U.G., and Vandoros, P., 1967, Basaltic rocks of the Paraná basin, *in* Bigarella, J.J., Becker, R.D., and Pinto, J.D., eds., Problems in Brazilian Gondwana geology: Curitiba, Brazil, International Symposium on Gondwana Stratigraphy and Paleontology, p. 207–231.

Coward, M.P., Purdy, E.G., Ries, A.C., and Smith, D.G., 1999, The distribution of petroleum reserves in basins of the South Atlantic margins, *in* Cameron, N.R., Bate, R.H., and Clure, V.S., eds., The oil and gas habitats of the South Atlantic: Geological Society [London] Special Publication 153, p. 101–131.

Daily, P., 2000, Tectonic and stratigraphic development of the Rio Muni Basin, Equatorial Guinea: The role of transform zones in Atlantic Basin evolution, *in* Mohriak, W.U., and Talwani, M., eds., Atlantic rifts and continental margins: American Geophysical Union Geophysical Monograph 115, p. 105–128.

Davison, I., 1999, Tectonics and hydrocarbon distribution along the Brazilian South Atlantic margin, *in* Cameron, N.R., Bate, R.H., Clure, V.S., eds., The oil and gas habitats of the South Atlantic: Geological Society [London] Special Publication 153, p. 133–151.

DePaolo, D.J., 1981, Trace element and isotopic effects of combined wallrock assimilation and fractional crystallisation: Earth and Planetary Science Letters, v. 53, p. 189–202.

Dias, J.L., 1993, Evolução da fase rift e a transição rift / drift nas bacias das margens leste e sudeste do Brasil: 3rd International Congress of the Brazilian Geophysical Society, Rio de Janeiro, Expanded Abstracts, v. 2, p. 1328–1332.

Dias, J.L., Sad, A.R.E., Latgé, M.A.L., and Silveira, D.P., 1994, Bacia de Pelotas: Estado da arte e perspectivas exploratórias: 2nd Seminário de Interpretação Exploratória, Petrobras, Departamento de Exploração, Rio de Janeiro, p. 270–275.

Drake, M.J., and Weill, D.F., 1975, Partition of Sr, Ba, Ce, Y, Eu^{+2}, Eu^{+3} and other REE between plagioclase feldspar and magmatic study: An experimental study: Geochimica et Cosmochimica Acta, v. 39, p. 689–712.

Eldholm, O., Skogseid, J., Planke, S., and Gladczenko, T.P., 1995, Volcanic margin concepts, *in* Banda, E., Tornè, M., and Talwani, M., eds., Rifted ocean-continent boundaries: Dordrecht, Kluwer, p. 1–16.

Erlank, A.J., Marsh, J.S., Duncan, A.R., Miller, R.McG., Hawkesworth, C.J., Betton, P.J., and Rex, D.C., 1984, Geochemistry and petrogenesis of the Etendeka volcanic rocks from SWA/Namibia: Geological Society of South Africa Special Publication 13, p. 195–245.

Etheridge, M.A., Symonds, P.A., and Lister, G.S., 1989, Application of the detachment model to reconstruction of conjugate passive margins, *in* Tankard, A.J., and Balkwill, H.R., eds., Extensional tectonics and stratigraphy of the North Atlantic margins: American Association of Petroleum Geologists Memoir, v. 46, p. 23–40.

Faure, G., 1986, Principles of isotope geology: New York, John Wiley and Sons, 590 p.

Fodor, R.V., 1987, Low- and high-TiO_2 flood basalts of Southern Brazil: Origin from picritic parentage and a common mantle source: Earth and Planetary Science Letters, v. 84, p. 423–430.

Fodor, R.V., and Vetter, S.K., 1984, Rift-zone magmatism: Petrology of basaltic rocks transitional from CFB to MORB, Southeastern Brazil margin: Contributions to Mineralogy and Petrology, v. 88, p. 307–321.

Fodor, R.V., McKee, E.H., and Asmus, H.E., 1983, K-Ar ages and the opening of the South Atlantic Ocean: Basaltic rocks from the Brazilian margin: Marine Geology, v. 54, M1–M8.

Fodor, R.V., Corwin, C., and Rosemberg, A., 1985a, Petrology of Serra Geral (Paraná) continental flood basalts, Southern Brazil: Crustal contamination, source material, and South Atlantic magmatism: Contributions to Mineralogy and Petrology, v. 91, p. 54–65.

Fodor, R.V., Corwin, C., and Sial, A.N., 1985b, Crustal signatures in the Serra Geral flood basalt province, Southern Brazil: O and Sr isotope evidence: Geology, v. 13, p. 763–765.

Fodor, R.V., McKee, E.H., and Asmus, H.E., 1983/1984, K-Ar ages and the opening of the South Atlantic Ocean: Basaltic rock from the Brazilian margin: Marine Geology, v. 54, p. M1–M8.

Fonck, J.M., Cramez, C., and Jackson, M.P.A., 1998, Role of subaerial volcanic rocks and major unconformities in the creation of South Atlantic margins, *in* Mello, M.R., and Yilmaz, P.O., eds., Extended Abstract Volume: American Association of Petroleum Geologists International Conference and Exhibition, Rio de Janeiro, Brazil, p. 38–39.

Fontana, R. L., 1996, SDR (seaward-dipping reflectors) e a transição crustal na Bacia de Pelotas: Anais do 39 Congresso Brasiliero de Geologia, Sociedade Brasileira de Geologia, Salvador–Bahia, v. 5, p. 425–430.

Francheteau, J., and Le Pichon, X., 1972, Marginal fracture zones as structural framework of continental margins in South Atlantic Ocean: American Association of Petroleum Geologists Bulletin, v. 56, p. 991–1007.

Fujimaki, H., Tatsumoto, M., and Aoki, K., 1984, Partition coefficients of Hf, Zr and REE between phenocrysts and groundmasses: Journal of Geophysical Research, v. 89 (suppl.), p. B662–B672.

Furlong, K.P., and Fountain, D.M., 1986, Lithospheric evolution with underplating: Thermal considerations and seismic-petrologic consequences: Journal of Geophysical Research, v. 91, p. 8285–8294.

Gallagher, K., and Brown, R., 1999, The Mesozoic denudation history of the Atlantic margins of southern Africa and southeast Brazil and the relationship to offshore sedimentation, *in* Cameron, N.R., Bate, R.H., and Clure, V.S., eds., The oil and gas habitats of the South Atlantic: Geological Society [London] Special Publication 153, p. 41–53.

Garland, F., Turner, S., and Hawkesworth, C., 1996, Shifts in the source of the Paraná basalts through time: Lithos, v. 37, p. 223–243.

Gerrard, I., Smith, G.C., 1982, Post-Paleozoic succession and structure of the Southwestern African continental margin, *in* Watkins, J.S., and Drake, C.L., eds., Studies in continental margin geology: American Association of Petroleum Geologists Memoir, v. 34, p. 49–62.

Gibson, S.A., Thompson, R.N., Dickin, A.P., and Leonardos, O.H., 1995, High-Ti and low-Ti mafic potassic magmas: Key to plume-lithosphere interactions and continental flood-basalt genesis: Earth and Planetary Science Letters, v. 136, p. 149–165.

Gladczenko, T.P., Hinz, K., Eldholm, O., Meyer, H., Neben, S., and Skogseid, J., 1997, South Atlantic volcanic margins: Journal of the Geological Society of London, v. 154, p. 465–470.

Gomes, P.O., Gomes, B.S., Palma, J.J.C., Jinno, K., and Souza, J.M., 2000, Ocean-continent transition and tectonic framework of the oceanic crust at the continental margin off NE Brazil: Results of LEPLAC Project, *in* Mohriak, W.U., and Talwani, M., eds., Atlantic rifts and continental margins: American Geophysical Union Geophysical Monograph 115, p. 261–291.

Groschel-Becker, H.M., 1996, Formational processes of oceanic crust at sedimented spreading centers: Perspectives from the West African continental margin and Middle Valley, Juan de Fuca Ridge [Ph.D. thesis]: Miami, Florida, University of Miami, 328 p.

Guardado, L.R., Gamboa, L.A.P., Lucchesi, C.F., 1989, Petroleum geology of the Campos Basin, a model for a producing Atlantic-type basin, *in* Edwards, J.D., and Santogrossi, P.A., eds., Divergent/passive margin basins: American Association of Petroleum Geologists Memoir, v. 48, p. 3–79.

Gudmundsson, A., 1995, Infrastructure and mechanics of volcanic systems in Iceland: Volcanology and Geothermal Research Journal, v. 64, p. 1–22.

Guiraud, R., and Maurin, J., 1992, Early Cretaceous rifts of Western and Central Africa: An overview: Tectonophysics, v. 213, p. 153–168.

Harry, D.L., and Sawyer, D.S., 1992, Basaltic volcanism, mantle plumes, and the mechanics of rifting: The Paraná flood basalt province of South America: Geology, v. 20, p. 207–210.

Hawkesworth, C.J., Marsh, J.S., Duncan, A.R., Erlank, A.J., and Norry, M.J., 1984, The role of continental lithosphere in the generation of the Karoo volcanic rocks: Evidence from combined Nd- and Sr-isotope studies: Geological Society of South Africa Special Publication 13, p. 341–354.

Hawkesworth, C.J., Gallagher, K., Kirstein, L., Mantovani, M.S.M., Peate, D.W., and Turner, S.P., 2000, Tectonic controls on magmatism associated with continental break-up: An example from the Paraná-Etendeka province: Earth and Planetary Science Letters, v. 179, p. 335–349.

Hawkesworth, C.J., Gallagher, K., Kelley, S., Mantovani, M., Peate, D.W., Regelous, M., and Rogers, N.W., 1992, Paraná magmatism and the opening of the South Atlantic, *in* Storey, B.C., Alabaster, T., and Pankhurst, R.J., eds., Magmatism and the causes of continental break-up: Geological Society [London] Special Publication 68, p. 221–240.

Henry, S.G., Mohriak, W.U., and Mello, M.R., 1996, South Atlantic sag basins: New petroleum system components [abs.], American Association of Petroleum Geologists International Conference and Exhibition, Caracas, Venezuela: American Association of Petroleum Geologists Bulletin, v. 80/8, p. 1300.

Hinz, K., 1981, A hypothesis on terrestrial catastrophes: Wedges of very thick oceanward dipping layers beneath passive continental margins: Their origin and paleoenvironmental significance: Geologische Jahrbuch, Reihe E, Geophysik, v. 22, p. 3–28.

Hinz, K., 1990, The Argentine eastern continental margin: Structure and geological development: American Association of Petroleum Geologists Bulletin, v. 74, p. 675–676.

Hinz, K., and Weber, J., 1976, Zum geologischen Aufbau des Norwegischen Kontinentalrandes und der Barents-See nach refexionsseismischen Messungen: Compendium, Erganzungsbad der Erdol und Kohle, v. 57/76, p. 3–29.

Hinz, K., Mutter, J.C., Zehnder, C.M., and the NGT Study Group, 1987, Symmetric conjugation of continent-ocean boundary structures along the Norwegian and east Greenland margins: Marine and Petroleum Geology, v. 4, p. 166–187.

Hinz, K., Neben, S., Schreckenberger, B., Roeser, H.A., Block, M., Souza, K.G., and Meyer, H., 1999, The Argentine continental margin north of 48°S: Sedimentary successions, volcanic activity during breakup: Marine and Petroleum Geology, v. 16, no. 1, p. 1–25.

Irvine, T.N., and Baragar, W.R.A., 1971, A guide to the chemical classification of common volcanic rocks: Canadian Journal of Earth Sciences, v. 8, p. 523–547.

Irving, A.J., and Frey, F.A., 1978, Distribution of trace elements between garnet megacrysts and host volcanic liquids of kimberlitic to rhyolitic composition: Geochimica et Cosmochimica Acta, v. 42, p. 771–787.

Jackson, M.P.A., Cramez, C., and Fonck, J.M., 2000, Role of subaerial volcanic rocks and mantle plumes in creation of South Atlantic margins: Implications for salt tectonics and source rocks: Marine and Petroleum Geology, v. 17, p. 477–498.

Jackson, M.P.A., Cramez, C., and Mohriak, W.U., 1998, Salt tectonics provinces across the continental–oceanic boundary in the Lower Congo and Campos Basins on the South Atlantic Margins [abs.]: American Association of Petroleum Geologists Bulletin, v. 82, p. 1926.

Karner, G.D., and Driscoll, N.W., 1998, Tectonic setting of late rift stage source rocks of the West African continental margin, *in* Mello, M.R. and Yilmaz, P.O., eds., American Association of Petroleum Geologists International Conference and Exhibition, Nov. 8–11, 1998, Extended Abstract Volume, p. 826.

Karner, G.D., and Driscoll, N.W., 1999, Tectonic and stratigraphic development of the West African and eastern Brazilian margins: Insights from quantitative basin modelling, *in* Cameron, N.R., Bate, R.H., and Clure, V.S., eds., The oil and gas habitats of the South Atlantic: Geological Society [London] Special Publication 153, p. 11–40.

Karner, G.D., Driscoll, N.W., McGinnis, J.P., Brumbaugh, W.D., and Cameron, N.R., 1997, Tectonic significance of syn-rift sediment packages across the Gabon-Cabinda continental margin: Marine and Petroleum Geology, v. 14, no. 7/8, p. 971–1000.

Keen, C.E., 1985, The dynamics of rifting: Deformation of the lithosphere by active and passive driving forces: Geophysical Journal of the Royal Astronomical Society, v. 80, p. 95–120.

Keen, C.E., and deVoogd, B., 1988, The continent-ocean boundary at the rifted margin off eastern Canada: New results from deep seismic reflection studies: Tectonics, v. 7, p. 101–124.

Keen, C.E., Courtney, R.C., Dehler, S.A., and Williamson, M.C., 1994, Decompression melting at rifted margins: Comparison of model predictions with distribution of igneous rocks on the Eastern Canadian margin: Earth and Planetary Science Letters, v. 121, p. 403–416.

Kusznir, N., Lehner, P., and DeRuiter, P.A.C., 1977, Structural history of the Atlantic margin of Africa: American Association of Petroleum Geologists Bulletin, v. 61, p. 961–981.

Larsen, H.C., and Jakobsdóttir, S., 1988, Distribution, crustal properties and significance of seawards-dipping sub-basement reflectors off E Greenland, *in* Morton, A.C., and Parson, L.M., eds., Early Tertiary volcanism and the opening of the NE Atlantic: Geological Society [London] Special Publication 39, p. 95–114.

Lase Study Group, 1986, Deep structure of the US East Coast passive margin from large aperture seismic experiments (LASE): Marine and Petroleum Geology, v. 3, p. 234–242.

Leg 152 Shipboard Party, 1994, Drilling unearths "fire and ice" at southeast Greenland margin: Eos (Transactions, American Geophysical Union), v. 75, no. 35, p. 401, 403, and 406.

Lehner, P., and de Ruiter, P.A.C., 1977, Structural history of Atlantic margin of Africa: American Association of Petroleum Geologists Bulletin, v. 61, p. 961–981.

LeMaitre, R.W., 1989, A classification of igneous rocks and glossary of terms, *in* Recommendations of the International Union of Geological Sciences Subcommission on the Systematics of Igneous Rocks: Oxford, Blackwell, 193 p.

LeRoex, A.P., Cliff, R.A., and Adair, B.J.I., 1990, Tristan da Cunha, South Atlantic: Geochemistry and petrogenesis of a basanite-phonolite lava series: Journal of Petrology, v. 31, p. 779–812.

Light, M.P.R., Maslanyj, M.P., and Banks, N.L., 1992, New geophysical evidence for extensional tectonics on the divergent margin offshore Namibia, *in* Storey, B.C., Alabaster, T., and Pankhurst, R.J., Magmatism and the causes of continental break-up: Geological Society [London] Special Publication 68, p. 257–270.

Lobo, J.T., 2000, Petrogênese dos basaltos do Cretáceo Inferior das bacias de Campos e Pelotas, SE do Brasil [M.S. thesis]: Rio de Janeiro, Universidade do Estado do Rio de Janeiro, Faculdade de Geologia, 97 p.

Lobo, J.T., Valente, S.C., Thomaz Filho, A., and Szatmari, P., 1999b, Diabásios da Serra do Mar e basaltos da Bacia de Campos: Comparação dos processos de AFC através de modelamento geoquímico quantitativo: 7th Simpósio de Geologia do Sudeste, São Pedro, São Paulo, Sociedade Brasileira de Geologia, Abstracts, p. 56.

Lohmann, H.H., Hoffman-Rohle, J., and Hinz, K., 1995, Argentina, *in* Kulke, H., ed., Regional petroleum geology of the world. Part II: Africa, America, Australia and Antarctica: Berlin-Stuttgart, Gebruder Borntraeger Verlagsbuchhandlung, p. 549–577.

Macedo, J.M., 1990, Evolução tectônica da Bacia de Santos e áreas continentais adjacentes, *in* Gabaglia, G.P.R., and Milani, E.J., eds., Origem e evolução de bacias sedimentares: Rio de Janeiro, Petrobrás, p. 361–376.

Marques, L.S., Dupré, B., and Piccirillo, E.M., 1999, Mantle source compositions of the Paraná magmatic province (southern Brazil): Evidence from trace element and Sr-Nd-Pb isotope geochemistry: Journal of Geodynamics, v. 28, p. 439–458.

Marton, L.G., Tari, G.C., and Lehmann, C.T., 2000, Evolution of the Angolan passive margin, West Africa, with emphasis on post-salt structural styles, in Mohriak, W.U., and Talwani, M., eds., Atlantic rifts and continental margins: American Geophysical Union Geophysical Monograph 115, p. 129–149.

Marzoli, A., Melluso, L., Morra, V., Renne, P.R., Sgrosso, I., D'Antonio, M., Morais, L.D., Morais, E.A.A., and Ricci, G., 1999, Geochronology and petrology of Cretaceous basaltic magmatism in the Kwanza basin (western Angola), and relationships with the Paranà-Etendeka continental flood basalt province: Journal of Geodynamics, v. 28, p. 341–356.

Maslanyj, M.P., Light, M.P.R., Greenwood, R.J., and Banks, N.L., 1992, Extension tectonics offshore Namibia and evidence for passive rifting in the South Atlantic: Marine and Petroleum Geology, v. 9, p. 590–601.

Matos, R.M.D., 1992, The northeast Brazilian rift system: Tectonics, v. 11, p. 766–801.

Matos, R.M.D., 1999, History of the northeast Brazilian rift system: Kinematic implications for the break-up between Brazil and West Africa, *in* Cameron, N.R., Bate, R.H., and Clure, V.S., eds., The oil and gas habitats of the South Atlantic: Geological Society [London] Special Publication 153, p. 55–73.

Matos, R.M.D., 2000, Tectonic evolution of the equatorial South Atlantic, *in* Mohriak, W.U., and Talwani, M., eds., Atlantic rifts and continental margins: American Geophysical Union Geophysical Monograph 115, p. 331–354.

McBride, J.H., White, R.S., Henstock, T.J., and Hobbs, R.W., 1994, Complex structure along a Mesozoic sea-floor spreading ridge: BIRPS deep seismic reflection, Cape Verde abyssal plain: Geophysical Journal International, v. 119, p. 453–478.

McCarthy, J., Mutter, J.C., Morton, J.L., Sleep, N.H., and Thompson, G.A., 1988, Relic magma chamber structures preserved within the Mesozoic North Atlantic crust?: Geological Society of America Bulletin, v. 100, p. 1423–1436.

McKenzie, D., 1978, Some remarks on the development of sedimentary basins: Earth and Planetary Science Letters, v. 40, p. 25–32.

Meyers, J.B., 1995, Rifted continental margin architecture off West Africa, as revealed by deep-penetrating multi-channel seismic reflection and potential field data [Ph.D. thesis]: Miami, Florida, University of Miami, 234 p.

Meyers, J.B., and Rosendahl, B.R., 1991, Seismic reflection character of the Cameroon volcanic line: Evidence for uplifted oceanic crust: Geology, v. 19, p. 1072–1076.

Meyers, J.B., Rosendahl, B.R., and Austin, J.A., 1996a, Deep-penetrating MCS images of the South Gabon Basin: Implications for rift tectonics and post-breakup salt remobilization: Basin Research, v. 8, p. 65–84.

Meyers, J.B., Rosendahl, B.R., Groschel-Becker, H., Austin, J.A., and Rona, P.A., 1996b, Deep-penetrating MCS imaging of the rift-to-drift transition, offshore Doula and North Gabon Basins, West Africa: Marine Petroleum Geology, v. 13, p. 793–836.

Milani, E.J., and Zalán, P.V., 1999, An outline of the geology and petroleum systems of the Paleozoic interior basins of South America: Episodes, v. 22, no. 3, p. 199–205.

Milner, S.C., Duncan, A.R., Whittingham, A.M., and Ewart, A., 1995, Trans-Atlantic correlation of eruptive sequences and individual silicic units within the Paraná-Etendeka igneous province: Journal of Volcanology and Geothermal Research, v. 69, p. 137–157.

Mizusaki, A.M.P., Thomaz Filho, A., and Cesero, P., 1998, Ages of the magmatism and the opening of the South Atlantic Ocean: Universidade Federal do Rio Grande do Sul, Instituto de Geociências, Pesquisas, v. 25, no. 2, p. 47–57.

Mizusaki, A.M.P., Petrini, R., Bellieni, G., Comin-Chiaramonti, P., Dias, J., DeMin, A., and Piccirillo, E.M., 1992, Basalt magmatism along the passive continental margin of SE Brazil (Campos basin): Contributions to Mineralogy and Petrology, v. 111, p. 143–160.

Mohriak, W.U., 1995, Salt tectonics structural styles: Contrasts and similarities between the South Atlantic and the Gulf of Mexico, *in* Travis, C.J., Harrison, H., Hudec, M.R., Vendeville, B.C., Peel, F.J., and Perkins, B.E., eds., Salt, sediment and hydrocarbons: Gulf Coast Section, SEPM (Society for Sedimentary Geology) Foundation 16th Annual Research Conference, Houston, Texas, p. 177–191.

Mohriak, W.U., 2001, Salt tectonics, volcanic centers, fracture zones and their relationship with the origin and evolution of the South Atlantic Ocean: Geophysical evidence in the Brazilian and West African margins, 7th International Congress of the Brazilian Geophysical Society, Salvador–Bahia, Brazil, Oct. 28–31, 2001, Expanded Abstract, p. 1594.

Mohriak, W.U., and Latge, M.A.L., 1991, Deep seismic survey of Brazilian passive margin basins: The southeastern region, *in* Congresso Brasileiro de Geofisica, 2, Salvador, Bahia, Sociedade Brasileira de Geofiscia: Salvador, Bahia, Boletim de Resumos Expandidos, p. 621–626.

Mohriak, W.U., Hobbs, R., and Dewey, J.F., 1990, Basin-forming processes and the deep structure of the Campos Basin, offshore Brazil: Marine and Petroleum Geology, v. 7, no. 2, p. 94–122.

Mohriak, W.U., Barros, M.C., Rabelo, J.H.L., and Matos, R.D., 1993, Deep seismic survey of Brazilian Passive Basins: The northern and northeastern regions, *in* Third International Congress of the Brazilian Geophysical Society: Rio de Janeiro, Expanded Abstracts, p. 1134–1139.

Mohriak, W.U., Rabelo, J.H.L., Matos, R.D., and Barros, M.C., 1995a, Deep seismic reflection profiling of sedimentary basins offshore Brazil: Geological objectives and preliminary results in the Sergipe Basin: Journal of Geodynamics, v. 20, p. 515–539.

Mohriak, W.U., Bassetto, M., and Vieira, I.S., 1995b, Deep seismic constraints on the crustal architecture of sedimentary basins in the Brazilian margin: Tectonic and exploratory implications: Boletim de Resumos Expandidos, 5th Simposio Nacional de Estudos Tectonicos-95, Gramado, p. 246–248.

Mohriak, W.U., Macedo, J.M., Castellani, R.T., Rangel, H.D., Barros, A.Z.N., Latgé, M.A.L., Ricci, J.A., Misuzaki, A.M.P., Szatmari, P., Demercian, L.S., Rizzo, J.G., and Aires, J.R., 1995c, Salt tectonics and structural styles in the deep-water province of the Cabo Frio region, Rio de Janeiro, Brazil, *in* Jackson, M.P.A., Roberts, D.G., and Snelson, S., eds., Salt tectonics: A global perspective: American Association of Petroleum Geologists Memoir, v. 65, p. 273–304.

Mohriak, W.U., Bassetto, M., and Vieira, I.S., 1998a, Crustal architecture and tectonic evolution of the Sergipe-Alagoas and Jacuipe basins, offshore northeastern Brazil: Tectonophysics, v. 288, p. 199–220.

Mohriak, W.U., Palagi, P.R., and Mello, M.R., 1998b, Tectonic evolution of South Atlantic salt basins: American Association of Petroleum Geologists Bulletin, v. 82, p. 1945.

Mohriak, W.U., Jackson, M.P.A., and Cramez, C., 1999, Salt tectonics provinces across the continental-oceanic boundary in the Brazilian and West African margins: 6th International Congress of the Brazilian Geophysical Society, CD with Abstracts, Sociedade Brasileira de Geofisica 191.

Mohriak, W.U., Bassetto, M., and Vieira, I.S., 2000a, Tectonic evolution of the rift basins in the northeastern Brazilian region, *in* Mohriak, W.U., and Talwani, M., eds., Atlantic rifts and continental margins, American Geophysical Union Geophysical Monograph 115, p. 293–315.

Mohriak, W.U., Rosendahl, B.R., and Palagi, P.R., 2000b, Comparisons of rift architecture and petroleum systems of the eastern Brazilian and west African sedimentary basins: Manchester, England, University of Manchester, Geoscience 2000, Abstracts, p. 144.

Montadert, L., Roberts, D.G., DeCharpal, D., and Guennoc, P., 1979, Rifting and subsidence of the northern continental margin of the Bay of Biscay, *in,* Initial Reports of the Deep Sea Drilling Project, Volume 48: Washington, D.C., U.S. Government Printing Office, p. 1025–1060.

Morgan, W.J., 1983, Hot spot tracks and the early rifting of the Atlantic: Tectonophysics, v. 94, p. 123–139.

Müller, R.D., Royer, J.Y., and Lawver, L.A., 1993, Revised plate motions relative to the hot spots from combined Atlantic and Indian Ocean hot spot tracks: Geology, v. 21, p. 275–278.

Müller, R.D., Roest, W.R., Royer, J.Y., Gahagan, L.M., and Sclater, J.G., 1997, Digital isochrons of the world's ocean floor: Journal of Geophysical Research, v. 102, no. B2, p. 3211–3214.

Mutter, J.C., Talwani, M., and Stoffa, P.L., 1982, Origin of seaward-dipping reflectors in oceanic crust off the Norwegian margin by "subaerial seafloor spreading": Geology, v. 10, p. 353–357.

Mysen, B.O., and Kushiro, I., 1977, Compositional variations of coexisting phases with degree of melting of peridotite in the upper mantle: American Mineralogist, v. 62, p. 843–865.

Nakamura, N., 1974, Determination of REE, Ba, Fe, Mg, Na and K in carbonaceous and ordinary chondrites: Geochimica et Cosmochimica Acta, v. 38, p. 757–775.

Nürnberg, D., and Müller, R.D., 1991, The tectonic evolution of the South Atlantic from Late Jurassic to present: Tectonophysics, v. 191, p. 27–53.

O'Connor, J. M., and Duncan, R.A., 1990, Evolution of the Walvis Ridge–Rio Grande Rise hot spot system: Implications for African and South American plate motions over plumes: Journal of Geophysical Research, v. 95, p. 17474–17502.

Palagi, P.R., 1998, Bacia do Atlântico Sul-Avaliação Regional: Petrobras–Exploração e Produção–Gerência de Explorção/Gerência de Interpretação de Novas Fronteiras, 86 p., atlas.

Pálmason, G., 1980, A continuum model of crustal generation in Iceland: Kinematic aspects: Journal of Geophysics, v. 47, p. 7–18.

Peate, D.W., 1997, The Paraná-Etendeka province, *in* Mahoney, J.J., and Coffin, M.F., eds., Large igneous provinces: Continental, oceanic, and planetary flood volcanism: American Geophysical Union Geophysical Monograph 100, 438 p.

Peate, D.W., and Hawkesworth, C.J., 1996, Lithospheric to asthenospheric transition in low-Ti flood basalts from southern Paraná, Brazil: Chemical Geology, v. 127, p. 1–24.

Peate, D.W., Hawkesworth, C.J., and Mantovani, M.S.M., 1992, Chemical stratigraphy of the Paraná lavas (South America): Classification of magma types and their spatial distribution: Bulletin of Volcanology, v. 55, p. 119–139.

Peate, D.W., Hawkesworth, C.J., Mantovani, M.S.M., and Shukovsky, W., 1990, Mantle plumes and flood basalt stratigraphy in the Paraná South America: Geology, v. 18, p. 1223–1226.

Piccirillo, E.M., and Melfi, A.J., 1988, The Mesozoic flood volcanism of the Paraná basin: Petrogenetic and geophysical aspects: São Paulo, Brazil, Instituto Astronomico e Geofisico da Universidade de São Paulo, 600 p.

Piccirillo, E.M., Bellieni, G., Cavazzini, G., Comin-Chiaramonti, P., Petrini, R., Melfi, A.J., Pinesi, J.P.P., Zantadeschi, P., and DeMin, A., 1990, Lower Cretaceous tholeiitic dyke swarms from the Ponta Grossa (southeast Brazil): Petrology, Sr-Nd isotopes and genetic relationships with the Paraná flood volcanics: Chemical Geology, v. 89, p. 19–48.

Planke, S., Symonds, P.A., Alvestad, E., and Skogseid, J., 2000, Seismic volcanostratigraphy of large-volume basaltic extrusive complexes on rifted margins: Journal of Geophysical Research, v. 105, no. B8, p. 19335–19351.

Pontes, C.E.S., Castro, F.C.C., Rodrigues, J.J.G., Alves, R.R.P., Castellani, R.T., Santos, S.F., and Monis, M.B., 1991, Reconhecimento tectônico e estratigráfico da Bacia de Sergipe-Alagoas em águas profundas: 2nd Congresso Internacional da Sociedade Brasileira de Geofísica, Resumos Expandidos, Salvador, Bahia, v. 2, p. 638–643.

Popoff, M., 1988, Du Gondwana a l'Atlantique sud: Les connexions du fosse de la Benoue avec les bassins du nord-est bresilien jusqu'a l'overture du golfe de Guinee au Cretace inferieur: Journal of African Earth Science, v. 7, p. 409–431.

Renne, P.R., Deckart, K., Ernesto, M., Ferand, G., and Piccirillo, E.M., 1996, Age of the Ponta Grossa dike swarm (Brazil), and implications to Paraná flood volcanism: Earth and Planetary Science Letters, v. 144, p. 199–211.

Renne, P.R., Ernesto, M., Pacca, I.G., Coe, R.S., Glen, J.M., Prevot, M., and Perrin, M., 1992, The age of Paraná flood volcanism, rifting of Gondwanaland, and the Jurassic–Cretaceous boundary: Science, v. 258, p. 975–979.

Roberts, D.G., Backman, J., Morton, A.C., Murray, J.W., Keene, J.B., 1984, Evolution of volcanic rifted margins: Synthesis of Leg 81 results on the west margin of Rockall Plateau, *in* Blackman J., ed., Initial reports of the Deep Sea Drilling Project, Volume 81: Washington, D.C., U.S. Government Printing Office, p. 883–911.

Rock, N.M.S., 1991, Lamprophyres: Glasgow, Blackie, 285 p.

Rosendahl, B.R., 1997, The seismic roots of the West African passive margin: Northern Cameroon to southern Gabon: 5th International Congress of the Brazilian Geophysical Society, Expanded Abstracts, v. 1, p. 9–10.

Rosendahl, B.R., and Groschel-Becker, H., 1999, Deep seismic structure of the continental margin in the Gulf of Guinea: A summary report, *in* Cameron, N.R., Bate, R.H., and Clure, V.S., eds., The oil and gas habitats of the South Atlantic: Geological Society [London] Special Publication 153, p. 75–83.

Rosendahl, B.R., and Groschel-Becker, H., 2000, Architecture of the continental margin in the Gulf of Guinea as revealed by reprocessed deep-imaging seismic data, *in* Mohriak, W.U., and Talwani, M., eds., Atlantic rifts and continental margins: American Geophysical Union Geophysical Monograph 115, p. 85–103.

Rosendahl, B.R., Meyers, J., Groschel, H., and Scott, D., 1992, Nature of the transition from continental to oceanic crust and the meaning of reflection Moho: Geology, v. 20, p. 721–724.

Rosendahl, B.R., Groschel-Becker, H., Meyers, J., and Kaczmarick, K., 1991, Deep seismic reflection study of a passive margin, southeastern Gulf of Guinea: Geology, v. 19, p. 291–295.

Rosendahl, B.R., Mohriak, W.U., and Turner, J.P., 2000, Fitting the West African and Brazilian margins back together: Geoscience 2000, University of Manchester, England, Abstracts, p. 145.

Sadowski, G.R., and Dias Neto, C.M., 1981, O lineamento sismo-tectônico do Cabo Frio: Revista Brasileira de Geociências, v. 11, no. 4, p. 209–212.

Sandwell, D., and Smith, W., 1997, Marine gravity anomaly from GEOSAT and ERS-1 satellite altimetry: Journal of Geophysical Research, v. 102, p. 10039–10054.

Schlumberger, 1983, WEC Afrique de L'Ouest—Well Evaluation Conference. 1. Geologie, p. I.1–I.69.

Schobbenhaus, C., Campos, D.A., Derze, G.R., and Asmus, H.E., 1984, Geologia do Brasil: MME/DNPM, Brasilia, 501 p.

Schuepbach, M.A., and Vail, P.R., 1980, Evolution of outer highs on divergent continental margins, *in* Studies in geophysics: Continental tectonics: Washington, D.C., National Academy of Sciences, p. 50–61.

Skogseid, J., and Eldholm, O., 1995, Rifted continental margin off mid-Norway, *in* Banda, E., Tornè, M., and Talwani, M., eds., Rifted ocean-continent boundaries: Dordrecht: Kluwer, p. 147–152.

Smith, W.H.F., and Sandwell, D.T., 1997, Global sea floor topography from satellite altimetry and ship depth soundings: Science, v. 277, p. 1956–1962.

Souza, K.G., Fontana, R.L., Mascle, J., Macedo, J.M., Mohriak, W.U., and Hinz, K., 1993, The southern Brazilian margin: An example of a South Atlantic volcanic margin: Third International Congress of the Brazilian Geophysical Society, Rio de Janeiro, v. 2, p. 1336–1341.

Stewart, K., Turner, S., Kelley, S., Hawkesworth, C.J., Kirstein, L., and Mantovani, M., 1996, 3-D, Ar/Ar geochronology in the Paraná continental flood basalt province: Earth and Planetary Science Letters, v. 143, p. 95–109.

Sun, S.S., 1980, Lead isotopic study of young volcanic rocks from mid-ocean ridges, ocean islands and island arcs: Royal Society of London Philosophical Transactions, ser. A, v. 297, p. 409–445.

Sun, S.S., and McDonough, W.F., 1989, Chemical and isotopic systematics of oceanic basalts: Implications for mantle composition and processes, *in* Saunders, A.D., and Norry, M.J., eds., Magmatism in the ocean basins: Geological Society [London] Special Publication 42, p. 313–345.

Szatmari, P., 1998, Tectonic habitat of petroleum along the South Atlantic margins: American Association of Petroleum Geologists International Conference, Extended Abstracts Volume, Rio de Janeiro, Brazil, p. 362–363.

Szatmari, P., Conceição, J.C.J., Destro, N., Smith, P.E., Evensen, N.M., and York, D., 2000, Tectonic and sedimentary effects of a hotspot track of alkali intrusions defined by Ar-Ar dating in SE Brazil: 31 International Geological Congress, Rio de Janeiro, Abstract Volume, CD-ROM.

Talwani, M., and Abreu, V., 2000, Inferences regarding initiation of oceanic crust formation from the U.S. East Coast margin and conjugate South Atlantic margins, *in* Mohriak, W.U., and Talwani, M., eds., Atlantic rifts and continental margins: American Geophysical Union Geophysical Monograph 115, p. 211–233.

Teisserenc, P., Villemin J., 1989, Sedimentary basin of Gabon: Geology and oil systems, *in* Edwards, J.D., Santogrossi P.A., eds., Divergent/passive margin basins: American Association of Petroleum Geologists Memoir, v. 48, p. 117–199.

Thompson, R.N., 1972, The 1-atmosphere melting patterns of some basaltic volcanic series: American Journal of Science, v. 272, p. 901–932.

Thompson, R.N., and Gibson, S.A., 1991, Subcontinental mantle plumes, hot spots and pre-existing thinspots: Journal of the Geological Society of London, v. 147, p. 973–977.

Thompson, R.N., Morrison, M.A., Hendry, G.L., and Parry, S.J., 1984, An assessment of the relative roles of crust and mantle in magma genesis: An elemental approach: Royal Society of London Philosophical Transactions, ser. A, v. 310, p. 549–590.

Turner, J.P., 1995, Gravity-driven structures and rift basin evolution: Rio Muni Basin, offshore Equatorial West Africa: American Association of Petroleum Geologists Bulletin, v. 79, p. 1138–1158.

Turner, J.P., 1999, Detachment faulting and petroleum prospectivity in the Rio Muni Basin, Equatorial Guinea, West Africa, *in* Cameron, N.R., Bate, R.H., and Clure, V.S., eds., The oil and gas habitats of the South Atlantic: Geological Society [London] Special Publication 153, p. 303–320.

Turner, S., Regelous, M., Kelley, S., Hawkesworth, C., and Mantovani, M., 1994, Magmatism and continental break-up in the South Atlantic: High precision $^{40}Ar/^{39}Ar$ geochronology: Earth and Planetary Science Letters, v. 121, p. 333–348.

Turner, S.P., Peate, D.W., Hawkesworth, C.J., and Mantovani, M.S.M., 1999, Chemical stratigraphy of the Paraná basalt succession in western Uruguay: Further evidence for the diachronous nature of the Paraná magma types: Journal of Geodynamics, v. 28, p. 459–469.

Ulbrich, H.H.G.J., and Gomes, C.B., 1981, Alkaline rocks from continental Brazil: Earth-Science Reviews, v. 17, p. 135–154.

Ussami, N., Karner, G.D., and Bott, M.H.P., 1986, Crustal detachment during South Atlantic rifting and formation of Tucano-Gabon basin system: Nature, v. 322, p. 629–632.

Valente, S.C., Ellam, R.L., Meighan, I.G., and Fallick, A.E., 1999, The Serra do Mar and Ponta Grossa dyke swarms: Dynamic melting and geodynamic models for the Early Cretaceous in Southeast Brazil: Abstracts of the 7th Simpósio Nacional de Estudos Tectônicos, Lençóis, Belo Horizonte, Sociedade Brasileira de Geologia, p. 102–103.

Wannesson, J., Icart, J.C., and Ravart, J., 1991, Structure and evolution of adjoining segments of the west African margin determined from deep seismic profiling, *in* Meissner, R., Brown, L., Durbaum, H.J., Franke, W., Fuchs, K., and Seifert, F., eds., Continental lithosphere: Deep seismic reflections: American Geophysical Union, Geodynamic Series, v. 22, p. 275–289.

Wernicke, B., 1985, Uniform-sense normal simple shear of the continental lithosphere: Canadian Journal of Earth Science, v. 22, p. 108–125.

White, R.S., and McKenzie, D., 1989, Magmatism at rift zones: The generation of volcanic continental margins and flood basalts: Journal of Geophysical Research, v. 94, no. B6, p. 7685–7729.

White, R.S., and McKenzie, D.M., 1995, Mantle plumes and flood basalts: Journal of Geophysical Research, v. 100, no. B9, p. 17543–17585.

Wisse, E., Scharroo, R., Naeije, M.C., and Wakker, K.F., 1994, Mean sea surface computation from ERS-1 data: Proceedings of the Second ERS-1 Symposium, Space at the Service of our Environment, Hamburg, Germany, 11–14 October 1993, ESA SP-361, p. 1053–1058.

Wood, B.J., and Fraser, D.G., 1976, Elementary thermodynamics for geologists: Oxford, Oxford University Press, 303 p.

Zalán, P.V., Wolf, S., Astolfi, M.A.M., Vieira, I.S., Conceição, J.C.J., Appi, V.T.T., Neto, E.V.S., Cerqueira, J.R., and Marques, A., 1990, The Paraná Basin, Brazil, *in* Leighton, M.W., Kolata, D.R., Oltz, D.S., and Eidel, J.J., eds., Interior cratonic basins: American Association of Petroleum Geologists Memoir, v. 51, p. 681–701.

Manuscript Accepted by the Society October 15, 2001

Printed in the U.S.A.

Geological Society of America
Special Paper 362
2002

Volcanic passive margin of Namibia: A potential fields perspective

B. Corner
Corner Geophysics Namibia cc., P.O. Box 2055, Swakopmund, Namibia
J. Cartwright
Department of Earth Sciences, Cardiff University, P.O. Box 914, Cardiff F10 3YE, UK
R. Swart
National Petroleum Corporation of Namibia (NAMCOR (Pty) Ltd.), Private Bag 13196, Windhoek, Namibia

ABSTRACT

Namibia is now regarded as one of the type areas for studying the development of a passive volcanic margin. Recent work has concentrated on interpreting seismic data, and little use has been made of existing magnetic or gravity data. New aeromagnetic data that were recently acquired offshore Namibia by NAMCOR have been merged with existing onshore and offshore aeromagnetic surveys, as well as with ship-borne data. In addition, satellite-derived offshore free air gravity data have been merged with onshore Bouguer gravity data.

A number of new features pertinent to the evolution of the volcanic margin are apparent from the new combined data sets. The most striking is that the G, M2, and M4 magnetic anomalies are better defined and are seen to swing coastward toward northern Namibia and southern Angola, in contrast to studies by earlier researchers in which these anomalies were interpreted to swing away from the coast. This has important implications for the mechanism and rate of breakup, for the age of the volcanics, and for the age of sedimentation along the coastal margin.

Some interpreted seismic lines have also been modeled with coincident magnetic and gravity profiles to test whether the interpretation of the entire package of seaward-dipping reflectors (SDR) as being basaltic flows is correct. The magnetic data indicate that at least part of this package is likely to be filled with sediment, while the gravity data are less conclusive in this respect due to unknown and hence ambiguous density contributions from deeper crustal levels. These SDR packages of interpreted basalts interbedded with sediments are seen as suboutcrops along the prominent edge gravity high, which correlates closely with both the hinge zone and the G magnetic anomaly.

Westward (seaward) of the hinge zone the edge gravity anomaly loses amplitude over the underlying basalts. Bearing the sources of ambiguity in mind, forward modeling nevertheless shows that the gravity anomaly can be completely neutralized, in the first order, by the overlying low-density sedimentary pile, which thickens westward. This suboutcrop of the SDR package models well as the source of the dipolar G magnetic anomaly, and contributes significantly to the shelf edge gravity high. No field reversals were required to obtain a good fit in the forward modeling, although their presence cannot be discounted.

Corner, B., Cartwright, J., and Swart, R., 2002, Volcanic passive margin of Namibia: A potential fields perspective, *in* Menzies, M.A., Klemperer, S.L., Ebinger, C.J., and Baker, J., eds., Volcanic Rifted Margins: Boulder, Colorado, Geological Society of America Special Paper 362, p. 203–220.

Individual volcanic complexes can also be seen in the data sets; e.g., the Phoenix high, which was originally recognized and interpreted by Norsk Hydrogeologists. New volcanic complexes have now also been recognized. A number of interpreted major onshore structures, mostly of Pan-African age at their earliest manifestation, are seen to displace the edge gravity and G, M2, and M4 magnetic anomalies. This implies reactivation of these structures into the offshore domain at least until the late Mesozoic. Two of the structures, the Opuwa lineament and Omaruru lineament zone, continue westward as the Rio Grande Rift and Walvis Fracture Zone, respectively, having probably constituted zones of weakness with favorable orientations for initiation of these fracture zones at the time of breakup.

INTRODUCTION

The Namibian continental margin is in many respects a type area for studying the development of a volcanic margin basin (Gladczenko et al., 1997). The margin occupies a relatively linear stretch of the eastern border of the South Atlantic (Fig. 1), and is relatively free from the tectonic complexities related to more complex plate boundaries. The margin is located in the path of influence of the Tristan hotspot, and this provides an opportunity to study interactions between possible plume impact and the lithospheric processes operative during continental breakup. The margin is flanked by a major physiographic feature of continental dimensions (the Great Escarpment; King, 1951), and is also associated with a major flood basalt province, the Etendeka (Fig. 1) (Milner et al., 1994), both of which provide additional dimensions for an integrated analysis of the evolution of the margin. A final point of interest arises from the structure of the continental crust flanking the margin. The margin has developed in the north along the north-northwest–trending coastal branch of the Pan-African Damara mobile belt (Martin, 1973; Clemson et al., 1999), and is segmented along its strike farther south by structures that appear to be related to the interior, or inland, northeast–trending branch of the Damara mobile belt (Clemson et al., 1999). In the extreme south the margin has developed along the trend of the Pan-African Gariep Belt. The Namibian margin is thus also ideally situated for a study of the influence of lithospheric heterogeneity on volcanic margin style and evolution.

Much previous work has concentrated on interpreting seismic data (Clemson, 1997; Gladczenko et al., 1997; Aizawa et al., 2000). Magnetic and gravity data interpretation, e.g., some forward modeling of gravity and magnetic ship-track profiles offshore Namibia, was conducted by Bauer et al. (2000), whereas Stewart et al. (2000) conducted gravity modeling and a rigorous subsidence analysis offshore Namibia. In this chapter, we present an interpretation of newly acquired offshore aeromagnetic data that have been merged with the existing national onshore aeromagnetic and gravity data sets as well as with satellite-derived offshore gravity data and ship-track magnetic data. These data provide an invaluable basis for mapping regional crustal structure. The specific aim of this study is to interpret these potential field data sets, together with the seismic data, in order to elucidate offshore structure and the links thereof with onshore structures. Emphasis is placed on mapping and modeling the important features associated with the volcanic passive margin such as the G, M2, and M4 magnetic anomalies (Gladczenko et al., 1997), the offshore edge gravity high, and the seaward-dipping reflectors (SDR) (Clemson, 1997).

A particular problem of direct relevance to characterization of the volcanic margin style (Planke et al., 2000) is to quantify the extent of the offshore lava suites. At present, vast thicknesses of layered reflective sequences are uncalibrated by drilling, and their lithology is unknown. SDR sequences in the deeper offshore domain have been universally interpreted by previous workers as being extrusive igneous rocks, but the status of a 10–14-km-thick package located closer inshore is a point of current debate (Clemson, 1997; Gladczenko et al., 1997). Major onshore structures and tectonostratigraphic domains extending into the offshore domain were mapped in this study, and a number of new structural features and intrusions are identified and discussed.

REGIONAL GEOLOGIC SETTING

The Namibian continental margin consists of a substantial thickness (3–5 km) of mainly clastic postrift sediments overlying basement rocks of unknown affinity (Clemson, 1997). The innermost part of the margin is likely to be floored by stretched continental crust (Clemson et al., 1999), but the outermost part is floored by oceanic crust. The location and geometry of the continent-ocean boundary are poorly defined. The synrift sequence has barely been calibrated offshore Namibia (possibly the uppermost few hundred meters). The bulk of the synrift is thought to consist of Karoo Supergroup sediments overlain by Late Jurassic to Early Cretaceous siliciclastics and volcanics (Clemson, 1997; Clemson et al., 1999).

The postrift sequence along the outer portion of the margin overlies a sequence of SDRs that are interpreted to be the result of a huge subaerial basaltic extrusive event associated with breakup (Gladczenko et al., 1997; Clemson, 1997). One of the major unknowns of this margin is the seaward extent of the onshore Etendeka flood basalt province, and the link, if any, with the offshore SDR province (Clemson, 1997).

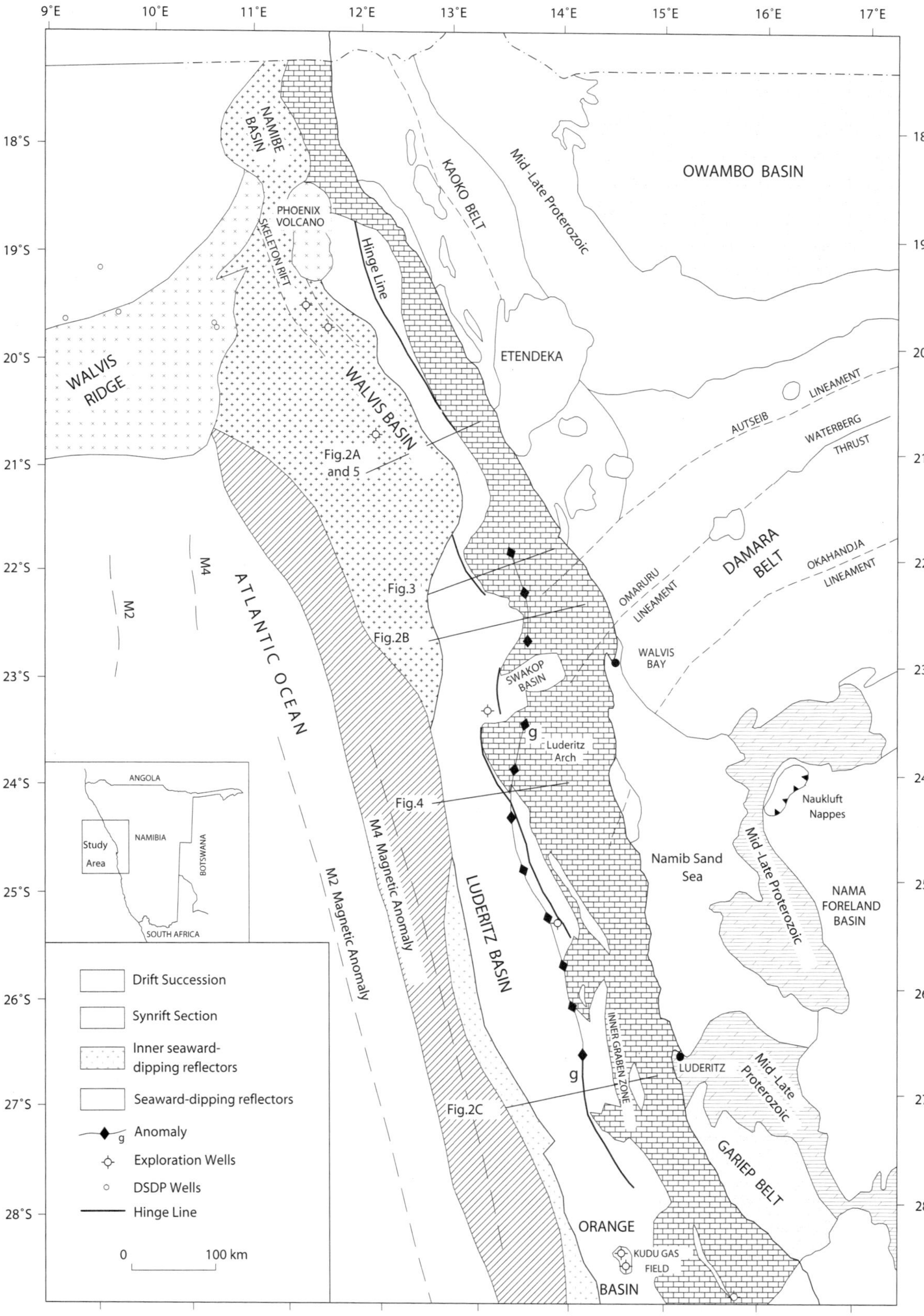

Figure 1. Map showing major basins and some of the related tectonic features offshore and onshore Namibia that were known before potential field studies (covered in this chapter) were started. Northern position of M2 and M4 has since changed and relationship between onshore and offshore structures has since become clearer. DSDP, Deep Sea Drilling Project.

The general structure of the margin and the geometry of the main stratigraphic units, as known prior to this study, are shown schematically in Figure 1, which includes the earlier locations of the M2 and M4 magnetic anomalies (Rabinowitz and LaBreque, 1979). The geoseismic cross sections (e.g., Fig. 2) are based on regional multichannel reflection seismic (MCS) data acquired in the late 1980s and early 1990s. From these sections, it is evident that there is limited upper crustal extension along the margin, although the limited penetration of the data means that the deeper structure is ambiguous. The dominant rotation of the older units is oceanward. The post-Karoo synrift sequence thickens dramatically beneath the modern slope, in a structural unit widely referred to as the inner half-graben.

Based on detailed interpretation of ~8000 line kilometers of MCS data, we have interpreted the basal units of this thick synrift package (i.e., directly above the Karoo in Fig. 2) as being the offshore equivalent of the Etendeka Group extrusive series. This interpretation is driven by the following key observations.

1. Stratigraphic relationships are exposed on land where Etendeka Group basalts erupted onto an unconformity marking the upper boundary of the Karoo Supergroup (Milner et al., 1994).

2. Similar stratigraphic and topographic relationships are at a particular level in the offshore MCS data (Fig. 3). The unit is correlated along the entire margin, a distance of ~1300 km; it has a consistent stratigraphic position relative to fault-bounded underlying sequences (interpreted as Karoo), and overlying seaward-thickening packages developed within the inner half-graben (Clemson, 1997; Clemson et al., 1999).

This interpretation differs considerably from that of Gladczenko (1994), who interpreted the oceanward-thickening wedge of the inner half-graben as a "pre-breakup sedimentary complex" consisting of "Paleozoic (?) and Mesozoic prerift and synrift sediments covered by Etendeka basalts" (p. 130). This definition would thus place the Etendeka at the upper boundary of what we show as synrift (Fig. 2), rather than at the basal contact with the Karoo as shown in our interpretation. This difference is highly significant. The Etendeka is dated as 132 ± 1 Ma (Renne et al., 1996). In our interpretation, there is an ~6–8 m.y. interval between the extrusion of the Etendeka Group and the deposition of shallow-marine limestones (dated as Barremian in two offshore wells) that effectively form the upper boundary to the synrift. In Gladczenko's (1994) interpretation, however,

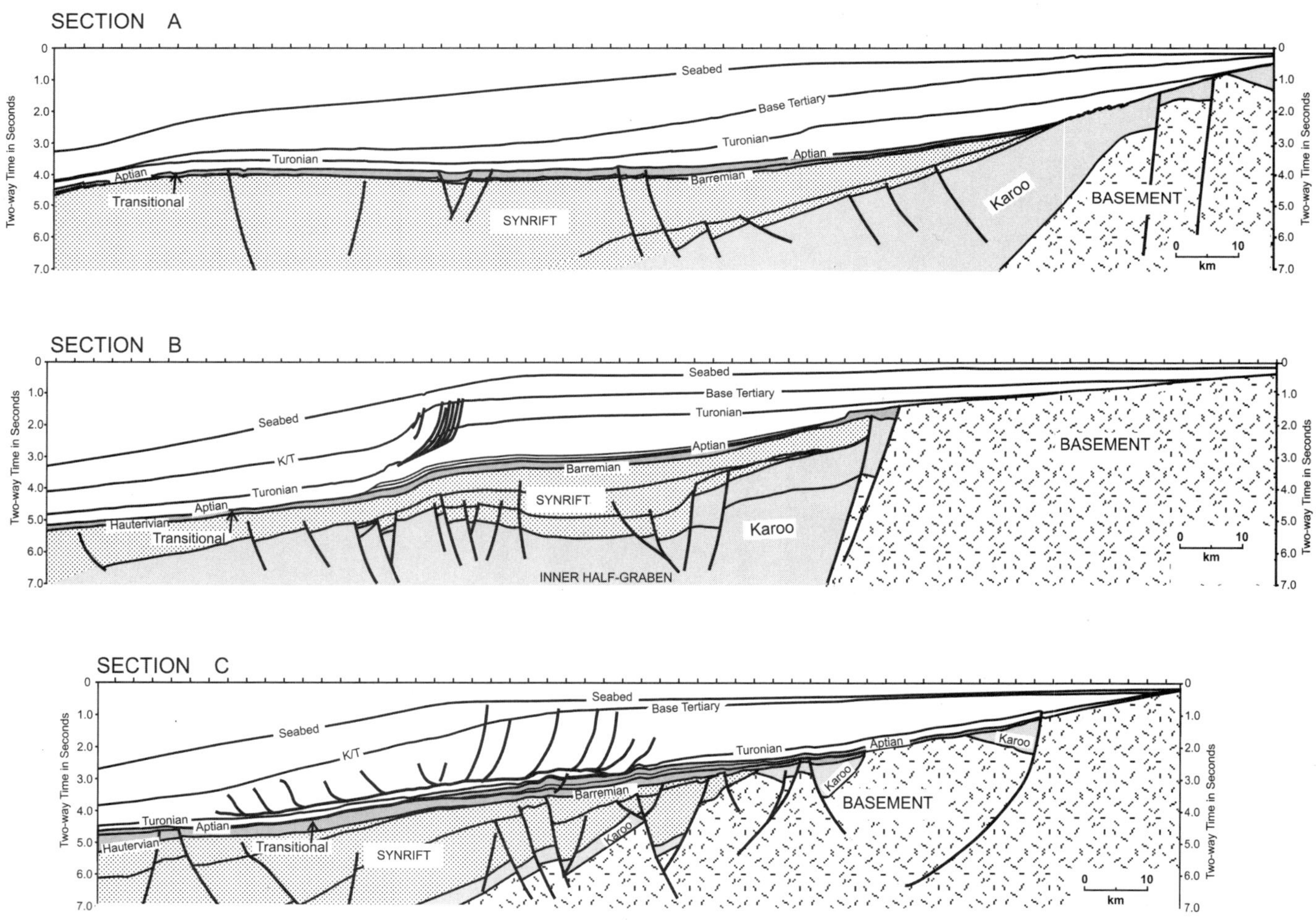

Figure 2. Three geoseismic sections illustrating variations in tectonostratigraphic style of margin. Section localities shown in Figure 1.

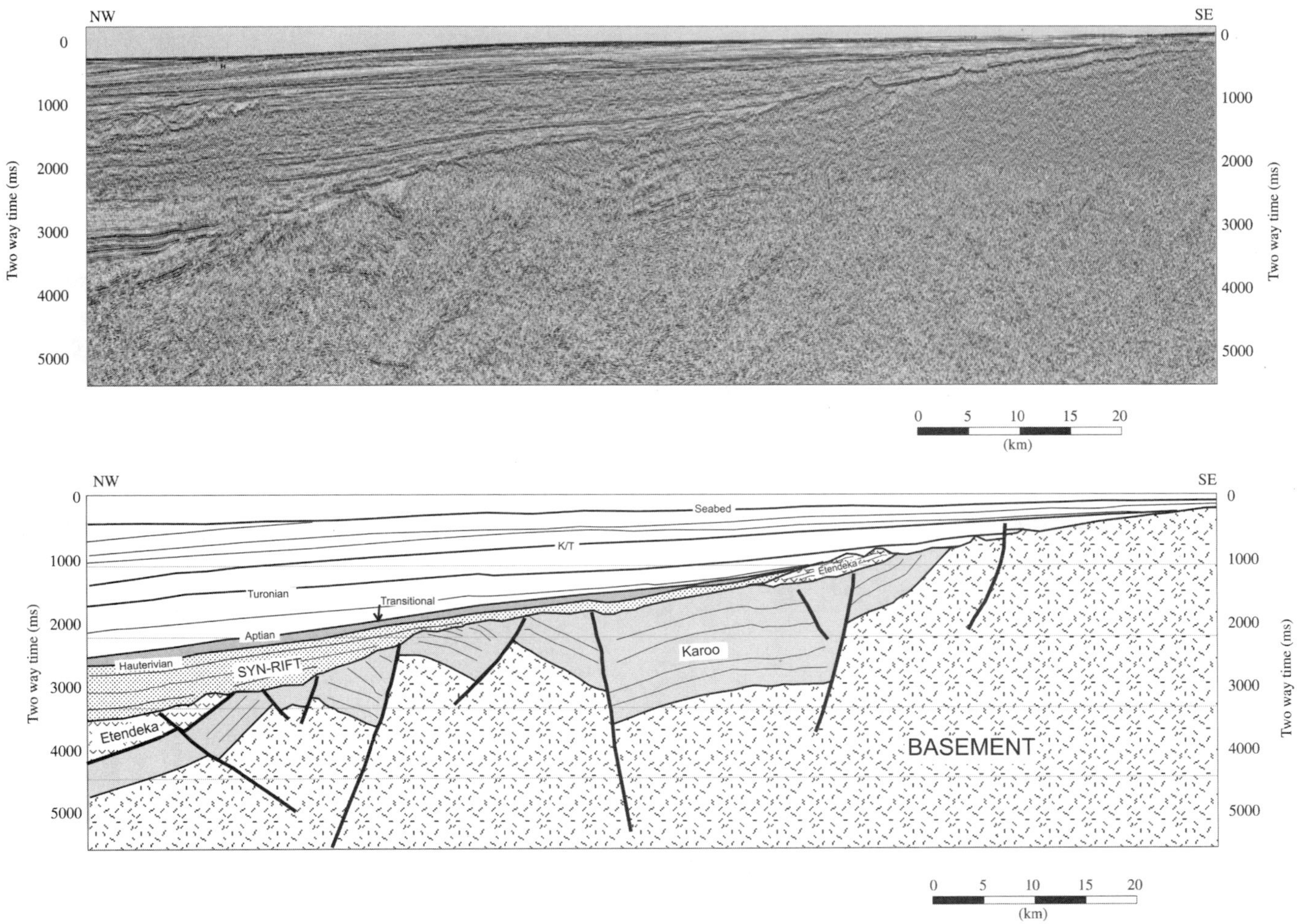

Figure 3. Seismic profile (top) and geoseismic section (bottom) from Walvis Basin, illustrating nature of basal and upper Etendeka surfaces in offshore, and pronounced landward onlap of Cretaceous units. Section locality shown in Figure 1.

there is an unspecified (but much longer) period representing the units occupying the inner half graben, and an implication that they are mainly sedimentary rather than igneous. If our interpretation is correct, it raises the extremely important question of the composition of the inner half-graben fill, the maximum thickness of which ranges from values of 6 to 14 km within the Walvis Basin (Clemson, 1997). It is to address this question that the magnetic modeling described herein was directed.

DATA SETS AND PROCESSING

The regional onshore aeromagnetic data set of Namibia, acquired mostly at a 1 km line spacing, has been significantly enhanced in recent years in two respects, i.e., by the ongoing National Programme of the Geological Survey of Namibia to improve the resolution to a 200 m line spacing, and by the addition of an offshore data set extending ~250 km out to sea. The latter was acquired at a square-grid line spacing of 25 km under contract to the National Petroleum Corporation of Namibia (Pty) Ltd. (NAMCOR). Onshore gravity coverage is much more sparse in terms of data density, stations having been acquired at coarse intervals of 1 km or more, mostly along the main roads. The offshore satellite-derived gravity data, acquired by NAMCOR from Geophysical Exploration Technology, University of Leeds, by comparison, shows excellent resolution.

A number of combinations of the above data sets have been used in the interpretation, including: (1) offshore aeromagnetic and ship-track magnetic data merged with the onshore aeromagnetic data; (2) offshore aeromagnetic and ship-track magnetic data merged with the onshore Bouguer and free air gravity data; and (3) offshore satellite-derived gravity data (both free air and trend-corrected Bouguer data) merged with both the onshore aeromagnetic and Bouguer gravity data.

The merged images are too numerous to include in this text. We present two of these that show the most salient features of all the data sets, i.e., the merged offshore and onshore magnetic data, presented as Figure 4, and the satellite-derived offshore free air gravity data merged with the onshore Bouguer gravity,

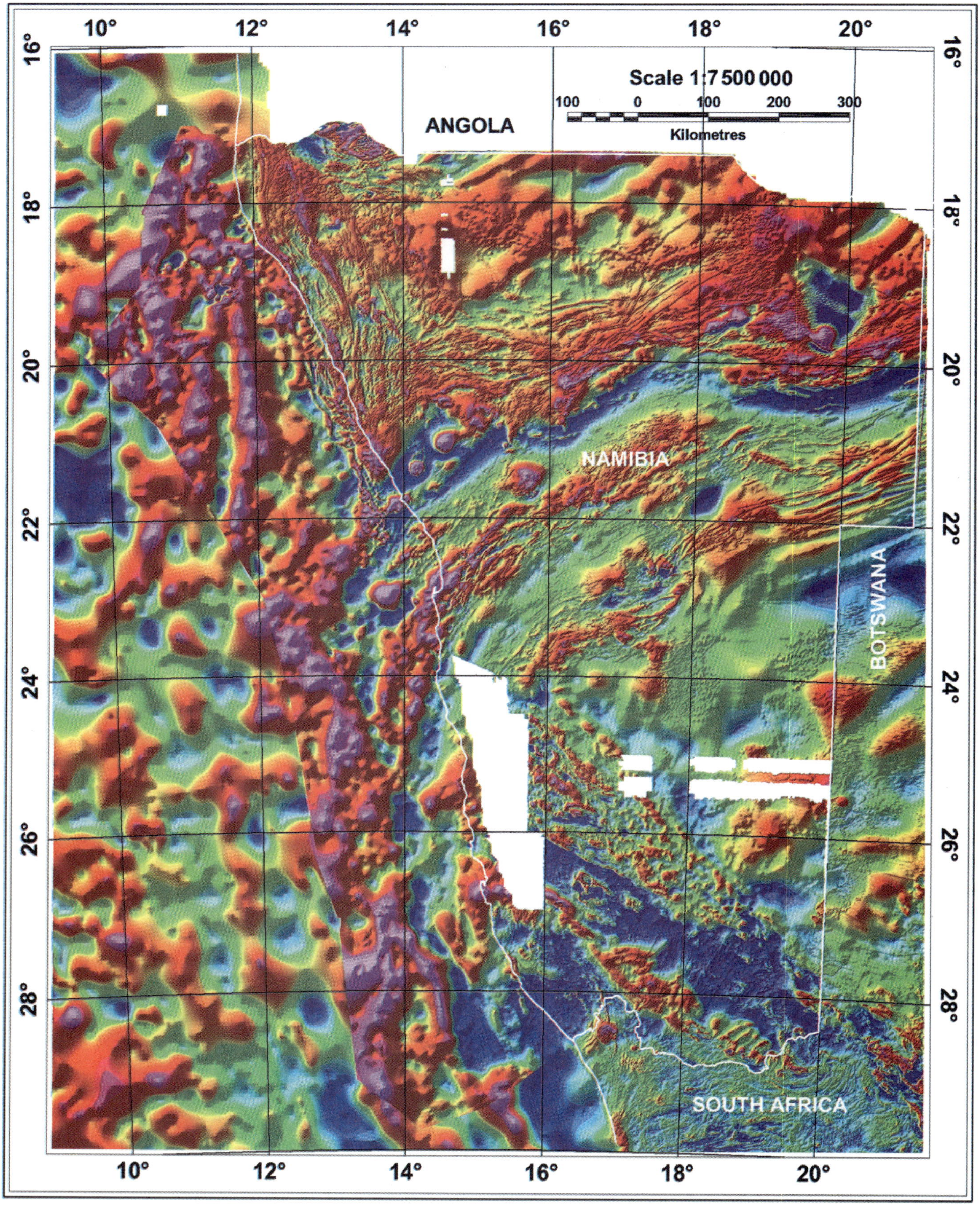

Figure 4. Namibian onshore and offshore total-field magnetic data (with northwest sun shading).

presented as Figure 5. Because an isostatic anomaly map of Namibia is not currently available, a first-order long-wavelength trend has been removed from the Bouguer gravity data set in Figure 5 so as to reduce the artifact of the continental margin Bouguer gradient.

A number of transformations and filters have been applied to the data sets in order to enhance features of interest. These transformations include pole reduction, analytical signal, first vertical derivative, and trend removal. Real-time imaging (apparent sun shading and color palette rotation) of these data sets was applied in order to facilitate identification of structures. Figures 4 and 5 are presented here at a highly reduced scale of 1:7 500 000 for the sake of economy in this volume. In addition, the presented images reflect only one sun-shaded direction and one color palette. Thus many of the features that helped identify certain of the lineaments and structures may not necessarily be seen easily in these images. A study with similar objectives and methodologies to those of the present study is that of Smethurst (2000), who used magnetic and gravity data to study land and offshore tectonic links between western Norway and the North Sea.

FORWARD MODELING

The process of forward or inverse modeling of potential field data is inherently ambiguous, allowing a large variety of combinations of depth, attitude, geometry, and physical property contrast (density or magnetic susceptibility), and direction of magnetization (in the case of magnetics) to yield equally good solutions. In the absence of the control of deep crustal seismic data in this study, modeling was confined to the upper crustal region covered by the shallower MCS data, i.e., limited to the upper ~10 km. The gravitational effects of deeper crustal and upper mantle sources were assumed to contribute to a longer wavelength regional field in the data. The gravitational effects of these sources were studied by Watts and Fairhead (1999), Bauer et al. (2000), and Stewart et al. (2000). In this study we have quantified the magnetic and gravitational effects of the basalt sequences using the control of the relatively shallow marine seismic lines. The regional gravity field was removed using an interactive spline function fitted to the data, incorporating the effect of the water layer from a model thereof. Such regional removal is a source of ambiguity in view of the mixing of frequency content of shallow and deep sources. Nevertheless, it was felt that the residual anomalies arising from shallower crustal levels were adequately isolated. Two seismic lines, ECL 89-32 and ECL 91-405 (Figs. 6 and 7), showing the major offshore structural and stratigraphic elements, were selected for forward modeling of the corresponding gravity and magnetic data sets. Depth conversion of these two seismic lines (M. Aizawa, 1998, personal commun.) provided lateral contact and depth data for the modeling process. Velocities for the depth conversion were obtained from checkshot data from five nearby boreholes, and deeper into the profiles from stacking velocities. These boreholes also provided the necessary lithological data to interpret the seismic lines and the geological control required for the modeling process. A number of objectives were addressed with the forward modeling interpretation, including (1) to test the nature and origin of the G, M2, and M4 magnetic anomalies, and coast-parallel gravity highs, in comparison to previous interpretations (Rabinowitz and LaBrecque, 1979; Light et al., 1992; Gladczenko, 1994; Clemson et al., 1999; Watts and Fairhead, 1999; Bauer et al., 2000, Stewart et al., 2000); and (2) to test whether the inner half-graben is dominantly basaltic flows, dominantly siliciclastic rocks, or an intercalation of the two.

The modeled sections (Figs. 8 and 9) were interpreted using the forward modeling software of Professional Geophysical Services (Witwatersrand University, Johannesburg). A regional gradient and effect of the water layer was estimated, as described above, and subtracted from the gravity profiles. Only a minor regional was removed from the magnetic data due to the relatively poor depth of penetration (having a fall-off rate one inverse power of source distance higher than for an equivalent gravity case, thus reducing considerably contributions from deeper sources). An important consideration was the possible effect of basaltic sequences having a reversed magnetization. Classically, where marine magnetic anomalies display a primarily negative anomaly form, a magnetization reversal is invoked (e.g., Vine and Matthews, 1963). The G magnetic anomaly comprises both positive and negative components. These components could be explained by a classic normal and reversed scenario. Bauer et al. (2000) invoked an arbitrary shaped body of reversed magnetization in their modeling of the G anomaly offshore Namibia to fully explain the negative anomaly. Carruthers et al. (2001) similarly used reversely magnetized units to model the negative anomaly. In our alternative scenario both the positive and negative anomaly components could be easily explained by a single north-south–striking normally magnetized body with a very shallow dip to the west, i.e., from the suboutcrop of the feather edge of the magnetic units. This is precisely the structural style of the SDRs as interpreted from the seismic sections. In the first instance we therefore assumed normal (induced) magnetization before testing for reversals and, using the depth and geometry control from the seismic data, successfully modeled the magnetic profiles. Thus, although the presence of some reversed magnetization may enhance the negative component of the anomaly, the magnetic anomalies studied can also be explained by induced magnetization using the well-controlled geometry derived from interpretation of the SDRs. Such are the limits of ambiguity of the modeling process. The susceptibility values are somewhat high, but still within the range of published values for basalts (Clark, 1997). The susceptibilities could be reduced by slightly decreasing the depth to the basalts. This would be dependant on the errors (unknown) associated with the seismic depth conversions. During the forward modeling process an attempt was made to fill the entire inner half-graben with basalt. This could not be supported by the modeling,

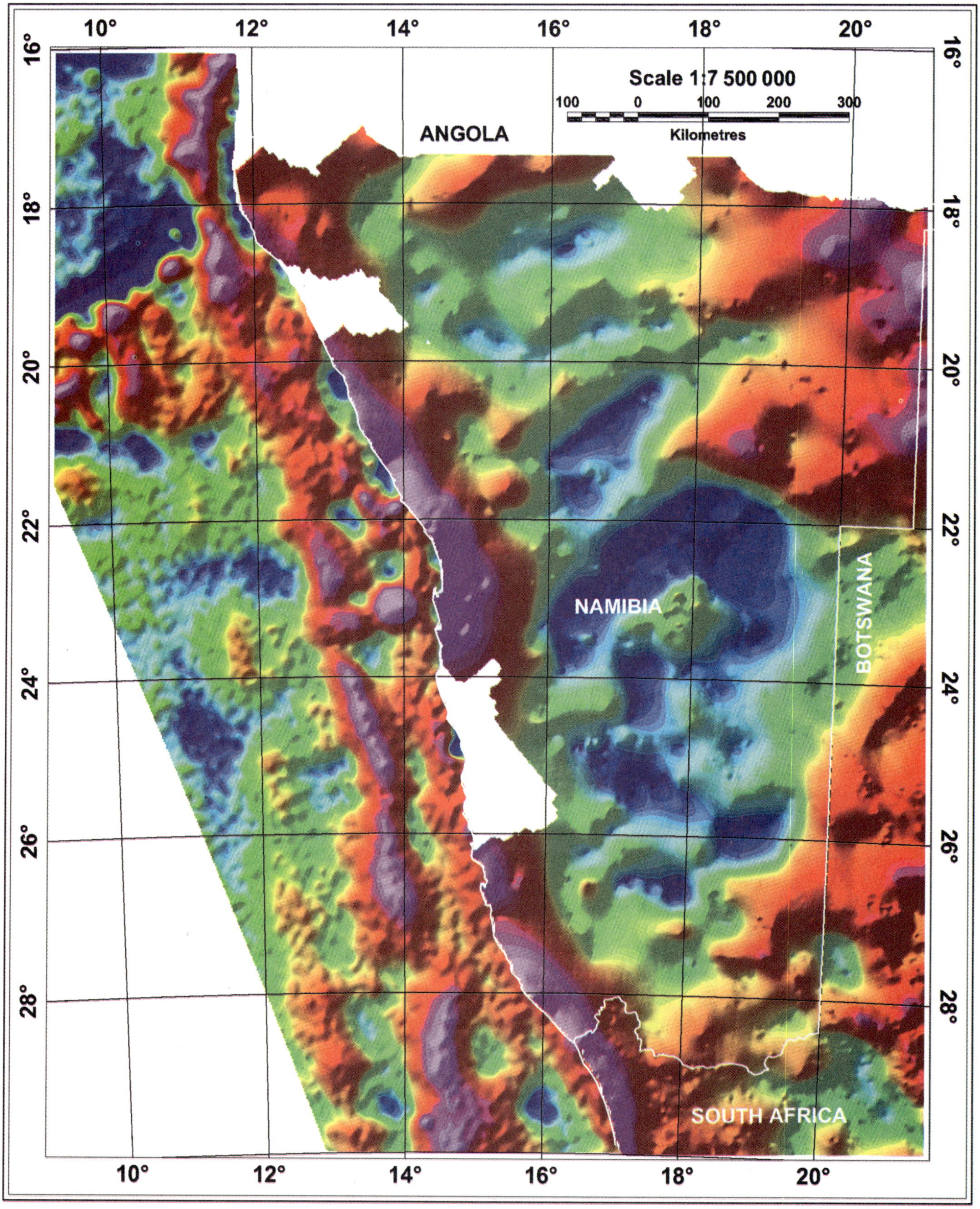

Figure 5. Namibian onshore Bouguer and offshore free air gravity data (first-order trend corrected Bouguer data, with northwest sun shading).

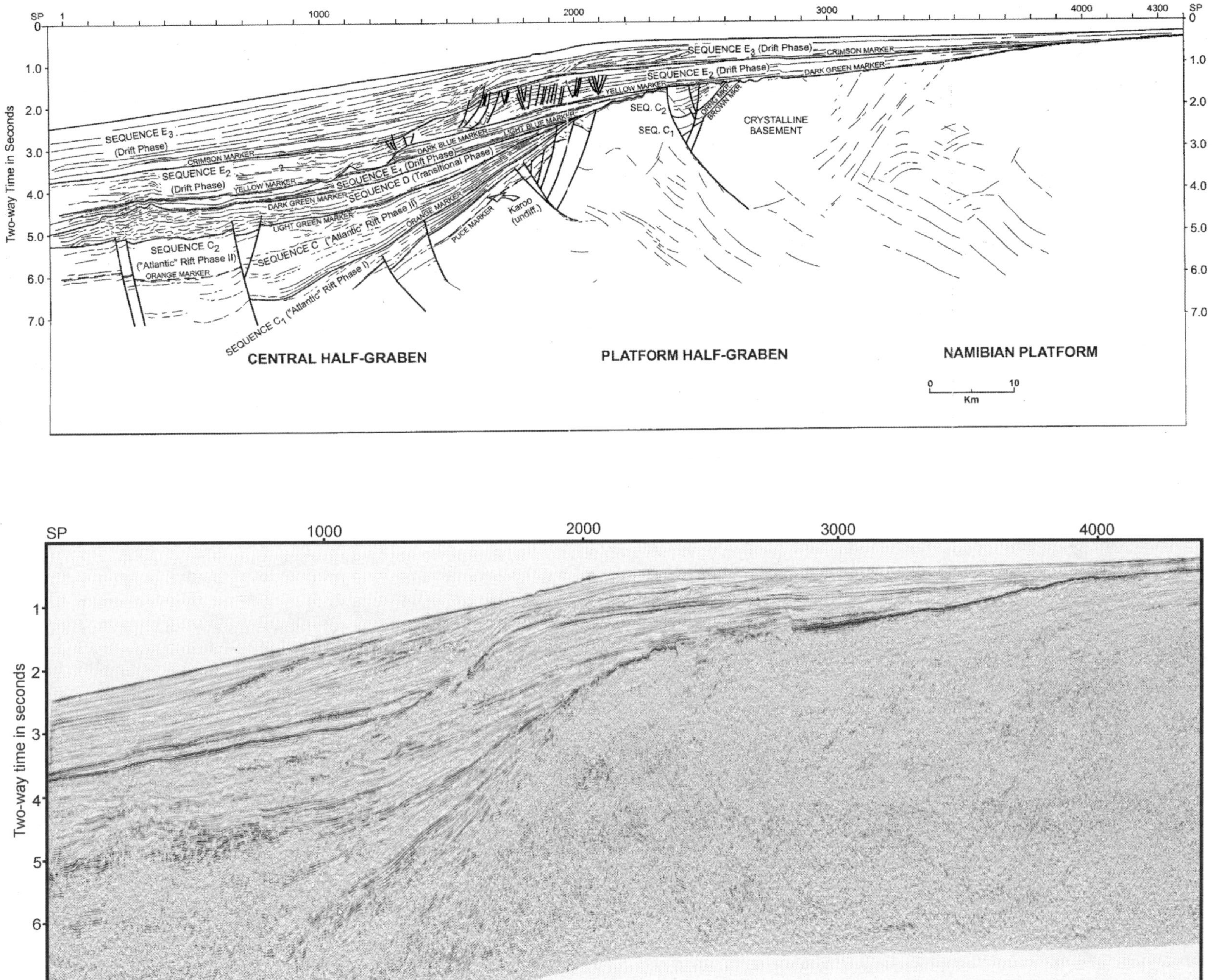

Figure 6. Interpreted line diagram of seismic line ECL-89-32 (after Clemson, 1997).

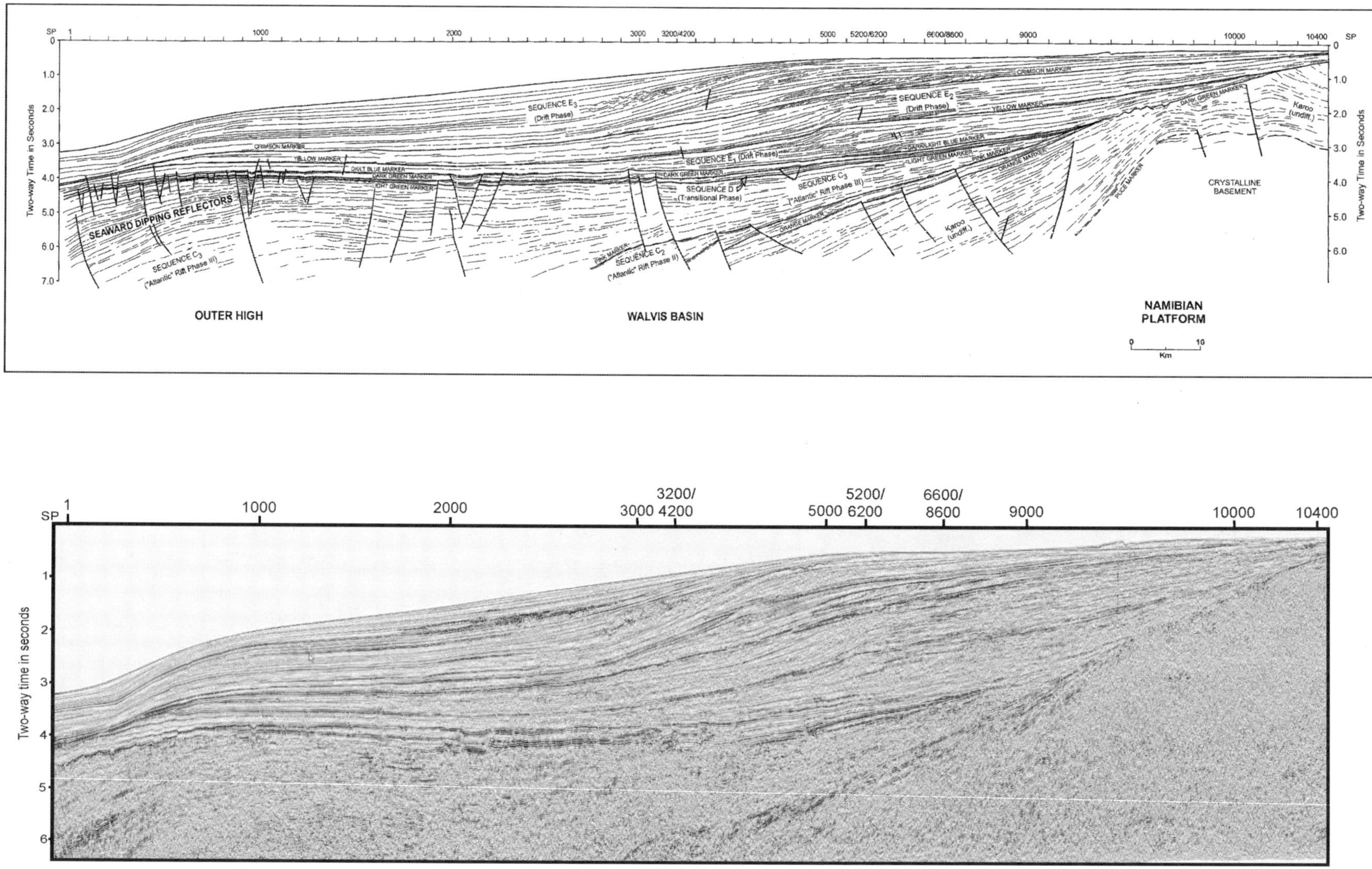

Figure 7. Interpreted line diagram of seismic line ECL-91-405 (after Clemson, 1997).

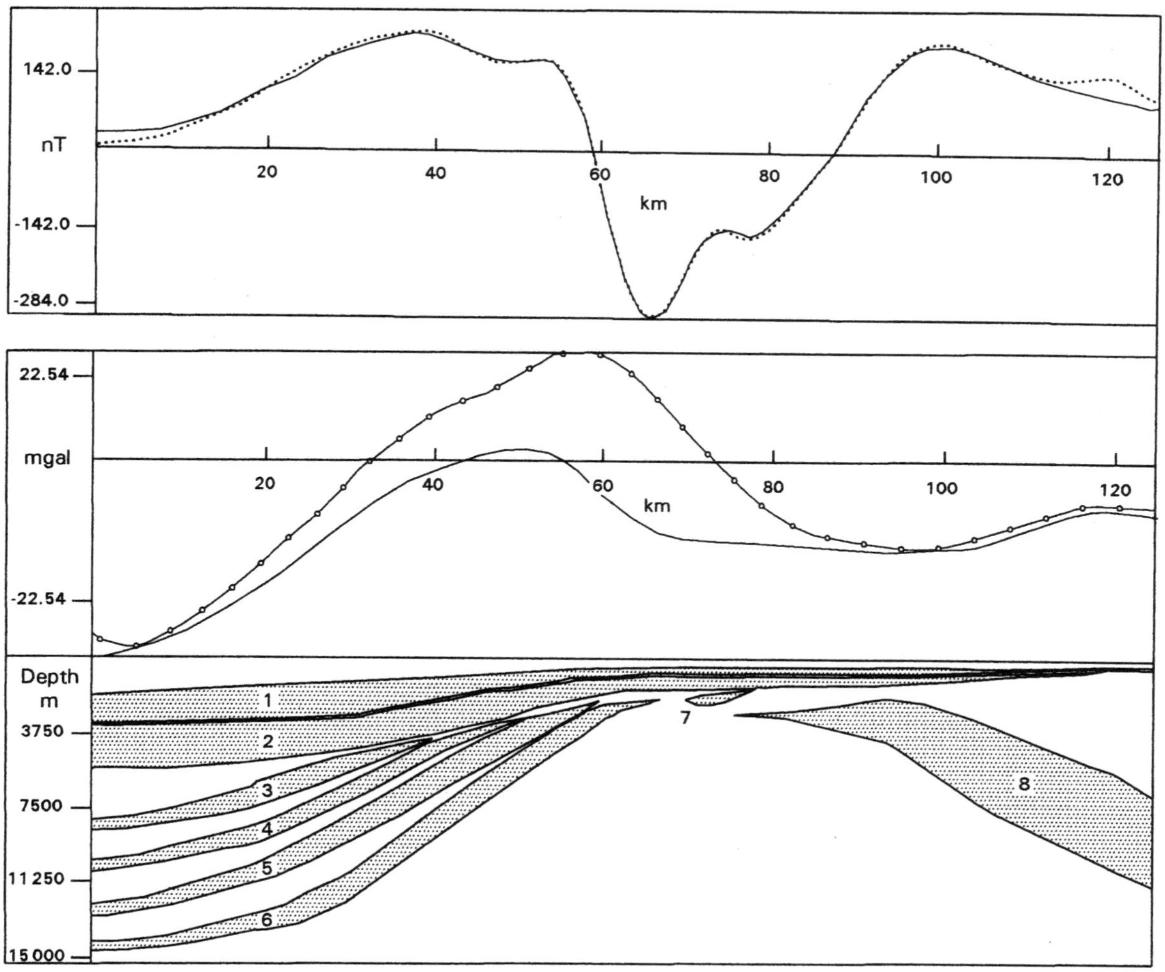

Figure 8. Line 89-32: forward model section is solution for magnetic data (top). Lower profile is equivalent gravity solution for this model. Dotted and solid curves are observed and theoretical, respectively. Physical property contrasts with respect to crystalline basement (susceptibility = 0 SI; density = 2.65 gm/cm^3) are given. See Table 1.

TABLE 1. LINE 89-32: PHYSICAL PROPERTIES

Body	Susceptibility contrast (SI)	Density contrast (gm/cm^3)	Unit
1	0.0	–0.45	Barremian and younger
2	0.0	–0.35	Barremian and younger
3	0.10	0.25	Basalt SDR
4	0.13	0.25	Basalt SDR
5	0.13	0.35	Basalt SDR
6	0.29	0.40	Basalt SDR
7	0.07	0.00	Damaran
8	0.15	0.00	Damaran

Note: The white areas are of zero magnetic and density contrast, i.e., beneath bodies 6 and 8 is crystalline basement, in between bodies 2 and 3, 3 and 4, 4 and 5, and 5 and 6 are possible intercalated sediments.

The vertical exaggeration is 3.3.

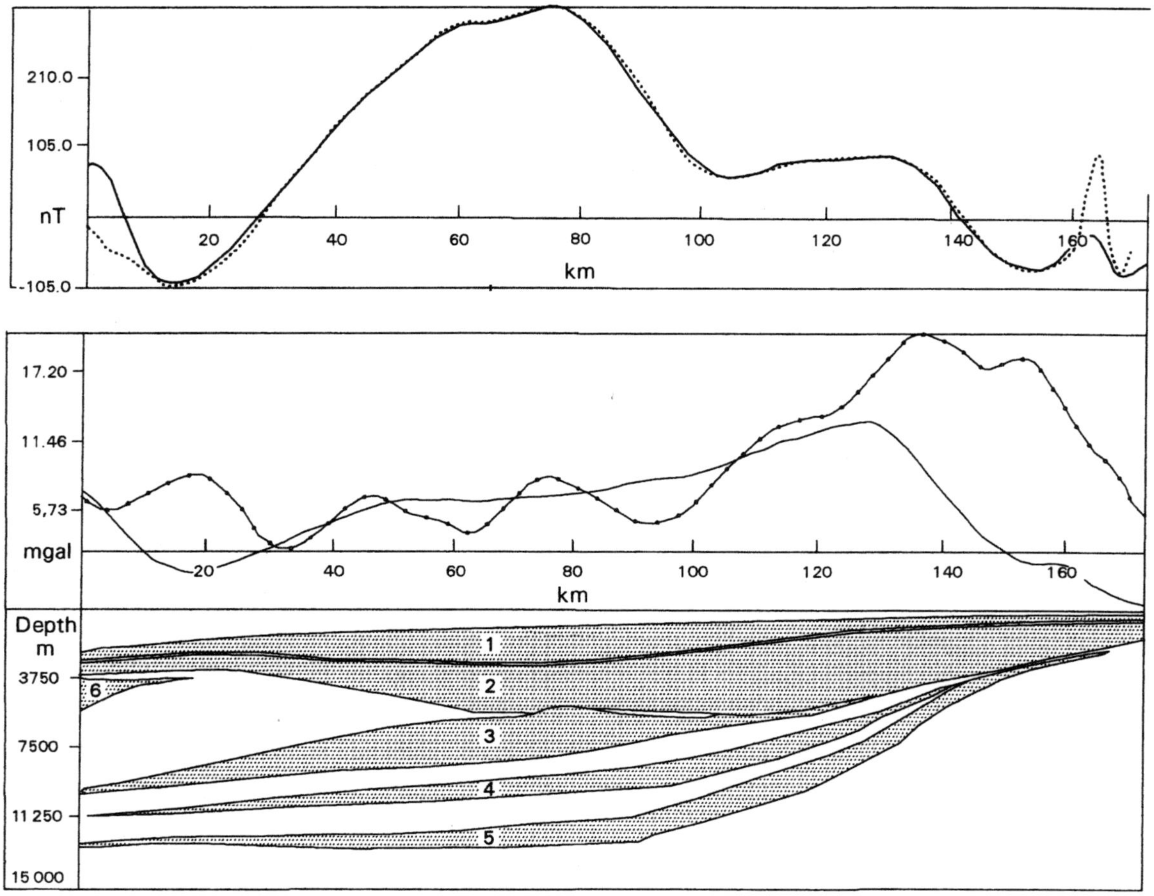

Figure 9. Line 91-405: forward model section is solution for magnetic data (top). Lower profile is equivalent gravity solution for this model. Dotted and solid curves are observed and theoretical, respectively. Physical property contrasts with respect to crystalline basement (susceptibility = 0 SI; density = 2.65 gm/cm^3) are given. See Table 2.

TABLE 2. LINE 91-405: PHYSICAL PROPERTIES

Body	Susceptibility contrast (SI)	Density contrast (gm/cm^3)	Unit
1	0.0	–0.30	Barremian and younger
2	0.0	–0.05	Barremian and younger
3	0.35	0.10	Basalt SDR
4	0.28	0.35	Basalt SDR
5	0.25	0.35	Basalt SDR
6	0.30	0.30	Basalt SDR

Note: The white areas are of zero magnetic and density contrast, i.e., beneath body 5 is crystalline basement, in between bodies 6 and 3, 3 and 4, and 4 and 5 are possible intercalated sediments.

The vertical exaggeration is 2.6.

leading to the conclusion that intercalated sediments must also be present. The magnetic models thus demonstrate that the basalt sequences associated with the SDRs give rise to the observed anomalies, regardless of whether they are normally or reversely magnetized.

The modeled magnetic sections were replicated without geometric modification to the bodies in order to assess their gravitational field. Densities, either estimated typical values or derived from borehole logs, were ascribed instead of susceptibilities (Figs. 8 and 9). Without any further forward modeling of the gravity field, the models show that the gravity high arising from the basalts corresponds both in position and wavelength to the observed edge effect gravity high (we refer to the edge effect anomaly as the "100 km gravity high" herein in order to avoid usage of a term which suggests a genetic origin). The modeled amplitudes are, however, slightly lower and their peaks are slightly farther seaward. This could be caused by a number of factors, the most important being an incorrect estimation of the regional field associated with crustal stretching and underplating. Other factors could include an incorrect choice of densities, too great a depth of burial (supported by the discussion on susceptibilities and depths above), too few basaltic bodies, and density heterogeneities in the upper crust. Because none of these factors could be quantified, no further gravity modeling was conducted. The objective, however, of calculating the gravity field associated with the magnetically modeled units was in part fulfilled by demonstrating that the SDRs contribute to the observed gravity anomalies.

In the magnetically modeled section of line ECL 89-32 (Figs. 6 and 8) a magnetic unit within the Damaran basement to the east must be included to obtain a good fit. This is not inconsistent with the known magnetic mineralogy of the Damara belt (Corner, 1983). Line ECL 91-405 (Figs. 7 and 9) also covers the suboutcrop of a further set of SDRs at its western extremity. The gravity profile along this line suggests a considerable amount of density heterogeneity that is not clear in the magnetic data and is thus not accommodated in the modeling.

We thus conclude that the G and M series of magnetic anomalies, and the 100 km gravity high, comprise essentially two components, i.e., relatively near surface contributions associated with the basalt-related SDRs, and deeper contributions seen primarily in the gravity data related to crustal extension and magmatic underplating. The deeper crustal contributions were successfully modeled gravimetrically by Bauer et al. (2000), whereas Stewart et al. (2000) were not able to fully model the observed gravity data using rigorous subsidence analysis and an underplating model. Stewart et al. concluded that the residual gravity highs of the 100 km gravity high may be related to the Etendeka volcanics, in view of their close spatial relationship. Our modeling fully supports and in part quantifies this conclusion. With the magnetic data we have demonstrated that an excellent fit can be obtained using a simple geological model, based on the seismic sections, of an inner half-graben fill comprising intercalated basalts and siliciclastic rocks with a shallow westward-dipping wedge geometry, thinning eastward by convergence to a pinchout at the hinge zone. Although our modeling demonstrates that the basaltic sequences of reversed polarity are not essential in obtaining a good fit using the above geological model, they are nevertheless likely to occur. The degree to which they may be present cannot, however, be estimated from the magnetic data alone in view of the ambiguity of the modeling process.

REGIONAL INTERPRETATION OF THE POTENTIAL FIELD DATA

The regional crustal framework of onshore Namibia has recently been interpreted from aeromagnetic and gravity data (Corner, 2000). The most significant large-scale lineaments and ring structures mapped in that study, tied in with the tectonostratigraphic terranes as delineated by Petzel and Schreiber (1999), are indicated in the interpretation map (Fig. 10). The interpretation of the offshore domain presented here is an extension of previous interpretations by Corner and Swart (1997, 1999).

It is important to note that what is shown as a lineament in Figure 10, for presentational simplicity, is a line representing only the approximate central locus of the lineament's swath. The nature of a lineament was succinctly defined by Richards (2000): "Large-scale crustal lineaments are recognized as corridors of aligned geological, structural, geomorphological, or geophysical features . . . they are commonly difficult to observe on the ground, the scale of the features and their interrelationships being too large to map except on a regional scale. They are therefore most easily identified from satellite and geophysical (gravity and magnetic) maps" (p. 1). The lineaments shown in Figure 10 conform with this definition, having swaths up to 50 km wide.

Continental margin

The margin of the unextended continental crust, occurring at the hinge zone in the region west of Walvis Bay down to Lüderitz, falls sharply into focus in the offshore aeromagnetic data set (Figs. 4 and 10). East of the hinge zone the offshore trend of the Damara orogeny is clearly seen for the first time. The data have allowed the offshore extensions of the Northern Central zone, Southern Central zone, and Southern zone terranes to be mapped (Fig. 10).

Phoenix volcanic province

A major linear northeast-trending series of gravity highs flanks the Walvis ridge in the north, immediately south of the Rio Grande Fracture Zone (Figs. 5 and 10), and is interpreted to be associated with transcurrent faulting and mafic intrusion. The Turonian-Cenomanian Phoenix volcano is where this trend intersects the 100 km offshore gravity high, forming one of a number of intrusions constituting a local volcanic province, here termed the Phoenix volcanic province (Holtar and Forsberg, 2000).

Offshore regional-scale faulting

A significant result of this study is that the offshore gravity and magnetic anomalies associated with the Mesozoic volcanics have been clearly offset in places along the offshore continuations of a number of Pan-African and Irumide lineaments; e.g., including inter alia the Autseib, Omaruru, Welwitschia, Transkalahari, and Pofadder lineaments. This suggests that the early Paleozoic to Late Proterozoic structures not only determined the architecture of the extended crust, but were probably also reactivated during the late Mesozoic. Recent to present-day movement has occurred along the seismically active Hebron

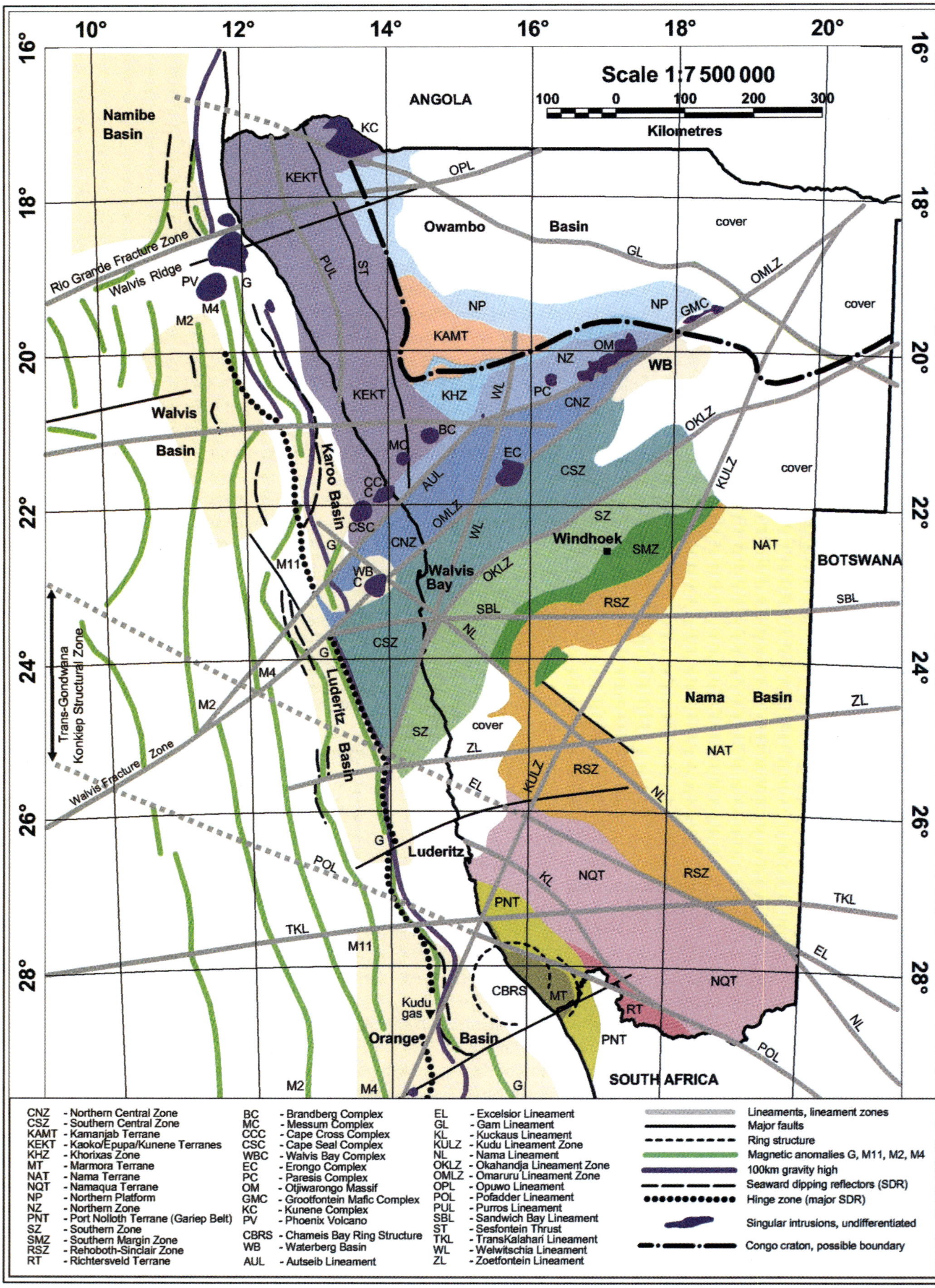

Figure 10. Interpreted Namibian offshore and onshore structure, basins, and terranes. (Tectono-stratigraphic terranes are simplified and modified after Petzel and Schreiber, 1999.)

fault, which forms part of the Nama lineament. An important implication arising from these conclusions is that these structures may have provided potential pathways for the larger drainage systems, and hence the focus of major offshore sedimentation, controlling the evolution of the offshore basins (see Clemson et al., 1999).

Omaruru and Autseib lineaments

The Pan-African Omaruru lineament (Corner, 1983) and the Autseib lineament (Miller, 1983) are interpreted (Corner, 2000) to be associated parallel features, constituting a broader Omaruru lineament zone. The zone, probably the largest regional structure seen to extend from the onshore into the offshore domain, is also clearly evident in the gravity data as the southern boundary of a major linear onshore regional gravity high (Figs. 5 and 10). Major structural change is seen across the Omaruru lineament zone offshore; e.g., the Damaran domains extend farther seaward to the south and hallmark the change in regional Damaran structural direction from the northeast branch to the north-northwest coast-parallel branch. The 100 km gravity high and G and M11 magnetic anomalies show right-lateral displacement across the Omaruru lineament zone. The left-lateral displacement of the M4 and M2 anomalies along the Omaruru lineament zone appears to be of smaller scale and may reflect a later event. The Omaruru lineament zone is clearly seen in the ship-track data (beyond the western border of Fig. 4) to continue southwestward as the Walvis Fracture Zone. This phenomenon is interpreted to arise from the likelihood that the Pan-African lineaments constituted a favorable direction at the time of breakup, initiating the Walvis Fracture Zone transform fault as the continents moved apart (Clemson, 1997). The Walvis Fracture Zone in turn reactivated the Autseib and Omaruru lineaments as recently as the Cretaceous, as evidenced in part by the Waterberg fault, which bounds the Waterberg Basin (Fig. 10). Clemson et al. (1999), in their study of basement structure of the Namibian passive margin, also noted the evidence from offshore seismic data that the Omaruru and Autseib lineaments were active in post-Karoo time. A similar structural relationship is considered to apply between the Rio Grande Fracture Zone (Walvis ridge) and the onshore Opuwo lineament.

The well-known line of intrusive complexes, Brandberg, Messum, Cape Cross, and the offshore complex to the southwest of Cape Cross, newly identified in this study and here named the Cape Seal Complex, all lie along the northern flank of the Omaruru lineament zone (Autseib lineament, Fig. 10). Farther northeastward, the Paresis Complex, the Otjiwarongo Massif, and the Grootfontein Mafic Complex are also situated along the Omaruru lineament zone.

The intrusive complex, evident offshore of Walvis Bay as a major gravity and magnetic anomaly (Corner and Swart, 1997) and named here the Walvis Bay Complex, as well as the Erongo Complex, both occur along the southern flank of the Omaruru lineament zone. These two complexes are interpreted to be similar in that both are capped by Etendeka basalts. Evidence for a major confined Karoo basin being located above the Walvis Bay Complex is seen in the offshore seismic data (Clemson, 1997; Clemson et al., 1999). In both cases the presence of intersecting lineaments may have played an important role in the emplacement of the intrusions.

TransGondwana Konkiep structural zone

Corner and Swart (1997) termed the zone flanked by the Pofadder and Excelsior lineaments the TransGondwana Konkiep structural zone. It encompasses the Namaqua metamorphic terrane and in part the Port Nolloth terrane. Within a Gondwana framework the Pofadder lineament was recognized by Tankard et al. (1995; their Tantalite-Valley-Pofadder lineament zone) as continuing into South America. They linked the Alto Paraguay and Rio de la Plata cratons tectonically to the Namaqua Province in the Middle Proterozoic. Corner and Swart (1997) were of the opinion that the Tantallite-Valley-Pofadder lineament zone was somewhat restrictive laterally. They thus invoked the broader TransGondwana Konkiep structural zone, encompassing the Namaqua terrane in Namibia (Gordonia Province of Tankard et al., 1995), as being the southern African continuation of the tectonized belt that separates the two South American cratons.

Kudu lineament zone

The Kudu lineament zone has been recognized for some time in the mineral exploration community (Corner, 2000). Offshore, in the vicinity of the Kudu gas field, major apparent right-lateral displacement is seen to occur of the 100 km gravity high, the hinge zone, and of the G magnetic anomaly along the Kudu lineament zone. A series of north-northeast–trending faults, evident in the magnetic data inland, are associated with its swath. In view of its parallelism to the Welwitschia lineament (Corner, 1983), the Kudu lineament zone is interpreted to be of similar age, at least at its earliest limit, i.e., late Damaran, as it cuts across the northeast Damaran trend. This study suggests that it was active at least until the late Mesozoic.

Chameis Bay ring structure

The Chameis Bay ring structure (Figs. 4 and 10) was recognized for the first time by Corner and Swart (1999). Arcuate structures are seen flanking the Kudu gas field in the offshore gravity, magnetic, and seismic data. These are seen in the magnetic data (Figs. 4 and 10) to form part of a ring structure continuing onshore and encompassing the Marmora and Port Nolloth terranes. The origin is uncertain, but it correlates with the locality of a long-lived mantle plume proposed by Frimmel et al. (1996). It is suggested that the Chameis Bay ring structure is a manifestation of deep-seated plutonism partly affecting local structural style, and may be associated with hydrothermal activity and intrusives. The phenomenon of ring structures and associated mineralizing fluids, although new to Namibia, is not new globally. O'Driscoll published extensively on the subject with respect particularly to Australian mineral deposits (e.g., O'Driscoll and Campbell, 1997). The question may be asked whether it is coincidence that the Kudu gas field is situated on, or close to, the Chameis Bay ring structure and Kudu lineament zone or whether these structures aided the thermal regime required for hydrocarbon maturation.

Offshore semilinear gravity highs

At least four major coast-parallel gravity highs are seen in the offshore data (Fig. 5). The most prominent of these occurs ~100 km offshore (Figs. 5 and 10) and continues along the entire length of the Namibian coastline. It is referred to here as the 100 km gravity high as defined in the preceding modeling section (also referred to elsewhere as the shelf edge anomaly; e.g., Bauer et al., 2000). The gravity highs are structurally disrupted along seaward extensions of major fault lineaments identified in the onshore data, signifying either later reactivation of these lineaments or control by these early lineaments of the architecture of the extended crust. It is likely that a combination of both of these scenarios prevailed (Clemson, 1997; Clemson et al., 1999). Light et al. (1992) first reported on the 100 km gravity high, detected at that time by relatively sparse marine gravity profiles. They favored a model in which the anomaly arises from a continuous mantle wedge displacing less dense continental material. They did not, however, discount the possibility of an intrusive or extrusive source. Gladczenko (1994), however, proposed that this gravity high represents the shelf edge, particularly in view of a close correlation with the 500 m bathymetric contour defining the shelf edge. This does not explain the other linear gravity highs and cannot be maintained by forward modeling (Corner and Swart, 1999).

As discussed in the forward modeling section, the 100 km gravity high is proposed to arise in the first order by the updip convergence and pinchout of the major package of basaltic SDRs defining the hinge zone. The 100 km gravity high is also clearly associated with the G magnetic anomaly along the entire length of the hinge zone (cf. Hinz et al., 1999, on the conjugate margin). The suboutcrop position of the SDRs at the hinge zone is delineated in this study from the seismic profiles and is thus interpreted to be that of a major sequence of basalts, in line with the earlier findings of Corner and Swart (1999). The modeling further suggests that the basalts are interlayered with sediments (see forward modeling section), a conclusion supported by the parallel modeling studies of Bauer et al. (2000) and Carruthers et al. (2001). The 100 km gravity high occurs either sympathetically over the hinge zone or immediately downdip, as the basaltic sequence starts to thicken. Our modeling has shown, however, that it loses amplitude completely farther seaward as the overlying low-density sedimentary pile neutralizes the gravity high due to the basalts, which are nevertheless still interpreted in the seismic sections as being present at depth. A similar origin could be invoked in part for the lower order gravity highs farther out to sea, i.e., correlating with further pulses of basalt extrusion expressed as the SDRs. Stewart et al. (2000), however, showed that these gravity highs can be fully explained by an underplating model.

The preceding interpretation is thus in contrast to that of Watts and Fairhead (1999), who applied a process-orientated approach to modeling edge effect anomalies such as the 100 km gravity high. The processes they considered include rifting, sedimentation, and magmatic underplating. By quantifying these they are able to model the edge effect anomalies. Bauer et al. (2000) used a similar approach to that of Watts and Fairhead (1999) to model two gravity profiles offshore Namibia. They converted P-velocities, derived from their deep seismic data, to densities and demonstrated a good fit between observed and theoretical gravity profiles. Nevertheless, they imply that dense material at higher levels in the crust is necessary to achieve the good agreement. Stewart et al. (2000) similarly concluded that the residual gravity highs of the 100 km gravity high may be related to the Etendeka volcanics, in view of their close spatial relationship. These three studies do not directly consider the density effect of the basaltic seaward-dipping reflectors, nor do they take full cognizance of the associated magnetic anomalies in this respect. In our view, the most exact and meaningful solution should incorporate a quantitative combination of the basalt-origin model (Corner and Swart, 1999; this study) and the deeper crustal and magmatic underplating models of Watts and Fairhead (1999), Bauer et al. (2000), and Stewart et al. (2000).

Offshore magnetic anomalies

The G and M series of magnetic anomalies identified by a number of previous workers as seafloor-spreading–related features (e.g., Rabinowitz and LaBrecque, 1979; Gladczenko, 1994) are readily evident in the recent offshore data and are thus now more definitively positioned in Figure 10. Of major significance are the newly mapped trends of the M2 and M4 anomalies, which are seen to converge toward the northern Namibian and Angolan coastlines, in contrast to their earlier placement much farther out to sea in the northern area (Fig. 1). This has potential major implications: for the relative ages of the offshore sedimentary sequences within the inner half-graben from south to north along the margin (northward propagation of rifting envisaged, e.g., by Clemson, 1997), for relative rates of extension from the southern to northern offshore areas (Corner and Swart, 1997), and for rotation about the Tristan hotspot at this time (C. Reeves, 2001, personal commun.).

The preceding results imply that, at least within the confines of the offshore area bounded by the M2 anomaly in the west, clear oceanic crust cannot be seen within the present data because responses are masked by the SDRs in all data sets. More likely, the region constitutes extended and massively intruded continental crust, passing laterally westward into true oceanic crust.

A result of possibly major significance derives from the modeling interpretations presented in Figures 8 and 9. For volcanic margins with thick sequences of SDRs, the classical concept of the M series of magnetic anomalies arising from a progressively seaward-younging series of mid-oceanic styled sets of sheeted dikes (Vine and Mathews, 1963) may be inapplicable at least in the nearshore areas, either because of the masking effect of the extrusive SDRs or because they generate anomalies of similar character. This study clearly suggests that laterally extensive and strikingly linear anomalies with all the characteristics of classical seafloor-spreading anomalies can arise from the updip featheredges of the wedge-shaped SDR units (cf. anomaly G in Hinz et al., 1999). The modeling of Bauer et al. (2000), conducted in parallel to this study as part of the MAMBA experiment, also supports this conclusion. In terms of relative ages the implications are similar, i.e., younging occurs progressively seaward. However, whether each M anomaly in the region of extended crust can be uniquely matched in age with its accepted counterpart on the opposite side of the Atlantic is open to question. We suspect that a similar misinterpretation of seafloor-spreading anomalies and their age assignments might apply to many other volcanic margins.

CONCLUSIONS

Interpretation of the recently acquired offshore aeromagnetic data, together with offshore seismic data and other marine, land, and satellite gravity and magnetic data sets, has been extremely beneficial in understanding the evolution and structure of the continental margin. Structural, extrusive, and intrusive features are now more clearly understood or mapped, if not identified, for the first time. The G magnetic anomaly and what have been mapped as the M2 and M4 magnetic anomalies are better defined, swinging coastward toward northern Namibia and southern Angola, in contrast to earlier studies in which these anomalies were interpreted to swing away from the coast. This has important implications for the breakup mechanism, for the relative ages of the offshore sedimentary sequences within the inner half-graben from south to north along the margin, for relative rates of extension from the southern to northern offshore areas (e.g., northward propagation of rifting), and for rotation about the Tristan hotspot.

Forward magnetic modeling of the G and M magnetic anomalies has shown that an excellent fit can be obtained using a simple geological model, based on depth-converted seismic sections, of an inner half-graben fill comprising intercalated basalts and siliciclastic rocks with a shallow westward-dipping wedge geometry, thinning eastward by convergence to a pinchout at the hinge zone. Both the positive and negative anomaly components of the G magnetic anomaly are simply explained by this structural style assuming induced magnetization. Although basaltic sequences of reversed polarity cannot be discounted to enhance the negative components of the magnetic anomalies, there was no need to invoke this in the modeling. The corresponding gravity response of the magnetically modeled basalt sequence correlates in the first order in position and wavelength with the 100 km edge effect gravity high associated with the hinge zone. It is thus concluded that the G magnetic anomaly and the 100 km gravity high comprise essentially two components; primarily a relatively near surface component of sympathetic responses associated with the suboutcrop of basalt-related SDRs at the hinge zone, and a secondary deeper crustal component seen largely in the gravity data related to crustal extension and magmatic underplating.

A further result of possible major significance also derives from the modeling interpretations. For the Namibian margin, with its thick sequences of SDRs, the classical concept of the M series of magnetic anomalies arising from a progressively seaward-younging series of mid-oceanic styled sets of sheeted dikes appears to be inapplicable at least in the nearshore areas, perhaps including the M4 and M2 anomalies, because the extrusive SDRs generate anomalies of similar character. This study clearly suggests that laterally extensive and strikingly linear anomalies with all the characteristics of classical seafloor-spreading anomalies can arise from the updip featheredges of the wedge-shaped SDR units. This observation may apply to other volcanic margins where SDRs are well developed, and raises the question of whether each M anomaly in the region of nearshore extended crust can be uniquely matched in age with its accepted counterpart on the opposite side of the Atlantic.

It is also concluded that a number of interpreted major onshore structures, mostly of Pan-African age, which are seen to displace the 100 km offshore gravity and G, M2, and M4 magnetic anomalies, were reactivated along their offshore extensions at least until the late Mesozoic. Two of these structures, the Opuwa lineament and Omaruru lineament zone, are seen to continue westward as the Rio Grande Rift and Walvis Fracture Zone, respectively, having probably constituted zones of weakness with favorable orientations for initiation of these fracture zones at the time of breakup.

ACKNOWLEDGMENTS

We express our gratitude to the National Petroleum Corporation of Namibia (NAMCOR) for funding this project and to the Geological Survey of Namibia for allowing the use of their onshore data sets.

REFERENCES CITED

Aizawa, M., Bluck, B., Cartwright, J.A., Milner, S.C., Swart, R., and Ward, J.D., 2000, Constraints on the geomorphological evolution of Namibia from the offshore stratigraphic record: Communications of the Geological Survey of Namibia, v. 12, p. 337–346.

Bauer, K., Neben, S., Schrekenberger, B., Emmerman, R., Hinz, K., Jokat, W., Schulze, A., Trumbull, R.B., and Weber, K., 2000, Deep structure of the

Namibia continental margin as derived from integrated geophysical studies: Journal of Geophysical Research, v. 105, no. B11, p. 25829–25853.

Carruthers, R., Ritchie, D., and Richards, P., 2001, Integrated seismic and potential field modelling in the Walvis Basin region, offshore Namibia: British Geological Survey Report, confidential, CR/01/50.

Clark, D.A., 1997, Magnetic petrophysics and magnetic petrology: Aids to geological interpretation of magnetic surveys: Australian Geological Survey Organisation, AGSO Journal of Australian Geology and Geophysics, v. 17, 2, p. 83–104.

Clemson, J., 1997, Segmentation of the Namibian continental margin [Ph.D. thesis]: London, University of London, 347 p.

Clemson, J., Cartwright, J., and Booth, J., 1997, Structural segmentation and the influence of basement structure on the Namibian passive margin: Journal of the Geological Society of London, v. 154, p. 477–482.

Clemson, J., Cartwright, J., and Swart, R., 1999, The Namib Rift: A rift system of possible Karoo age, offshore Namibia, *in* Cameron, N., ed., Oil and gas habitats of the South Atlantic: Geological Society of London Special Publication 153, p. 381–402.

Corner, B., 1983, An interpretation of the aeromagnetic data covering the western portion of the Damara Orogen in South West Africa/Namibia: Geological Society of South Africa Special Publication 11, p. 339–354.

Corner, B., 2000, Crustal framework of Namibia derived from magnetic and gravity data: Communications of the Geological Survey of Namibia, v. 12, p. 13–19.

Corner, B., and Swart, R., 1997, Structural insights gained from a comparison of offshore Namibia satellite data with onshore magnetic and gravity data, poster presentation, *in* Proceedings of the 5th Technical Meeting of the South Africa Geophysical Association: Swakopmund, Namibia, p. 173–174.

Corner, B., and Swart, R., 1999, Interpretation of combined onshore and offshore magnetic and gravity data sets of Namibia, poster presentation, *in* Proceedings of the 6th Technical Meeting of the South Africa Geophysical Association: Cape Town, South Africa, p. P11–P12.

Frimmel, H.E., Hartnady, C.J.H., and Koller, F., 1996, Geochemistry and tectonic setting of magmatic units in the Pan-African Gariep Belt, Namibia: Chemical Geology, v. 130, p. 101–121.

Gladczenko, T.P., 1994, Crustal structure and composition of selected transient large igneous provinces, [Ph.D. thesis]: Oslo, Norway, University of Oslo, Applied Geophysics, 220 p.

Gladczenko, T.P., Hinz, K., Eldholm, O., Meyer, H., and Skogseid, J., 1997, South Atlantic volcanic margins: Journal of the Geological Society of London, v. 154, p. 465–470.

Hinz, K., Neben, S., Schrekenberger, H.A., Roeser, M., Block, K., and Meyer, H., 1999, The Argentine continental margin north of 48S: Sedimentary successions, volcanic activity during breakup: Marine and Petroleum Geology, v. 16, p. 1–25.

Holtar, E., and Forsberg, A.W., 2000, The post-rift development of the Walvis Basin, Namibia: Results from the exploration campaign in Quadrant 1911, *in* Mello, M.R., and Katz, B.J., eds, Petroleum systems of South Atlantic margins: American Association of Petroleum Geologists Memoir 73, p. 429–446.

King, L., 1951, South African scenery: London, Oliver and Boyd, 308 p.

Light M.P.R., Maslanyj, M.P., and Banks, N.L., 1992, New geophysical evidence for extensional tectonics on the divergent margin offshore Namibia, *in* Storey, B.C., Alabaster, T., and Pankhurst, R.J., eds., Magmatism and the causes of continental break-up: Geological Society [London] Special Publication 68, p. 257–270.

Martin, H., 1973, The Atlantic margin of southern Africa between latitude 17° S and the Cape of Good Hope, *in* Nairn, A.E.M., and Stehli, F.G., eds., The ocean basins and margins, Volume 1, The South Atlantic: New York, Plenum Publishing, p. 277–300.

Miller, R.McG., 1983, The Pan-African Damara Orogen of South West Africa/Namibia, *in* Miller, R. McG., ed., Evolution of the Damara Orogen: Geological Society of South Africa Special Publication 11, p. 431–515.

Milner, S.C., Duncan, A.R., Ewart, A., and Marsh, J.S., 1994, Promotion of the Etendeka Formation to group status: A new integrated stratigraphy: Communications of the Geological Survey of Namibia, v. 9, p. 5–12.

Müller, R.D., Roest, W.R., Royer, J.-Y., Gahagan, L.M., and Sclater, J.G., 1997, Digital isochrons of the world's ocean floor: Journal of Geophysical Research, v. 102, no. B2, p. 3211–3214.

O'Driscoll, E.S.D., and Campbell, I.B., 1997, Mineral deposits related to Australian continental ring and rift structures with some terrestrial and planetary analogies: Global Tectonics and Metallogenesis, v. 6, p. 83–93.

Petzel, V., and Schreiber, U., 1999, Simplified tectonostratigraphic map of Namibia [unpublished preliminary map], scale 1:4 000 000: Geological Survey of Namibia, 1 sheet.

Planke, S., Symonds P.A., Alvestad, E., and Skogseid, J., 2000, Seismic volcanostratigraphy of large-volume basaltic extrusive complexes on rifted margins: Journal of Geophysical Research, v. 105, no. B8, p. 19335–19351.

Rabinowitz, P.D., and LaBrecque, J., 1979, The Mesozoic South Atlantic ocean and evolution of its continental margins: Journal of Geophysical Research, v. 84, no. B11, p. 5973–6002.

Renne, P.R., Glen, J.M., Milner, S.C., and Duncan, A.R., 1996, Age of Etendeka flood volcanism and associated intrusions in southwestern Africa: Geology, v. 7, p. 659–662.

Richards, J.P., 2000, Lineaments revisited, Society of Economic Geologists Newsletter, v. 42, p. 13–20.

Smethurst, M.A., 2000, Land-offshore tectonic links in western Norway and the northern North Sea: Journal of the Geological Society of London, v. 157, p. 769–781.

Stewart, J., Watts, A.B., and Bagguley, J.G., 2000, Three-dimensional subsidence analysis and gravity modelling of the continental margin offshore Namibia: Geophysical Journal International, v. 141, p. 724–746.

Tankard, A.J., Uliana, M.A., Welsink, H.J., Ramos, V.A., Turic, M., Franca, A.B., Milani, E.J., de Brito Neves, B.B., Eyles, N., Skarmeta, J., Santa Ana, H., Wiens, F., Cirbian, M., Lopez Paulsen, O., Germs, G.J.B., De Wit, M.J., Machacha, T., and Miller, R. McG., 1995, Structural and tectonic controls of basin evolution in southwestern Gondwana during the Phanerozoic, *in* Tankard, A.J., Suarez, R., and Welsink, H.J., eds., Petroleum basins of South America: American Association of Petroleum Geologists Memoir, v. 62, p. 5–52.

Vine, F.J., and Matthews, D.H., 1963, Magnetic anomalies over ocean ridges: Nature, v. 199, p. 947–949.

Watts, A.B., and Fairhead, J.D., 1999, A process-oriented approach to modelling the gravity signature of continental margins: The Leading Edge, v. 18, p. 258–263.

Manuscript Accepted by the Society October 15, 2001

Printed in the U.S.A.

Geological Society of America
Special Paper 362
2002

Petrophysical modeling of high seismic velocity crust at the Namibian volcanic margin

R.B. Trumbull, S.V. Sobolev, and K. Bauer
GeoForschungsZentrum Potsdam, Telegrafenberg, 14473 Potsdam, Germany

ABSTRACT

Two recent onshore-offshore seismic transects across the Namibian passive margin reveal a thick (to 20 km) prism of material at the base of the crust with high seismic velocity (*Vp* = 7.1–7.6 km/s). To better understand the nature of this material and the processes that formed it, we estimate the bulk chemical composition of the high-velocity crust by relating its seismic velocity to a petrophysical model that links basalt composition and conditions of partial melting of peridotite.

Observed average seismic velocities in the igneous crust are consistent with basaltic material with ~14–18 wt% MgO. This conclusion is not affected by the presence of cumulate minerals because it integrates over the full thickness of the body; however, the highest *Vp* values of 7.6 km/s are consistent with velocities expected for cumulate minerals produced by fractional crystallization of a 14%–18% MgO parental melt. The subsolidus growth of garnet is unlikely to be a significant factor for the crustal velocity above the Moho depth of 30 km. Garnet growth in a magnesium-rich basaltic crust can be expected to limit the crustal thickness to ~30 km because bulk densities at deeper levels may exceed those of the peridotite mantle.

The relationship between MgO content of partial melts and the potential temperature of a fertile peridotite source suggests that the estimated 14–18 wt% MgO basalts were generated from mantle at ~1440–1560 °C potential temperature, which may be a good estimate for the potential temperature of the ancestral Tristan mantle plume at the Namibian margin. The igneous crust has the greatest volume, highest MgO contents, and highest inferred mantle potential temperatures at the location of the northern transect, which is closest to the Walvis Ridge hotspot trace. The mantle potential temperature estimated for the southern transect is 50–100 °C lower, suggesting cooling of the plume material during its flow southward.

INTRODUCTION

Volcanic rifted margins constitute one of the main categories of Earth's large igneous provinces (Mahoney and Coffin, 1997). These are the sites of extremely intense magmatism rooted in the mantle, which forms in the course of continental breakup and the birth of ocean basins. Understanding the scope and origin of this volcanic rifted margin magmatism is an important goal in geoscience, in terms of the geodynamic processes involved, the role of volcanic rifted margins in crustal evolution, and their potential environmental impact. Continental flood basalts are the most obvious expression of volcanic rifted margin magmatism, and they have been thoroughly studied in many places. However, mature rifted margins are, by their very nature, partially submerged, and a large proportion of the igneous rocks produced are located offshore. Seismic surveys have proven effective in detecting some of the unexposed components of volcanic rifted margin magmatism. In particular, two

Trumbull, R.B., Sobolev, S.V., and Bauer, K., 2002, Petrophysical modeling of high seismic velocity crust at the Namibian volcanic margin, *in* Menzies, M.A., Klemperer, S.L., Ebinger, C.J., and Baker, J., eds., Volcanic Rifted Margins: Boulder, Colorado, Geological Society of America Special Paper 362, p. 221–230.

associations are common. First, extensive basalt flow sequences on the continental shelf show up as seaward-dipping reflector sequences (Hinz, 1981; White and McKenzie, 1989; Oh et al., 1995). Second, the lower crust at the continent-ocean boundary on many margins has high seismic velocities ($Vp > 7$ km/s), which suggests that a large proportion of it consists of basaltic igneous intrusions underplating the crust (White and McKenzie, 1989; Holbrook et al., 1994; Kelemen and Holbrook, 1995; Leitch et al., 1998; Talwani and Abreu, 2000).

In contrast to the seaward-dipping reflector sequences, the underplated igneous crust has never been directly sampled. Based on an adiabatic decompression melting model for dry mantle peridotite and experimental data from melting experiments, McKenzie and Bickle (1988) and White and McKenzie (1989, 1995) showed that systematic relationships exist between the potential temperature of the mantle and pressure interval of melting on the one hand, and the thickness and bulk composition of igneous crust produced on the other. These relationships allow one to estimate the composition of igneous crust from observed seismic velocities and thickness, and to make inferences on the potential temperature of the mantle involved. Applications have been made in studies of volcanic margins in the North Atlantic (e.g., Kelemen and Holbrook, 1995; Barton and White, 1997).

In this chapter we extend the approaches of McKenzie and Bickle (1988) and White and McKenzie (1989) and derive new relationships linking the chemical composition of the magmatic underplate, its seismic velocities, and mantle potential temperature based on petrophysical and thermodynamic modeling of recent experimental melting data. We apply these relationships to the first offshore-onshore wide-angle dataset available from the South Atlantic (Bauer et al., 2000) and present estimates of the nature of high-velocity igneous crust under the margin of Namibia and of the potential temperature of the ancestral Tristan mantle plume.

GEOLOGIC SETTING

The South Atlantic volcanic rifted margins were produced in the Early Cretaceous (ca. 130 Ma) by continental breakup of western Gondwana. The onshore igneous record of breakup and rifting comprises flood basalts and associated large-volume felsic volcanics in the Paraná-Etendeka province, and a large number of subvolcanic intrusive complexes. There is a voluminous literature on these rocks; compilations of recent references can be found in the reviews of the Paraná-Etendeka province by Peate (1997) and Marsh et al. (2001), and in summaries of the Namibian intrusive complexes by Harris (1995) and by Trumbull et al. (2000).

The structure of the Namibian continental margin is known from seismic reflection surveys (summarized in Gladczenko et al., 1997) and from two onshore-offshore seismic refraction transects (Bauer et al., 2000). The main features of the margin, according to the latter study, are as follows (Fig. 1).

1. Oceanic crust of ~8 km thickness overlain by 3–4 km of sediments is found at the seaward ends of the transects (400 km offshore). The thickness of oceanic crust here is slightly higher than the global average value of 7.1 ± 0.8 km (White et al., 1992), suggesting some influence of the mantle plume.

2. The continent-ocean boundary is located ~150–100 km offshore and is marked by the landward end of a wedge of seaward-dipping reflector sequences in the upper crust and by prominent gravity and magnetic anomalies.

3. Beginning at the continent-ocean boundary, and extending ~200 km seaward, is a zone of thick (to 20 km) igneous crust with high seismic P-wave velocities (between 7.1 km/s and 7.6 km/s). The boundary to continental crust on the landward side is abrupt and steep. There appears to be no transition zone from oceanic to continental crust and no evidence for major crustal extension prior to breakup. Talwani and Abreu (2000) reported essentially the same findings from seismic sections of the eastern United States margin.

4. The Moho depth at the coast is 35 km. The velocity structure of the continental crust is uniform in transect 2, whereas transect 1 reveals high-velocity anomalies beneath two prominent Cretaceous intrusive complexes, Cape Cross and Messum.

The total volume of basaltic magmatism in the South Atlantic igneous province is enormous. Peate (1997) estimated the volume of terrigenous Paraná-Etendeka flood basalts to be 2×10^6 km^3. To this must be added the volume of basalt represented by seaward-dipping reflector sequences, which are known to occur over a length of 2000 km on both sides of the Atlantic (Gladczenko et al., 1997), and the volume of thick igneous crust at the continent-ocean boundary. It has often been stated that a mantle plume now centered under the island of Tristan da Cunha is responsible for the Walvis Ridge and Rio Grande Rise hotspot tracks on the ocean floor (e.g., White and McKenzie, 1989; O'Connor and LeRoex, 1992) and for much of the basaltic magmatism in the province as a whole. However, the precise role of the Tristan plume is not clear because on the one hand, the wedges of seaward-dipping reflectors extend along the margin farther south than the expected reach of even a large mantle plume (White and McKenzie, 1989). On the other hand, most flood basalts in the Paraná-Etendeka province do not have isotopic and geochemical signatures consistent with a plume source (Hawkesworth et al., 1999; Peate, 1997; Turner et al., 1996), whereas the mafic alkaline intrusive complexes and carbonatites in Namibia do (Milner and LeRoex, 1996; Harris et al., 1999; Trumbull et al., 2000).

METHODS AND RESULTS

Our objective is to estimate the bulk composition of the thick igneous crust at the continent-ocean boundary based on the observed seismic velocities. The approach taken is to extract a set of model chemical compositions (major element oxides) for basaltic magma from experimental data on partial melting of peridotite under different conditions of pressure and temperature.

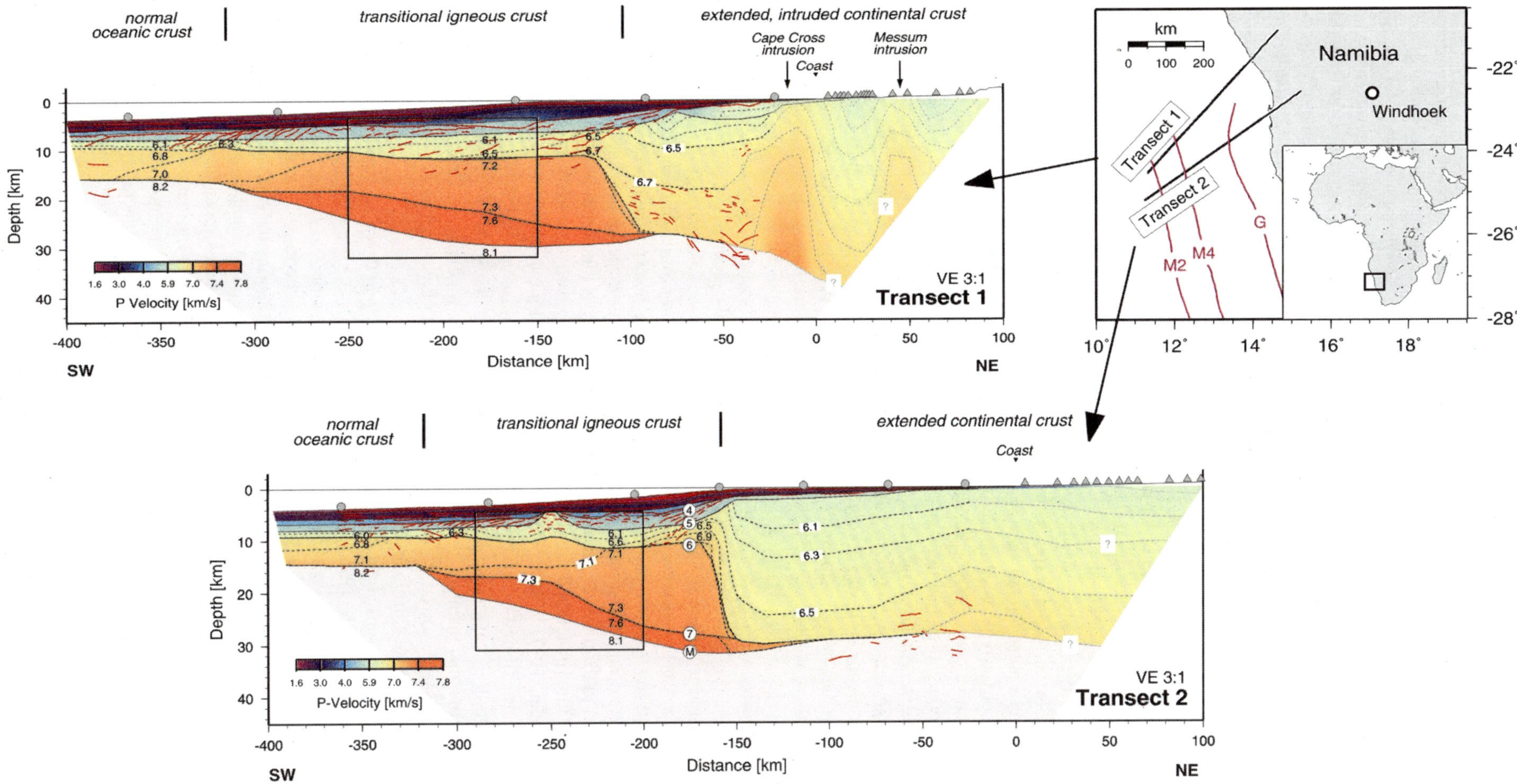

Figure 1. Interpreted two-dimensional crustal depth sections across Namibian margin based on wide-angle and reflection seismic data from MAMBA experiment (Bauer et al., 2000). Ocean-continent boundary is characterized by landward end of seaward-dipping reflector wedges in upper crust and by abrupt velocity gradient in middle and lower crust. Seafloor magnetic lineaments shown in inset map are from Rabinowitz and LaBrecque (1979). Boxes on both sections outline limits of petrophysical models of high-velocity crust. *V*p-boundary labels on section for transect 2 (4, 5, 6, 7, M) were omitted from transect 1 for clarity.

We then use an equilibrium thermodynamic model (Sobolev and Babeyko, 1994) to calculate the solid mineral assemblage corresponding to each bulk composition after crystallization and cooling to ambient conditions appropriate for the Namibian margin. Petrophysical modeling of this model crust then generates seismic P-wave velocity values as a function of bulk composition, the basis for predicting composition based on the observed velocity values.

The observed velocities in the lower crustal body are at the heart of this study and their uncertainties are therefore critical. We performed an analysis of uncertainty for the specific regions of high-velocity crust shown in Figure 1 (i.e., between boundary 6 and Moho in the boxed areas of each transect). This was done by a method similar to that of Holbrook et al. (1994), where we vary the best-fit average velocity in the regions of interest by fixed increments and then assess the model response to these variations by calculating the root-mean-square (RMS) traveltime misfit for all raypaths through the regions. A measure of uncertainty is the amount by which velocities can be varied without critically worsening the models. If one assumes a value of 0.1 s for an acceptable RMS misfit, which we consider to be an upper limit, the range of average velocity uncertainty is ±0.06 km/s for transect 2 and ±0.08 km/s for transect 1. From this we conclude that the uncertainty in the average velocity of the underplated crust is smaller than these values.

Estimating magma bulk composition

The partial melting of mantle peridotite is a subject of intense current research and there exists a large body of high-quality experimental data as well as several parameterization schemes based thereon, which are designed to predict melt composition as functions of temperature, pressure, and degree of melting (e.g., McKenzie and Bickle, 1988; Langmuir et al., 1992; Kinzler and Grove, 1993; Kinzler, 1997; Walker, 1999). However, even for the well-studied case of mid-ocean ridge basalts, the problem of reliably estimating the aggregate bulk magma composition to be expected from incremental decompression melting of upwelling mantle is not yet solved. The thermodynamic MELTS program (Ghiorso and Sack, 1995) offers the only potential way to realistically model polybaric fractional melting and melt aggregation, but Hirschmann et al. (1998) showed that the method is not sufficiently accurate for forward modeling of the melting process, particularly for the near-solidus region.

We envision that the thick igneous crust at the Namibian margin formed from accumulation of basaltic melts generated by decompression partial melting of ascending mantle. In detail, the melting process is likely to involve polybaric, near-fractional melting as suggested for the mid-ocean ridges, and is therefore difficult to model realistically. However, our intention in this study is not to investigate the melting process, but to define the average composition of a large body of igneous crust (to ~20 km thick, see Fig. 1) from seismic velocity data. For this purpose, it seems preferable to use a simple and robust estimate of melt compositions based directly on self-consistent, high-quality experimental data rather than to employ complex and model-dependent parameterization of the polybaric melting process.

Our estimates of magma composition are based on the experimental results of melting dry, fertile peridotite at 1.0 and 1.5 GPa pressures by Kushiro (1996). We performed a linear regression analysis on 13 reported experimental compositions (6 at 1.0 GPa and 7 at 1.5 GPa) generated at temperatures between 1200 and 1450 °C, and with a degree of melting between 10% and 38%. The results showed that the major oxide compositions and the temperature of melting can be expressed with good accuracy as a linear function of the melt MgO content (see Table 1). The predicted melt compositions in Table 1 are strictly valid only for the particular source composition shown in the table. However, Hirose and Kushiro (1993) showed experimentally that the overall major oxide composition of lherzolite partial melts is insensitive to changes in starting composition (within the limits of dry lherzolite examined). In particular, the melting experiments showed that MgO and SiO_2 contents are entirely independent of source composition. Other individual oxides showed compositional dependency, but the overall melt composition, in terms of normative minerals, is insensitive to source variation. Hirose and Kushiro (1998) showed experimentally that there is little difference between the bulk composition of an aggregate, accumulated melt produced by incremental melting and the composition of a single-batch melt of the same source at the same overall degree of melting (although there are differences in the source productivity between the two melting processes; see following). Therefore, we feel confident that the range of compositions listed in Table 1 is a fair representation of the range to be expected from varying degrees of partial melting of typical fertile peridotite. The choice of dry peridotite to model this process is thought to be reasonable because of the volumes of melt involved. It is likely that the first melting of upwelling mantle will involve volatile and alkali-enriched metasomatized material, for which the melting experiments cited do not apply. However, at the scale of melting required to produce a 20-km-thick layer, we expect that metasomatized portions of the mantle will not play a significant role.

Relationship between seismic velocity and bulk composition

The calculation of compressional wave velocities (Vp) to be expected for the range of bulk compositions shown in Table 1 requires estimating the corresponding bulk mineral assemblages for each chemical composition as a function of pressure and temperature ranges dictated by the geologic history and geometric structure of the margin. White and McKenzie (1989) simply took the CIPW norm for a given bulk composition as an estimate of the solid mineral assemblage and calculated seismic velocities from the aggregate. We take a more realistic approach and calculate the thermodynamically stable assemblage for each bulk composition using the program of Sobolev and Babeyko (1994), which accounts for nonideality of solid solutions and has

TABLE 1. MODEL BASALT COMPOSITION PREDICTED FROM EXPERIMENTAL PARTIAL MELTING OF PERIDOTITE

Source composition (wt%)		Regression parameters* a	b	r^2	Predicted melt composition at a given MgO (in wt %) 8	10	12	14	16	18	20	Error† (%)
Peridotite melting at 1.5 Gpa (7 experiments, degree of melting 10%–38%)												
SiO_2	43.70	0.207	47.149	0.76	48.80	49.22	49.63	50.05	50.46	50.87	51.29	0.84
TiO_2	0.25	−0.110	2.533	0.99	1.66	1.44	1.22	1.00	0.78	0.56	0.34	0.06
Al_2O_3	2.75	−0.986	26.259	0.96	18.37	16.40	14.42	12.45	10.48	8.51	6.53	1.02
FeO_T	10.05	0.273	5.704	0.93	7.89	8.43	8.98	9.52	10.07	10.61	11.16	0.31
MgO	37.22	n.a.	n.a.	n.a.	8.00	10.00	12.00	14.00	16.00	18.00	20.00	n.a.
CaO	3.26	−0.301	14.693	0.85	12.29	11.68	11.08	10.48	9.88	9.28	8.67	0.63
Na_2O	0.33	−0.189	4.159	0.94	2.65	2.27	1.90	1.52	1.14	0.76	0.39	0.16
K_2O	0.14	−0.104	2.086	0.98	1.25	1.04	0.83	0.62	0.42	0.21	0.00	0.05
Peridotite melting at 1.0 Gpa (6 experiments, degree of melting 12%–30%)												
SiO_2	43.70	0.151	49.122	0.70	50.33	50.63	50.93	51.23	51.54	51.84	52.14	0.39
TiO_2	0.25	−0.062	1.751	0.78	1.26	1.13	1.01	0.89	0.77	0.64	0.52	0.08
Al_2O_3	2.75	−0.982	25.692	0.98	17.83	15.87	13.91	11.94	9.98	8.01	6.05	0.43
FeO_T	10.05	0.312	4.938	0.94	7.43	8.05	8.68	9.30	9.92	10.54	11.17	0.32
MgO	37.22	n.a.	n.a.	n.a.	8.00	10.00	12.00	14.00	16.00	18.00	20.00	n.a.
CaO	3.26	−0.512	17.123	0.99	13.03	12.01	10.98	9.96	8.94	7.91	6.89	0.21
Na_2O	0.33	−0.127	2.957	0.83	1.94	1.68	1.43	1.17	0.92	0.66	0.41	0.23
K_2O	0.14	−0.097	1.812	0.86	1.03	0.84	0.64	0.45	0.25	0.06	0.00	0.21

Note: n.a. denotes "not applicable"; FeO_T denotes total Fe as FeO. Source of experimental data: Kushiro (1996).
* Linear regression parameters fit to Y = aX + b, where X is MgO and Y other oxide components; r^2 is the sum of squared residuals.
† Uncertainty of the prediction envelope for the linear regression expressed at the 95% confidence level.

been shown to closely match subsolidus equlibria in mafic rocks reported by high-pressure, high-temperature (*P, T*) experimental studies. This program also calculates average density, *V*p, and *V*s for each mineral assemblage using the classical Hashin-Shtrikman procedure (Hashin and Shtrikman 1963), in which end members of solid solutions are treated as mechanical components and an average value is calculated as an arithmetic mean of the corresponding Hashin-Shtrikman bounds. Elastic moduli and densities of the end members are taken from the laboratory measurements on single crystals compiled in Duffy and Anderson (1983) and Sobolev and Babeyko (1994).

Because ~130 m.y. have elapsed since breakup-related magmatism on land (e.g., Milner et al., 1995) and the onset of seafloor spreading offshore (Gladzcenko et al., 1997), we can assume that the thermal anomaly has decayed completely and the Namibian margin at the continent-ocean boundary has reached a steady-state geothermal gradient typical for oceanic lithosphere of this age. However, it is unrealistic to expect that equilibrium in dry basaltic material can be maintained during cooling down to the present-day temperatures of the Moho (~500 °C). Instead, mineral reactions will cease at some blocking temperature and a metastable mineral assemblage will remain. For a dry basaltic system lacking active deformation, the blocking temperature will likely be controlled by the volume diffusion of Al as the rate-determining step of metamorphic reactions, and is probably above 800 °C, depending on cooling rate (Sobolev and Babeyko, 1989). Therefore, we use the range 800–900 °C and not the ambient temperatures for calculating the equilibrium phase assemblage. The more sophisticated approach of Sobolev and Babeyko (1989), which estimates reaction kinetics as a function of a cooling model, leads to essentially the same result. Note, however, that the ambient temperature is important for its effect on the seismic velocity (i.e., d*V*p/d*T*).

Figure 2 shows the dependence of calculated *V*p values on the bulk composition of basaltic rocks listed in Table 1, using the MgO concentration as a compositional index. The plotted *V*p values represent the average integrated over a *P-T* range corresponding to the depth interval from 7 to 30 km. The error bar shown (±1 standard deviation) reflects the uncertainties in bulk composition derived in the following way. For each step of MgO shown (e.g., 10, 12, 14 wt%) we randomly generated 100 bulk compositions using the predicted values and prediction error for each oxide from Table 1. Equilibrium mineral assemblages were then calculated for these compositions along a vertical section through the igneous crust, taking 800 °C and 900 °C as equilibrium temperatures (blocking temperatures) throughout and pressures corresponding to the depths indicated by the seismic profile (Fig. 1). Density and seismic velocity values for each mineral assemblage were then calculated as a function of the *P-T* conditions appropriate for the present-day igneous crust assuming an average oceanic geotherm for >100 Ma lithosphere. Average values of *V*p and density for each particular bulk composition were calculated by integration over the depth interval from 7 to 30 km. As seen in Figure 2, typical errors on the average *V*p values calculated by this procedure are on the order of ±0.04 km/s. The

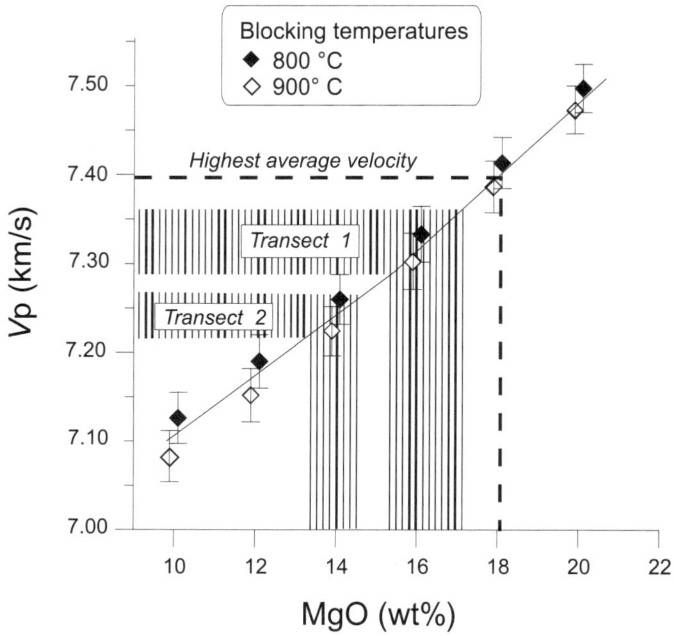

Figure 2. Relationship between bulk composition of igneous crust (plotted as MgO wt%) and its average compressional seismic velocity (*Vp*). *Vp* values were calculated using modeling approach of Sobolev and Babeyko (1994) and bulk compositions of melts from Table 1 for two blocking temperatures of 800 °C (filled diamonds) and 900 °C (open diamonds) (see text for details). Error bars shown (±1 standard deviation) reflect effect of uncertainties in bulk compositions from Table 1. Ranges of observed average velocities within underplated crust along two seismic transects are hachured. Observed seismic velocities correspond to ~15.5–17 wt% MgO on transect 1 and 13.5–14.5 wt% MgO on transect 2. Highest average velocity (7.39 km/s) is encountered in high-velocity body between kilometer –230 and –190 on transect 1 (shown as dashed line). This velocity corresponds to ~18 wt% MgO.

maximal error due to deviation of actual rock fabric from the ideal equigranular, isotropic fabric (the largest half difference between Hashin-Shtrikman bounds) is 0.06 km/s.

It is clear from the figure that the seismic velocity of the thick igneous crust is largely controlled by its MgO content and is insensitive to compositional uncertainties. Although the *Vp* values in Figure 2 were calculated from composition-MgO regressions based on the 1.5 GPa experiments, almost the same numbers result from calculations based on the 1.0 GPa data. Thus the results of this analysis are robust with respect to reasonable uncertainties in the pressure of melting.

The observed average seismic velocities of the thick igneous crust from the two seismic transects are shown as ruled bands in Figure 2. These average values are the integrated seismic velocities over the crustal volumes between kilometers –250 and –150 at transect 1, and between kilometers –290 and –200 at transect 2 (boxed areas in Fig. 1). The range indicated by the ruled pattern represents the difference between velocity averages for the underplate proper (interval between boundaries 6 and M in Fig. 1) and averages for the basement as a whole (boundaries 4 to M). In calculating the latter, we use the velocity at boundary 6 for the crust above it because seismic velocities in the upper crust are likely to be lowered by fracturing and alteration processes (a similar approximation was used by Kelemen and Holbrook, 1995). The dashed line in the figure represents the maximal average velocity for the igneous crust (7.4 km/s), which is the integrated seismic velocity for the region on transect 1 between boundaries 6 and M and between kilometers –230 and –190. This maximum average velocity value corresponds to basaltic rock with ~18 wt% MgO and the total range of observed velocities can be matched by bulk compositions corresponding to 15.5–17 wt% MgO at transect 1 and 13.5–14.5 wt% MgO at transect 2. The maximum uncertainty of observed velocities of ±0.06–0.08 km/s translates to an uncertainty in estimated MgO contents of ~1–1.5 wt%.

DISCUSSION

Effect of cumulates

The base of the thick igneous crust in Figure 1 is characterized by values of *Vp* between 7.3 and 7.6 km/s, considerably higher than the averages shown in Figure 2. A reasonable explanation for these high velocities is the accumulation of olivine and pyroxene from fractional crystallization of the basaltic melts. Farnetani et al. (1996) used the MELTS program of Ghiorso and Sack (1995) to demonstrate that mineral cumulates from fractional crystallization of basalt can explain the high seismic velocities (7.3–7.8 km/s) observed at the base of thick igneous crust below oceanic plateaus and hotspot islands such as Hawaii. We take the same approach here to calculate the expected result of fractionation crystallization from the 16% MgO basalt composition of Table 1. Fractional crystallization models were calculated for this bulk composition assuming an oxygen fugacity fixed at the QFM buffer and a pressure of 0.6 GPa, which corresponds roughly with mid-depth in the thick igneous crust. The bulk compositions of mineral fractionates and residual melts after various degrees of fractional crystallization were then used to calculate seismic *Vp* values by the approach described here. The results (Fig. 3) show that the expected cumulate mineralogy from a 16% MgO basalt could readily account for the high seismic velocities observed at the base of the igneous crust. Moreover, the figure shows that the average velocity of cumulate and residual melt together is nearly constant over a wide range of fractionation degree and is close to the observed average velocity in the thick crust. This is an important result because it means that the constraints on bulk composition from the average seismic velocities in Figure 2 are valid regardless of the effects of fractional crystallization. The overall conclusion remains that basaltic material with ~14–18 wt% MgO is required to explain the average velocities of thick igneous crust at the Namibian margin.

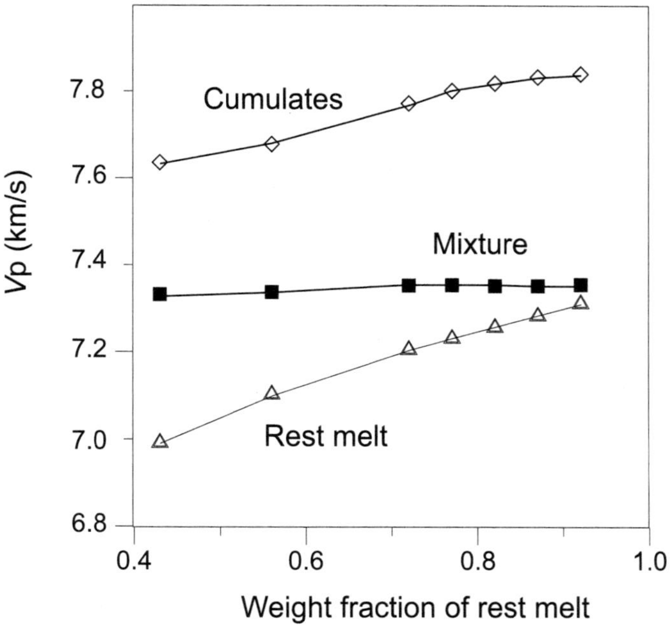

Figure 3. Calculated Vp for cumulates and residual melt at different degrees of fractional crystallization, calculated for bulk composition of 16 wt% MgO basalt (Table 1) using MELTS program of Ghiorso and Sack (1995). Open diamonds show velocities for cumulates, open triangles for residual liquid and filled boxes show average velocities from a mixture of both, calculated as average velocity of a two-component layered media. Note that average velocity is nearly constant over a wide range of fractionation.

Problem of garnet stability

The pressure in the lower part of the high-velocity crust at the Namibian margin is sufficient that garnet could form in basaltic bulk compositions, and the presence of garnet is potentially very important because its high density could upset the simple dependence of seismic velocity on bulk composition suggested in Figure 2. If it is assumed that equilibrium is maintained in the crust during cooling to 600–700 °C, then we can expect >20 vol% garnet for a mid-ocean ridge basalt (MORB)-like bulk composition (10 wt% MgO basalt), and this amount of garnet would raise the average P-wave velocity of basalt from 7.1 km/s to ~7.4–7.5 km/s, which is equivalent to the velocity of 18–20 wt% MgO basalt without garnet.

The effect of garnet is potentially serious, but it is unrealistic to expect that crystallized basalt will continue to equilibrate at temperatures much below the solidus unless the basalt is hydrous or subject to active deformation during cooling. Neither of these factors is likely to play a significant role in this geologic setting. Kelemen and Holbrook (1995) also invoked a kinetic barrier to reaction in their dismissal of garnet as an explanation for the high-velocity igneous crust under the United States east coast margin. For the garnet-forming reaction in a dry basaltic system, lattice diffusion of Al is likely to be the rate-determining step, and kinetic calculations based on a closing temperature of 800 °C suggest that <10% garnet will form in the deepest part of the crust (below 27 km). Therefore, garnet cannot be expected to contribute significantly to the observed crustal velocities. Note also that the zone with the highest seismic velocities at the base of the crust (>7.4 km/s) is not horizontal, as would be expected if velocity were controlled by the pressure limit of garnet stability.

Garnet is unlikely to play a significant role in the observed crust, but it can be significant in another way because the density of basaltic material in the garnet stability field may exceed that of the mantle, resulting in mechanical instability of the mafic crust. This process may put fundamental limitations on the thickness of the crust (Sobolev and Babeyko, 1989). According to our calculations, cumulates from 50% crystallization of basalt with 16 wt% MgO, if they equilibrated at a blocking temperature of 800 °C and depths >30 km, will be denser than the ultramafic mantle at this depth (3.27 g/cm^3). The maximum thickness of the igneous crust under the Namibian margin is ~30 km, and we suggest that some cumulate material may have been detached from the crust. If so, the total volume of igneous crust produced, and the average MgO content of the material, could have been greater than our estimates from the present-day crustal structure and velocity.

To summarize the arguments in this section, we have demonstrated above that the seismic P-wave velocities of thick crust at the continent-ocean transition beneath the Namibian margin require a magnesium-rich basaltic composition, with 14–18 wt% MgO. Referring to Figure 1, the part of margin to which our velocity and compositional estimates refer (boxed) is ~50 km seaward of the steep continent-ocean boundary. We consider it very likely that the entire crust here is igneous, but cannot exclude some minor contribution of extremely stretched continental crust above boundary 6. From a geologic standpoint, we expect that the underplated body will be heterogeneous and consist of some arrangement of sills, dikes, and laccoliths with cumulate layers and layers crystallized from more evolved melt. The internal structure cannot be resolved by the seismic observation (except for the more dense, probably cumulate layer at the base) nor can it be realistically modeled, and it does not affect our conclusions.

Calculation of mantle potential temperature

The relationship of seismic velocities and thickness of underplated basaltic crust to the potential temperature of its mantle source has been discussed in several papers (e.g., White and McKenzie, 1989). Kelemen and Holbrook (1995) used this approach to show that the thick igneous crust under the United States east coast margin requires a hot mantle source. In this section we discuss limits on the potential temperature of the mantle source for the Namibia high-velocity crust based on a new relation between mantle potential temperature and melt composition derived directly from experimental data.

Figure 4 shows the relationship between MgO contents in partial melts and the potential temperature based on published experimental studies of melting dry, fertile peridotite. The experimental data used are from Kushiro (1996) and Falloon et al. (1999) and represent the situation of batch melting. Each point on the diagram (open boxes and triangles) represents one experimental run and we have plotted data from all pressures studied together (0.5–2.0 GPa for Kushiro and 1.0–1.5 GPa for Falloon).

The potential temperature values (*T*p) were calculated from the reported experimental values of pressure, temperature, and degree of melting according to the definition of *T*p as the temperature of the solid at 1 bar pressure, which has the same entropy as the partially molten rock at *T* and *P* in question (McKenzie and Bickle, 1988). The calculation of *T*p used the following values for thermodynamic parameters, derived from the MELTS thermodynamic program (Ghiorso and Sack, 1995): entropy of fusion of 300 J/kg; heat capacities of 1.29 kJ/kg/K (solid), and 1.60 kJ/kg/K (melt). We assume an adiabatic gradient of 0.5 K/km (solid) and 1.0 K/km (melt) after McKenzie and Bickle (1988). Reasonable variation of these parameters (e.g., entropy of fusion by 50 J/kg) change the estimated potential temperature by only a few tens of degrees (see McKenzie and Bickle, 1988)

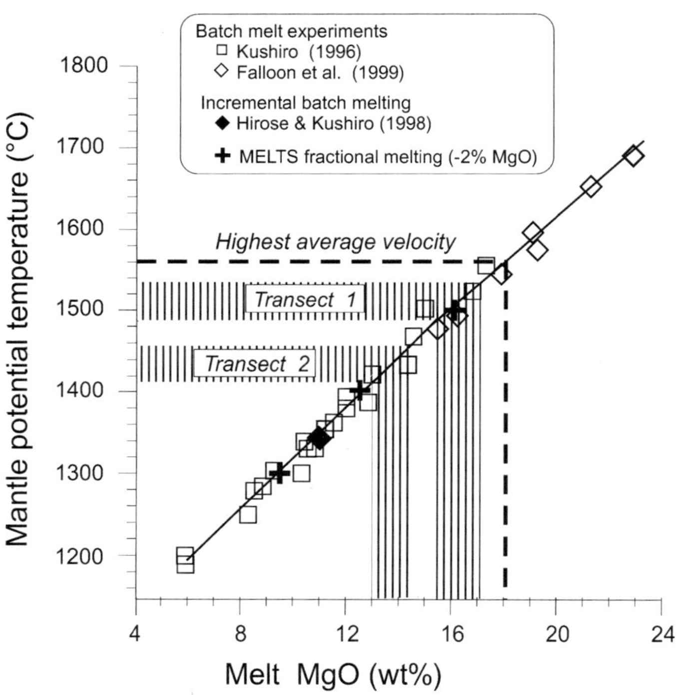

Figure 4. Relationship between MgO content in basaltic melts and potential temperature of peridotite mantle source. Experimental data on batch melting used are from Kushiro (1996) (open boxes) and Falloon et al. (1999) (open diamonds). Each point represents one experimental run. Potential temperatures (*T*p) were calculated from experimental values of pressure, temperature, and degree of melting achieved as explained in text. Filled diamonds show experimental results of polybaric incremental batch melting by Hirose and Kushiro (1998). Crosses are predictions for polybaric fractional melting calculated by MELTS (Asimow et al., 2001) corrected for 2 wt% systematic overestimate of MgO (Hirschmann et al., 1998). The 13.5–18 wt% range in MgO content corresponds to range in potential temperature of 1420–1560 °C.

In Figure 4, the *T*p-MgO points from batch melting experiments by Kushiro (1996) and Falloon et al. (1999) define a line hereafter referred to as the batch line. The filled diamonds represent experimental results of polybaric incremental batch melting by Hirose and Kushiro (1998), and these data also plot on the batch line. The crosses correspond to MgO contents predicted for polybaric fractional melting using the MELTS thermodynamic program (taken from Fig. 14d in Asimow et al., 2001). Previous studies of peridotite melting using this program revealed a systematic overestimate of MgO in melts by ~2 wt% (Hirschmann et al., 1998). When the predicted MgO values are corrected for this effect, they also fall on the batch line. From this comparison of experimental and theoretical melting studies in Figure 4 we conclude that there is a quasilinear relation between starting potential temperature and MgO content of the final melts, and that this relationship holds for isobaric batch melting, polybaric incremental batch melting (with 5 wt% batches), and polybaric near-fractional melting processes.

The *T*p-MgO diagram in Figure 4 predicts that a mantle potential temperature of ~1420–1560 °C would be required to generate basaltic melt with 13.5–18 wt% MgO, corresponding to the estimated compositional range of thick igneous crust at the Namibian margin. An uncertainty of 1 wt% for the calculated MgO contents corresponds to ~30 °C temperature uncertainty. These temperature estimates are in line with potential temperatures inferred from seismological observations for the Iceland (Shen et al., 1998) and Hawaii (Li et al., 2000) plumes, although slightly lower than the 1600 °C inferred for Hawaii. Gurenko et al. (1991) and Sobolev and Nikogosian (1994) estimated MgO contents of primary melts from the Iceland and Hawaii plumes to be 15 and 18.5 wt%, respectively, which are close to our estimates from the Namibian margin. Our estimate of the mantle potential temperature is considerably lower than the 1700 °C value for the core of the Tristan plume proposed by Thompson and Gibson (2000). Their estimate was derived from the inferred presence of unerupted komatiitic basalts with 24 wt% MgO, based on the discovery of high-Mg olivine crystals carried to the surface in basaltic dikes. While we do not dispute this finding, the seismic velocity evidence reported here indicates that such basaltic material can be only a small part of the total magmatism. The velocity expected for a 24% MgO basalt is >7.6 km/s (see Fig. 2), and this is clearly inconsistent with the bulk of unerupted basalts intruding the margin.

An interesting final point is that the mantle potential temperature appears to be ~50–100 °C higher at transect 1 than at transect 2. Because transect 1 is located ~100 km farther north and therefore closer to the Tristan hotspot track (Walvis Ridge), we speculate that there was a gradient of plume temperature from its center to the periphery. Barton and White (1997) reached a similar conclusion in their study of crustal thickness and subsidence history of the Edoras Bank, North Atlantic. They estimated that the mantle potential temperature fell by ~100 °C over a 100 km distance from the Iceland plume. Namibia transect 1 is ~300 km south of the Walvis Ridge crest and if the postulated temperature gradient is real, the potential temperature of the Tristan plume at the Walvis Ridge may have exceeded that of the Hawaii plume.

CONCLUSIONS

Bauer et al. (2000) showed that igneous crust at the continent-ocean transition on the Namibian margin is as thick as 20 km and has seismic P-wave velocities between 7.1 and 7.6 km/s. This study demonstrates that petrologic and petrophysical modeling, constrained by experimental data on peridotite melting, can be used to estimate the bulk composition of this material from the velocity data. Our main conclusion is that the average *Vp* values integrated over the depth range from 7 to 30 km requires a magnesium-rich basaltic composition, with 14–18 wt% MgO. This conclusion is not affected by the presence of cumulate minerals because it integrates over the full thickness of the body; however the highest *Vp* values of 7.6 km/s are consistent with velocities expected for cumulate minerals produced by fractional crystallization of the parental melt. The subsolidus growth of garnet is unlikely to be a significant factor affecting the crustal velocity above the Moho depth of 30 km. However, it can be shown that garnet formation in a magnesium-rich basaltic crust may limit the crustal thickness to ~30 km because bulk densities at deeper levels can exceed those of the peridotite mantle.

The relationship between MgO of partial melts and potential temperature of a fertile peridotite source suggests that basalts with 14–18 wt% MgO were generated from mantle at ~1450–1560 °C potential temperature, and this estimate for the potential temperature of the ancestral Tristan mantle plume is between those proposed for the Iceland and the Hawaii plumes. The highest seismic velocities, and greatest inferred MgO contents and mantle potential temperatures, are found at the northern transect, which is closer to the Tristan hotspot track at the Walvis Ridge. Mantle potential temperature estimated for the southern transect is 50–100 °C lower, suggesting cooling of the plume material during its flow southward.

ACKNOWLEDGMENTS

This paper is based on geophysical experiments funded by the GeoForschungsZentrum Potsdam, the Alfred Wegener Institute, Bremerhaven, and the Bundesanstalt für Geowissenschaften und Rohstoffe, Hannover. Equipment from the Geophysical Instrument Pool Potsdam at the GeoForschungsZentrum Potsdam made the onshore survey possible. We thank Paul Asimow for a preprint of his article and Ian Davidson and Robert White for their constructive reviews of the manuscript.

REFERENCES CITED

Asimow, P.D., Hirschmann, M.M., and Stolper, E.M., 2001, Calculation of peridotite partial melting from thermodynamic models of minerals and melts. 4. Adiabatic decompression and the composition and mean properties of mid-ocean ridge basalts: Journal of Petrology, v. 42, p. 963–998.

Barton, A.J., and White, R.S., 1997, Crustal structure of the Edoras Bank continental margin and mantle thermal anomalies beneath the North Atlantic: Journal of Geophysical Research, v. 102, p. 3109–3129.

Bauer, K., Neben, S., Schreckenberger, B., Emmermann, R., Hinz, K., Fechner, N., Gohl, K., Schulze, A., Trumbull, R.B., and Weber, K., 2000, Deep structure of the Namibia continental margin as derived from integrated geophysical studies: The MAMBA experiment: Journal of Geophysical Research, v. 105, p. 25829–25853.

Duffy, T.S., and Anderson, D.L., 1983, Seismic velocities in mantle minerals and the mineralogy of the upper mantle: Journal of Geophysical Research, v. 94, p. 1895–1912.

Falloon, T.J., Green, D.H., Danyushevsky, L.V., and Faul., U.H., 1999, Peridotite melting at 1.0 and 1.5 GPa: An experimental evaluation of techniques using diamond aggregates and mineral mixes for determination of near-solidus melts: Journal of Petrology, v. 40, p. 1343–1375.

Farnetani, C.G., Richards, M.A., and Ghiorso, M.S., 1996, Petrologic models of magma evolution and deep crustal structure beneath hotspots and flood basalt provinces: Earth and Planetary Sciences Letters, v. 143, p. 81–94.

Ghiorso, M.S., and Sack, R.O., 1995, Chemical mass-transfer in magmatic processes. 4. A revised and internally-consistent thermodynamic model for the interpolation and extrapolation of liquid-solid equilibria in magmatic systems at elevated temperatures and pressures: Contributions to Mineralogy and Petrology, v. 119, p. 197–212.

Gladczenko, T.P., Hinz, K., Eldholm, O., Meyer, H., Neben, S., and Skogseid, J., 1997, South Atlantic volcanic margins: Journal of the Geological Society, London, v. 154, p. 465–470.

Gurenko, A.A., Sobolev, A.V., and Kononkova, N.N., 1991, Petrology of the primary magma of the Reykjanes Peninsula rift tholeiites: Geochemistry International, v. 28, p. 58–71.

Harris, C., 1995, Oxygen isotope geochemistry of the Mesozoic anorogenic complexes of Damaraland, northwest Namibia: Evidence for crustal contamination and its effect on silica saturation: Contributions to Mineralogy and Petrology, v. 122, p. 308–321.

Harris, C., Marsh, J.S., and Milner, S.C., 1999, Petrology of the alkaline core of the Messum igneous complex, Namibia: Evidence for the progressively decreasing effect of crustal contamination: Journal of Petrology, v. 40, p. 1377–1397.

Hashin, Z., and Shtrikman, S., 1963, A variational approach to the elastic behavior of multiphase materials: Journal of Mechanics and Physics of Solids, v. 11, p 127–140.

Hawkesworth, C.J., Kelley, S., Turner, S., le Roex, A., and Storey, B., 1999, Mantle processes during Gondwana break-up and dispersal: Journal of African Earth Sciences, v. 28, p. 239–261.

Hinz, K., 1981, A hypothesis on terrestrial catastrophes: Wedges of very thick oceanward dipping layers beneath passive margins: Their origin and paleoenvironmental significance: Geologisches Jahrbuch, v. E22, p. 345–363.

Hirose, K., and Kushiro, I., 1993, Partial melting of dry peridotites at high pressure: Determination of compositions of melts segregated from peridotite using aggregates of diamond: Earth and Planetary Sciences Letters, v. 114, p. 477–489.

Hirose, K., and Kushiro, I., 1998, The effect of melt segregation on polybaric mantle melting: Estimation from the incremental melting experiments: Physics of Earth and Planetary Interiors, v. 107, p. 111–118.

Hirschmann, M.M., Ghiorso, M.S., Wasylenki, L.E., Asimow, P.D., and Stolper, E.M., 1998, Calculation of peridotite partial melting from thermodynamic models of minerals and melts. 1. Review of methods and comparison with experiments: Journal of Petrology, v. 39, p. 1091–1115.

Holbrook, W.S., Purdy, G.M., Sheridan, R.E., Glover, L. III, Talwani, M., Ewing, J., and Hutchinson, D., 1994, Seismic structure of the U.S. Mid-Atlantic continental margin: Journal of Geophysical Research, v. 99, p. 17871–17891.

Kelemen, P.B., and Holbrook, W.S., 1995, Origin of thick, high-velocity igneous crust along the U.S. east coast margin: Journal of Geophysical Research, v. 100, p. 10077–10094.

Kinzler, R.J., 1997, Melting of mantle peridotite at pressures approaching the spinel to garnet transition: Application to mid-ocean ridge basalt petrogenesis: Journal of Geophysical Research, v. 102, p. 853–874.

Kinzler, R.J., and Grove, T.J., 1993, Corrections and further discussion of the primary magmas of mid-ocean ridge basalts, 1 and 2: Journal of Geophysical Research, v. 98, p. 22339–22347.

Kushiro, I., 1996, Partial melting of a fertile mantle peridotite at high pressure: An experimental study using aggregates of diamond, *in* Basu, A., and Hart, S., eds., Earth processes: Reading the isotopic clock: American Geophysical Union Geophysical Monograph 95, p. 109–122.

Langmuir, C.H., Klein, E., and Plank, T., 1992, Petrological systematics of mid-ocean ridge basalts: Constraints on melt generation beneath ocean ridges, *in* Phipps Morgan, J., Blackman, D.K., and Sinton, J.M., eds., Mantle flow and melt generation at mid-ocean ridges: American Geophysical Union Geophysical Monograph 71, p. 183–280.

Leitch, A.M., Davies, G.F., and Wells, M., 1998, A plume head under a rifting margin: Earth and Planetary Science Letters, v. 161, p. 161–177.

Li, X., Kind, R., Priestley, K., Sobolev, S.V., Tilmann, F., Yuan, X., and Weber, M., 2000, Mapping the Hawaiian plume conduit with converted seismic waves: Nature, v. 405, p. 938–941.

Mahoney, J.J., and Coffin, M.F., editors, 1997, Large igneous provinces: Continental, oceanic, and planetary flood volcanism: American Geophysical Union Geophysical Monograph 100, 438 p.

Marsh, J.S., Ewart, A., Milner, S.C., Duncan, A.R., and Miller, R.McG., 2001, The Etendeka Igneous Province: Magma types and their stratigraphic distribution with implications for the evolution of the Paraná-Etendeka flood basalt province: Bulletin of Volcanology, v. 62, p. 464–486.

McKenzie, D.P., and Bickle, M.J., 1988, The volume and composition of melt generated by extension of the lithosphere: Journal of Petrology, v. 29, p. 625–679.

Milner, S.C., and le Roex, A.P., 1996, Isotope characteristics of the Okenyenya igneous complex, northwestern Namibia: Constraints on the composition of the early Tristan plume and the origin of the EM1 mantle component: Earth and Planetary Science Letters, v. 141, p. 277–291.

Milner, S.C., le Roex, A.P., and O'Connor, J.M., 1995, Age of Mesozoic igneous rocks in northwestern Namibia and their relationship to continental breakup: Journal of the Geological Society, London, v. 152, p. 97–104.

O'Connor, J.M., and le Roex, A.P., 1992, South Atlantic hot spot: Plume systems. 1. Distribution of volcanism in time and space: Earth and Planetary Science Letters, v. 113, p. 343–364.

Oh, J., Austin, J.A., Jr., Phillips, J.D., Coffin, M.F., and Stoffa, P.L., 1995, Seaward-dipping reflectors offshore the southeastern United States: Seismic evidence for extensive volcanism accompanying sequential formation of the Carolina Trough and Blake Plateau basin: Geology, v. 23, p. 9–12.

Peate, D.W., 1997, The Paraná–Etendeka Province, *in* Mahoney, J.J., and Coffin, M.F., eds., Large igneous provinces: American Geophysical Union Geophysical Monograph 100, p. 217–246.

Rabinowitz, P.D., and LaBrecque, J., 1979, The Mesozoic South Atlantic ocean and evolution of its continental margins: Journal of Geophysical Research, v. 84, p. 5973–6002.

Shen, Y., Solomon, S.C., Bjarnason, I.Th., and Wolfe, C.J., 1998, Seismic evidence for a lower-mantle origin of the Iceland plume: Nature, v. 395, p. 62–65.

Sobolev, A.V., and Nikogosian, I.K., 1994, Petrology of long–lived mantle plume magmatism: Hawaii, Pacific and Reunion Island, Indian Ocean: Petrology, v. 2, p. 111–144.

Sobolev, S.V., and Babeyko, A.Yu., 1989, Phase transformations in the lower continental crust and its seismic structure, *in* Mereu, R.S., Mueller, S., and Fountain, D.M., eds., Properties and processes of Earth's lower crust: American Geophysical Union Geophysical Monograph 51, p. 311–320.

Sobolev, S.V., and Babeyko, A.Yu., 1994, Modelling of mineralogical composition, density and elastic wave velocities in the anhydrous rocks: Surveys in Geophysics, v. 15, p. 515–544.

Talwani, M., and Abreu, V., 2000, Inferences regarding initiation of oceanic crust formation from the U.S. coast margin and conjugate South Atlantic margins, *in* Mohriak, W., and Talwani, M., eds., Atlantic rifts and continental margins: American Geophysical Union Geophysical Monograph 115, p. 211–233.

Thompson, R.N., and Gibson, S.A., 2000, Transient high temperatures in mantle plume heads inferred from magnesian olivines in Phanerozoic picrites: Nature, v. 407, p. 502–506.

Trumbull, R.B., Emmermann, R., Bühn, B., Gerstenberger, H., Mingram, B., Schmitt, A., and Volker, F., 2000, Insights on the genesis of the Cretaceous Damaraland igneous complexes in Namibia: The Nd- and Sr-isotopic perspective: Henno Martin Commemorative Issue, Communications of the Geological Survey of Namibia, v. 12, p. 313–324.

Turner, S.P., Hawkesworth, C.J., Gallagher, K., Stewart, K., Peate, D.W., and Mantovani, M.S.M., 1996, Mantle plumes, flood basalts and thermal models for melt generation beneath continents: Assessment of a conductive heating model and application to the Paraná: Journal of Geophysical Research, v. 101, p. 11503–11518.

Walker, M.J., 1999, Melting residues of fertile peridotite and the origin of cratonic lithosphere, *in* Fei, Y., Bertka, C.M., and Mysen, B.O., eds., Mantle petrology: Field observations and high-pressure experimentation: Houston, Texas, Geochemical Society Special Publication 6, p. 225–240.

White, R.S., and McKenzie, D.P., 1989, Magmatism at rift zones: The generation of volcanic continental margins and flood basalts: Journal of Geophysical Research, v. 94, p. 7685–7729.

White, R.S., and McKenzie, D.P., 1995, Mantle plumes and flood basalts: Journal of Geophysical Research, v. 100, p. 17543–17585.

White, R.S., McKenzie, D.P., and O'nions, R.K., 1992, Oceanic crustal thickness from seismic measurements and rare earth element inversions: Journal of Geophysical Research, v. 97, p. 19683–19715.

Manuscript Accepted by the Society October 15, 2001

Printed in the U.S.A.